U0358714

全国科学技术名词审定委员会

科学技术名词·工程技术卷（全藏版）

17

海峡两岸化学工程名词

海峡两岸化学工程名词工作委员会

国家自然科学基金资助项目

科学出版社

北　京

内 容 简 介

　　本书是由海峡两岸化学工程界专家会审的海峡两岸化学工程名词对照本,是在全国科学技术名词审定委员会公布的《化学工程名词》(1995)和由台湾化工学会提供的《化工名词》的基础上修订而成。共收词约 13 000条。供海峡两岸化学工程界和相关领域的人士使用。

图书在版编目(CIP)数据

───────────────────────────────

科学技术名词. 工程技术卷: 全藏版 / 全国科学技术名词审定委员会审定.
—北京: 科学出版社, 2016.01
　　ISBN 978-7-03-046873-4

　　I. ①科… II. ①全… III. ①科学技术–名词术语 ②工程技术–名词术语
IV. ①N-61 ②TB-61

───────────────────────────────

　　中国版本图书馆 CIP 数据核字(2015)第 307218 号

───────────────────────────────

　　　　责任编辑: 才 磊 王宝瑄 / 责任校对: 陈玉凤
　　　　　责任印制: 张 伟 / 封面设计: 铭轩堂

科 学 出 版 社 出版
北京东黄城根北街 16 号
邮政编码: 100717
http://www.sciencep.com
北京厚诚则铭印刷科技有限公司印刷
科学出版社发行　各地新华书店经销
*
2016 年 1 月第 一 版　　开本: 787×1092 1/16
2016 年 1 月第一次印刷　　印张: 44 1/2
字数: 1 037 000

定价: 7800.00 元(全 44 册)
(如有印装质量问题, 我社负责调换)

海峡两岸化学工程名词工作委员会委员名单

召 集 人：蒋楚生

委　　员：苏健民　　萧成基　　罗北辰

秘　　书：才 磊

召 集 人：石延平

委　　員：萬其超　　紀榮昌　　曾憲政　　蔡信行　　蔣孝澈

　　　　　劉清田　　賴君義

序

  科学技术名词作为科技交流和知识传播的载体,在科技发展和社会进步中起着重要作用。规范和统一科技名词,对于一个国家的科技发展和文化传承是一项重要的基础性工作和长期性任务,是实现科技现代化的一项支撑性系统工程。没有这样一个系统的规范化的基础条件,不仅现代科技的协调发展将遇到困难,而且,在科技广泛渗入人们生活各个方面、各个环节的今天,还将会给教育、传播、交流等方面带来困难。

  科技名词浩如烟海,门类繁多,规范和统一科技名词是一项十分繁复和困难的工作,而海峡两岸的科技名词要想取得一致更需两岸同仁作出坚韧不拔的努力。由于历史的原因,海峡两岸分隔逾50年。这期间正是现代科技大发展时期,两岸对于科技新名词各自按照自己的理解和方式定名,因此,科技名词,尤其是新兴学科的名词,海峡两岸存在着比较严重的不一致。同文同种,却一国两词,一物多名。这里称"软件",那里叫"软体";这里称"导弹",那里叫"飞弹";这里写"空间",那里写"太空";如果这些还可以沟通的话,这里称"等离子体",那里称"电浆";这里称"信息",那里称"资讯",相互间就不知所云而难以交流了。"一国两词"较之"一国两字"造成的后果更为严峻。"一国两字"无非是两岸有用简体字的,有用繁体字的,但读音是一样的,看不懂,还可以听懂。而"一国两词"、"一物多名"就使对方既看不明白,也听不懂了。台湾清华大学的一位教授前几年曾给时任中国科学院院长周光召院士写过一封信,信中说:"1993年底两岸电子显微学专家在台北举办两岸电子显微学研讨会,会上两岸专家是以台湾国语、大陆普通话和英语三种语言进行的。"这说明两岸在汉语科技名词上存在着差异和障碍,不得不借助英语来判断对方所说的概念。这种状况已经影响两岸科技、经贸、文教方面的交流和发展。

  海峡两岸各界对两岸名词不一致所造成的语言障碍有着深刻的认识和感受。具有历史意义的"汪辜会谈"把探讨海峡两岸科技名词的统一列入了共同协议之中,此举顺应两岸民意,尤其反映了科技界的愿望。两岸科技名词要取得统一,首先是需要了解对方。而了解对方的一种好的方式就是编订名词对照本,在编订过程中以及编订后,经过多次的研讨,逐步取得一致。

  全国科学技术名词审定委员会(简称全国科技名词委)根据自己的宗旨和任务,始终把海峡两岸科技名词的对照统一工作作为责无旁贷的历史性任务。近些年一直本着积极推进,增进了解;择优选用,统一为上;求同存异,逐步一致的精神来开展这项工作。先后接待和安排了许多台湾同仁来访,也组织了多批专家赴台参加有关学科的名词对照研讨会。工作中,按照先急后缓、先易后难的精神来安排。对于那些与"三通"有关的学科，以及名词混乱现象严重的学科和条件成熟、容易开展的学科先行开展名词对照。

在两岸科技名词对照统一工作中，全国科技名词委采取了"老词老办法，新词新办法"，即对于两岸已各自公布、约定俗成的科技名词以对照为主，逐步取得统一，编订两岸名词对照本即属此例。而对于新产生的名词，则争取及早在协商的基础上共同定名，避免以后再行对照。例如101～109号元素，从9个元素的定名到9个汉字的创造，都是在两岸专家的及时沟通、协商的基础上达成共识和一致，两岸同时分别公布的。这是两岸科技名词统一工作的一个很好的范例。

海峡两岸科技名词对照统一是一项长期的工作，只要我们坚持不懈地开展下去，两岸的科技名词必将能够逐步取得一致。这项工作对两岸的科技、经贸、文教的交流与发展，对中华民族的团结和兴旺，对祖国的和平统一与繁荣富强有着不可替代的价值和意义。这里，我代表全国科技名词委，向所有参与这项工作的专家们致以崇高的敬意和衷心的感谢！

值此两岸科技名词对照本问世之际，写了以上这些，权当作序。

2002 年 3 月 6 日

前　言

随着海峡两岸学术交流的不断加强,两岸化工名词不一致的问题,给两岸的学术交流带来的障碍逐渐显露出来。两岸化工界的同行们认为共同开展海峡两岸化工名词的交流和研讨,并出版对照本是非常必要的。在两岸化工名词对照研讨过程中两岸专家可以增进了解,取长补短,逐步达成共识,从而促进名词的统一。这项工作将惠及两岸化工领域的学术交流、知识传播以及相关文献的编纂和检索。鉴于此,全国科学技术名词审定委员会、中国化工学会和台湾李国鼎科技发展基金会、台湾化工学会的有关负责人协商,决定在两岸分别邀请有关专家组成"海峡两岸化工名词工作委员会"承担此项重任。该委员会决定以台湾化工学会提供的《化工名词》和全国科学技术名词审定委员会(原全国自然科学名词审定委员会)公布的《化学工程名词》(1995)作为对照蓝本,并由大陆专家先行完成《海峡两岸化工名词》补充对照工作,后通过电子邮件请台湾专家补充修订,完成对照初稿。

2005 年 11 月 18 日在北京召开了"海峡两岸化工名词研讨会"。专家们进一步明确了收词原则,对初稿中存疑的问题和不一致的定名进行了认真、细致地讨论。此外,还对学术上存在争议的名词或以其他学科为主定名的名词,也进行了讨论,在尊重习惯、择优选用、求同存异、逐步一致的原则指导下,专家的认识趋于接近。会后两岸专家又分别对相关部分进行了审核,完成了定稿。

通过对化工名词的对照研讨,两岸专家认识到,名词对照统一工作是一项长期而细致的工作,应该长期地进行下去。这项工作对海峡两岸的学术交流和知识传播都会起到积极的促进作用和支撑作用。今后两岸化工专家应加强交流与沟通,共同确定本学科领域的新名词,以减少新名词定名不一致的现象。我们的工作仅仅是一个开始,在化工名词的对照研讨及本书的编辑过程中,虽经两岸化工专家反复推敲、认真讨论,但不妥之处尚属难免,尤其是对照本中大陆使用的许多名词尚未审定公布,希望得到两岸读者的指正。

海峡两岸化学工程名词工作委员会

2007 年 1 月

编　排　说　明

一、本书是海峡两岸化学工程名词对照本。

二、本书分正篇和副篇两部分。正篇按汉语拼音顺序编排;副篇按英文的字母顺序编排。

三、[　]中的字使用时可以省略。

正篇

五、正名和异名分别排序,并在异名处用(=)注明正名。

六、对应的英文名为多个时(包括缩写词)用","分隔。

副篇

七、英文名对应多个相同概念的汉文名时用","分隔,不同概念的用① ② ③分别注明。

八、英文名的同义词用(=)注明。

九、英文缩写词排在全称后的( )内。

目　　录

序

前言

编排说明

正 篇

A

祖国大陆名	台湾地区名	英 文 名
阿伦尼乌斯方程	阿瑞尼斯方程式	Arrhenius equation
阿马加定律	阿馬加定律	Amagat law
安全玻璃	安全玻璃	safety glass
安全操作压力	安全操作壓力	safe working pressure
安全阀	安全閥	safety valve
安全放空	安全通氣孔	safety vent
安全工作应力	安全工作應力	safe working stress
安全减压阀	釋放閥	relief valve
安全检查[清]单	安全檢查報表	safety check list
安全门	安全門	safety door
安全系数	安全因數	safety factor
安全泄压阀	安全釋放閥	safety-relief valve
安全泄压面积	安全釋放面積	safety relief area
安全泄液阀	液體釋放閥	liquid relief valve
安全性试验	安全性試驗	safety testing
安全裕度	安全邊界	safety margin
安全炸药	安全炸藥	safety explosive
安全装置	安全裝置	safety device
安托万方程	安托因方程	Antoine equation
安息香	安息香	benzoin
安装	安裝	installation
安装成本	安裝成本	installation cost
氨饱和器	氨飽和器	ammonia saturator
氨苄青霉素	胺苄青黴素	ampicillin
氨萃取过程	氨萃取法	ammonia extraction process
氨反应器	氨反應器	ammonia reactor
氨合成	氨合成	ammonia synthesis
氨化肥料	氨化肥料	ammoniated fertilizer
氨化[作用]	氨化	ammoniation
氨基化	胺化	amination

祖国大陆名	台湾地区名	英 文 名
氨基树脂	胺基樹脂	amino resin
氨基酸	胺基酸	amino acid
氨碱法	氨鹼法	ammonia-soda process
氨解[作用]	氨解	ammonolysis
氨水	氨水	ammonia liquor, ammonia water, aqua ammonia, aqueous ammonia
氨洗涤器	洗氨器	ammonia washer
氨盐水	氨化鹽水	ammoniated brine
鞍点共沸物	鞍式共沸液	saddle azeotrope
铵矾	銨礬	ammonium alum
铵皂	銨皂	ammonium soap
昂萨格倒易关系	昂薩格倒易關係	Onsager reciprocal relation
凹槽	凹槽, V形槽	notch
凹槽式滤板	帶框濾板	recessed plate
凹痕	凹痕	sink mark
凹坑	坑, 槽, 池	pit
螯合剂	螯合劑, 鉗合劑	chelating agent, sequestering agent
奥长石	鈉鈣長石	oligoclase
奥纶	奧綸	orlon
奥斯陆蒸发结晶器	奧斯陸蒸發結晶器	Oslo evaporative crystallizer

B

祖国大陆名	台湾地区名	英 文 名
八田数	八田數	Hatta number
巴龙霉素	巴龍黴素	paromomycin
巴氏灭菌器	低溫殺菌器	pasteurizer
巴氏消毒法	低溫殺菌法, 巴氏殺菌法	pasteurization
巴西棕榈蜡	棕櫚蠟, 卡拿巴蠟	carnauba wax
拔顶蒸馏	直餾	topping
钯催化剂	鈀觸媒	palladium catalyst
靶管	靶管	target tube
靶效率	靶效率	target efficiency
白垩	白堊	chalk
白放线菌素	放線菌素	actinomycetin
白合金	白合金	white metal
白金	白金[含鉛金]	white gold

祖国大陆名	台湾地区名	英　文　名
白磷	白磷	white phosphorus
白热光	白熱光, 白熱燈	incandescent light
白鞣酸	白鞣酸, 明礬糊	tawing paste
白水泥	白水泥	white cement
白炭黑	白碳, 白煙	white carbon
白铁矿	白鐵礦	marcasite
白液	白液, 燒鹼液	white liquor
白云母	白雲母	muscovite
白云石	白雲石	dolomite
白云石灰石	白雲石灰石	dolomitic limestone
白云石石灰	白雲石石灰	dolomitic lime
白云石水泥	白雲石水泥	dolomite cement
白云石陶器	白雲石陶器	dolomite earthenware
百分绝对湿度	百分絕對濕度	percentage absolute humidity
百分湿度	百分濕度	percentage humidity
百叶板填充塔	填板塔	slat packed tower
百叶窗挡板	百葉窗擋板	louver type baffle
摆动连续结晶槽	擺動連續結晶槽	Wulff-Bock crystallizer
摆式黏度计	擺式黏度計	pendulum viscometer
摆式张力试验机	擺式張力試驗機	pendulum-type tension testing machine
摆旋鼓风机	擺旋鼓風機	cycloidal blower
[班伯里]密[闭式混]炼机	密閉式混煉機, 班伯里混煉機	Banbury mixer
斑点	斑漬, 斑點	spot
斑点反应	斑點反應	spot reaction
斑点分析	斑點分析	spot analysis
斑点试验	斑點試驗	spot test
斑染	斑染	speck dyeing
斑贴试验	布片試驗	patch test
斑铜矿	斑銅礦	bornite
板波纹填料	板波紋填充	Mellapak packing
板翅换热器	板翅換熱器	plate-fin heat exchanger
板框[式]过滤机	框板過濾機	frame and plate filter
板框组件	板框組件	plate-and-frame module
板式换热器	板式換熱器	plate heat exchanger
板式输送机	板式運送機, 板運機	slat conveyor
板式塔	層板塔	plate column, plate tower
板式蒸发器	板式蒸發器	plate-type evaporator
板数	板數	plate number

祖国大陆名	台湾地区名	英　文　名
板效率	板效率	tray efficiency
半八角形池	半八角形池	semi-octagonal cell
半补强炉黑	半補強爐黑	semireinforcing furnace black
半导体	半導體	semiconductor
半导体催化剂	半導體觸媒	semiconductor catalyst
半导体聚合物	半導體聚合物	semiconductive polymer
半导体涂料	半導體塗料	semiconductive coating
半对数坐标纸	半對數坐標紙	semi-logarithmic paper
半分馏	半分餾	semifractionating
半分批法	半批次法	semi-batch process
半分批反应器	半分批反應器	semi-batch reactor
半分批[式]操作	半批式操作	semi-batch operation
半分批选择性	半批式選擇性	semi-batch selectivity
半封闭流化态催化剂床	半封閉流化態觸媒床	semi-isolated fluidized catalyst bed
半封闭叶轮	半封閉葉輪	semi-enclosed impeller
半干性油	半乾性油	semi drying oil
半固体物料	半固體物料, 塑性物質	soft solid materials
半光涂料	半光澤塗料	semigloss coating
半合成物	半合成物	semisynthetic
半合成纤维	半合成纖維	semi-synthetic fiber
半糊状漆	半糊狀漆	semi-paste paint
半互穿透聚合物网络	半互穿聚合物網絡	semi-interpenetrating polymer network
半化学纸浆	半化學紙漿	semi-chemical pulp
半化学制浆法	半化學製漿法	semi-chemical pulping process
半极性键	半極性鍵	semipolar bond
半结晶聚合物	半結晶聚合物	semi-crystalline polymer
半连续过程	半連續程序	semicontinuous process
半连续聚合	半連續聚合	semi-continuous polymerization
半连续窑	半連續窯	semi-continuous kiln
半流动式反应器	半流動式反應器	semiflow reactor
半硫化	半硫化	semi-cure, semi-curing
半耐久性黏合剂	半耐久性黏合劑	semi-durable adhesive
半挠性链	半撓性鏈	semi-flexible chain
半强性促进剂	半強促進劑	semi-strong accelerator
半区间法	半區間法	interval halving
半生活性	半生活性	midlife activity
半衰期	半衰期	half life
半衰期法	半衰期法	half life method

祖国大陆名	台湾地区名	英　文　名
半水晶	半晶體	semi-crystal
半体	半體	half body
半透明度	①半透明度②半透明性	translucence
半透膜	半透膜	semipermeable membrane
半透性分子	半透性分子	semipermeable molecule
半微量分析	半微量分析	semi-microanalysis
半无光	半無光	semi-dull
半无烟煤	半無煙煤	semi-anthracite
半纤维素	半纖維素	hemicellulose
半悬浮聚合	半懸浮聚合	semi-pearl polymerization
半烟煤	半煙煤	semi-bituminous coal
半硬泡沫	半硬性泡沫	semi-rigid foam
半硬质纤维板	半硬質纖維板	semi-hard board
半煮法	半煮法	semi-boiling process
半自动系统	半自動系統	semi-automatic system
半自动压机	半自動壓縮積機)	semi-automatic press
拌和机	拌和機	mingler
棒磨机	棒磨機	rod mill
磅达(英制力的单位)	磅達	poundal
包藏	①吸著②包藏	occlusion
包埋	包埋	entrapment
包装	包裝	packaging
包装纸	包裝紙	packing paper
胞内酶	胞內酶	intracellular enzyme
胞腔模型	細胞模型	cell model
胞外酶	胞外酶	extracellular enzyme
饱和	飽和, 飽和度	saturation
饱和度	飽和度	degree of saturation
饱和极限	飽和極限	saturation limit
饱和聚酯	飽和聚酯	saturated polyester
饱和空气	飽和空氣	saturated air
饱和器	飽和器	saturator
饱和曲线	飽和曲線	saturation curve
饱和溶液	飽和溶液	saturated solution
饱和色	飽和色	saturated color
饱和湿度	飽和溼度	saturated humidity
饱和态	飽和態	saturated state
饱和烃聚合物	飽和碳氫聚合物	saturated hydrocarbon polymer

祖国大陆名	台湾地区名	英 文 名
饱和温度	飽和溫度	saturation temperature
饱和压力	飽和壓力	saturation pressure
饱和液	飽和液	saturated liquid
饱和蒸气	飽和蒸氣	saturated vapor
饱和蒸气压	飽和蒸氣壓	saturated vapor pressure
饱和蒸汽	飽和蒸汽	saturated steam
保留(＝截留)		
保留期间	停留期間	retention period
保留体积	停留體積	retention volume
保温材料	保溫材料, 隔熱材料	insulating material
保险	保險	insurance
保证质量	保證品質	guaranteed quality
报告	報告	report
报警器	警報器	alarm
刨纹	切削條痕	sheeter line
爆聚[合]	爆聚合	explosive polymerization
爆裂	爆裂	burst
爆裂强度	爆裂強度	bursting strength
爆裂压力	爆裂壓力	bursting pressure
爆燃	爆燃	deflagration
爆燃器	爆燃器	deflagrator
爆炸	爆炸	explosion
爆炸化合物	爆炸化合物	blasting compound
爆炸极限	爆炸界限	explosion limit
爆炸时间	爆炸時間	detonation time
爆炸势	爆炸潛力, 爆炸勢	explosion potential
爆炸危险	爆炸傷害	explosion hazard
爆炸压[力]	爆炸壓力	explosion pressure
爆炸指数	爆炸指數	explosion index
爆震试验	震爆試驗	knock characteristic test
杯混合平均组成	杯混合平均組成	cup-mixing composition
北极鲸蜡油	抹香鯨油, 北極鯨蠟油	arctic sperm oil
备择假设	備擇假設	alternative hypothesis
备择投资	交替投資	alternative investment
背压	反壓, 背壓	back pressure
钡矾	鋇礬	barium alum
倍示压力计	倍示壓力計	multiplying manometer
倍增时间	倍增時間	doubling time
焙烧炉	①焙燒爐②風箱	roaster

祖国大陆名	台湾地区名	英 文 名
本构方程	本質常數	constitutive equation
本森[溶解度]系数	本森係數	Benson’s solubility coefficient
本体聚合[反应]	本體聚合[反應]	mass polymerization
本体模量	整體模數	bulk modulus
本体温度	整體溫度	bulk temperature
本体相	整體相	bulk phase
本体性质	整體性質	bulk property
本征动力学	固有動力學	intrinsic kinetics
苯胺橙	苯胺橙	aniline orange
苯胺点	苯胺點	aniline point
苯胺黑	苯胺黑	aniline black, nigrosine
苯胺红	苯胺紅	aniline red
苯胺蓝	苯胺藍	aniline blue
苯胺绿	苯胺綠	aniline green
苯胺染料	苯胺染料	aniline dye
苯胺紫	苯胺紫	aniline violet
苯并呋喃	苯并呋喃, 薰草哢	benzofuran, coumarone
苯并呋喃-茚树脂	苯并呋喃樹脂	coumarone-indene resin
苯乙烯	苯乙烯	styrene
苯乙烯-丙烯腈共聚物	苯乙烯-丙烯腈共聚物	styrene acrylonitrile copolymer
苯乙烯-丙烯酸酯[共聚]涂料	苯乙烯-丙烯酸酯塗料	styrene acrylate copolymer coating
苯乙烯-二乙烯基苯共聚物	苯乙烯-二乙烯基苯共聚物	styrene divinylbenzene copolymer
苯乙烯基化油	苯乙烯基化油	styrenated oil
苯乙烯-甲基丙烯酸甲酯树脂	苯乙烯-甲基丙烯酸甲酯樹脂	styrene methyl methacrylate resin
苯乙烯树脂	苯乙烯樹脂	styrene resin
苯乙烯-顺丁烯二酸酐共聚物	苯乙烯-順丁烯二酸酐共聚物	styrene maleic anhydride copolymer
苯乙烯橡胶	苯乙烯橡膠	styrene rubber
泵	泵	pump
泵功率	泵功率	pump horsepower
泵送功	泵功	pumping work
泵送损耗	泵抽損失	pumping loss
泵特性	泵特性	pump characteristics
泵效率	泵效率	pump efficiency
泵压降	泵壓[力]降	pump pressure drop
泵轴	泵軸	pump shaft

祖国大陆名	台湾地区名	英 文 名
泵转子	泵轉子	pump rotor
比表面	比面, 表面係數	specific surface
比表面积	比表面積	specific surface area
比冲量	比衝量, 比推力	specific impulse
比例-重调-速率作用	比例-重整-速率作用	proportional plus reset plus rate action
比例-重调作用	比例-重整作用	proportional plus reset action
比电阻	比阻力	specific resistance
比反应速率	比反應速率	specific reaction rate
比刚度	比剛度, 剛性-重量比	stiffness-weight ratio
比刚性	比剛性	specific rigidity
比焓	比焓	specific enthalpy
比极化度	比極化度, 比偏光度	specific polarization
比较器	比較器	comparator
比绝缘电阻	比絕緣電阻	specific insulation resistance
比例带	比例帶	proportional band
比例带调整	比例帶調整	proportional band adjustment
比例换算(=定标)		
比例-积分控制	比例-積分控制	proportional-integral control
比例-积分-微分作用	比例-積分-微分作用	proportional-integral-derivative action
比例-积分作用	比例-積分作用	proportional-integral action
比例控制	比例產物	proportional control
比例控制偏离	比例控制的偏離	bias in proportional
比例控制器	比例控制器	proportional controller
比例灵敏度	比例靈敏度	proportional sensitivity
比例-微分控制	比例-微分控制	proportional-derivative control
比例-微分作用	比例-微分作用	proportional-derivative action
比例元件	比例元件	proportional element
比例增益	比例增益	proportional gain
比例作用	比例作用	proportional action
比滤饼阻力	比濾餅阻力	specific cake resistance
比黏度	比黏度	specific viscosity
比浓对数黏度	固有黏度	inherent viscosity
比强度	比強度, 強度係數, 比抗張力, 強度-重量比	specific strength, specific tenacity, strength-to-weight ratio
比热	比熱	specific heat
比容[积]	比容	specific volume
比色计	比色計	chromometer, colorimeter
比色密度计	比色密度計	color densitometer

祖国大陆名	台湾地区名	英　文　名
比色器	比色器	color comparator
比熵	比熵	specific entropy
比生长速率	比生長速率	specific growth rate
比湿度	比濕度	specific humidity
比速率	比速率	specific rate
比速率系数	比速率係數	specific rate coefficient
比推力	比推力	specific thrust
比消光系数	比消光係數, 吸光係數	specific extinction coefficient
比旋光度	光轉偏極係數	specific rotatory power
比重	比重	specific gravity, specific weight
API 比重计	美制[石油]比重計	API hydrometer
比重瓶	比重瓶	pycnometer, specific gravity bottle
API 比重指数	美制[石油]比重	API gravity
比转速	特定速率	specific speed
比浊法	濁度測定法	turbidimeter
比阻尼容量	比衰減容量	specific damping capacity
吡唑啉酮染料	吡唑啈染料	pyrazolone dye
毕奥数	畢奧數	Biot number
闭工叶轮	封閉葉輪	enclosed impeller
闭环	閉環	closed loop
闭环传递函数	閉環循環函數	closed loop transfer function
闭环控制	閉環控制	closed loop control
闭环频率应答	閉環頻率應答	closed loop frequency response
闭环稳定性	閉環穩定性	closed loop stability
闭环系统	閉環系統	closed loop system
闭路	閉路	closed circuit
闭路压碎	閉路壓碎, 循環壓碎	closed circuit crushing
闭路研磨	閉路研磨, 循環研磨	closed circuit grinding
闭式边界	閉式邊界	closed boundary
闭式容器	密閉容器	closed vessel
闭锁式料斗	閉鎖式料斗	lock hopper
蓖麻油	蓖麻油	castor oil
壁厚	壁厚	wall thickness
壁摩擦角	壁摩擦角	angle of wall friction
壁湍流	壁紊流	wall turbulence
壁效应	壁效應	wall effect
边际利润	邊際利潤	marginal profit
边界层	邊界層	boundary layer
边界层方程	邊界層方程式	boundary layer equation

祖国大陆名	台湾地区名	英 文 名
边界层浓度	邊界層濃度	boundary layer concentration
边界条件	邊界條件	boundary condition
边界相	邊限相位	margin phase
边界增益	邊限增益	margin gain
边缘效应	邊緣效應	edge effect
编码	代號, 規範, 簡碼	code
鞭毛	鞭毛	flagella
变动成本	變動成本	variable cost
变动系数	變異係數	coefficient of variation
变分学	變分學, 變分法	variational calculus, calculus of variation
变换	變換式	transform
变换反应	[水煤氣]轉化反應	shift reaction
变换炉	[水煤氣]轉化器	shift converter
变阶反应	變階反應	shifting-order reaction
变量	變數	variable
变送器	傳送器, [信號]發設機	transmitter
变送器增益	傳送氣增益	transmitter gain
变速泵	變速泵	variable-speed pump
变体	變異體	variant
变温吸附	變溫吸附	temperature swing adsorption
变形	變形	deformation
变形点	變形點	deformation point
变形范围	變形範圍	deformation range
变形功	變形功	deformation work
变形共熔物	變形共熔物	deformation eutectic
变形速率	變形速率	deformation rate
变形坐标	變形座標	deformation coordinates
变性	變性	denaturation
变性剂	變性劑	denaturant
变性酒精	變性酒精, 加甲醇酒精	denatured alcohol, methylated spirit
变压器	變壓器	transformer
变压吸附	變壓吸附	pressure swing adsorption(PSA)
变异性	變異性	variability
标称值	標稱值	nominal value
标度律	①比例定律②放大定律	scaling law
标度因子	結垢係數	scale factor
标度指示器	標度指示器	scale indicator
标量	純量	scalar

祖国大陆名	台湾地区名	英　文　名
标准	標準	standard
标准操作	標準操作	standard operation
标准操作步骤	標準操作步驟	standard operation procedure
标准差	標準偏差	standard deviation
标准稠度	標準稠度	standard consistency
标准大气压	標準大氣壓[力]	standard atmospheric pressure
标准电池	標準電池	standard cell
标准反应焓变化	標準反應焓改變量	standard enthalpy change of reaction
标准反应热	標準反應熱	standard heat of reaction
标准反应条件	標準反應條件	standard reaction condition
标准符号	標準符號	standard symbol
标准化	標準化	standardization
标准化合物	標準化合物	standard compound
标准化流程图	標準化流程圖	standardization flow-sheet
标准浇口	標準澆口	standard gate
标准孔口	標準孔口	standard orifice
标准量规	標準規	standard gage
标准螺纹弯头	標準螺旋彎頭	standard screwed elbow
标准命名法	標準命名法	standard nomenclature
标准喷嘴	標準注嘴	standard nozzle
标准燃烧热	標準燃燒熱	standard heat of combustion
标准溶液	標準溶液	standard solution
标准乳浊玻璃	標準乳濁玻璃	standard opal glass
标准筛	標準篩	standard screen, standard sieve
标准生成焓变化	標準生成焓改變量	standard enthalpy change of formation
标准生成热	標準生成熱	standard heat of formation
标准试验	標準試驗	standard test
标准试验筛	標準試驗篩	standard testing sieve
标准态	標準狀態	standard state
标准态逸度	標準狀態逸壓	standard state fugacity
标准温度	標準溫度	standard temperature
标准温度和压力	標準溫度和壓力	standard temperature and pressure
标准温度计	標準溫度計	standard thermometer
标准误差	標準誤差	standard error
标准纤维素	標準纖維素	standard cellulose
标准线性固体	標準線性固體	standard linear solid
标准状态	標準狀態	standard conditions
表	表	gauge
表观比热	視比熱	apparent specific heat

祖 国 大 陆 名	台 湾 地 区 名	英 文 名
表观比重	視比重	apparent specific gravity
表观纯度	視純度	apparent purity
表观活度	視活性	apparent activity
表观活化能	視活化能	apparent activation energy
表观剪切率	視剪率	apparent shear rate
表观剪切应力	視剪應力	apparent shear stress
表观阶[数]	視階	apparent order
表观孔隙率	視孔隙度	apparent porosity
表观扩散系数	視擴散係數	apparent diffusivity
表观密度(＝视密度)		
表观黏度	視黏度	apparent viscosity
表观溶解度	視溶解度	apparent solubility
表观速度	表觀速度	superficial velocity
表观体积	視體積	apparent volume
表观应力	視應力	apparent stress
表观质量	視質量	apparent mass
表观重力	視重力	apparent gravity
表观重量	視重量	apparent weight
表观组成	視組成	apparent composition
表面层	表面層	surface layer
表面处理	表面處理	surface treatment
表面粗糙度	表面糙度	surface roughness
表面粗化处理	表面粗化處理	surface roughening treatment
表面催化	表面催化[作用]	surface catalysis
表面电导	表面傳導	surface conductance
表面电荷	表面電荷	surface charge
表面电位	表面電位, 表面勢	surface potential
表面电阻	表面電阻	surface resistance
表面电阻率	表面電阻率	surface resistivity
表面发酵	表面發酵	surface fermentation
表面发射率	表面發射係數	surface emissivity
表面反应阻力	表面反應阻力	surface reaction resistance
表面覆盖度	表面覆蓋率	surface coverage
表面改性	表面改性	surface modification
表面改性剂	表面改性劑	surface modifier
表面改性纤维	表面改性纖維	surface modified fiber
表面更新理论	表面更新理論	surface renewal theory
表面更新因数	表面更新因數	surface renewal factor
表面工程	表面工程	surface engineering

祖国大陆名	台湾地区名	英 文 名
表面横向结晶	表面横列結晶, 表面跨晶結晶	surface transcrystallinity
表面化学	界面化學, 表面化學	surface chemistry
表面活性	表面活性, 界面活性	surface activity
表面活性剂	表面活性劑	surface-active agent, surfactant
表面活性剂液泛	界面活性劑泛流[法]	surfactant flooding
表面积	表面積	surface area
表面接枝	表面接枝	surface grafting
表面结构	表面結構	surface structure
表面卷取机	表面捲取機	surface winder
表面科学	表面科學	surface science
表面孔隙度	表面孔隙度	surface porosity
表面扩散	表面擴散	surface diffusion
表面冷凝器	表面冷凝器	surface condenser
表面力	表面力	surface force
表面裂纹	表面裂紋	surface crack
表面流变学	表面流變學	surface rheology
表面流动	表面流動	surface flow
表面轮廓	表面輪廓	surface profile
表面络合物	表面錯合物	surface complex
表面密度	表面密度	surface density
表面摩擦	表面摩擦	skin friction
表面磨蚀	表面磨蝕, 表面磨耗	surface abrasion
表面能	表面能	surface energy
表面黏度	表面黏度	surface viscosity
表面黏度计	表面黏度計	surface viscometer
表面浓度	表面濃度	surface concentration
表面培养	表面培養	surface cultivation
表面缺陷	表面缺陷	surface imperfection
表面施胶	表面上膠	surface sizing
表面水分	表面水分	surface moisture
表面速率	表面速率	surface rate
表面酸性部位	表面酸性部位	surface acid site
表面涂层	表面塗層	surface coating
表面位错	表面移位	surface dislocation
表面温度	表面溫度	skin temperature, surface temperature
表面吸附	表面吸附	surface adsorption
表面吸收器	表面吸收器	surface absorber
表面现象	表面現象	surface phenomenon

祖国大陆名	台湾地区名	英　文　名
表面形态	表面形態	surface morphology
表面压力	表面壓[力]	surface pressure
表面曳引	表面阻力	surface drag
表面硬化	表面硬化	case hardening, surface hardening
表面粘接	表面黏接	surface bonding
表面张力	表面張力	surface tension
表面张力计	表面張力計	surface tensiometer
表面制备	表面製備	surface preparation
表面状况	表面狀況	surface condition
表面自由能	表面自由能	surface free energy
表皮效应	表面效應	skin effect
表压	錶壓	gage pressure
宾厄姆流体	賓漢流體	Bingham fluid
冰醋酸	冰醋酸	glacial acetic acid
冰点	冰點	ice point
冰晶石	冰晶石	cryolite
冰糖	冰糖	rock sugar
冰浴	①冰浴②冰槽	ice bath
丙二酸	丙二酸	malonic acid
丙酸纤维素	丙酸纖維素	cellulose propionate, propionate cellulose
丙烯低聚物	低聚丙烯	propylene oligomer
丙烯腈	丙烯腈	acrylonitrile
丙烯腈-丁二烯-苯乙烯[树脂]	丙烯腈-丁二烯-苯乙烯[樹脂]	acrylonitrile-butadiene-styrene resin
丙烯酸树脂	壓克力樹脂, 丙烯酸酯樹脂	acrylic resin
丙烯酸酯	丙烯酸酯	acrylic ester
丙烯酸酯橡胶	丙烯酸酯橡膠	acrylic ester rubber
并发反应	併發反應	concurrent reaction
并联补偿	並聯補償	parallel compensation
并联电路	並聯電路	parallel circuit
并联运行	並聯運轉	parallel running
并列复式涡轮机	並列複式渦輪機	parallel compound turbine
并流操作	同向流操作	cocurrent operation
并流过程	同流程序	cocurrent process
并流热交换	並流熱交換	paraflow heat exchange
并流, 同向流	同向流動	cocurrent flow, paraflow
并行操作	並聯操作	parallel operation
并行过程	並聯程序	parallel processes

祖国大陆名	台湾地区名	英 文 名
病毒	病毒	virus
病原体	病原體	pathogen
波长	波長	wavelength
波传播	波之傳播	wave propagation
波动力学	波動力學	wave mechanics
波数	波數	wave number
波纹板	浪板	corrugated sheet
波纹管	波形管	corrugated tube
波纹管式流量计	伸縮囊式流量計	bellows type flowmeter
波纹管式压力计	伸縮囊壓力計	bellows manometer
波纹塔板	波動穿流板	ripple tray
波状流	波狀流	wavy flow
波状运动	波形運動	wave motion
玻耳兹曼分布	波茲曼分布	Boltzmann distribution
玻化瓷器	玻化瓷器	vitreous china
玻化相	玻化相	vitreous phase
玻璃	玻璃	glass
玻璃衬里	玻璃襯裡	glass lining
玻璃吹制	玻璃吹製	glass blowing
玻璃电极	玻璃電極	glass electrode
玻璃[固]化	玻化	vitrification
玻璃[固]化期	玻化期	vitrification period
玻璃[固]化砖	玻化磚	vitrified brick
玻璃管	玻璃管	glass pipe, glass tube
玻璃管及配件	玻璃管子及配件	glass pipe and fitting
玻璃过滤器	玻璃過濾器	glass filter
玻璃拉制	玻璃拉製	glass drawing
玻璃料	玻璃料	frit
玻璃棉	玻璃絨, 玻璃綿	glass wool
玻璃强化塑料	玻璃強化塑膠	glass reinforced plastics
玻璃区	玻璃區	glassy zone
玻璃态	玻璃狀態	glassy state
玻璃搪瓷	玻璃搪瓷	glass enamel
玻璃陶瓷	玻璃陶瓷	glass ceramics
玻璃温度计	玻璃溫度計	glass stem thermometer
玻璃纤维	玻璃纖維	glass fiber, glass textile
玻璃纤维强化塑料	玻璃纖維強化塑膠	glass fiber reinforced plastics (GFRP)
玻璃液位计	玻璃液位計, 玻璃量規	glass gage
玻璃转变温度	玻璃轉移溫度	glass transition temperature

祖国大陆名	台湾地区名	英　文　名
玻色-爱因斯坦分布	玻式愛因斯坦分布	Bose-Einstein distribution
剥离	剝離, 去皮	peeling
伯努利方程	白努利方程	Bernoulli equation
泊	泊	poise
铂重整	鉑媒重組	platforming
铂重整产品	鉑媒重組油	platformate
铂重整过程	鉑媒重組程序	platforming process
铂电阻球	鉑電阻球	platinum resistance bulb
铂电阻温度计	鉑電阻溫度計	platinum resistance thermometer
铂坩埚	鉑坩堝, 白金坩堝	platinum crucible
铂海绵	鉑海綿	platinum sponge
铂黑催化剂	鉑黑觸媒	platinum black catalyst
铂接触过程	鉑接觸法	platinum contact process
铂-镍催化剂	鉑鎳觸媒	platinum nickel catalyst
铂网	鉑網	platinum gauze
铂氧化铬催化剂	鉑鉻氧觸媒	platinum-chromia catalyst
铂氧化铝催化剂	附鋁氧觸媒	platinum-alumina catalyst
博登施泰数	博登施泰數	Bodenstein number
[博拉德]双层网环	雙層網環	Borad ring
薄层	薄片, 薄層, 一板	lamina
薄层色谱[法]	薄層分析[法]	thin layer chromatography
薄膜吹制法	薄膜吹製法	sheet blowing method
薄膜分离	薄膜分離	membrane separation
薄膜分离技术	薄膜分離技術	membrane separation technology
薄膜干燥机	薄膜乾燥器	film drier
薄膜闪蒸器	[急]驟[薄]膜蒸發器	flash film evaporator
薄膜蒸发器	薄膜蒸發器	Luwa evaporator, thin-film evaporator
薄片	薄片	slice
薄荷脑	薄荷腦	menthol
薄荷油	薄荷油	peppermint oil
簸动	簸選	jigging
簸动筛	簸選篩, 選礦篩	jigging screen
补偿	①補償②調整	compensation
补偿器	補償器	compensator
补偿温度计系统	補償溫度計系統	compensated thermometer system
补偿效应	補償效應	compensation effect
补强剂	補強劑	strengthening agent
捕获剂	捕獲劑	scavenger
捕集效率	捕集效率	collection efficiency

祖国大陆名	台湾地区名	英 文 名
捕液器	液阱	liquid trap
不摆应答	不攞應答	dead beat response
不饱和	不飽和	unsaturation
不饱和聚酯	不飽和聚酯	unsaturated polyester
不饱和烃	不飽和烴	unsaturated hydrocarbon
不饱和脂肪酸	不飽和脂肪酸	unsaturated fatty acid
不变时间系统	不變時間系統	time invarying system
不变式	不變式	invariant
不对称流动	不對稱流動	asymmetrical flow
不合格产品	不合規格產品	off-spec product
不互溶流动	不互溶流動	immiscible flow
不互溶溶剂	不互溶溶劑	immiscible solvent
不互溶系统	不互溶系統	immiscible system
不互溶相	不互溶相	immiscible phase
不互溶性	不互溶性	immiscibility
不互溶液体	不互溶液體	immiscible liquid
不挥发性	不揮發性	nonvolatility
不挥发油	不揮發油	fixed oil
不均匀性	不匀性	heterogeneity
不可逆催化	不可逆催化[作用]	irreversible catalysis
不可逆电池	不可逆電池	irreversible cell
不可逆反应	不可逆反應	irreversible reaction
不可逆功	不可逆功	irreversible work
不可逆过程	不可逆程序	irreversible process
不可逆[过程]热力学	不可逆熱力學	irreversible thermodynamics
不可逆化学反应	不可逆化學反應	irreversible chemical reaction
不可逆性	不可逆性	irreversibility
不可压缩流动	不可壓縮流動	incompressible flow
不可压缩流体	不可壓縮流體	incompressible fluid
不可压缩滤饼	不可壓縮濾餅	incompressible filter cake
不可压缩性	不可壓縮性	incompressibility
不冷凝气体	不冷凝氣體	incondensable gas
不良分布	不良分布	maldistribution
不凝气体	不冷凝氣體	noncondensable gas
不确定型决策	不確定型決策	decision making under uncertainty
不溶解化	不溶解化	insolubilization
不渗透性	不透性	impermeability
不透明度	乳白度	opacity
不透明釉	不透明釉	opaque glaze

祖国大陆名	台湾地区名	英 文 名
不透性隔板	不透性隔板	impermeable partition
不透性膜	不透性薄膜	impermeability membrane
不稳定操作点	不穩定操作點	unstable operating point
不稳定常数	不穩定常數	instability constant
不稳定区	易變區[域]	labile region, labile zone
不稳定系统	不穩定系統	unstable system
不稳定响应	不穩定應答	unstable response
不稳定性	不穩定性, 不安定性	instability
不锈钢	不銹鋼	stainless steel
不锈钢纤维	不銹鋼纖維	stainless-steel fiber
不皂化物	未皂化物	unsaponified matter
布朗扩散	布朗擴散	Brownian diffusion
布滤机	布濾機	cloth filter
布罗伊登法	布羅伊登法	Broyden method
布置	布置	layout
布置检查表	布置檢查表	layout check list
布置图	布置圖	layout diagram
步进电机	步進馬達	step motor
钚反应器	鈽反應器	plutonium reactor
部分反应	部分反應	partial reaction
部分分隔流动	部分隔離流動	partially segregated flow
部分分解	部分分解	partial decomposition
部分分离	部分分離	partial separation
部分分式	部分分式	partial fraction
部分互溶	部分互溶	partial miscibility
部分互溶系统	部分互溶系統	partially miscible system
部分回流	部分回流	partial reflux
部分甲基化[反应]	部分甲基化[反應]	partial methylation
部分解耦	部分解偶	partial decoupling
部分冷凝	部分冷凝	partial condensation
部分裂化	部分裂解	partial cracking
部分燃烧	部分燃燒	partial combustion
部分寿命	部分壽命	fractional life
部分衰减法	部分衰減法	fractional-life method
部分氧化	部分氧化	partial oxidation
部分氧化反应	部分氧化反應	partial oxidation reaction
部分游离	部分游離	partial ionization
部分再循环过程	部分循環程序	partial recycle process
部分真空	部分真空	partial vacuum

祖国大陆名	台湾地区名	英 文 名
部分转化	部分轉化	partial conversion

C

祖国大陆名	台湾地区名	英 文 名
擦洁剂	擦潔劑	abrasive cleaner
材料成本	材料成本	material cost
材料成本指数	材料成本指數	material cost index
材料工程	材料工程	materials engineering
材料合成	材料合成	material synthesis
材料加工	材料加工	material processing
材料科学	材料科學	materials science
材料设计	材料設計	material design
材料制造	材料製造	material manufacturing
财产税	財產稅	property tax
财务会计成本	財務會計成本	financing cost
采样控制	取樣數據控制,抽樣數據控制	sampled data control
采样控制系统	取樣數據控制系統,抽樣數據控制系統	sampled-data control system
采样数据	取樣數據	sampled data
采样数据系统	取樣數據系統,抽樣數據系統	sampled-data system
采样数据信号	取樣數據信號,抽樣數據信號	sampled-data signal
菜[子]油	菜子油,菜油	rape oil, rape seed oil
参比标准	參考標準	reference standard
参比电极	參考電池	reference electrode
参比态,参考态	參考狀態	reference state
参比条件	參比條件	reference condition
参比温度	參考溫度	reference temperature
参考接点	參考接點	reference junction
参考输入	參考輸入	reference input
参考态(=参比态)		
参考值	參考值	reference value
参考坐标	參考坐標	frame of reference
参数	參數	parameter
参数泵	參數泵	parametric pump

祖国大陆名	台湾地区名	英　文　名
参数法分离	參數法分離	parametric pumping
参数集总	塊集	lumping
参数灵敏度	參數靈敏度	parameter sensitivity
参数推算	參數推算	coaptation
参数选择	參數選擇	parameter selection
残差	殘留誤差	residual error
残留物	殘留產物	residual product
残液, 釜液	殘渣, 殘留物	residue
残余贡献, 剩余贡献	剩餘貢獻	residual contribution
残余焓, 剩余焓	殘留焓	residual enthalpy
残[余价]值	殘餘價值	salvage value
残余馏分	殘留餾分	residual fraction
残余气体	殘留氣體	residual gas
残余熵, 剩余熵	殘留熵	residual entropy
残余体积, 剩余体积	殘留體積	residual volume
残余项, 剩余项	殘留項	residual term
残余性质, 剩余性质	殘留性質	residual property
残余应变	殘留應變	residual strain
藏量(=[库]存量)		
操作变量	操作變量	operational variable
操作标准	操作標準	operation standard
操作参量	操作參數	operating parameter
操作成本	操作成本	operation cost
操作窗	操作窗	operating window
操作点	操作點	operating point
操作放大器	操作放大器	operation amplifier
操作界限	操作界限	operating limit
操作流程图	操作流程圖	operation flowsheet
操作频率	操作頻率	operating frequency
操作曲线	操作曲線	operating curve
操作时间	操作時間	operating time, operation length
操作手册	操作手冊	operating manual
操作数据	操作數據	operating data, operation data
操作弹性	操作彈性	operating flexibility, turndown ratio
操作条件	操作條件	operating condition, operation condition
操作图	操作圖	operating diagram, operation diagram
操作温度	操作溫度	operating temperature
操作稳定性	操作穩定性	operating stability
操作线	操作線	operating line, operation line

祖国大陆名	台湾地区名	英　文　名
操作效率	操作效率	operation efficiency
操作性能图	性能圖	performance chart
操作压力	操作壓力	operating pressure
操作原理	操作原理	operating principle
操作再生循环	操作再生循環	operation regeneration cycle
操作周期	操作周期	operation cycle
槽	槽	channel
槽[法炭]黑	槽黑	channel black
槽炉	槽爐	tank furnace
槽式反应器	槽式反應器	tank reactor
槽式结晶器	槽式結晶器	tank crystallizer
槽式流量计	槽式流量計	channel type flowmeter
草纸浆	草紙漿	straw pulp
侧回流	側回流	side reflux
侧基	側基	side group
侧冷却器	側冷卻器	side cooler
侧链基	側鏈基	side chain radical
侧链运动	側鏈運動	side chain motion
侧[链]支	側支	side branch
侧流	側流, 支流	side stream
侧流(＝旁路)		
侧馏分	側流餾出物	sidecut, sidecut distillate
侧线[馏分]汽提塔	側流汽提塔	side stream stripper, side stripper
侧线塔	側流塔	sidestream column
侧向应力	側向應力	lateral stress
测尘器	測塵器	konimeter
测定	測定	determination
测量	量測, 量度, 測量	measurement
测量方法	量測方法	measuring means
测量元件	量測元件	measuring element
测量滞后	量測落後	measurement lag, measuring lag
测量装置	量測裝置	measuring device
测黏流动	測黏[度]流動	viscometric flow
测声计	測音計	acousimeter
测湿法	測溼法	hygrometry
测试步骤	測試步驟	test procedure
测试点	測試點	test point
测试仪	試驗器	tester
测微计	測微計, 分厘卡	micrometer

祖国大陆名	台湾地区名	英 文 名
测温元件	測溫元件	temperature measuring element
测压孔	壓力接頭	pressure tap
测压压头	壓力高差	manometric head
测压液	測壓液	manometric fluid
测振仪	振動計	vibrometer
层间应力	層間應力	interlaminar stress
层离黏土	剝層黏土	delaminated clay
层流	流線流動, 層流	streamline flow
层流边界层	層流邊界層	laminar boundary layer
层流底层	層流次層	laminar sublayer
层流反应器	層流反應器	laminar flow reactor
层流区	層流區[域]	laminar region
层流式喷流吸收器	層式噴流吸收器	laminar jet absorber
层流, 滞流	層流, 直線流動	streamlined flow, laminar flow
层析法(＝色谱法)		
层压	積層	lamination
层压安全玻璃	積層安全玻璃	laminated safety glass
层压板壁	積層板壁	laminated wall
层压材料	積層板	laminate
层压塑料	層壓塑料	stratified plastic
层压压制机	積層壓製機	lamination press
差动阀	差壓閥	differential valve
差分方程	差分方程式	difference equation
差降	差降	drop
差角	差角	angle of difference
差热分析法	微差熱分析	differential thermal analysis
差示分析仪	微分分析樓	differential analyzer
差示热机械分析法	差示熱機械分析法	differential thermomechanical analysis
差示热重分析法	差示熱重分析法	differential thermogravimetric analysis
差示扫描量热法	差示掃描量熱法	differential scanning calorimetry
差示筛析	微分篩析	differential screen analysis
差示吸附	微分吸附	differential adsorption
差示吸收	微分吸收	differential absorption
差示 U 形管	微差 U[形]管	differential U-tube
差示压力计	微差壓力計	differential manometer
差式风压计	差式壓力計	draft gage
差速离心[分离]	差速離心分離	differential centrifugation
差压池	微壓差計	differential pressure cell
差压传感器	微差壓力轉換器	differential pressure transducer

祖国大陆名	台湾地区名	英 文 名
差压界限	微差壓力極限	differential pressure limit
差异压头	微差高差	differential head
柴油	柴油燃料	diesel fuel
柴油爆震	柴油震爆	diesel knock
柴油机	柴油機	diesel engine
掺合	掺合, 掺配	blend
掺合槽	掺配槽	blending tank
掺合过程	掺配程序	blending process
掺合机	掺合機	blender
掺合剂	掺合劑	blending agent
掺合式混合器	掺合式混合器	blending mixer
掺合物	掺合物	blend
掺合系统	掺配系統	blending system
掺气流动	通氣流動, 曝氣流動	aerated flow
掺入	掺雜	dope
掺杂	掺雜	doping
掺杂剂	掺雜劑	dopant
缠结	纏繞	entanglement
产量(＝通过量)		
产品成本	產品成本	product cost
产品储存	產品儲存	product storage
产品分布变异	產物分布畸變	product distribution distortion
产品分离	產物分離	product separation
产品分离器	產物分離器	product separator
产品工程	產品工程	product engineering
产品规格	產品規格	product specification
产品规划	產品規劃	product planning
产品回收	產物回收	product recovery
产品开发	產品開發	product development
产品设计	產品設計	product design
产品物流	產物流	product stream
产品运送检查[清]单	產品運送檢查報表	product shipment check list
产品质量	產品品質	product quality
产物分布	產物分布	product distribution
产物抑制	產物抑制[作用]	product inhibition
长管蒸发器	長管蒸發器	long tube evaporator
长宽比	縱橫比	aspect ratio
长石	長石	feldspar
长石釉	鈉鈣長石	feldspathic glaze

祖国大陆名	台湾地区名	英 文 名
长条记录纸	長條記錄紙	strip chart
长弯头	長彎頭	long sweep elbow
长网[成型]机	長綱[製紙]機	fourdrinier machine
长效促进剂	長效催速劑	persistent accelerator
长效剂	長效劑	persistent agent
肠蛋白酶	腸(肽)酶	erepsin
常产玻璃纤维	玻璃短纖維	staple glass fiber
常数	常數	constant
常微分方程	常微分方程式	ordinary differential equation
常压操作	常壓操作	atmospheric operation
常压分馏	常壓分餾器, 常壓分餾塔	atmospheric fractionator
常压平衡蒸馏	空氣平衡蒸餾	air equilibrium distillation
常压闪蒸塔	常壓驟餾塔	atmospheric flash tower
常压芽浆糖化	常壓製醪	atmospheric mashing
常压蒸馏	空氣蒸餾	air distillation
常压蒸汽	常壓蒸汽	atmospheric steam
厂址	廠址	plant location
厂址选择	廠址選擇	plant site selection
敞开系统	開放系統	open system
钞票纸	鈔票紙	bank-note paper
超薄膜	超薄膜	ultrathin film
超重整	超重組	hyperforming
超大型集成电路	超大型積體電路	very large scale integrated circuit
超导薄膜	超導薄膜	superconducting thin film
超导发电器	超導發生器	superconducting generator
超导合金	超導合金	superconducting alloy
超导聚合物	超導聚合物	superconductive polymer
超导态	超導態	superconducting state
超导碳氮化物	超導碳氮化物	superconducting carbonitride
超导特性	超導特性	superconducting characteristic
超导体	超導體	superconductor
超导性	超導性	superconductivity
超导转变温度	超導轉變溫度	superconducting transition temperature
超电势	過電位	overpotential
超额函数	過剩函數	excess function
超额焓	過剩焓	excess enthalpy
超额化学势	過剩化學勢	excess chemical potential
超额吉布斯自由能	過剩吉布斯自由能	excess Gibbs free energy

祖国大陆名	台湾地区名	英 文 名
超额浓度法	過量濃度法	excess concentration technique
超额熵	過剩熵	excess entropy
超额体积	過剩體積	excess volume
超额性质	過剩性質	excess property
超分子	超分子	supermolecule
超分子结构	超分子結構	supermolecular structure, supramolecular structure
超分子有序	超分子有序	supermolecular order
超分子织态结构	超分子織[態結]構	supermolecular texture
超分子转变	超分子轉變	supermolecular transition
超高聚物	超聚合物	superpolymer
超过滤	超微過濾	hyperfiltration
超级计算机	超級計算機	supercomputer
超加速剂	超催速劑	ultra accelerator
超晶格	超晶格	superlattice
超净化	超淨化, 超純化	ultrapurification
超拉伸	超拉伸	super drawing
超离子导电聚合物	超離子導電聚合物	super ion-conductive polymer
超临界萃取	超臨界萃取	supercritical extraction
超临界流	超臨界流動	supercritical flow
超临界气体	超臨界氣體	supercritical gas
超流体	超流體	superfluid
超滤	超過濾	ultrafiltration
超滤机	超過濾器	ultrafilter
超前补偿	超前補償	lead compensation
超声波沉淀	超音波沈澱	ultrasonic precipitation
超声波分解	超音波分解	sonolysis
超声波计	超音波計	ultrasonic meter
超声波流动	超音波流動	ultrasonic flow
超声波黏度计	超音波黏度計	ultrasonic viscometer
超声波学	超音波學	ultrasonics
超声波振动器	超音波振動器	ultrasonic vibrator
超声附聚	超日波黏聚	ultrasonic agglomeration
超声净化	超音波清潔	ultrasonic cleaning
超声速喷嘴	超音速噴嘴	supersonic nozzle
超速离心	超速離心	ultracentrifugation
超[速]离心机	超離心機	ultracentrifuge
超弹性	超彈性	superelasticity
超弹性材料	超彈性物質	hyperelastic material

祖国大陆名	台湾地区名	英 文 名
超调	超越量	overshoot
超微量化学	超微量化學	ultra microchemistry
超吸附[法]	超吸附[法]	hypersorption
超吸附器	超吸附器	hypersorber
超细研磨机	超細研磨機	ultragrinder
超显微法	超顯微法	ultra microscopy
超显微镜	超顯微鏡	ultra microscope
超限保护	超限保護	over range protection
超小型计算机	超級小型計算機	superminicomputer
超氧化物	超氧化物	superoxide
超越控制	超越控制	override control
潮解	潮解	deliquescence
潮霉素	溼球菌黴素	hygromycin
沉底酵母	沈底酵母	bottom yeast
沉淀	沈澱, 沈積	precipitation
沉淀槽	沉積槽	sedimentation chamber
沉淀池	沉積池	sedimentation basin
沉淀滴定[法]	沈澱滴定[法]	precipitation titration
沉淀反应	沈澱反應	precipitation reaction
沉淀聚合	沈澱聚合[反應]	precipitation polymerization
沉淀瓶	沉積瓶	sedimentation bottle
沉淀器	沈澱器	precipitator
沉淀物	沈澱物	precipitate
沉积槽	沉積槽	sedimentation tank
沉积电位	沉降電位	sedimentation potential
沉积高岭土	沉積高嶺土	sedimentary kaolin
沉积黏土	沉積黏土	sedimentary clay
沉积速率	沉積速率	sedimentation rate
沉积物	沉積物	sediment
沉积岩	沉積岩	sedimentary rock
沉降	沉積	sedimentation, settling
沉降槽	沈降槽	settling tank
沉降常数	沉積常數	sedimentation constant
沉降池	沈降池	settling basin
沉降法	沉積法	sedimentation method
沉降过滤	沈降過濾	settling filtration
沉降盘	沉積盤	sedimentation pan
沉降平衡	沉積平衡	sedimentation equilibrium
沉降平衡法	沉積平衡法	sedimentation equilibrium method

祖国大陆名	台湾地区名	英　文　名
沉降平均分子量	沉积平均分子量	sedimentation average molecular weight
沉降器	沈降器, 沈降槽	settler
沉降区	沈降區	settling zone
沉降时间	沈降時間, 安定時間	settling time
沉降室	沈降室	settling chamber
沉降速度	沈降速度, 沉積速度	settling velocity, sedimentation velocity
沉降速度法	沉積速度法	sedimentation velocity method
沉降系数	沉積係數	sedimentation coefficient
沉降柱	沈降柱	settling column
沉降柱分析	沈降柱分析	settling column analysis
沉料	沉料	jetsam
陈化	陳化, 老化	aging
衬玻璃管	玻璃襯管	glass-lined pipe
衬里	襯裏	lining
衬里管	襯管	lined pipe
成本	成本	cost
成本工程师	成本工程師	cost engineer
成本估计	成本估計值	cost estimate, cost estimation
成本会计	成本會計	cost accounting
成本曲线	成本曲線	cost curve
成本效益	成本效益	cost effectiveness
成本效益分析	成本效益分析	cost benefit analysis, cost-effective analysis
成本效益管理	成本效益管理	cost-effective management
成本因子	成本因數	cost factor
成本指数	成本指數	cost index
成对(＝对偶)		
成核	成核	nucleation
成核结晶	成核結晶	nucleation crystallization
成气反应	成氣反應	gas-forming reaction
成球	成球	balling, prilling
成酸菌	生酸菌	acid-forming bacteria
成套设备	套裝裝置	packaged equipment
成套设计文件	成套設計文件	design package
承插接头	承插接頭	bell and spigot joint
城市垃圾	都市廢棄物	municipal waste
城市污水	都市污水	municipal wastewater
乘法器	相乘器	multiplier
程长	路徑長度	path length
程序控制	程式化控制	programmed control

祖国大陆名	台湾地区名	英文名
程序控制器	程式化控制器	programmed controller
程序升温	溫度規劃	temperature programming
程序升温图	溫度規劃圖	temperature programmer
澄清	澄清[作用]	clarification, defecation
澄清过程	澄清程序	clarification process
澄清过滤	澄清過濾	clarification filtration
澄清过滤器	澄清器	clarifier filter
澄清能力	澄清能力	clarification capacity, clarifying capacity
澄清器	澄清器	clarifier, defecator
澄清区	澄清區	clarification zone
澄清汁	澄清汁	defecated juice
弛豫法(=松弛法)		
弛豫模量	鬆弛模數	relaxation modulus
弛豫时间	鬆弛時間	relaxation time
池沸腾	池沸騰	pool boiling
池沸腾曲线	池沸騰曲線	pool boiling curve
持气率	持氣率	gas holdup
尺寸比	尺寸比	size ratio
尺寸稳定性	尺寸穩定性	dimensional stability
尺寸因子	大小因數	size factor
齿轮	齒輪	gear
齿轮泵	齒輪泵	gear pump
齿轮式混合器	齒輪式混合器	gear mixer
齿轮组	齒輪系	gear train
齿隙	背隙[齒輪螺紋], 空檔	backlash
赤磷	赤磷	red phosphorus
赤铁矿	赤鐵礦	hematite
赤铜矿	赤銅礦	cuprite
翅管式热交换器	鰭片管熱交換器	finned-tube heat exchanger
翅片	鰭片	fin
翅片表面	鰭片表面	finned surface
翅片管	鰭片管	finned pipe, finned tube
充分发展流	完全發展流動	fully developed flow
充气	通氣	aeration
充气槽	通氣槽, 曝氣槽	aeration tank
充气管	通氣管	aeration tube
充气搅拌	空氣攪拌	air agitation
充气搅拌器	空氣攪拌器	air agitator
充气率	通氣速率	aeration rate

祖国大陆名	台湾地区名	英 文 名
充气轮	通氣輪	aeration wheel
充气器	通氣器	aerator
充气室	充氣室	plenum chamber
充气数	曝氣數, 通氣數	aeration number
充气装置	通氣裝置	aeration device
充填率	填充分率	packing fraction
充氧器	充氧器	oxygenizer
冲程	衝程	stroke
冲击	衝擊	impact
冲击波	震波	shock wave
冲击波反应器	沖擊波反應器	shock wave reactor
冲击波聚合	沖擊波聚合	shock wave polymerization
冲击腐蚀	衝射腐蝕	impingement corrosion
冲击凝固	沖擊凝固	shock coagulation
冲击器	衝擊機	impactor
冲击强度	衝擊強度	impact strength
冲击式涤气器	衝射洗氣器	impingement scrubber
冲击式离心分离器	衝射離心分離器	impingement centrifugal separator
冲击式破碎机	衝擊式壓碎機	impact crusher
冲击试验	耐衝擊試驗	impact test
冲击数	衝擊數	impact number
冲击弹性	沖擊彈性	shock elasticity
冲击压力	衝擊壓[力]	impact pressure
冲击压头	衝擊高差	impact head
冲击载荷	震動負載	shock load
冲角	攻角	angle of attack
冲量	衝量	impulse
冲洗	沖盡	washout
重氮化	重氮化[反應]	diazotization
重氮化合物	重氮化合物	diazo compound
重氮染料	重氮染料	diazo dye
重叠操作	重疊操作	overlapping operation
重沸器(=再沸器)		
重复操作	重覆操作	repetitive operation
重复性聚合物	重覆性聚合物	repetitive polymer
重合研光机	重合壓延機	doubling calendar
重整	重組[反應]	reforming
重整催化剂	重組觸媒	reforming catalyst
重整反应	重組反應	reforming reaction

祖国大陆名	台湾地区名	英 文 名
重整汽油	重組汽油	reformed gasoline
重整器	重組器	reformer
重组	重組	recombination
重组 DNA	重組 DNA	recombinant DNA
抽丝	抽絲	spinning
臭氧	臭氧	ozone
臭氧龟裂	臭氧紋裂	ozone cracking
臭氧化	臭氧化[反應]	ozonization
臭氧计	臭氧計	ozonometer
出口	出口	outlet
出口阀	出口閥	outlet valve
出口气体	出口氣體	exit gas
出口温度	出口溫度	outlet temperature
初步设计	初步設計	preliminary design
初沸点	初沸點	initial boiling point
初级沉降槽	初級沈降槽, 初級澄清器	primary settler
初级澄清器	初級澄清器, 初級沈降槽	primary clarifier
初级净化	初級純化	primary purification
初级蒸馏	初級蒸餾	primary distillation
初馏塔	初餾塔	primary tower
初滤机	初濾機	preliminary filter
初始松弛系统	初始鬆弛系統	initial relaxed system
初始条件	初始條件	initial condition
初[始状]态	初始狀態	initial state
初速[度]	初速度	initial velocity
初速率	初速率	initial rate
初速率法	初速率法	method of initial rate
初值定理	初值定理	initial value theorem
除草剂	除草劑	herbicide, weed killer
除尘器	集塵器	precipitator
除虫菊酯	除蟲菊精	pyrethrin
除臭	除臭	deodorization
除臭剂	除臭劑	deodorant, deodorizer
除焦	除焦	decoking
除焦机	除焦機	coke knocker
除焦时间	除焦時間	decoking time
除气[作用]	除氣	outgassing

祖国大陆名	台湾地区名	英　文　名
除湿器	除濕器	dehumidifier
除雾器	除霧器	demister
除锈剂	除銹劑	rust remover
储槽	儲存槽, 儲存容器	storage tank, storage vessel
储存	儲存	storage
储存量	儲存容量	storage capacity
储存设备	儲存裝置	storage equipment
储存设施	儲存設施	storage facility
储存损失	儲存損失	storage loss
储存温度	儲存溫度	storage temperature
储存装置流程图	儲存裝置流程圖	storage equipment flowsheet
储料堆	儲料	stock pile
储能函数	儲能函數	stored energy function
储热器	貯熱器	heat reservoir
储油罐	油槽	oil tank
处理工厂	處理工廠	treatment plant
处理过程	①處理程序②處理法	treatment process
处理设备	處理設備	treatment device
处置	①排列②配置③處理, 處置	disposal
触变胶	①觸變膠②流動減黏膠	thixotrope
触变性	觸變性	thixotropy
触变性流体	觸變流體	thixotropic fluid
触变性质	觸變性質	thixotropic property
穿孔	①打洞②孔, 洞	perforation
穿流塔板	雙流塔板	dual-flow tray
穿流栅板	渦輪格子板	turbo-grid tray
穿透点	失效點	breakthrough point
穿透概率	穿透機率	penetration probability
穿透理论	穿透理論	penetration theory
穿透曲线	失效曲線	breakthrough curve
穿透容量	失效時容量	breakthrough capacity
穿透时间	失效時間	breakthrough time
穿透性	穿透性	penetrability
传代时间	傳代時間	generation time
传导	傳導	conduction
传导性	傳導性	conductibility
传递	傳送, 轉移	transfer

祖国大陆名	台湾地区名	英 文 名
传递成型	轉移模製	transfer mold
传递单元	傳送單元	transfer unit
传递单元数	傳送單元數	number of transfer unit (NTU)
传递函数	轉移函數, 轉換函數	transfer function
传递扩散系数	輸送擴散係數	transport diffusivity
传递系数	傳送係數	transfer coefficient
传递现象	輸送現象	transport phenomenon
传递性质	輸送性質	transport property
传递滞后	傳送落後	transfer lag, transmission lag
传感器	感測器	sensor
传热	熱傳送	heat transfer
传热单元数	熱傳單元數	number of heat transfer unit
传热面积	熱傳面積	heat transfer area
传热膜系数	薄膜熱傳係數	film heat transfer coefficient
传热设备	熱傳裝置	heat transfer equipment
传热水力半径	熱傳水力半徑	hydraulic radius for heat transfer
传热速率	熱傳速率	heat transfer rate
传热污垢因子	熱傳積垢因素	scaling factor for heat transfer
传热系数	熱傳係數	heat transfer coefficient
传热 j 因子(= j_H 因子)		
传热阻力	熱傳阻力	heat transfer resistance
传输线路	傳送線路	transmission line
传氧速率	傳氧速率	oxygen transfer rate
传氧系数	傳氧係數	oxygen transfer coefficient
传质	質量傳送	mass transfer
传质操作	質傳操作	mass transfer operation
传质单元高度	質傳單元高度	height of a mass transfer unit
传质控制	質傳控制	mass transfer control
传质区	質傳區	mass transfer zone (MTZ)
传质速率	質傳速率	mass transfer rate
传质系数	質傳係數	mass transfer coefficient
传质 j 因子(= j_D 因子)		
传质装置	質傳裝置	mass transfer equipment
传质阻力	質傳阻力	mass transfer resistance
船底漆	船底塗料	ship bottom paint
串级	①串級②階式	cascade
串级补偿	串級補償	cascade compensation
串级分馏板	串級分餾板	cascade-type plate
串级控制	串級控制	cascade control

祖国大陆名	台湾地区名	英 文 名
串级冷冻	串級冷凍	cascade refrigeration
串级冷却器	串級冷卻器	cascade cooler
串级浓缩板	串級濃縮板	cascade concentration plate
串级浓缩器	串級濃縮器	cascade concentrator
串级塔	串級塔	cascade tower
串级系统	串級系統	cascade system
串级蒸发器	串級蒸發器	cascade evaporator
串级组合	串級組合	cascade combination
串联复式涡轮机	串列複式渦輪機	tandem compound turbine
床层塌落技术	床層塌落技術	bed-collapsing technique
床层支架	床支架	bed support
床栅支架	床柵支架	bed grid support
吹管	吹管	blow pipe
吹模	吹模	blow mold
吹扫扩散	掃掠擴散	sweep diffusion
吹扫气	沖洗氣	purge gas
吹塑[成型]	吹氣成型法	blow molding
吹制玻璃	吹製玻璃	blown glass
吹制油	吹煉油	blown oil
垂直筛板	垂直篩板	vertical sieve tray
锤[式]磨[碎]机	鎚磨機	hammer mill
锤[式破]碎机	鎚碎機	hammer crusher
春材	春材,早材[木材]	spring wood
纯度	純度	purity
纯绿宝石	純綠寶石,祖母綠,翡翠	emerald
纯氧活性污泥法	純氧活性污泥法	unox process
纯组分	純成份	pure component
醇解	醇解	alcoholysis
醇酶	酒精酶	alcoholase
醇酸树脂	酸醇樹脂	alkyd resin
醇质清漆	酒精清漆	spirit varnish
瓷	瓷	porcelain
瓷坩埚	瓷坩堝	porcelain crucible
瓷绝缘子	瓷絕緣體	porcelain insulator
瓷漆	瓷漆	enamel paint
瓷漆干燥室	瓷漆乾燥室	enamel drier
瓷[器]	瓷,瓷器	china, chinaware
瓷土	瓷土	china clay

祖国大陆名	台湾地区名	英 文 名
瓷釉	瓷釉	porcelain glaze
瓷砖	磁磚	ceramic tile
磁场	磁場	magnetic field
磁带	磁帶	magnetic tape
磁导率	磁導率	permeability
磁分离	磁力分離	magnetic separation
磁化率	磁化率	magnetic susceptibility
磁化学	磁化學	magnetochemistry
磁黄铁矿	磁黄鐵礦	pyrrhotite
磁矩	磁矩	magnetic moment
磁力分离器	磁力分離器	magnetic separator
磁力过滤	磁力過濾	magnetic filtration
磁力流态化	磁力流態化	magneto fluidization
磁流体动力学	磁性流體力學	magnetohydrodynamics
磁盘	磁碟	magnetic disk
磁式数据储存	磁式資料儲存	magnetic data storage
磁铁矿	磁鐵礦	magnetite
磁稳流化床	磁穩流化床	magnetically stabilized fluidized bed
磁性不纯物	磁性不純物	magnetic impurity
次层	次層	sublayer
次级测量值	次量測值	secondary measurement
次级产品	次產物	secondary product
次级弛豫温度	次級鬆弛溫度	secondary relaxation temperature
次级矩	次級矩	second moment, second order moment
次级蠕变	次級蠕變	secondary creep
次级转变	次級轉變	secondary transition
次氯酸盐漂白	次氯酸鹽漂白	hypochlorite bleaching
次生胞壁	次生胞壁	secondary cell wall
次烟煤	次煙煤	subbituminous coal
次优控制	次最適控制	suboptimal control
次优控制器	次最適控制器	suboptimal controller
次优性	次最適性	suboptimality
次正应力系数	次正應力係數	secondary normal stress coefficient
刺激物	刺激物	irritant
粗糙标度	粗糙標度	roughness scale
粗糙度	粗糙度	roughness
粗糙因子	粗糙因數	roughness factor
粗滴乳状液	巨乳液	macro emulsion
粗级压碎机	粗級壓碎機	coarse crusher

祖国大陆名	台湾地区名	英 文 名
粗颗粒	粗顆粒	coarse particle
粗滤	粗濾	straining
粗滤床	粗濾床	roughing bed, roughing filter
粗滤过程	粗濾程序	straining process
粗略分离	粗略分離	sloppy separation
粗筛[网]	粗篩	coarse screen
粗筛选	初級篩選	preliminary sizing
粗悬浮液	粗懸浮液	coarse suspension
促进化催化剂	促進化觸媒	promoted catalyst
促进剂的早熟性	催速劑早熟性	precocity of accelerator
醋酸菌	醋酸菌	acetic acid bacteria
醋酸人造丝	醋酸嫘縈, 乙酸嫘縈	acetate rayon
醋酸纤维素	醋酸纖維素	acetyl cellulose, cellulose acetate
醋酸酯薄膜	醋酸酯膠膜, 乙酸酯膠膜	acetate film
催干剂	乾燥劑	drier, dryer
催化部位	催化部位	catalytic site
催化重整	觸媒重組	catalytic reforming
催化重整过程	觸媒重組法, 媒組法	catalytic reforming process
催化促进剂	觸媒促進劑, 助催化劑	catalytic promoter
催化动力学	催化動力學	catalytic kinetics
催化反应	催化反應, 觸媒反應	catalytic reaction
催化反应器	觸媒反應器	catalytic reactor
催化分解	催化分解	catalytic decomposition
催化过程	催化程序	catalytic process
催化化学吸附	催化化學吸附	catalytic chemisorption
催化活性	催化活性	catalytic activity
催化剂	觸媒, 催化劑	catalyst, catalytic agent
催化剂床[层]	觸媒床	catalyst bed
催化剂活化	觸媒活化	catalyst activation
催化剂活性	催化活性	catalyst activity
催化剂[几何]形状	觸媒[幾何]形狀	catalyst geometry
催化剂架	觸媒架	catalyst support
催化剂接收器	觸媒接收器	catalyst receiver
催化剂结焦	觸媒結焦	catalyst coking
催化剂浸渍	觸媒[之]浸漬	catalyst impregnation
催化剂孔隙度	觸媒孔隙度	catalyst porosity
催化剂粒子	觸媒粒子	catalyst particle
催化剂磨损	觸媒磨耗	attrition of catalyst, catalyst attrition

祖国大陆名	台湾地区名	英 文 名
催化剂黏结剂	觸媒黏結劑	catalyst binder
催化剂配方	觸媒[之]配方	catalyst formulation
催化剂球粒	觸媒粒	catalyst pellet
催化剂烧尽釜	觸媒清燒蒸餾器	catalyst burn-off still
催化剂失活	觸媒去活化	catalyst deactivation
催化剂寿命	觸媒壽命	catalyst life
催化剂衰变	觸媒衰變	catalyst decay
催化剂网	觸媒網	catalytic gauze
催化剂网反应器	觸媒網反應器	catalytic-gauze reactor
催化剂稳定剂	觸媒安定劑	catalyst stabilizer
催化剂污染	觸媒積垢	catalyst fouling
催化剂污染指数	觸媒污染指數	catalyst contamination index
催化剂稀释剂	觸媒稀釋劑	catalyst diluent
催化剂细孔结构	觸媒細孔結構	catalyst pore structure
催化剂循环	觸媒循環	catalyst circulation
催化[剂]抑制剂	觸媒抑制劑	catalyst inhibitor
催化剂有效性	觸媒有效性	catalyst effectiveness
催化剂载体	觸媒載體, 催化劑載體	catalyst carrier, catalyst support, catalytic carrier
催化剂再生	觸媒再活化, 觸媒再生	catalyst reactivation, catalyst regeneration
催化剂中毒	觸媒中毒	catalyst poisoning
催化剂转化	觸媒轉化	catalytic conversion
催化剂转化器	觸媒轉化器	catalytic converter
催化金属	催化性金屬	catalytic metal
催化聚合	催化聚合[反應]	catalytic polymerization
催化裂化	[觸]媒裂[解]法	catalytic cracking
催化裂化过程	觸媒裂解法, 煤裂法	catalytic cracking process
催化面	催化表面	catalytic surface
催化面积	催化面積	catalytic area
催化系统	觸媒系統	catalytic system
催化循环	催化循環	catalytic cycle
催化氧化	催化氧化[反應]	catalytic oxidation
催化增效剂	觸媒增效劑	catalyst extender
催化专一性	催化特定性	catalytic specificity
催化[作用]	催化[作用]	catalysis
催化作用	觸媒作用, 催化作用	catalytic action
催化作用机理	觸媒作用機構, 催化機構	mechanism of catalysis
催泪毒气	催淚毒氣	tear gas

祖国大陆名	台湾地区名	英 文 名
脆点	脆化點	brittle point
脆性	脆度, 脆性	brittleness, friability
脆性物料	脆性物料	brittle material
萃取	萃取	extraction
萃取超滤	萃取性超過濾	extractive ultrafiltration
萃取过程	萃取程序	extraction process
萃取结晶	萃取結晶	extractive crystallization
萃取结晶器	萃取結晶器	extractive crystallizer
萃取率	萃取速率	extraction rate
萃取器	萃取器	extractor
萃取器组	萃取器組	extraction battery
萃取塔	萃取塔	extraction column, extraction tower
萃取液	萃取液	extract
萃取因子	萃取因數	extraction factor
萃取蒸馏	萃取蒸餾	extractive distillation
萃余物, 抽余液	萃餘物	raffinate
存储单元	記憶格	memory cell
错流	横流, 交叉流動, 交流	crosscurrent, crossflow
错流操作	横流操作, 交流操作	crosscurrent operation, crossflow operation
错流萃取	横流萃取	crosscurrent extraction
错流吸附器	交叉流動吸附器	crossflow adsorber

D

祖国大陆名	台湾地区名	英 文 名
搭接接头	搭接	lap joint
打浆机	打漿機, 攪合機	beater, hollander, pulp beater
打泡	攪打	whipping
打印墨	打印墨	stamp pad ink
大肠杆菌	大腸桿菌	Escherichia coli
大豆蛋白质	大豆蛋白質	soybean protein
大豆纤维	大豆蛋白纖維	soybean fiber
大豆油	黃豆油	soybean oil
大分子	巨分子	macromolecule
大分子(=高分子)		
大孔	巨觀細孔	macropore
大气扩散	大氣擴散	atmospheric diffusion
大气冷凝器	氣壓冷凝器	barometric condenser

祖国大陆名	台湾地区名	英 文 名
大气排放管	大氣排泄管	barometric discharge pipe
大气品位	空氣品質	air quality
大气品位指数	空氣品質指數	air quality index
大气腿	氣壓真空柱, 氣壓管	barometric leg, barometric pipe
大气压	大氣壓力	atmospheric pressure, barometric pressure
大小比值	大小比值, 量比值	magnitude ratio
大小量度	大小量度	magnitude scaling
大中取大判据	大中取大判據	maximax criterion
大中取小遗憾判据	大中取小遺憾判據	minimax regret criterion
大宗化学品	大宗化學品	commodity chemical
大宗聚合物	大宗聚合物	commodity polymer
代谢控制	代謝控制	metabolic control
代谢速率	代謝速率	metabolic rate
代谢物	代謝物	metabolite
代谢[作用]	代謝作用, 新陳代謝	metabolism
带静电[作用]	帶靜電[作用]	static electrification
带宽	頻帶寬度	bandwidth
带式干燥器	帶式乾燥器	belt dryer
带式过滤机	帶濾機	belt filter
带状矩阵	帶式矩陣	banded matrix
袋滤器	袋濾器	bag filter
袋模塑	袋式成型	bag moulding
袋式过滤分离器	袋形過濾分離器	bag filter separator
袋式集尘器	袋式集塵器	bag type dust collector
袋式研磨机	袋式研磨機	pocket grinder
单部位机理	單部位機構	single-site mechanism
单程操作	單程操作	once-through operation
单程过程	單程程序	once-through process
单程加热器	單程加熱器	single-pass heater
单程收率	單程產率	yield per path
单程转化率	單程轉化率	single pass conversion
单冲程预塑机	單衝程預塑機	single-stroke preforming press
单纯浸渍法	單純浸漬法	straight dipping process
单纯形法	簡式法	simplex method
单纯形算法	簡式運算[法], 簡式演算[法]	simplex algorithm
单道分析仪	單波道分析儀	single channel analyzer
单底物反应	單受質反應	single substrate reaction
单点测定法	單點測定法	single-point determination

祖国大陆名	台湾地区名	英 文 名
单垫过滤机	單墊濾機	monopad filter
单动泵	單動泵	single acting pump
单反应	單一反應	single reaction
单分布组分	單分布成分	single distributed component
单分隔点	單分隔點	single split point
单分散	單分散	monodisperse
单分散性	單分散	monodispersion
单分子反应	單分子反應	monomolecular reaction, unimolecular reaction
单缸往复泵	單缸往復泵	simplex reciprocating pump
单股聚合物	單股聚合物	single-strand polymer
单辊压碎机	單輥壓碎機	single roll crusher
单环路控制	單環路控制	single-loop control
单级操作	單階操作	single stage operation
单级萃取	單階萃取	single stage extraction
单级导向阀	單級導引閥	single stage pilot valve
单级过程	單階程序	single-stage process
单级离心泵	單級離心泵	single stage centrifugal pump
单级离心机	單級離心機	single stage centrifuge
单级蒸馏	單階蒸餾	single stage distillation
单极点	單極點	simple pole
单晶	單晶	single crystal
单晶纤维	單晶纖維	single-crystal fiber
单克隆抗体	單克隆抗體	monoclonal antibody
单利	單利	simple interest
单零点	單零點	simple zero
单螺杆混合机	單軸混合器	single-screw mixer
单螺杆挤出机	單軸壓出機	single-screw extruder
单螺线螺杆	單絲螺桿	single flighted screw
单面涂布	單面塗佈	single spread
单宁	單寧, 鞣酸	tannin
单宁酸	單寧酸, 鞣酸	tannic acid
单偶氮染料	單偶氮染料	monoazo dye
单腔模具	單穴模	single cavity mold
单桥聚合物	單橋聚合物	single-bridged polymer
单球菌	單球菌	monococcus
单色发射能力	單色發射能力	monochromatic emissive power
单纱	單捻紗	single yarn
单纱强力试验机	單紗強度試驗儀	single thread tester

祖国大陆名	台湾地区名	英　文　名
单输入单输出系统	單輸入單輸出系統	single-input-single-output system
单丝	單絲, 單絲纖維, 絲狀纖維	filament,monofilament
单丝加捻	單捻線, 單紗捻度	single twist
单糖	單醣	monosaccharide
单体	單體	monomer
单位成本	單位成本	unit cost
单位成本估计	單位成本估計	unit cost estimation
单位回馈	單位回饋	unit feedback
单细胞蛋白	單細胞蛋白	single cell protein
单细胞生长	單細胞生長	unicellular growth
单相反应器	單相反應器	single-phase reactor
单相系统	單相系統	single-phase system
单向阀(=止逆阀)		
单向反应	單向反應, 不可逆反應	unidirectional reaction
单向解偶	單向解偶	one-way decoupling
单向狭缝	單向狹縫	unilateral slit
单效蒸发器	單效蒸發器	single effect evaporator
单型腔模	單模	single impression mold
单一稳态	單一穩態	unique steady state
单一液体近似法	單一液體近似法	single liquid approximation
单元操作	單元操作	unit operation
单元过程	單元程序, 單元方法	unit process
单元计算	單元計算	unit computation
单元折旧率	單一貶值	single unit depreciation
单轴应力	單軸向應力	uniaxial stress
单组分	單成分	single component
单组分系统	單成分系統	one-component system
胆固醇	膽固醇	cholesterol
胆甾状态	膽固醇型液晶態	cholesteric state
蛋白黑素	類黑精	melanoidin
蛋白酶	蛋白酶	protease, proteinase
蛋白水解	蛋白質水解	proteolysis
蛋白[水解]酶	蛋白解酶	proteolytic enzyme
蛋白质	蛋白質	protein
蛋白质分级	蛋白質分級	protein fractionation
蛋白质工程	蛋白質工程	protein engineering
氮肥	氮肥	nitrogenous fertilizer
氮化硅	氮化矽	silicon nitride

祖国大陆名	台湾地区名	英 文 名
氮芥子气	氮芥子氣	nitrogen mustard gas
氮气清除	氮氣沖洗	nitrogen purge
氮循环	氮循環	nitrogen cycle
氮氧需要量	氮氧需要量	nitrogen oxygen demand
弹内[加氧]陈化	彈内[加氧]老化	bomb aging
弹式反应器	彈式反應器	bomb reactor
弹式量热器	彈卡計	bomb calorimeter
当量	當量	equivalent
当量长度	相當長度	equivalent length
当量电导	當量導電度	equivalent conductance
当量粒径	相當粒徑	equivalent particle diameter
当量收缩管长	縮管相當管長	contraction equivalent pipe length
当量直径	相當直徑	equivalent diameter
挡板	擋板	flapper
挡板痕	擋板痕	baffle mark
挡板混合器	擋板式混合器, 擋板柱混合器	baffle mixer, baffle plate column mixer
挡板喷嘴机构	擋板噴嘴機構	flapper nozzle mechanism
挡板喷嘴系统	擋板噴嘴系統	baffle nozzle system
挡板室	擋板室	baffle chamber
挡板塔	擋板塔	baffle column, baffle plate tower, baffle tower
挡板塔混合器	擋板塔混合器	baffle column mixer
挡板塔盘	擋板盤	baffle pan
档板,折流板	擋板,折流板,緩衝板	baffle
导出单位	導出單位	derived unit
导电高分子	導電高分子	electroconductive polymer
导电计	導電計	electrolytic conductance meter
导电聚合物	導電聚合物	conductive polymer
导管	導管	duct
导管	導管	conduit
导流筒	通風管,流通管	draft tube
导流筒挡板	通風管擋板	draft tube baffle
导流筒挡板结晶器, DTB 结晶器	導流擋板結晶器	draft-tube-baffled crystallizer
导流筒混合器	流通管混合器	draft tube mixer
导热率(＝导热系数)		
导热系数	熱傳導係數	thermal conductivity, thermal conductivity coefficient

祖国大陆名	台湾地区名	英　文　名
导数	導數	derivative
导数膨胀计测定法	導數膨脹測定法	derivative dilatometry
导数热重分析法	導數熱重分析法	derivative thermogravimetric analysis, DTGA, derivative thermogravimetry
导体催化剂	導體觸媒	conductor catalyst
导温系数(＝热扩散系数)		
导线	導線	lead wire
导线误差	導線誤差	lead wire error
导向阀	導引閥	pilot valve
倒流	反流	backflow
倒数标绘	倒數圖	reciprocal plot
倒数法则	倒數法則	reciprocal rule
倒数关系	倒數關係	reciprocal relation
倒焰窑	倒焰窯	downdraft kiln
倒易律	互反律	reciprocating law
捣矿杆组	搗礦杆組	stamp battery
捣磨机	搗碎機	stamp mill
到达角	到達角	angle of arrival
灯黑	燈煙, 燈黑	lamp black
灯笼填函盖	燈籠填函蓋	lantern gland
等边三角图	等邊三角圖	equilateral triangular diagram
等沉降粒子	等沉降粒子	equal-settling particle
等当点	當量點	equivalent point
等电沉淀	等電沈澱	isoelectric precipitation
等电点	等電點	isoelectric point
等电点分离	等電點分離	isoelectric separation
等电聚焦	等電聚焦	isoelectric focusing
等动力学点	等動力點	isokinetic point
等动力学温度	等動力溫度	isokinetic temperature
等规聚合物(=全同立构聚合物)		
等焓过程	等焓程序	isenthalpic process
等焓膨胀	等焓膨脹	isenthalpic expansion
等焓压缩	等焓壓縮	isenthalpic compression
等离点	等離點	isoionic point
等离子加强蚀刻	電漿加強蝕刻	etching plasma-enhanced
等离子蚀刻	電漿蝕刻[法]	plasma etching
等离子蚀刻[法]	電漿蝕刻法	etching plasma

祖国大陆名	台湾地区名	英文名
等离子体	電漿	plasma
等离子增强蚀刻	電漿加強蝕刻[法]	plasma-enhanced etching
等[理论]板高度, 理论板当量高度	理論板相當高度	height equivalent of a theoretical plate
等摩尔逆向扩散	等莫爾逆向擴散	equilmolar counter diffusion
等浓度曲线	等濃度曲線	isoconcentration curve
等容过程	等容程序	isochoric process, isometric process
等容流程图	等距流程圖, 等角流程圖	isometric flowsheet
等容热容	等容熱容量	heat capacity at constant volume
等容线	等容線	isochore, isometrics
等熵过程	等熵程序	isentropic process
等熵流动	等熵流動	isentropic flow
等熵膨胀	等熵膨脹	isentropic expansion
等熵压缩	等熵壓縮	isentropic compression
等渗溶液	等滲壓溶液	isotonic solution
等渗压浓度	等滲壓濃度	isotonic concentration
等式约束	等式約束	equality constraint
等势线	等位線	equipotential line
等速电泳	等速電泳	isotachophoresis
等温操作	等溫操作	isothermal operation
等温多重性	等溫多重性	isothermal multiplicity
等温反应	等溫反應	isothermal reaction
等温反应器	等溫反應器	isothermal reactor
等温过程	等溫程序	isothermal process
等温环境	等溫環境	isothermal environment
等温扩散	等溫擴散	isothermal diffusion
等温流动	等溫流動	isothermal flow
等温膨胀	等溫膨脹	isothermal expansion
等温热重量分析法	等溫熱重量法	isothermal thermogravimetry
等温温度相当值	等溫溫度相當值	equivalent isothermal temperature
等温吸附	等溫吸附	isothermal adsorption
等温线	等溫線	isotherm, isothermal line
等温线迁移	等溫線遷移	isotherm migration
等温压缩	等溫壓縮	isothermal compression
等温压缩性	等溫壓縮性	isothermal compressibility
等温有效性	等溫有效度	isothermal effectiveness
等温蒸馏	等溫蒸餾	isothermal distillation
等相角曲线	等相角曲線	constant phase angle curve

祖国大陆名	台湾地区名	英　文　名
等相位轨迹	等相位軌跡	locus of constant phase
等效自由沉降直径	等效自由沉降直徑	equivalent free-falling diameter
等斜线	等斜率線	isocline
等压法	等蒸汽壓法	isopiestic method
等压过程	等壓程序	isobaric process
等压过热	等壓過熱	isobaric superheating
等压热容	等壓熱容量	heat capacity at constant pressure
等压线	等壓線	isobar
等值管长度	相當管長	equivalent pipe length
等值线图表	輪廓圖	contour plot
低共熔点	共熔點	eutectic point
低共熔合金	共熔合金	eutectic alloy
低共熔平衡	共熔平衡	eutectic equilibrium
低共熔温度	共熔溫度	eutectic temperature
低共熔物	共熔混合物, 共熔物	eutectic mixture, eutectics
低共熔系统	共熔系統	eutectic system
低共熔状态	共熔狀態	eutectic state
低聚反应	低[元]聚合[反應]	oligomerization
低聚糖(＝寡糖)		
低聚物	低聚合物	oligomer
低硫天然气	脫臭氣	sweet gas
低密度聚乙烯	低密度聚乙烯	low density polyethylene
低热值煤气	低熱量氣體	low-BTU gas
低渗溶液	低滲壓溶液	hypotonic solution
低通滤波器	低頻濾波器	low pass filter
低温工业	低溫工業	cryogenic industry
[低温]焊料	軟焊料, 焊錫	solder
低温碳化	低溫碳化[法]	low temperature carbonization
低选择器开关	低選擇器開關	low selector switch
低于额定值	①低越量②低越[現象]	undershoot
滴	滴	drop
滴滴涕	滴滴梯	dichloro-diphenyl-trichloroethane (DDT)
滴定	滴定[法]	titration
滴定[分析]法	滴定[分析]法	titrimetric method, titrimetry
滴定计	微滴定	titrimeter
滴干法	滴乾	drip dry
滴口	滴口	drip
滴流	滴流	trickling

祖国大陆名	台湾地区名	英 文 名
滴流床反应器	滴流床反應器	trickle-bed reactor
滴流床, 涓流床	滴流床	trickle bed
滴滤	滴濾[法]	trickling filtration
滴滤池	滴濾池	trickling filter
滴相	滴相	drop phase
滴状冷凝	滴式冷凝	dropwise condensation
底板	底板	bottom plate
底部产物	底部産物	bottom product
底部出料	底部卸料	bottom discharge
底部液位控制	塔底液位控制	bottom-level control
底部蒸汽	塔底蒸汽	bottom steam
底阀	底閥, 塔底閥腳閥	foot valve, bottom valve
底流	底流	underflow
底流速度	底流速度	underflow velocity
底流速率	底流速率	underflow rate
底漆	底漆, 底塗	primer, undercoat
底物, 基质	底物, 基質, 基材	substrate
底物抑制	受質抑制[作用]	substrate inhibition
底相, 下相	底相	bottom phase
底座	底座, 拖架	pedestal
芪	1,2-二苯乙烯	stilbene
芪染料	二苯乙烯染料	stilbene dye
地蜡	地蠟	earth wax
地下水	地下水	groundwater
地下水处理	地下水處理	groundwater treatment
地下水污染	地下水污染	groundwater pollution
递归	遞回	recursion
递归公式	遞回公式	recursion formula
递归关系[式]	遞回關係[式]	recursive relation
递归函数	遞回函數	recursive function
递减均衡法	遞減均衡法	declining balance method
递阶控制	階梯控制	hierarchical control
第二维里系数(=第二位力系数)		
第二位力系数, 第二维里系数	第二位力係數	second virial coefficient
第三维里系数(=第三位力系数)		

祖国大陆名	台湾地区名	英　文　名
第三位力系数, 第三维里系数	第三維里係數	third virial coefficient
缔合溶液模型	締合溶液模型	associated solution model
蒂勒模数	蒂勒模組	Thiele modulus
点焊	點焊	spot welding
点焊机	點焊機	spot welder
点汇座	點匯座	point sink
点火时间	點火時間	ignition time
点胶合	點膠合, 局部膠合	spot gluing
点蚀	麻點腐蝕	pitting corrosion
点效率	點效率	point efficiency
点选择性	點選擇性	point selectivity
点源	點源	point source
点阵	格子, 晶格	lattice
碘值	碘價, 碘值	iodine value
电槽室	電槽室	cell house
电场	電場	electric field
电沉积	電沉積[法]	electrodeposition
电沉积物	電積層[法]	electrodeposit
电瓷	絕電瓷	electric porcelain
电磁阀	電磁閥	solenoid valve
电磁分离[法]	電磁分離[法]	electromagnetic separation
电磁辐射	電磁輻射	electromagnetic radiation
[电]磁搅拌器	磁力攪拌器	magnetic stirrer
电磁离析器	電磁分離器	electromagnetic separator
电磁流量计	電磁流量計, 磁力流量計	electromagnetic flowmeter, magnetic flowmeter
电磁谱	電磁[波]譜	electromagnetic spectrum
电导	電導度	conductance
电导池	導體電池	conductivity cell
电导滴定	電導滴定法	conductometric titration
电导计	熱導計, 電導計	conductometer
电导率	導電係數, 傳導度	electric conductivity , conductivity
电导系数	導電係數	electric conductivity coefficient
电导性	電傳導性	electric conductivity
电动传送	電動傳送	electric transmission
电动传送器	電動傳送器	electric transmitter
电动机	馬達, 電動機	motor
电动机常数	馬達常數	motor constant

祖国大陆名	台湾地区名	英 文 名
电动机驱动器	馬達引動器	motor actuator
电动控制	電動控制	electric control
电动-气动引动器	電動氣動[式]引動器	electropneumatic actuator
电动势	電動勢	electromotive force (EMF)
电动天平	電動天平	electrobalance
电动信号	電動信號	electric signal
电动序	電動勢序	electromotive series
电动液压引动器	電動液壓[式]引動器	electrohydraulic actuator
电动引动器	電動引動器	electric actuator
电镀	電鍍	electroplating
电镀工程	電鍍工程	electroplating engineering
电镀工业	電鍍工業	plating industry
电分散	電分散	electrodispersion
电浮选法	電浮選[法]	electroflotation
电弧炉	電弧爐	arc furnace
电化当量	電化學當量	electrochemical equivalent
电化学	電化學	electrochemistry
电化学反应	電化學反應	electrochemical reaction
电化学反应工程	電化學反應工程	electrochemical reaction engineering
电化学反应器	電化學反應器	electrochemical reactor
电化学工业	電化學工業	electrochemical industry
电化学势	電化學勢	electric chemical potential
pH 电极	pH 電極	pH electrode
电极板除尘器	電板集塵器	electric plate precipitator
电解	電解	electrolysis
电解槽	電解槽	electrolysis bath, electrolyzer
电解池	電解槽	electrolytic cell
电解分离	電解分離	electrolytic separation
电解分析	電分析	electroanalysis
电解氟化	電氟化[反應]	electrofluorination
电解工业	電解工業	electrolytic industry
电解过程	電解法	electrolytic process
电解加氢二聚合	電解加氫二聚合[反應]	electrohydrodimerization
电解氧化	電解氧化	electrolytic oxidation
电解质	電解質	electrolyte
电控制器	電動式控制器	electric controller
电类比	電類比	electric analogy
电离	游離	ionization

祖国大陆名	台湾地区名	英　文　名
电离常数	游離常數	ionization constant
电离电位	游離電位	ionization potential
电离度	游離度	degree of ionization
电离能	游離能	ionization energy
电量滴定	電量滴定法	coulometric titration
电流滴定[法]	電流滴定[法]	amperometric titration
电流作用	電流作用	galvanic action
电氯化反应	電氯化[反應]	electrochlorination
电毛细管现象	電毛細管現象	electrocapillarity
电能	電能	electric energy
电黏效应	電黏性效應	electroviscous effect
电偶腐蚀	電流腐蝕	galvanic corrosion
电气管道除尘器	電管集塵器	electric pipe precipitator
电气设施	電力裝置	electric installation
电迁移	電移	electromigration
电桥	電橋	bridge circuit
电亲和性	電親和性	electroaffinity
电清洗	電解清洗	electrocleaning
电热反应器	電熱反應器	electric-impedance heated reactor
电热过程	電熱法	electrothermal process
电容	①電容②容率	capacitance
电容脉动式转速计	電容器脈衝[式]轉速計	capacitor impulse type tachometer
电容器	電容器	condenser
电容器纸	積電紙	condenser paper
电容湿度计	電容[式]溼度計	capacitance moisture meter
电容探针	電容探針	capacitance probe
电容压力转换器	電容[式]壓力轉換器	capacitance pressure transducer
电容液面计	電容[式]液面計, 電容[式]液位計	capacitance liquid level gage
电容元件	電率元件	capacitance element
电融合	電融合	electrofusion
电色谱	電色層分析	electrochromatography
电渗	電滲透	electroosmosis
电渗析	電透析[作用]	electrodialysis
电渗析过程	電透析法	electrodialysis process
电渗现象	電滲透	electric osmosis
电势	電位	potential
电势落	勢降	potential drop

祖 国 大 陆 名	台 湾 地 区 名	英 文 名
电势梯度	勢梯度	potential gradient
电枢族	電樞	armature
电涂	電塗, 電塗層	electrocoating
电涂层	電鑄	electrocasting
电涂装	電鑄	electrocasting
电位滴定[法]	電位滴定[法]	potentiometric titration
电位法	電位法	potentiometric method
电位分析[法]	電位分析	potentiometric analysis
电位计	電位計, 電勢計	potentiometer
电稳分散作用	電穩分散[作用], 電穩分散液	electrocratic dispersion
电锡镀	鍍錫	tin plating
电压	電壓	voltage
电冶金	電冶金學	electrometallurgy
电泳	電泳	electrophoresis
电致变色显示	電色顯示	electrochromic display
电致发光	電發光	electroluminescence
电铸	電鑄	electroforming
电铸莫来石	電鑄富鋁紅柱石	electrocast mullite
电子材料	電子材料	electronic material
电子放大器	電子放大器	electronic amplifier
电子计算机	電子計算機	electronic computer
电子控制	電子控制	electronic control
电子流量计	電子流量計	electric type flowmeter, electronic flowmeter
电子模拟器	電子模擬器	electronic simulator
电子偶产生	對偶生成	pair creation
电子偶生成	对偶生成	pair production
电子配分函数	電子分配函數	electronic partition function
电子讯号	電子訊號	electronic signal
电阻	電阻	resistance
电阻高温计	電阻高溫計	electric resistance pyrometer
电阻炉	電阻爐	resistance furnace
电阻温度计	電阻溫度計	resistance thermometer
垫料	填料	padding
垫片材料	墊片材料	gasket material
垫圈槽	墊圈溝	gasket groove
淀粉	澱粉, 漿糊	starch
淀粉碘化物试验	澱粉碘化物試驗	starch iodide test
淀粉碘试纸	澱粉碘試紙	starch iodine paper

祖国大陆名	台湾地区名	英 文 名
淀粉基聚合物	澱粉基聚合物	starch based polymer
淀粉酶	澱粉酶	amylase
淀粉黏合剂	澱粉黏合劑	starch adhesive
淀粉试纸	澱粉試紙	starch paper
淀粉水解	澱粉水解	amylolysis
淀粉糖化酶	澱粉酶	diastase
靛白	靛白	indigo white
靛蓝	靛藍	indigo blue
靛染	靛染	indigo dyeing
靛染缸	靛染缸	indigo dye vat
靛棕	靛棕	indigo brown
吊篮式反应器	籃式反應器	basket reactor
迭代法	疊代法	iterative method
迭式反应器	疊式反應器	stack-up reactor
叠加	疊加	superposition
叠加原理	疊加原理	superposition principle
蝶[形]阀	蝶型閥	butterfly valve
丁苯无规共聚物	丁苯無規共聚物	styrene butadiene random copolymer
丁基纤维素	丁基纖維素	butyl cellulose
丁基橡胶	丁基橡膠	butyl rubber
丁腈橡胶	腈橡膠	nitrile rubber
丁钠橡胶	鈉聚丁二烯橡膠, 鈉橡膠, 丁鈉橡膠	sodium polybutadiene rubber, sodium rubber, sodium-butadiene rubber
丁子香酚	丁香酚	eugenol
顶板	頂板	top plate
顶冷凝器	頂冷凝器	head condenser
顶相,上相	頂相	top phase
定比定律	定比定律	law of definite proportions
定标,比例换算	定標, 比例化, 標度化	scaling
定常运动	恆穩運動, 穩定運動	steady motion
定氮仪	氮量計	azotometer
定点观察	定點察視	spot survey
定理	定理	theorem
定量分析	定量分析	quantitative analysis
定律	定律, 律	law
定时开关	定時開關	time switch
定速泵	恆速泵	constant speed pump
定态	固定狀態, 穩態	stationary state, steady-state
定位槽	定高差槽	constant head tank

祖国大陆名	台湾地区名	英　文　名
定向	定向, 定位, 方位	orientation
定向分布	方向分布	orientation distribution
定向平面	定向平面	orientation flat
定向吸附	位向吸附	oriented adsorption
定性分析	定性分析	qualitative analysis
定域粒子系集	定域粒子系集	assembly of localize particles
定子	定子	stator
锭剂	研棒	pastille
锭子油	錠子油	spindle oil
冬化	低溫化, 冬化	winterizing
动电力学	電動力學	electrokinetics
动电势	電動位	electrokinetic potential
动电现象	電動現象	electrokinetic phenomenon
动力参数	動力參數	kinetic parameter
动力发动机	能量產生器, 發電机	power generator
动力方程	動態方程式	dynamical equation
动力分析	動力分析	kinetic analysis
动力模式	動力模式	kinetic model
动力黏度	動態黏度	dynamic viscosity
动力性质	動力性質	kinetic property
动力学	動力學	kinetics
动力学控制	動力學控制	kinetic control
动力学理论	動力理論	kinetic theory
动力学式	動力式	kinetic expression
动量	動量	momentum
动量传递	動量傳送	momentum transfer
动量传递系数	動量傳遞係數	momentum transfer coefficient
动量方程	動量方程式	momentum equation
动量分离器	動量分離器	momentum separator
动量矩守恒原理	動量力矩守恒原理	principle of conservation of moment of momentum
动量扩散系数	動量擴散係數	momentum diffusivity
动量平衡	動量均衡	momentum balance
动量守恒	動量守恒	momentum conservation
动量守恒原理	動量守恒原理	principle of conservation of momentum
动量输送	動量輸送	momentum transport
动量通量	動量通量	momentum flux
动量原理	動量原理	momentum principle
动能	動能	kinetic energy

祖国大陆名	台湾地区名	英 文 名
动态补偿	動態補償	dynamic compensation
动态反射光谱[法]	動態反射光譜學, 動態反射光譜法	dynamic reflectance spectroscopy
动态分析	動態分析	dynamic analysis
动态刚性	動態剛性	dynamic rigidity
动态规划	動態規劃	dynamic programming
动态过程	動態程序	dynamic process
动态模量	動態模數	dynamic modulus
动态模拟	動態模擬	dynamic simulation
动态模型	動態模式	dynamic model
动态平衡	動態平衡	dynamic equilibrium
动态去偶器	動態解偶器	dynamic decoupler
动态热机械法	動態熱機械法	dynamic thermomechanometry
动态热重量分析法	動態熱重量法	dynamic themogravimetry
动态试验	動態試驗	dynamic test
动态误差	動態誤差	dynamic error
动态误差系数	動態誤差係數	dynamic error coefficient
动态系统	動態系統	dynamic system
动态相似	動態相似性	dynamic similarity
动态响应	動態應答	dynamic response
动态行为	動態行為	dynamic behavior
动态学	①動態②動態學③動力學	dynamics
动态优化	動態最適化	dynamic optimization 翁
动态元素	動態元素	dynamic element
动态滞后	動態落後	dynamic lag
动物胶	動物膠	animal glue
动物厩肥	動物廄肥	animal manure
动物皮	動物皮	animal hide
动物油	動物油	animal oil
动压强	動壓[力]	dynamic pressure
动压头	動壓頭	kinetic head
冻凝	凍凝	congelation
冻凝点	凍凝點	congealing point, congelation point
冻凝温度	凍凝溫度	congealing temperature
胨	腖	peptone
胨化	腖化[作用]	peptonization
斗式提升机	斗式升降機	bucket elevator
斗式运输机	斗式運送機, 斗運機	bucket conveyor

祖国大陆名	台湾地区名	英 文 名
毒气	毒氣	toxic gas
毒素	毒素	toxin
毒物	毒物	poison
毒性	毒性	toxicity
毒性化学品	毒性化學品	toxic chemical
毒烟	毒煙	toxic smoke
独居石	獨居石	monazite
独立反应	獨立反應	independent reaction
独立粒子系集	獨立粒子系集	assembly of independent particles
独立失活	獨立失活	independent deactivation
堵塞	阻塞	plugging
杜里龙耐酸硅铁	高矽鐵	duriron
镀铂碳电极	鍍鉑碳[電]極	platinized carbon electrode
镀铬	鍍鉻	chrome plating
镀镍	鍍鎳	nickel plating
镀锡	鍍錫	tinning
镀锡板	馬口鐵	tin plate
镀锌	鍍鋅	galvanization, zinc plating
镀锌保护	鍍鋅保護	galvanic protection
短杆菌酪肽	短桿菌酪	tyrocidin
短杆菌素	短桿菌素, 混合短桿菌	tyrothricin
短杆菌肽	短桿菌素, 滅革蘭菌素	gramicidin
短链支化	短鏈支化	short-chain branching
短期熟成	短期成熟	short ripening
短纤纱	短纖紗, 細紗, 精紡紗	spun yarn
短纤维	短纖維, 定長纖維, 人造棉	staple fiber, short fiber
短纤维复合材料	短纖維複合材料	short fiber composite
短纤维增强	短纖維增強	short fiber reinforcement
短纤维增强塑料	短纖維增強塑料	short fiber reinforced plastic
短效加速剂	短效催速劑	fugitive accelerator
短支链	短支鏈	short-chain branch
断键	斷鍵	scission
断开	撕裂	tearing
断裂表面能	斷裂表面能	surface energy of fracture
断裂力学	破壞力學	fracture mechanics
断裂强度	破壞強度	fracture strength
断裂统计学	斷裂統計學	statistics of rupture
断裂应力	破壞應力, 斷裂應力	fracture stress, rupture stress

祖国大陆名	台湾地区名	英 文 名
煅富铁黄土	煆富鐵黃土	burnt sienna
煅烧	煆燒	calcination
煅烧焦	煆燒焦	calcined coke
煅烧炉	煆燒機	calcinator, calciner
煅烧黏土	煆燒黏土	calcined clay
煅烧区	煆燒區	calcining zone
煅石膏	煆石膏	calcined plaster
煅赭石	煆赭石	burnt ocher
堆垛	堆垛	stacking
堆垛效应	堆垛效應	stacking effect
堆肥	堆肥	compost
堆肥法	堆肥法	composting
堆[积]浸[取]	堆積浸取	heap and dump leaching
堆密度	體密度	bulk density
堆芯活性区	反應器核心, 反應器爐心	reactor core
对苯乙烯磺酸	對苯乙烯磺酸	p-styrene sulfonic acid
对比饱和蒸汽压	對比飽和蒸汽壓	reduced saturated vapor pressure
对比变量	對比變數	reduced variable
对比分子量	折合分子量	reduced molecular weight
对比密度	對比密度	reduced density
对比黏度	折合對比黏度	reduced viscosity
对比态原理(＝对应态原理)		
对比体积	對比體積	reduced volume
对比温度	對比溫度	reduced temperature
对比性质	對比性質, 折合性質	reduced property
对比压力	對比壓力	reduced pressure
对比状态方程	對比狀態方程式	reduced equation of state
对比浊度	對比濁度	reduced turbidity
对比坐标	對比坐標	reduced coordinates
对标换算因数	時間換算因數	time scale change factor
对称膜	對稱膜	symmetric membrane
对称性	對稱性	symmetry
对称中心	對稱中心	symmetry center
对称轴	對稱軸	axis of symmetry
对分搜索	二分法搜尋	dichotomous search
对接	對接	butt joint
对流	對流	convection

祖国大陆名	台湾地区名	英 文 名
对流传热	熱對流	convection heat transfer, convective heat transfer
对流传热系数	熱對流係數	convective heat transfer coefficient
对流传质	對流質傳	convection mass transfer
对流导数	對流導數	convective derivative
对流段	對流段	convection section
对流加速度	對流加速度	convective acceleration
对流通量	對流通量	convective flux
对硫磷	巴拉松	parathion
对偶, 成对	對偶, 成對	pairing
对偶变量	對偶變數	pairing variables
对偶[性]	對偶	duality
对数平均	對數平均	logarithmic mean
对数平均半径	對數平均半徑	logarithmic mean radius
对数平均面积	對數平均面積	logarithmic mean area
对数平均摩尔分率	對數平均莫耳分率	logarithmic mean mole fraction
对数平均温差	對數平均溫差	logarithmic mean temperature difference
对数生长期	對數生長期	logarithmic phase
对数图	對數圖	logarithmic plot
对数正态分布	對數常態分布函數	log normal distribution function
对应态	對應狀態	corresponding state
对应态律	對應狀態定律	law of corresponding state
对应态原理, 对比态原理	對應狀態原理	principle of corresponding state
对映体选择性	鏡像對應異構選擇性	enantiometric selectivity
对峙反应	可逆反應	opposing reactions
钝化	鈍化	passivation
钝化铁	不活性鐵, 鈍鐵	passive iron
钝态	鈍態, 不活性, 鈍性	passive state, passivity
多变比热	多變比熱	polytropic specific heat
多变常数	多變常數	polytropic constant
多变过程, 多方过程	多變程序	polytropic process
多变量控制系统	多變數控制系統	multivariable control system
多变量系统	多變數系統	multivariable system
多变量最优化	多變數最適化	multivariable optimization
多变膨胀	多變膨脹	polytropic expansion
多变性	多變性	polytropism
多变指数	多變指數	polytropic index
多层次优化法	多層次優化法	multilevel method of optimization

祖国大陆名	台湾地区名	英　文　名
多层吸附理论	多層吸附理論	multilayer adsorption theory
多巢控制环路	巢式控制環路	nested control loop
多程换热器	多程熱交換器	multipass exchanger
多程加热器	多程加熱器	multipass heater
多程冷凝器	多程冷凝器	multipass condenser
多重产物塔	多重產物塔	multiple product column
多重反应	多重反應	multiple reactions
多重反应器系统	多反應器系統	multiple reactor system
多重进料流	多重進料流	multiple feed stream
多重输出控制	多重輸出控制	multiple output control
多重输出系统	多[重]輸出系統	multiple output system
多重输入系统	多[重]輸入系統	multiple-input system
多重态	多重態, 多重性	multiplicity, multiplet
多重稳态	多重穩態	multiple stability, multiple steady state
多重稳态性	多穩態性	steady-state multiplicity
多重性因子	倍增因數	multiplication factor
多重再沸器	多重再沸器	multiple reboiler
多床反应器	多床反應器	multibed reactor, multiple bed reactor
多次萃取	多次萃取	multiple extraction
多带掺合机	複帶摻合機	multiribbon blender
多道分析仪	多通道分析儀	multichannel analyzer
多点记录器	多點記錄器	multipoint recorder
多段过程	多階程序	multistage process
多段体系	多階系統	multistage system
多方过程(＝多变过程)		
多分布组分	多分布成分	multiple distributed component
多分隔点	多分隔點	multiple split point
多釜串联模式	串聯槽模式	tanks-in-series model
多功能催化剂	多功能觸媒	multifunctional catalyst
多功能酶	多功能酶	multifunctional enzyme
多管反应器	多管反應器	multitube reactor
多管系统	多管系統	multiple pipe system
多管旋风分离器	多旋風分離器	multicyclone
多辊压延机	強度軋光機	supercalender
多回路控制系统	多環路控制系統	multiloop control system
多回路系统	多環路系統	multiloop system, multiple loop system
多级过程	分階程序	stage process
多级扩散	多階擴散	multistage diffusion
多级离心泵	多階離心泵	multistage centrifugal pump

祖国大陆名	台湾地区名	英 文 名
多级离心机	多階離心機	multiple stage centrifuge
多级流化床	多階流化床	multistage fluidized bed
多级模型	分階模式	stagewise model
多级喷射泵	多階射出器	multistage ejector
多级喷射器	多級射出器	stage ejector
多级批式萃取	多階批式萃取	multiple batch extraction
多级汽提	分階汽提	stage stripping
多级系统	分階系統	stage system
多级压缩	多階壓縮	multiple stage compression, multistage compression
多级压缩机	多階壓縮機	multiple stage compressor, multistage compressor
多级研磨	多階研磨	multistage grinding
多级蒸馏	多階蒸餾	multistage distillation
多价螯合作用	螯合作用, 鉗合作用	sequestration
多降液管塔板	多降液管塔板	multidowncomer tray, MD tray
多金属催化剂	多金屬觸媒	polymetallic catalyst
多进出口系统	多進出口系統	multiport system
多孔板	多孔板	perforated plate
多孔板筛	多孔[板]篩	perforated screen
多孔壁管	多孔壁管	porous-wall tube
多孔玻璃过滤器	燒結玻璃過濾器	fritted glass filter
多孔固体	多孔[性]固體	porous solid
多孔管	多孔管	perforated pipe
多孔假底	多孔假底	perforated false bottom
多孔介质	多孔介質	porous medium
多孔粒	多孔粒	porous pellet
多孔粒脉冲反应器	多孔粒脈衝反應器	porous pellet pulse reactor
多孔膜	多孔薄膜	porous membrane
多孔塑料	海綿塑膠	sponge plastic
多孔塔板	多孔板	perforated tray
多孔填料	多孔填充物	porous packing
多孔铁阴极	多孔鐵陰極	perforated iron cathode
多孔性材料	多孔[性]材料	porous material
多孔性催化剂	多孔性觸媒	porous catalyst
多孔砖	多孔磚	perforated brick
多流体理论	多流體理論	poly-fluid theory
多硫聚合物	多硫聚合物	polysulfide polymer
多目标规划	多目標規劃	multi-objective programming

祖国大陆名	台湾地区名	英 文 名
多黏菌素	多黏菌素	polymyxin
多嵌段共聚物	多段共聚物	segment copolymer, segmental copolymer, segmented copolymer
多区模型	多區模型	multi-region model
多室槽式反应器	分倉式反應器	compartmented tank reactor
多室反应器	複床爐	multiple hearth furnace
多室离心机	多室離心機	multichamber centrifuge
多室涡轮机	多室渦輪機	multicasing turbine
多输入多输出系统	多輸入多輸出系統	Multiple-input-multiple-output-system
多态[现象]	多形性	polymorphism
多糖	多糖	polysaccharide
多糖酶	多醣分解酶	polysaccharidase
多烷基化作用	多烷化[反應]	polyalkylation
多稳态性	多穩態性	multiplicity of steady states
多相反应器	多相反應器	multiphase reactor
多相流	多相流	multiphase flow
多相平衡	不勻相平衡	heterogeneous equilibrium
多相系统	多相系統	multiphase system
多相性	多相性	heterogeneity
多效蒸发	多效蒸發	multiple evaporation, multieffect evaporation
多效蒸发器	多效蒸發器	multiple evaporator
多叶混合器	多葉混合器	multiple blade mixer
多叶片风机	多葉風扇	multiblade fan
多叶片混合器	多葉混合器	multiblade mixer
多元醇	多元醇	polyalcohol
多元分离序列	多成分分離順序, 多成分分離定序	multicomponent separation sequence, multicomponent separation sequencing
多元混合物, 多组分混合物	多成分混合物	multicomponent mixture
多元系[统], 多组分系统	多成分系統	multicomponent system
多轴应力	多軸應力	multiaxial stress
多锥管分离器	多錐管分離器	multicone separator
多组分	多成分, 多元	multicomponent
多组分共沸物	多成分共沸液	multicomponent azeotrope
多组分混合物(=多元混合物)		
多组分扩散	多成分擴散	multicomponent diffusion

祖国大陆名	台湾地区名	英　文　名
多组分吸收	多成分吸收	multicomponent absorption
多组分系统(＝多元系[统])		
惰化剂	惰化劑	inactivator
惰性气体	惰性氣體	inert gas
惰性物	惰性物	inert
扼流	阻流	choke flow
恶臭污染物	氣味污染物	odor pollutant

E

祖国大陆名	台湾地区名	英　文　名
颚[式破]碎机	顎碎機	jaw crusher, jaw breaker
蒽醌还原染料	蒽醌缸染料	anthraquinone vat dye
蒽醌染料	蒽醌染料	anthraquinone dye
蒽油	蒽油	anthracene oil
二次成核	第二次成核	secondary nucleation
二次成核作用	二次成核作用	secondary nucleation
二次反应	次要反應, 後續反應	secondary reaction
二次夹带	二次夾帶	re-entrainment
二次胶合	二次膠合	secondary gluing
二次色散	二次色散	secondary dispersion
二次石油回收	二次石油回收	secondary oil recovery
二次蒸发器	二次蒸發器	second evaporator
二道底漆	二道底漆	surfacer
二级胺	二級胺	secondary amine
二级标准	二級標準	secondary standard
二级沉降槽	二級沉降槽	secondary settler
二级成核作用	二級成核作用	second order nucleation
二级澄清器	二級澄清器, 二級沉降槽	secondary clarifier
二级处理	二級處理	secondary treatment
二级滴滤池	二級滴濾池	secondary trickling filter
二级电离电位	二級電離	second order ionization potential
二级加料器	二級進料器	secondary feeder
二级结晶	二級結晶	second order crystallization
二级聚合反应	二級聚合反應	secondary polymerization reaction
二级转变	二級轉變	second order transition

祖国大陆名	台湾地区名	英 文 名
二级转变温度	二級轉變溫度	second order transition temperature
二极管	二極體	diode
二极函数发生器	雙極函數產生器	diode function generator
二甲苯	二甲苯	xylene
二阶流体	二階流體	second-order fluid
二阶系统	二階系統	second-order system
二阶响应	二階應答	second-order response
二聚	二聚合作用	dimerization
二肽酶	二肽酶	dipeptidase
二维模型	二維模式	two-dimensional model
二相传热	兩相熱傳	two phase heat transfer
二向色谱法	二維層析[法]	two-dimensional chromatography
二行程引擎	二行程引擎	two-cycle engine
二氧化锆	氧化鋯	zirconia
二氧化硅	矽石, 二氧化矽	silica
二氧化钛	氧化鈦	titania
二氧化碳测定计	二氧化碳測定計	calcimeter
二元催化剂	二元觸媒	binary catalyst
二元共沸液	二元共沸液	binary azeotrope
二元混和物, 双组分混合物	二元混和物	binary mixture
二元系[统], 二组分系统	二元系, 雙成分系統	binary system
二元蒸馏	二元蒸餾	binary distillation
二组分系统(＝二元系[统])		

F

祖国大陆名	台湾地区名	英 文 名
发电厂	發電廠	power plant
发电厂循环	動力廠循環	power plant cycle
发光	發光	luminescence
发光菌	發光菌	photogenic bacteria
发光染料	發光染料	luminescent dye
发光涂料	發光漆	luminous paint
发汗工艺	發汗法	sweating process
发汗(增塑剂渗出)	發汗(增塑劑滲出)	sweating, sweat out

祖国大陆名	台湾地区名	英 文 名
发酵	發酵[法]	fermentation
发酵槽	發酵缸, 發酵槽	fermentation vat, fermentation tank
发酵管	發酵管	fermentation tube
发酵管试验	發酵管試驗	fermentation tube test
发酵罐	發酵槽, 缸式發酵槽	fermenter, fermentor
发酵过程	發酵程序	fermentation process
发酵机理	發酵機構	fermentation mechanism
发酵技术	發酵技術	fermentation technology
发酵窖	發酵窖	fermentation cellar
发酵模式	發酵方式	fermentation pattern
发酵培养基	發酵基, 發酵介質	fermentation medium
发酵器械	發酵裝置	fermentation apparatus
发酵室	發酵室	fermentation chamber
发酵途径	發酵途徑	fermentation pathway
发酵效率	發酵速率	fermentation efficiency
发明	發明	invention
发泡	發泡	foaming
发泡剂	發泡劑	blowing agent, foam agent, inflating agent
发散冷却	蒸散冷卻	transpiration cooling
发色体	色素原	chromogen
发色团	發色團, 色基	chromophore
发射光谱	發射光譜	emission spectrum
发射光谱仪	發射分光計	emission spectrometer
发射率	發射率	emissivity
发射能力	發射能力	emissive power
发射体	發射體	emitter
发射因数	排放因數, 發射係數	emission factor
发生炉煤气	發生爐氣	producer gas
发烟弹	煙幕彈	smoke shell
发烟罐	發煙罐	smoke pot
发烟试验	發煙試驗	smoking test
罚函数	處罰函數	penalty function
阀当量管长	閥相當管長	valve equivalent pipe length
阀范围性	閥範圍性	valve rangeability
阀杆	閥桿	valve stem
阀鉴定编码	閥鑑定簡碼	valve identification code
阀控制	閥控制	valve control
阀门定位器	閥定位器	valve positioner
阀驱动器	閥引動器	valve actuator

祖国大陆名	台湾地区名	英 文 名
阀容量	閥容量	valve capacity
阀塞	閥塞	valve plug
阀损失	閥損失	valve loss
阀特性	閥特性	valve characteristics
阀体	閥主體	valve body
阀响应	閥應答	valve response
阀序列	閥定序	valve sequencing
阀增益	閥增益	valve gain
阀滞后	閥遲滯	valve hysteresis
阀阻力	閥阻力	valve resistance
法定责任	法定責任	legal liability
ASOG 法	ASOG 法	analytical solution of group contribution method
法拉第	法拉第	faraday
法兰	凸緣法蘭	flange
法兰接头	凸緣法蘭接頭	flange coupling
法兰连接	凸緣法蘭連接	flange connection
法兰联管节	凸緣法蘭接頭	flange joint
法兰旋塞	凸緣分接頭	flange tap
法向应力	正應力, 法向應力	normal stress
法向应力差	正應力差	normal stress difference
法坐标	正規座標	normal coordinates
发状弹簧	髮狀彈簧	hair spring
发夹管	髮夾管	hair-pin tube
珐琅釉	琺瑯釉	enamel glaze
凡士林	凡士林	vaseline
矾土	礬土	alumina
钒钾铀矿	釩酸鉀鈾礦	carnotite
钒接触法	釩接觸法	vanadium contact process
繁殖槽	繁殖槽	propagator
反常黏度	反常黏度	anomalous viscosity
反萃取	逆萃取	reverse extraction
反电动势	反電動勢	counter emf
反杠杆法则	反摃桿法則	inverse lever arm rule
反回作用	反作用	return action
反胶团萃取	逆膠團萃取	reverse micelle extraction
反截留	逆陷	backtrapping
反竞争抑制	不競爭抑制[作用]	uncompetitive inhibition
反馈	回饋	feedback

祖国大陆名	台湾地区名	英　文　名
反馈补偿	回饋補償	feedback compensation
反馈回路	回饋環路	feedback loop
反馈机理	回饋機構	feedback mechanism
反馈控制	回饋控制	feedback control
反馈控制器	回饋控制器	feedback controller
反馈控制系统	回饋控制系統	feedback control system
反馈伸缩囊	回饋伸縮囊	feedback bellow
反馈系统	回饋系統	feedback system
反馈信号	回饋信號	feedback signal
反馈元件	回饋元件	feedback element
反扩散	逆擴散	reverse diffusion
反流	逆向流動	reverse flow
反流态化	去流體化	defluidization
反凝胶萃取	反凝膠萃取	antigelation extraction
反乳化	去乳化[作用]	demulsification
反射分光光度计	反射分光光度計	reflection spectrophotometer
反射光谱法	①反射光譜學②反應光譜法	reflectance spectroscopy
反射计	反光計	reflectometer
反射聚光器	反射聚光器	reflecting condenser
反射率	反射係數	reflectivity
反渗透	逆滲透	reverse osmosis
反式加成	反式加成[反應]	transaddition
反式[异构体]	反式	trans-form
反微分控制	反微分控制	inverse derivative control
反洗	反洗, 逆洗	backflushing, backwashing
反洗速率	逆洗速率	backwash rate
反相乳液	反乳液	inverse emulsion
反向泵	反向泵	reverse-running pump
反向传播算法	反向傳播算法	back-propagation algorithm
反向电流	逆流	reverse current
反向扩散	回擴散	back diffusion
反向压力计	反向壓力計	inverted manometer
反硝化细菌	脫氮[化]菌	denitrifying bacteria
反硝化作用	脫氮化[反應]	denitrification
反应	反應	reaction
反应步骤	反應步驟	reaction step
反应萃取	反應萃取	reactive extraction
反应答	反應答	inverse response

祖 国 大 陆 名	台 湾 地 区 名	英 文 名
反应动力学	反應動力學	reaction kinetics
反应对	反應對	reaction pair
反应分子数	①分子度②分子性	molecularity
反应釜	反應釜	reaction kettle
反应概率	反應機率	reaction probability
反应工程	反應工程	reaction engineering
反应过程	反應程序, 反應性程序	reaction process, reactive process
反应焓	反應焓	enthalpy of reaction
反应机理	反應機制	reaction mechanism
反应级	反應階	order of reaction
反应级数	反應級數	reaction order
反应溅射蚀刻法	反應性噴濺蝕刻[法]	reactive sputtering etching
反应进度	反應程度	extent of reaction
反应进行变量	反應進行變數	reaction progress variable
反应扩散网络	反應擴散網路	reaction diffusion network
反应离子蚀刻	反應性離子蝕刻[法]	reactive ion etching
反应灵敏度	反應靈敏度	reaction sensitivity
反应络合物	反應物錯合物	reactant complex
反应器	反應器	reactor
反应器安全	反應器安全性	reactor safety
反应器参数	反應器參數	reactor parameter
反应器操作	反應器操作	reactor operation
反应器操作特性	反應器性能	reactor performance
反应器动力学	反應器動力學	reactor kinetics
反应器动态学	反應器動態[學]	reactor dynamics
反应器分析	反應器分析	reactor analysis
反应器工程	反應器工程	reactor engineering
反应器构型	反應器組態	reactor configuration
反应器规模	反應器容量	reactor capacity
反应器控制	反應器控制	reactor control
反应器理论	反應器理論	reactor theory
反应器模型	反應器模式	reactor model
反应器设计	反應器設計	reactor design
反应器网络	反應器網絡	reactor network
反应器稳定性	反應器穩定性	reactor stability
反应器系统	反應器系統	reactor system
反应器优化	反應器最適化	reactor optimization
反应器中毒	反應器中毒	reactor poisoning
反应器装配	反應器配置	reactor setup

祖国大陆名	台湾地区名	英　文　名
反应器组	反應器組	reactor train
反应亲和力	反應親和力	reaction affinity
反应区	反應區	reaction zone
反应曲线	反應曲線	reaction curve
反应曲线法	反應曲線法	reaction curve method
反应热	反應熱	heat of reaction
反应时间	反應時間	reaction time
反应式分配器	作用式分配器	reaction-type distributor
反应室	反應室	reaction chamber
反应速度	反應速度	reaction velocity
反应速率	反應速率	reaction rate
反应速率常数	反應速率常數	reaction rate constant
反应速率理论	反應速率理論	reaction rate theory
反应塔	反應塔	reaction tower
反应图	反應圖	reaction diagram
反应途径	反應途徑	reaction path
反应网络	反應網絡	reaction network
反应网络分析	反應網路分析	analysis of reaction network, reaction network analysis
反应温度	反應溫度	reaction temperature
反应物	反應物	reactant
反应物比例	反應物比例	reactant ratio
反应系统	反應系統	reaction system
反应相	反應相	reacting phase
反应性	①反應性②反應度	reactivity
反应循环	反應循環	reaction cycle
反应蒸馏	反應蒸餾	distillation with chemical reaction, reactive distillation
反应中心	反應中心	reactive center
反应周期	反應周期	reaction cycle time
反应专一性	反應特異性	reaction specificity
反应坐标	反應坐標	reaction coordinates
返回率	報酬率	return rate
返混反应器	返混反應器	backmix reactor
返位时间	重整時間	reset time
泛点	氾流點	flooding point
泛点速度	氾流速度	flooding velocity
泛函方程式	泛函方程式	functional equation
泛速度分布	泛速度分布	universal velocity distribution

祖国大陆名	台湾地区名	英 文 名
范德瓦耳斯常数	凡得瓦常數	van der Waals constant
范德瓦耳斯方程	凡得瓦方程式	van derWaals equation
范德瓦耳斯力	凡得瓦力	van der Waals force
范德瓦耳斯吸附	凡得瓦吸附	van der Waals adsorption
范德瓦耳斯状态方程	凡得瓦狀態方程式	van der Waals equation of state
范拉尔方程	凡得拉方程	van Laar equation
范托夫定律	凡特荷夫定律	van't Hoff's law
范围	範圍	scope
方波	正方形波	square wave
方程	方程式	equation
BET 方程	BET 方程	Brunauer-Emmett-Teller equation
BWR 方程	BWR 方程	Benedict-Webb-Rubin equation, BWR equation
MESH 方程组	MESH 方程組	MESH equations
NRTL 方程, 非随机两液方程	NRTL 方程, 非随機兩液方程	non-random two liquid equation, NRTL equation
RK 方程	RK 方程	Redlich-Kwong equation(RK equation)
方法	方法	method
方解石	方解石	calcite
方块三对角矩阵	方塊三對角矩陣	block tridiagonal matrix
方框流程图	程序方塊圖	block flow diagram
方镁石	方鎂石	periclase
方硼石	方硼石	boracite
方石英	白矽石	cristobalite
芳构化	芳香[烴]化	aromatization
芳[香]化合物	芳香烴	aromatics
芳香油	香料油	perfume oil
防爆	防爆	explosion proof
防冲板	擋板, 折流板	baffle plate
防喘振控制	抗波動控制	antisurge control
防冻	抗凍劑	antifreeze
防冻剂	①防凍器②防凍劑	antifreezer, antifreezing agent
防腐	防腐	preservation
防腐剂	防腐劑	preservative
防腐蚀	耐蝕, 防蝕	anticorrosion, corrosion prevention
防护催化剂	防護觸媒	guard catalyst
防护罩	通風櫥, 排氣罩	hood
防火	火災防護, 防火	fire protection
防火漆	防火漆	fire-retarding paint

祖国大陆名	台湾地区名	英　文　名
防溅挡板	防濺板	splash plate
防溅盘	防濺盤	splash disc
防焦剂	防焦劑	scorch retarder
防结块剂	防阻塞劑, 防結塊劑	antiblock agent, anticaking agent
防沫剂	防沫劑	antifoam agent
防黏剂	防黏著劑	abherent
防软剂	軟化防止劑	antisoftener
防声涂料	防聲塗料	sound-proof coating
防蚀剂	防銹劑, 防蝕劑	anticorrosive agent
防蚀漆	防銹劑, 防蝕漆	anticorrosive paint
防水处理	①防水[處理]②防水劑	water proofing
防水润滑脂	防水潤滑脂	water proof grease
防水水泥	防水水泥	water proof cement
防缩模	防縮模	shrinkage jigs
防缩器	防縮器	shrink fixture
防缩温度	防縮溫度	shrinkage temperature
防缩楦模	縮裂	shrinkage block jigs
防缩整理	防縮整理	shrink resistant finish
防污染剂	防汙劑	antistaining agent
防锈剂	防銹脂	antirusting grease
防盐雾性	防鹽霧性	salt-fog resistance
防震涂料	防震塗料	shock proof lacquer
仿羊皮纸	玻璃紙	parchmyn paper
仿真(＝模拟)		
纺前染色	紡前染色	spun-colored
纺前染色丝	紡紗	spun-dyed yarn
纺丝	紡絲	spinning
纺丝仓	紡絲倉, 紡絲[機]部位	spinning cell
纺丝罐	紡絲罐	spinning can, spinning pot
纺丝机	紡絲機	spinning machine
纺丝溶液	紡絲溶液	spinning solution
[纺丝]丝饼	[紡絲]絲餅	spinning cake
纺丝酸	紡絲酸	spinning acid
纺丝头	紡絲帽, 噴絲帽	spinning nozzle
纺丝头拉伸	紡絲頭拉伸	spinning draft
纺丝信道	紡絲通道	spinning cabinet
纺丝甬道	紡絲甬道	spinning channel
纺丝油	紡絲油	spinning oil

祖国大陆名	台湾地区名	英　文　名
纺丝浴	紡絲浴	spinning bath
纺丝浴拉伸	紡絲浴拉伸	spinning bath stretch
纺织品	①紡織品②纖維	textile
纺织纤维	紡織纖維	textile fiber
放大	放大, 按比例放大	amplification, scale up
放大法则	放大法則	scale-up rule
放大器	放大器, 增輻器	amplifier
放大问题	放大問題	scale-up problem
放空罐	泄料槽	blow tank
放气阀	釋放閥	release valve
放热反应	放熱反應, 放熱反應	exothermic reaction, heat-liberating reaction
放热过程	放熱程序	exothermic process
放热性	放熱度	exothermicity
放射化学	放射化學	radiochemistry
放射性	放射性	radioactivity
放射性尘	放射性塵	radioactive dust
放射性毒物	放射性毒	radioactive poison
放射性废物	放射性廢棄物	radioactive waste
放射性核	放射性核	radioactive nucleus
放射性核素	放射性核種	radioactive nuclide
放射性裂变	放射性分裂	fission radioactive, radioactive fission
放射性落尘	放射性落塵	radioactive fallout
放射性示踪剂	放射性示蹤劑	radiotracer, radioactive tracer
放射性衰变	放射性衰變	radioactive decay
放射性同位素	放射性同位素	radioactive isotope
放射性蜕变	放射性蛻變	radioactive disintegration
放射性污染	放射性污染	radioactive contamination
放射性污染物	放射性污染物	radioactive pollutant
放射性元素	放射性元素	radioactive element
放射性指示剂	放射性示蹤劑	radioactive indicator
放射源	放射[性]源	radioactive source
放线菌素	放線菌	actinomycin
放线菌酮	環己醯亞胺素	cycloheximide
放泄阀	排洩閥	drain valve
飞溅	濺鍍	spatter
飞温	溫度失控	temperature runaway
非重叠操作	非重疊操作	nonoverlapping operation
非重复聚合物	非重複性聚合物	nonrepetitive polymer
非催化反应	非催化反應	noncatalytic reaction

祖国大陆名	台湾地区名	英文名
非等温反应器	非等溫反應器	nonisothermal reactor
非等温吸收	非等溫吸收	non-isothermal absorption
非等温性	非等溫性	nonisothermality
非定常流	非穩態流動	unsteady flow
非定态	非恆穩狀態	unsteady state
非定态操作	非穩態操作	unsteady state operation
非定态传热	非穩態熱傳	unsteady state heat transfer
非定态过程	非穩態程序	unsteady state process
非定域粒子系集	非定域粒子系集	assembly of non-localized particles
非独立粒子系集	非獨立粒子系集	assembly of interacting particles
非对称膜	非對稱膜	asymmetric membrane
非多孔颗粒吸收器	非孔性粒吸收器	nonporous pellet absorber
非多孔膜	非多孔膜	nonporous membrane
非分离反应器	非隔離反應器	nonsegregated reactor
非分离返混	非隔離逆混	nonsegregated backmixing
非分离混合	非隔離混合	nonsegregated mixing
非干性油	不乾性油	nondrying oil
非惯性坐标系统	非慣性坐標系統	noninertial coordinate system
非基本反应	非基本反應	nonelementary reaction
非结构模型	末結構化模式	unstructured model
非结合水分	非結合水	unbound water
非金属	非金屬	nonmetal
非晶二氧化硅	非晶[形]矽石	amorphous silica
非晶形[现象]	非晶體	amorphism
非晶性的	非晶	noncrystalline
非竞争性抑制	非競爭性抑制[作用]	noncompetitive inhibition
非绝热反应	非絕熱反應	nonadiabatic reaction
非绝热反应器	非絕熱反應器	nonadiabatic reactor
非均相催化	不勻相催化(作用)	heterogeneous catalysis
非均相反应	不勻相反應	heterogeneous reaction
非均相反应器	不勻相反應器	heterogeneous reactor
非均相共沸混合物	不均勻相共沸液, 不勻相共沸液	heteroazeotrope, heterogeneous azeotrope
非均相聚合	不勻相聚合(反應)	heterogeneous polymerization
非均相系统	不勻系統	heterogeneous system
非均相状态	不勻相狀態	heterogeneous state
非均一流化床	不勻相流體化床	heterogeneously fluidized bed
非均匀流	非均勻流動	nonuniform flow
非均匀性	不勻性, 不均勻性	inhomogeneity, nonhomogeneity

祖国大陆名	台湾地区名	英　文　名
非离子洗涤剂	非離子清潔劑	nonionic detergent
非离子型表面活性剂	非離子界面活性劑	nonionic surface active agent, nonionic surfactant
非理想表面	非理想表面	nonideal surface
非理想反应器	非理想反應器	nonideal reactor
非理想流动	非理想流動	nonideal flow
非理想气体	非理想氣體	nonideal gas
非理想溶液	非理想溶液	nonideal solution
非连锁反应	非連鎖反應	nonchain reaction
非炼焦煤	非煉焦煤	noncoking coal
非黏流动	無黏性流動	inviscid flow
非牛顿流体	非牛頓流體	non-Newtonian fluid
非平衡级模型	非平衡級模型	non-equilibrium stage model
非平衡热力学, 不可逆过程热力学	不平衡熱力學	non-equilibrium thermodynamics
非平衡系统	非平衡系統	non-equilibrium system
非破坏性试验	非破壞性試驗	nondestructive test
非亲和吸附	非親和吸附	non-affinity adsorption
非侵入性遥感器	非侵入性感測器	noninvasive sensor
非侵入性仪器	非侵入性儀器	noninvasive instrument
非润湿表面	非潤溼表面	nonwetting surface
非时变系统	非時變系統	time invariant system
非丝状细菌	非絲狀細菌	nonfilamentous bacteria
非随机两液方程 (=NRTL 方程)		
非稳态	不穩定狀態	unstable state
非稳态过程	非穩態程序	nonsteady state process
非洗式板框	非洗式板框	nonwashing plate
非线性	非線性	nonlinearity
非线性动力学	非線性動力學	nonlinear kinetics
非线性规划	非線性規劃	nonlinear programming
非线性回归分析	非線性回歸分析	nonlinear regression analysis
非线性控制阀	非線性控制閥	nonlinear control valve
非线性控制器	非線性控制器	nonlinear controller
非线性控制系统	非線性控制系統	nonlinear control system
非线性黏弹性	非線性黏彈性	nonlinear viscoelasticity
非线性稳定性	非線性穩定性	nonlinear stability
非线性系统	非線性系統	nonlinear system
非圆管当量直径	非圓管相當直徑	equivalent non-circular duct diameter

祖国大陆名	台湾地区名	英 文 名
非再生能源	非再生性能源	nonrenewable energy source
非正常反应	異常反應	abnormal reaction
非正规溶液	非規則溶液	nonregular solution
非正式报告	非正式報告	informal report
非质子溶剂	無質子溶劑	aprotic solvent
非自发反应	非自發反應	nonspontaneous reaction
非自发过程	非自發程序	non-spontaneous process
非最小相位落后	非最小相位落後	nonminimum phase lag
菲克定律	費克定律	Fick's law
肥料	肥料	fertilizer
斐波那契搜索法	費布納西搜尋法	Fibonacci search method
榧子油	榧子油	kaya oil
翡翠绿	翡翠綠	emerald green
废碱	廢鹼	spent caustic
废碱液	廢鹼液	spent lye
废气	廢氣	waste gas
废弃物	廢棄物, 廢料	waste
废弃物陆上处理	[廢棄物]陸上處理[法]	land disposal
废燃料	廢燃料, 廢油	waste fuel
废热	廢熱	waste heat
废热干燥器	廢熱乾燥器	waste heat drier
废热锅炉	廢熱鍋爐	waste heat boiler
废热回收	廢熱回收	waste heat recovery
废水	廢水, 污水	waste water
废水处理	廢水處理	waste water treatment
废酸	廢酸	waste acid
废糖蜜中的糖	含蜜糖	molasses sugar
废物	廢棄物	refuse
废物处理	廢棄物處理	waste disposal, waste treatment
废物管理	廢棄物管理	waste management
废物坑	廢棄物坑	refuse pit
废橡胶, 杂胶	廢橡膠, 雜膠	scrap rubber
废液	廢液	spent liquor, waste effluent, slop
废渣埋填	[廢棄物]掩埋	landfill
沸点	沸點	boiling point
沸点测定器	沸點測定器	boiling point apparatus
沸点降低	沸點下降	boiling point depression, boiling point lowering
沸点曲线	沸點曲線	boiling point curve

祖国大陆名	台湾地区名	英 文 名
沸点升高	沸點上升	boiling point elevation, boiling point rise
沸点图	沸點圖	boiling point diagram
沸点指数	沸點指數	boiling point index
沸点-组成图	沸點-組成圖	boiling point-composition diagram
沸石	沸石	zeolite
沸石催化剂	沸石催化劑	zeolite catalyst
沸腾	沸騰	boiling, ebullition
沸腾床	沸騰床, 氣泡床	ebullating bed, bubbling bed
沸腾床反应器	沸騰床反應器	ebullated bed reactor, ebullating bed reactor
沸腾淀粉	糊化澱粉	boiling starch
沸腾范围	沸點範圍	boiling range
费米-狄拉克分布	費米-笛拉克分布	Fermi-Dirac distribution
分瓣模	組合模, 瓣合塑模	split mold
分贝	分貝	decibel (dB)
分辨能力	解析能力	resolving power
E 分布	E 分布, 滯留時間分布	E distribution
分布板	分配板	distribution plate
分布参数模型	分布參數模型	distributed-parameter model
分布参数系统	分布參數系統	distributed-parameter system
分布成分	分布成分	distributed component
分布函数	分布函數	distribution function
分布器	分布器	distributor
分布器臂	分配器臂	distributor arm
分布曲线	分布曲線	distribution curve
分布系统	分布系統	distributed system
分布性质	分布性質	distributed property
分布滞后	分布落後	distributed lag
分步结晶	分段結晶	fractional crystallization
分步膨胀	分步膨脹	fractional expansion
分层	剝層[作用], 成層[作用]	delamination, stratification
分层流	分層流動	stratified flow
分叉分析	分歧分析	bifurcation analysis
分叉, 分支	分歧	bifurcation
分程调节	分開範圍控制	split-range control
分段操作	分階操作	stage operation
分段床反应器	分段床反應器	segmental-bed reactor
分段分析	分段分析	fractional analysis
分段梯形聚合物	分段梯形聚合物	step ladder polymer

祖 国 大 陆 名	台 湾 地 区 名	英 文 名
分隔	分隔	partitioning
分汞器	分汞器	mercury decomposer
分光光度计	分光光度計, 光譜儀	spectrophotometer
分光计	分光計	spectrometer
分光镜	分光鏡	spectroscope
分光浊度滴定	分光濁度滴定	spectroturbidimetric titration
分级	類析[法], 分類	classification
分级沉淀	分級沈澱	fractional precipitation
分级模型	分階模式	stage model
分级器	類析器	classifier
分级渗析	分級透析	fractional dialysis
分级吸附	分吸附	fractional adsorption
分级吸收	分吸收	fractional absorption
分级效率	分級效率	fractional efficiency
分解	分解	decomposition
分解程度	分解程度	extent of decomposition
分解代谢	細胞分解作用	catabolism
分解电势	分解勢	decomposition potential
分解电压	分解電壓	decomposition voltage
分解剂	分解劑	decomposer
分解器	分解器	decomposer
分解热	分解熱	heat of decomposition
分解效率	分解效率	decomposition efficiency
分解协调法	分解協調法	decomposition-coordination method
分解压力	分解壓[力]	decomposition pressure
分解蒸馏	分解蒸餾	destructive distillation
分界表面, 界面相	界面相	dividing surface
分界线	分模線	parting line
分馈吸附	分饋吸附	split feed adsorption
分离	分離	separation
分离池	分離池	separation cell
分离点	分離點	separation point
分离度	解析度	resolution
分离高度	分離高度	transport disengaging height
分离鼓	液氣分離器	knockout drum
分离过程	分離程序	separation process
分离技术	分離技術	separation technology
分离技术规格表	分離規格表	separation specification table
分离空间(=自由空间)		

祖 国 大 陆 名	台 湾 地 区 名	英 文 名
分离流	分離流	separated flow
分离膜	分離膜	separative membrane
分离器	分離器	separator
分离因子	分離因數	separation factor
分裂	裂解	splitting
分流	分叉流動	split flow
分流器	分裂機, 分離機	splitter
分馏	分餾, 選擇性蒸發	fractional distillation, fractionation, selective evaporation
分馏萃取	分[步]萃取	fractional extraction
分馏器	分餾器	fractionator
分馏塔	分餾塔	fractional column, fractionating tower
分馏柱	分餾塔	fractionating column
分馏装置	分餾裝置	fractionation assembly
分率	分率	fraction
分凝管	分凝管	fractional condensing tube
分凝器	分凝器, 部分冷凝器	fractional condenser, fractional condenser
分凝作用	分級冷凝	fractional condensation
分配比	分配比	distribution ratio
分配常数	分配常數	partition constant
分配定律	分配定律	distribution law
分配律	分配律	distributive law
分配色谱法	分配層析[法]	partition chromatography
分配系数	分布係數, 分配係數	distribution coefficient, partition coefficient
分批培养	批式培養	batch culture
分期偿还	分期償還	amortization
分散	分散	dispersion
分散本领	分散能力	dispersive power
分散度	分散度	degree of dispersion
分散剂	分散劑	dispersant, disperser, dispersion reagent
分散节涌流	分散塞流	dispersed slug flow
分散介质	分散介質	dispersed medium
分散控制	分區控制	decentralized control
分散流, 弥散流	分散流	dispersed flow
分散磨	分散磨	dispersion mill
分散系统	分散系統	dispersed system
分散相	分散相	dispersed phase, disperse phase
分散性	分散性	dispersibility
分散[性]染料	分散染料	dispersed dye

祖国大陆名	台湾地区名	英　文　名
分时	分時	time sharing
分数	分數	fraction
分数级反应	分數階次反應	fractional-order reaction
分析器	分析儀, 分析器	analyzer
分析天平	分析天平	analytical balance
分析型	分析型	analysis mode
分析蒸馏	分析蒸餾	analytical distillation
分销成本	分銷成本	distribution cost
分压力	分壓	partial pressure
分支(＝分叉)		
分支定界法	分支定界法	branch and bound method
分子	分子	molecule
分子参数	分子參數	molecular parameter
分子重排	分子重排	molecular rearrangement
分子动态法	分子動態法	molecular dynamic method, MD method
分子对称性	分子對稱性	molecular symmetry
分子分散	分子分散	molecular dispersion
分子构型	分子組態	molecular configuration
分子间力	分子間力	intermolecular force
分子结构	分子結構	molecular structure
分子扩散	分子擴散	molecular diffusion
分子扩散系数	分子擴散係數	molecular diffusivity
分子量	分子量	molecular weight
分子量分布	分子量分布	molecular weight distribution (MWD)
分子模拟	分子模擬	molecular simulation
分子内力	分子內力	intramolecular force
分子配分函数	分子配分函數	molecular partition function
分子热力学	分子熱力學	molecular thermodynamics
分子筛	分子篩	molecular sieve
分子筛催化剂	分子篩觸媒	molecular-sieve catalyst
分子筛沸石	分子篩沸石	molecular-sieve zeolite
分子生物学	分子生物學	molecular biology
分子输送	分子輸送	molecular transport
分子束沉积法	分子束沈積法	molecular beam deposition
分子束外延	分子束磊晶	molecular beam epitaxy
分子速度	分子速度	molecular velocity
分子物种	分子物種	molecular species
分子相互作用	分子相互作用	molecular interaction
分子蒸馏	分子蒸餾	molecular distillation

祖国大陆名	台湾地区名	英　文　名
分子转换	分子轉換	molecular transformation
分子自集	分子自集	molecular self-assembly
芬斯克方程	芬斯基方程式	Fenske's equation
芬斯克填料(＝螺线圈填料)		
酚红	酚紅	phenol red
酚醛树脂	①酚醛樹脂,酚甲醛樹脂,酚數脂②電木	bakelite, phenol formaldehyde resin, phenolic resin
酚醛塑料	酚醛樹脂	phenolics
酚树脂黏合剂	酚樹脂黏合劑	phenol resin adhesive
酚酞	酚酞	phenolphthalein
焚烧炉	焚化爐	incinerator
粉尘	塵,灰塵,粉塵	dust
粉尘分离器	粉塵分離器	dust separator
粉尘粒子	粉塵粒子	dust particle
粉尘粒子大小	粉塵粒子大小	dust particle size
粉尘旋风分离器	粉塵旋風分離器	dust cyclone separator
粉尘装置	粉塵裝置	dust equipment
粉煤	粉煤	pulverized coal
粉磨机	粉碎機	pulverizer, pulverizing mill
粉末冶金	粉末冶金[學]	powder metallurgy
粉碎	碎解,磨碎,粉碎	size reduction
粉碎比	磨碎程度	size reduction ratio
粉碎机	粉碎機	comminuting machine, comminutor
粉[体]	①粉②火藥	powder
粉体工程(＝粉体技术)		
粉体技术, 粉体工程	粉體技術	powder technology
粉体密度	粉體密度	powder density
粉状燃料	粉狀燃料	pulverized fuel
丰度	豐度	abundance
丰度曲线	①豐度曲線②含量曲線	abundance curve
风洞	風洞	wind tunnel
风干	風乾	air drying, loft drying
风干失重	風乾損失	air-drying loss
风干收缩	風乾收縮	air shrinkage
风化	風化	efflorescence, weathering
风帽分布板	風帽分布板	tuyere distributor
风煤气	風煤氣	air gas

祖国大陆名	台湾地区名	英 文 名
风蚀试验	風蝕試驗	slacking test
风速计	①流速計②風速計	anemometer
风速仪	風向計	anemoscope
风险	風險	risk
风险分析	風險分析	risk analysis, venture analysis
风险函数	風險函數	risk function
风险利润	風險利潤	venture profit
风险收益率	風險收益率	risk earning rate
风险型决策	風險型決策	decision making under risk
风险因子	風險因素	risk factor
风箱	①風箱②伸縮囊	bellows
封闭导管	封閉導管	closed conduct
封闭式叶轮	隱閉葉輪	closed impeller
封闭系统	封閉系統	closed system
封口	封合	sealing
封口机	封口機	sealing machine
峰度	峰態, 峭度	kurtosis
峰负荷	尖峰負載	peak load
峰共振	尖峰共振	peak resonance
峰流量	尖峰流量	peak flow
峰时间	尖峰時間	peak time
峰增益	尖峰增益	peak gain
峰增益比	尖峰增益比	peak gain ratio
峰值	尖峰	peak
锋面分析	鋒面分析	frontal analysis
蜂蜡	蜂蠟	bee wax
蜂窝状对流	蜂巢形對流	cellular convection
冯卡门边界层理论	馮卡門邊界層理論	von Karman boundary layer theory
冯卡门积分法	馮卡門積分法	von Karman integral method
冯卡门类比	馮卡門類比	von Karman analogy
冯卡门数	馮卡門數	von Karman number
冯卡门涡街	馮卡門渦列	von Karman vortex street
冯韦曼方程式	馮韋曼方程式	von Weiman equation
缝编织物	縫編織物	stitch bonded fabric
缝焊	縫焊	seam welding
缝焊机	縫焊機	seam welder
缝隙腐蚀	裂紋腐蝕	crevice corrosion
呋喃树脂	呋喃樹脂	furan resin
呋喃塑料	呋喃塑造	furan plastic

祖国大陆名	台湾地区名	英　文　名
麸酸球菌	麸酸球菌	glutamicus micrococcus
弗洛里-哈金斯理论	佛洛里-哈金斯理論	Flory Huggins theory
伏特计	電壓計, 伏特計	voltmeter
氟磷灰石	氟磷灰石	fluorapatite
氟氯烷	氟氯烷	freon
氟碳化物	氟碳化物	fluorocarbon
氟碳树脂	氟碳樹脂	fluorocarbon resin
浮标	浮子, 浮桶, 浮標	float
浮标指示计	浮標指示計	float indicator
浮雕漆	浮彫漆	relief paint
浮顶槽	①浮頂油槽②浮頂槽	floating roof tank
浮动控制	浮動控制	floating control
浮动速率	浮動速率	floating rate, floating speed
浮动压力操作	浮動壓力操作	floating pressure operation
浮动压力控制	浮動壓力控制	floating pressure control
浮动作用	浮動作用	floating action
浮阀板	浮閥板	floating valve tray
浮阀塔板	閥板	valve tray
浮法玻璃	飄浮玻璃	float glass
浮盖消化槽	浮[動]蓋消化槽	floating-cover digester
浮力	浮力	buoyancy, buoyant force
浮力效应	浮力效應	buoyant effect
浮[力中]心	浮力中心	center of buoyancy
浮料	浮料	floatsam
浮球阀	浮閥	float valve
浮石	浮石	pumice
浮式压力计	浮標式壓力計	float-type manometer
浮式钟型压力计	浮式鐘型壓力計	floating type bell gage
浮水皂	浮皂	floating soap
浮头换热器	浮[動]頭熱交換器	floating head heat exchanger
浮选	浮選[法]	flotation
浮选槽	浮選槽	flotation tank
浮选过程	浮選法	flotation process
浮渣	浮渣	scum
浮渣收集装置	浮渣收集設備	scum-collecting device
浮钟压力计	鐘式壓力計	bell manometer
符号	符號	symbol
符号规定	符號規定	sign convention
福尔马林	福馬林, 甲醛液	formalin

祖国大陆名	台湾地区名	英 文 名
辐射	輻射	radiation
辐射安全	輻射安全性	radiation safety
辐射测量仪	輻射計	radiation meter
辐射常数	輻射常數	radiation constant
辐射穿透	輻射穿透	radiation penetration
辐射传热	輻射熱傳	radiation heat transfer
辐射定律	輻射定律	radiation law
辐射段	輻射段	radiation section
辐射防护	輻射防護	radiation protection
辐射高温计	輻射高溫計	radiation pyrometer
辐射隔屏	輻射屏	radiation screen
辐射沟槽滤板	輻射溝狀濾板, 輻溝濾板	radial grooved filter plate
辐射化学	輻射化學	radiation chemistry
辐射剂量	輻射劑量	radiation dose
辐射剂量计	輻射劑量計	radiation dosimeter
辐射帽	輻射帽	radiation bonnet
辐射密度	輻射密度	radiation density
辐射能	輻射能	radiant energy, radiation energy
辐射能发射	輻射能發射	radiant energy emission
辐射能级	輻射能階	radiation level
辐射屏蔽	輻射遮蔽	radiation shield
辐射器	輻射器	radiator
辐射强度	輻射強度	radiation intensity
辐射热	輻射熱	radiation heat
辐射热计	輻射熱測定器	bolometer
辐射束	輻射束	radiant beam
辐射损耗	輻射損失	radiation loss
辐射损伤	輻射損害	radiation damage
辐射危害	輻射危害	radiation hazard
辐射系数	輻射發射係數	radiant emissivity
辐射消毒	輻射滅菌	radiation sterilization
辐射压[强]	輻射壓[力]	radiation pressure
辐射引发的固态聚合釜液(=残液)	固態輻射引發聚合	solid-state radiation-initiated polymerization
辅酶	輔酵素	coenzyme
辅因子	輔因子	cofactor
辅[助]电极	輔助電極	auxiliary electrode
辅助吸收器	輔助吸收器	auxiliary absorber

祖国大陆名	台湾地区名	英　文　名
辅助增塑剂	助增塑劑	secondary plasticizer
辅助周期	不生產期	nonproductive period
腐浆防治	腐漿控制	slime control
腐蚀	腐蝕	corrosion
腐蚀剂	腐蝕質	corrodent
腐蚀疲劳	腐蝕疲勞	corrosion fatigue
腐蚀试验	腐蝕試驗	corrosion test
腐蚀速率	腐蝕速率	corrosion rate
腐蚀裕量	腐蝕容許量	corrosion allowance
腐蚀作用	腐蝕作用	corrosive action
腐殖质	堆肥	humus
负催化剂	負觸媒	negative catalyst
负催化作用	負催化[作用]	negative catalysis
负电阻	負電阻	negative resistance
负反馈	負回饋	feedback negative, negative feedback
负荷变动	負荷變化	load change
负荷不足	負荷不足	underloading
负荷搅动	負荷擾動	load disturbance
负荷效应	負荷效應	loading effect
负化学促进剂	負化學促進劑	negative chemical promoter
负偶合	負偶合	negative coupling
负偏差	負偏離	negative deviation
负吸附	負吸附	negative adsorption
负压	真空壓力	subatmospheric pressure
负载	負載, 負荷	load
负载因数	負載因數	duty factor
负增长反应	去增長反應	depropagation reaction
负值产物	負值產物	negative valued product
附铂浮石	附鉑浮石	platinized pumice
附铂硅胶催化剂	附鉑矽膠觸媒	platinized silica gel catalyst
附铂石棉	附鉑石綿	platinized asbestos
附铂石棉催化剂	附鉑石綿觸媒	platinized asbestos catalyst
附着热	附著熱	heat of adhesion
复变量	複變數	complex variable
复分解	複分解	double decomposition, metathesis
复分解反应	複分解反應	metathetical reaction
复合壁	複合壁	composite wall
复合材料	複合物, 複合材料	composite, composite material
复合传递系数	複合傳送係數	composite transfer coefficient

祖国大陆名	台湾地区名	英 文 名
复合管道系统	複合管路系統	complex pipe system
复合过程	複合程序	complex process
复合膜	複合膜	composite membrane
复合平均方程式	複合平均方程式	composite-averaged equation
复合球磨机	複合球磨機	compound ball mill
复合形法	複合形法	complex method
复合样品	複合樣品	composite sample
复合液体	複合液體	composite liquids
复利	複利率法	compound interest
复利因子	複利率因數	compound interest factor
复平面	複數平面	complex plane
复式簿记	複式簿記	double entry book keeping
复式精馏塔	複式精餾塔	compound rectifying column
复式蒸馏器	複式蒸餾器	compound distillating apparatus
复数共轭根	複數共軛根	complex conjugate root
复数黏度	複數黏度成分	complex viscosity
复位	重整	reset
复位角	重整角	reset corner
复位势	複勢	complex potential
复位速率	重整速率	reset rate
复位响应	重整應答	reset response
复位终结	重整繞緊	reset windup
复位作用	重整作用	reset action
复写纸	複寫紙	carbon paper
复杂反应	複雜反應	complex reaction
复杂反应网络	複雜反應網路	complex reaction network
复杂性	摻雜	sophistication
复制[品]	繁殖	reproduction
副产物	副産物	by-product
副反应	副反應	side reaction
副环路	副環路	secondary loop
副回路	次環路	minor loop
副效应	副效應	side effect
傅里叶数	傅利葉數	Fourier number
富氮海鸟粪	富氮海鳥糞	nitrogen guano
富过磷酸钙	重過磷酸鈣	double superphosphate
富集	增濃	enrichment
富集培养	增濃培養	enrichment culture
富集因子	增濃因數	enrichment factor

祖国大陆名	台湾地区名	英　文　名
富气	富氣	rich gas
富相	富相	rich phase
覆盖率	覆蓋率	fraction of coverage

G

祖国大陆名	台湾地区名	英　文　名
改性醇酸树脂	改質醇酸樹脂	modified alkyd resin
改性淀粉	改質澱粉	modified starch
改性聚丙烯腈纤维	副丙烯腈纖維	modacrylic fiber
钙玻璃	鈣玻璃	lime glass
钙长石	鈣長石	lime feldspar
钙芒硝	鈣芒硝	glauberite
钙镁橄榄石	鈣橄欖石	monticellite
钙皂	鈣皂	calcium soap
概率	機率, 或然率	probability
概率标绘图	機率圖	probability plot
概率分布	機率分布	probability distribution
概率分布函数	機率分布函數	probability distribution function
概率密度	機率密度	probability density
概率密度函数	機率密度函數	probability density function
概率曲线	機率曲線	probability curve
干电池	乾電池	dry cell
干纺 `	乾紡[絲]	dry spinning
干馏	乾餾	dry distillation
干馏釜(＝甑)		
干滤器	乾式過濾器	drying filter
干凝胶	乾凝膠	xerogel
干球温度	乾球溫度	dry bulb temperature
干球温度计	乾球溫度計	dry bulb thermometer
干扰素	干擾素	interferon
干热灭菌	乾燥滅菌	dry heat sterilization
干涉图样	干涉圖型	interference pattern
干湿比	溼度計量比	psychrometric ratio
干湿陈化试验	乾濕老化試驗	admiralty test
干湿球湿度	乾濕球濕度計	wet and dry bulb hygrometer
干湿球温度计	乾濕球溫度計, 溼度計	dry and wet thermometer, psychrometer
干式燃烧	乾式燃燒	dry combustion

祖国大陆名	台湾地区名	英文名
干式氧化	乾式氧化法	dry oxidation
干性油	乾性油	drying oil
干燥	乾燥	drying
干燥橱	乾燥櫥	drier cabinet, drying cabinet
干燥房	乾燥房	drying house
干燥过程	乾法	dry process
干燥剂	乾燥劑	desiccant, drying agent
干燥架	乾燥架	drying hack
干燥炉	乾燥爐	drying stove
干燥面	乾燥面	dry surface
干燥盘	乾燥盤	drying tray
干燥棚	乾燥棚	drying shed
干燥器	乾燥器, 乾燥機, 乾燥室	desiccator ,drier, dryer
干燥器皿	乾燥器	desiccation apparatus
干燥区	乾燥區	drying zone
干燥室	乾燥室	drying room
干燥收缩	乾燥收縮	drying shrinkage
干燥速率	乾燥速率	drying rate
干燥速率曲线	乾燥速率曲線	drying-rate curve
干燥塔	乾燥塔	drying tower
干燥箱	乾燥箱	drier bin
干燥窑	烘窯	drying kiln
干燥转鼓	乾燥桶	drying drum
干蒸汽	乾蒸汽	dry steam
甘汞电池	甘汞電池	calomel cell
甘汞电极	甘汞電極	calomel electrode
甘酞树脂	酞酸樹脂	glyptal resin
甘油	甘油	glycerine
甘油单酯	單甘油酯	monoglyceride
甘油二酯	二甘油酯	diglyceride
甘油松香酯	松香甘油酯, 松香硬酯	ester gum
甘油酯	甘油酯	glyceride
甘蔗	甘蔗	sugarcane
甘蔗蜡	蔗蠟	sugarcane wax
杆菌	桿菌	bacillus
杆菌肽	桿菌	bacitracin
肝糖	肝糖	glycogen
感光玻璃	感光玻璃	photoform glass

祖国大陆名	台湾地区名	英　文　名
感光减敏剂	感光減敏劑	photographic desensitizer
[感光]密度计	感光密度計	densitometer
感光乳剂	感光乳液	photographic emulsion
感光纸	感光紙	sensitized paper
感胶液晶	向液性液晶	lyotropic liquid crystal
感应炉	感應爐	induction furnace
橄榄仁油	橄欖仁油	olive-kernel oil
橄榄石	橄欖石	olivine
橄榄油	橄欖油	olive oil
刚度	剛性	rigidity
刚度试验	勁度試驗	stiffness test
刚铝石	剛[鋁]石	alundum
刚铝石过滤器	剛[鋁]石過濾器	alundum filter
刚性方程	剛性方程式	stiff equation
刚性模量	剛性模量	stiffness modulus
刚性模数	剛性模數	rigidity modulus
缸内晶种	罐內晶種	pan seed, pan seeding
缸式发酵槽	缸式發酵槽	jar fermentor
钢	鋼	steel
钢管	鋼管	steel pipe
钢化玻璃	回火玻璃, 強化玻璃, 韌化玻璃	tempered glass, toughened glass
钢水包	澆斗	ladle
钢纤维	鋼纖維	steel fiber
杠杆安全阀	槓桿安全閥	lever safety valve
杠杆法则	槓桿法則	lever arm rule
杠杆原理	槓桿原理	lever arm principle
高苯乙烯橡胶	高苯乙烯橡膠	high styrene rubber
高差效应	高差效應	head effect
高分子, 大分子	聚合物, 高分子, 聚合體	polymer
高分子半导体	高分子半導體	semiconducting polymer
高分子材料	聚合材料, 高分子材料	polymeric material
高分子共混物	聚摻合物	polyblend
高硅铁	高矽鐵	high silica iron
高硅氧玻璃	耐熱玻璃	shrunk glass, vycor glass
高级处理	高級處理	advanced treatment
高级废水处理	高級廢水處理[法]	advanced wastewater treatment
高级糖蜜	高級糖蜜	high test molasses

祖国大陆名	台湾地区名	英 文 名
高径比	縱橫比, 細長度, 長徑比	aspect ratio, slenderness ratio
高聚物	聚合物, 高分子化合物	high polymer
高抗爆性燃料	高抗震爆燃料	high antiknocking fuel
高抗冲聚苯乙烯	耐衝擊聚苯乙烯	high impact polystyrene
高抗拉强度	高抗張力	high tensile strength
高岭石	高嶺石	kaolinite
高岭土	高嶺土	kaolin
高炉	高爐	blast furnace
高炉焦	高爐焦	blast furnace coke
高炉煤气	高爐氣, 鼓風爐氣	blast furnace gas
高炉渣	高爐渣, 鼓風爐渣	blast furnace slag
高铝玻璃	高鋁玻璃	high alumina glass
高铝耐火材料	高鋁耐火物	high alumina refractory
高铝黏土	高鋁黏土	high alumina clay
高铝水泥	高鋁水泥	aluminous cement, high alumina cement
高铝砖	高鋁磚, 礬士磚	alumina brick, high alumina brick
高锰酸盐[滴定]值	過錳酸鹽值	permanganate value
高锰酸盐值	過錳酸鹽數	permanganate number
高密度聚乙烯	高密度聚乙烯	high density polyethylene
高能化分子	高能化分子	energized molecule
高强硅盐水泥	高強度[卜特蘭]水泥	high strength portland cement
高热量气体	高熱量氣體	high-Btu gas
高热值	高熱值	high heating value
高渗溶液	高滲壓溶液	hypertonic solution
高斯消元法	高斯消去法	Gaussian elimination
高速离心机	超速離心機	supercentrifuge
高速率消化槽	高速率消化槽	high-rate digester
高速碳钢	高速碳鋼	high speed carbon steel
高速蒸发器	高速蒸發器	high speed evaporator
高铁[卜特兰]水泥	高鐵[卜特蘭]水泥	high iron Portland cement
高通滤波器	高通濾波器	high-pass filter
高透光玻璃	高透光玻璃	high transmission glass
高温法	高溫法	hot process
高温分解	熱解, 高溫分解	pyrogenic decomposition
高温计	高溫計	pyrometer
高温橡胶	熱製橡膠	hot rubber
高效型洗涤剂	強力清潔劑	heavy duty detergent
高效液相色谱仪	高性能液相層析儀	high performance liquid chromatograph

祖国大陆名	台湾地区名	英　文　名
高选择器开关	高選擇器開關	high selector switch
高真空操作	高真空操作	high vacuum operation
高真空蒸馏	高真空蒸餾	high vacuum distillation
锆石	鋯英石	zircon
锆石瓷	鋯瓷	zircon porcelain
锆英石耐火材料	鋯英石耐火物	zircon refractory
割缝	狹縫	slit
格拉斯霍夫数	葛瑞斯何夫數	Grashof number
格雷茨数	格雷茲數	Graetz number
格筛	柵篩	grizzly
隔板	隔板	dummy plate
隔离法	隔離法, 單離法	method of isolation
隔离剂	防黏著劑	dusting·agent
隔离流动	隔離流動	segregated flow
隔离流反应器	隔離流反應器	segregated reactor
隔离系统, 孤立系统	隔離系統	isolated system
隔膜	隔膜, 隔板	septum
隔膜泵	隔膜泵	diaphragm pump
隔膜操作阀	隔膜操作閥	diaphragm operated valve
隔膜电解槽	隔膜[電解]槽	diaphragm cell
隔膜阀	隔膜閥	diaphragm valve
隔膜盒	隔膜盒	diaphragm box
隔膜盒液位计	隔膜盒液位計	diaphragm box level gauge
隔膜控制阀	隔膜控制閥	diaphragm control valve
隔膜片	隔膜片	diaphragm
隔膜式压力转换器	隔膜壓力轉換器	diaphragm type pressure transducer
隔膜式应变计	隔膜應變計	diaphragm type strain gage
隔膜室	隔膜室	diaphragm chamber
隔膜微压差	隔膜微壓差	diaphragm D/P cell
隔膜压缩机	隔膜壓縮機	diaphragm compressor
隔热	保溫, 隔熱	insulation, thermal insulation, heat insulation
隔热材料	保溫材料	heat insulating material
隔热层	保溫層	lagging
隔热衬里	絕緣襯裏	insulating lining
隔热管	保溫管	lagged pipe
隔热面	保溫面	lagged surface
隔热[蒸馏]塔	隔熱柱, 保溫柱	insulated column
隔声材料	隔音材料	sound-insulating material

祖国大陆名	台湾地区名	英 文 名
隔声黏合剂	隔音黏合劑	sound-insulating adhesive
镉电池	鎘電池	cadmium cell
镉黄	鎘黃	cadmium yellow
镉皂	鎘皂	cadmium soap
个别沉降	個別沉降	discrete settling
个人计算机	個人計算機, 個人電腦	personal computer
各向同性	各向同性	isotropy
各向同性固体	各向同性固體	isotropic solid
各向同性流动	各向同性流動	isotropic flow
各向同性湍流	各向同性紊流	isotropic turbulence
各向异性	各向異性	anisotropy
各向异性膜	各向異性膜	anisotropic membrane
各向异性湍流	非各向同性紊流	nonisotropic turbulence
各向异性吸收	[各]向異性吸收	anisotropic absorption
铬橙	鉻橙	chrome orange
铬钢	鉻鋼	chrome steel
铬革	鉻革, 鞣皮	chrome leather
铬红	鉻紅	chrome red
铬黄	鉻黃	chrome yellow
铬绿	鉻綠	chrome green
铬媒染色	鉻媒染色	chrome mordant dyeing
铬镁砖	鉻美磚	chrome-magnesite brick
铬染料	鉻媒染料	chrome dye
铬鞣	鉻鞣	chrome tannage
铬鞣制	鉻鞣製	chrome tanning
铬颜料	鉻顏料	chrome pigment
铬质耐火材料	鉻[質]耐火物	chrome refractory
铬砖	鉻磚	chrome brick
根轨迹	根軌跡	root locus
根轨迹法	根軌跡法	root locus method
根轨迹方法	根軌跡法	root locus technique
根轨迹角	根軌跡角	root locus angle
根轨迹图	根軌跡圖	root locus plot
更迭成本	更迭成本	changeover cost
更迭时间	更迭時間	changeover time
更换费用	置換成本	replacement cost
更换价值	汰換價值	replacement value
更换评估	汰換評估	replacement evaluation
工厂布置	工廠布置	plant layout

祖国大陆名	台湾地区名	英 文 名
工厂成本估算	工廠成本估計	plant cost estimate
工厂资产	工廠資產	plant asset
工程	①工程②工程學	engineering
工程材料	工程材料	engineering material
工程分析	工程分析	engineering analysis
工程流程图	機械流程圖	mechanical flow diagram, mechanical flowsheet
工程设计	工程設計	engineering design
工程设计成本	工程設計成本	engineering design cost
工程实验	工程實驗	engineering experiment
工程塑料	工程塑膠	engineering plastic
工况	操作方式	operation mode
工时	人時, 工時	man hour
工时学	工時學	time and motion study
工业	工業	industry
工业反应器	工業反應器	industrial reactor
工业废料	工業廢棄物	industrial waste
工业废水	工業廢水	industrial wastewater
工业废水处理	工業廢水處理	industrial wastewater treatment
工业过程	工業程序	industrial process
工业化学	工業化學	industrial chemistry
工业化学计量学	工業化學計量學	industrial stoichiometry
工业酒精	工業酒精	industrial alcohol
工业煤气	工業用氣體	industrial gas
工业色谱	工業色譜	process-scale chromatography
工业水处理	工業用水處理	industrial water treatment
工业仪器	工業儀器	industrial instrument
工业用水	工業用水	industrial water
工业用洗涤剂	工業用清潔劑	industrial-use detergent
工艺安全	程序安全性	process safety
工艺布置图	程序布置	process layout
工艺参数	程序參數	process parameter
工艺单元	程序單元	process unit
工艺负荷	程序負載	process load
工艺负荷变化	程序負載改變	process load change
工艺工程师	程序工程師, 方法工程師	process engineer
工艺规划	程序規劃	process planning
工艺过程设计	程序設計	process design

祖国大陆名	台湾地区名	英 文 名
工艺计算	程序計算	process calculation
工艺技术	程序技術	process technology
工艺流程图	程序圖, 程序流程圖	process chart, process flow diagram, process flowsheet
工艺设备	程序裝置	process equipment
工艺设计工程师	程序設計工程師	process design engineer
工艺设计师	程序設計師	process designer
工艺条件	程序條件	process condition
工艺物流	程序流	process stream
工艺系统	程序系統	process system
工艺性能	程序性能	process performance
工艺用水	程序用水	process water
工艺用真空[系统]	真空程序	process vacuum
工艺蒸汽	程序蒸汽	process steam
工资	工資	wage
工资单	薪資單	payroll
工作单	工作單	worksheet
工作流体	工作流體	working fluid
工作特性	性能特性	performance characteristics
工作应力	工作應力	working stress
弓形降液管	弓形下導排管	segmental downtake calandria
弓形孔口	弓形孔口	segmental orifice
弓形孔口板	弓形孔口板	segmental orifice plate
公称	標稱	nominal
公称尺寸	標稱尺寸	nominal size
公称管径	標稱管徑	nominal pipe diameter
公共设施	共用設施	common utility
公共设施检查报表	公用設施檢查報表	utility check list
公共设施流程图	公用設施流程圖	utility flowsheet
公式	[公]式	formula
公用服务事业	公用設施	utility service
公用设施	公用設施	utility
公制	公制, 米制	metric system
功函数	功函數	work function
功力消耗	功率消耗量	power consumption
功率	功率	power
功率数	功率數	power number
功率损失	功率損失	power loss
功率效率	功率效率	power efficiency

祖 国 大 陆 名	台 湾 地 区 名	英 文 名
功率需要量	功率需要量	power requirement
功率因数	功率因數	power factor
功能高分子	功能性聚合物	functional polymer
功需要量	功需要量	work requirement
功指数	功指數	work index
供给压力	供給壓[力]	supply pressure
供氧	供氧	oxygen supply
供应	供應	supply
汞浮子压力计	汞浮子壓力計	mercury float manometer
汞扩散泵	汞擴散泵	mercury diffusion pump
汞齐	汞齊	amalgam
汞污染	汞污染	mercury pollution
汞阴极电解槽	汞陰極電解槽	mercury cathode cell
拱砖	拱磚	arch brick
共萃取	共萃取	coextraction
共存方程	共存方程式	coexistence equation
共轭对	共軛對	conjugate pair
共轭溶液	共軛溶液	conjugate solution
共轭深度	共軛深度	conjugate depth
共轭相	共軛相	conjugate phase
共反应剂	共反應物	coreactant
共沸	共沸, 恆沸	constant boiling
共沸参数	共沸參數	azeotropic parameter
共沸点	共沸點, 恆沸點	azeotropic point, constant boiling point
共沸范围	共沸範圍	azeotropic range
共沸干燥	共沸乾燥	azeotropic drying
共沸混合物	共沸混合液	azeotropic mixture
共沸塔	共沸塔	azeotrope tower
共沸温度	共沸溫度	azeotropic temperature
共沸物, 恒沸物	共沸物	azeotrope
共沸系统	共沸系統	azeotropic system
共沸现象	共沸現象	azeotropic phenomenon
共沸线	共沸線	azeotropic line
共沸[性]	共沸法	azeotropy
共沸性质	共沸性質	azeotropic property
共沸蒸馏, 恒沸蒸馏	共沸蒸餾	azeotropic distillation
共沸[蒸馏]过程	共沸程序	azeotropic process
共沸状态	共沸狀態	azeotropic state
共沸组成	共沸組成	azeotropic composition

祖国大陆名	台湾地区名	英　文　名
共混聚合物	高分子掺合物, 聚合體掺合物	polymer blend
共价电子对	共用電子對	shared electron pair
共聚合	共聚合反應	copolymerization
共聚物	共聚物	copolymer
共离子	共離子	co-ion
共溶点, 褶点	褶點	plait point
共生	共生, 共棲	symbiosis
共生现象	①共棲②共存	commensalism
共形映射	保形映射, 保角映射	conformal mapping
共振	共振	resonance
共振峰	共振尖峰	resonance peak, resonant peak
共振积分	共振積分式	resonance integral
共振频率	共振頻率	resonance frequency, resonant frequency
沟缝速度	溝縫速度	slot velocity
沟缝液体密封	溝縫液體密封	slot liquid seal
沟流	溝流	channeling
沟流效应	溝流效應	channeling effect
构象	構形	conformation
构型性质(=位形性质)		
估计方差	估計變異數	estimated variance
估计量	估計值	estimate
估计器	估計器	estimator
估算(=估值)		
估值, 估算	估計	estimation
孤立系统(=隔离系统)		
箍	捏縮	pinch
箍缩效应	捏縮效應	pinch effect
古伊-斯托多拉定理	高伊-斯托多拉定理	Gouy Stodola theorem
骨架振动	骨架振動	skeletal vibration
骨炭	骨黑, 骨碳	bone black, bone char
鼓风	①鼓風②爆炸	blast
鼓风机	鼓風扇, 強制通風扇, 送風機	forced draft fan, air blower, blast blower, blower
鼓风炉	鼓風爐	blast furnace
鼓风通气机	鼓風通氣器	forced draft aerator
鼓泡	氣泡	bubbling
鼓泡反应器	氣泡反應器	sparged reactor
鼓泡流化床	氣泡流體化床	bubbling fluidized bed

祖国大陆名	台湾地区名	英 文 名
鼓泡流态化	氣泡流態化	bubbling fluidization
鼓泡气	噴佈氣	sparged gas
鼓泡器	噴佈器	sparger
鼓泡式吸收器	氣泡吸收器	bubbling absorber
鼓泡塔	鼓風塔	bubble column
鼓泡塔盘	泡罩板	bubble tray
鼓式标度指示计	鼓式標[度指]示計	drum type scale indicator
鼓式分离器	桶式分離器	drum separator
鼓式干燥器	桶式乾燥器	drum drier
鼓式计数器	鼓式計數器	drum type counter
鼓式记录器	鼓式記錄器	drum type recorder
鼓型过滤机	桶式過濾器	drum filter
固醇(=甾醇)		
固氮菌	固氮菌	azotobacteria
固氮[作用]	氮固定	nitrogen fixation
固定	固定	fixation
固定成本	固定成本	fixed cost
固定成长系统	固定成長系統	fixed-growth system
固定床	固定床	fixed bed
固定床反应器	固定床反應器	fixed bed reactor
固定床过程	固定床法	fixed bed process
固定催化剂床	固定觸媒床	fixed catalyst bed
固定氮	固定氮	fixed nitrogen
固定费用	固定費用	fixed charge
固定管板换热器	固定管板熱交換器	fixed tube-sheet heat exchanger
固定化技术	固定化技術	immobilization technology
固定化酶	固定酶	immobilized enzyme
固定化细胞反应器	固定化細胞反應器	immobilized cell reactor
固定剂	固定劑	fixed agent
固定台	固定台	stationary platen
固定浴	固定浴	fixation bath
固定资本	固定資本	fixed capital
固定资本投资额	固定資本投資額	fixed capital investment
固固相反应	固固反應	solid-solid reaction
固-固相转变	固-固相轉變	solid-solid transition
固化计	熟化計	cure meter
固化酒精	固化酒精	solidified alcohol
固化汽油	固化汽油	solidified gasoline
固化热	固化熱	heat of solidification, solidification heat

祖国大陆名	台湾地区名	英 文 名
固化速率	固化速率	solidification rate
固化速率参数	固化速率參數	solidification rate parameter
固溶体	固溶體	solid solution
固溶体合金	固溶體合金	solid solution alloy
固溶线	[固]溶線	solvus
固态	固態	solid state
固态高分子电解质	固態高分子電解質	solid-state polyelectrolyte
固态化学	固態化學	solid state chemistry
固态扩散	固態擴散	solid-state diffusion
固态物理	固態物理	solid state physics
固体	固體	solid
固体超载	固體超載	solid overload
固体催化反应	固體催化反應	solid-catalyzed reaction
固体点阵	晶體晶格	solid lattice
固体废弃物处理	固體廢棄物處理	solid waste disposal
固体废弃物掩埋场	固體廢棄物掩埋場	solid waste landfill
固体废物	固體廢棄物	solid waste
固体聚合法	固相聚合	solid polymerization
固体力学	固體力學	solid mechanics
固体粒子	固體粒子	solid particle
固体浓度	固體濃度	solid concentration
固体燃料	固體燃料	solid fuel
固体润滑剂	固體潤滑劑	solid lubricant
固体输送	固體輸送	solid transport
固体填充塑料	固體填充塑料	solid filled plastic
固体通量	固體通量	solid flux
固体通量理论	固體通量理論	solid flux theory
固体推进剂	固體推進劑	solid propellant
固体悬浮体	固體懸浮物	solid suspension
固相	固相	solid phase
固相反应	固相反應	solid phase reaction
固相含量	固體含量	solid content
固相合成	固相合成	solid phase synthesis
固相化	固定化	immobilization
固相聚合	固相聚合, 固態聚合	solid phase polymerization, solid-state polymerization
固相缩聚	固相縮聚	solid phase polycondensation
固相线	固相線	solidus
固液分离器	固液分離器	solid-liquid separator

祖国大陆名	台湾地区名	英 文 名
固有安全	固有安全	intrinsic safety
固有活化能	固有活化能	intrinsic activation energy
固有角动量	固有角能量	intrinsic angular momentum
固有能	固有能	intrinsic energy
固有热	蘊熱, 固有熱	intrinsic heat
固有速率	固有速率	intrinsic rate
故障形式和影响分析	故障形式和影響	failure mode and effect analysis
故障诊断	故障診斷	failure diagnosis, fault diagnosis
刮板	刮刀	scraper
刮板式换热器	刮刀式熱交換器	scraper heat exchanger
刮板输送机	梯板運送機, 梯運機, 刮運機	flight conveyor, scraper conveyor
刮铲角	刮鏟角	angle of spatula
刮刀工去皮机	刀式去皮機	knife barker
刮刀式冷却器	刮刀式冷凍器	scraper chiller
刮痕	刮痕	scratch
刮痕硬度	刮痕硬度	scratch hardness
刮涂机	刮塗機	spread coater
寡糖, 低聚糖	寡糖	oligosaccharide
拐点	反曲點, 拐點	inflection point
拐角流	角形流動	corner flow
拐折空气加热器	交錯空氣加熱器	staggered air heater
关闭高差	關閉高差	shut-off head
关键字	關鍵字	key word
关键组分	關鍵成分, 主成分	key component
关联, 关联式	相關式	correlation
关联矩阵	關聯矩陣	incidence matrix
关联式(=关联)		
关税	關稅	tariff
观测次序	測得階	observed order
观察孔	視孔	peep hole
官能团	官能基, 功能基	functional group
管	管	pipe
管板	管板	tube sheet
管壁厚系列号	管分類號, 分類號	pipe schedule number, schedule number
管程	管程	tube[side]pass
管道流程图	配管流程圖	piping flowsheet
管沟	溝	trench
管过滤器	管過濾器	pipe filter

祖国大陆名	台湾地区名	英 文 名
管件	管配件	pipe fitting
[管]接头	接頭, 接合	joint, coupling joint
管接头	管接頭	pipe joint
管接头密封	管密封	pipe sealing
管径	管徑	pipe diameter
管理费用	管理費	overhead cost
管流	管流	pipe flow
管路(=配管)		
管路混合器	管路混合器	in-line mixer
管路摩擦	管路摩擦	pipe friction
管路网络	管路網絡	pipeline network
管路系统	管路系統, 配管系統	piping system
管路-仪表流程图	管線及儀器圖	piping and instrument diagram(PID)
管磨机	管磨機	tube mill
管排	管排	tube bank
管式反应器	管式反應器	tubular reactor
管式干燥器	管式乾燥器	tube drier
管式加热器	管式加熱器	tubular heater
管式冷凝器	管式冷凝器	tubular condenser
管式炉	管式爐	tube furnace
管式气压计	管式氣壓計	tube gage
管式蒸发器	管式蒸發器	tubular evaporator
管式蒸馏釜	管馏器	pipe still,tube still,tubular still
管式组件	管式組件	tubular module
管束	管束	tube bundle
管线	管線	pipeline
管线表	管線表	pipe line list
管线过滤器	管線過濾器	in-line filter
管线应力	管線應力	piping stress
管中心距	節距	pitch
管子尺寸	管子尺寸	pipe size
管子粗糙度	管粗糙度	pipe roughness
管子相对粗糙度	管子相對粗糙度	pipe relative roughness
惯性	慣性	inertia
惯性沉降	慣性沉降	inertial settling
惯性离心分离器	慣性離心分離器	inertial centrifugal separator
惯性力	慣力	inertial force
惯性装置	慣性裝置	inertial device
灌注培养	灌注培養	perfusion culture

祖国大陆名	台湾地区名	英 文 名
罐头厂废水	罐頭廠廢水	cannery waste water
光催化	光催化[作用]	photocatalysis
光催化剂	光觸媒	photocatalyst
光[导]纤[维]	光纖[維]	optical fiber
光电, 光电学	光電學	photoelectricity
光电比色计	光比色法, 光電比色計	photocolorimetry, photoelectric colorimeter
光电材料	光導電材料	photoconductive material
光电池	光電管	photocell
光电池[管]	光電管	photoelectric cell
光电导性	光電導性	photoconductivity
光电分光光度计	光電光譜儀	photoelectric spectrophotometer
光电管	光電管	phototube
光电计	光電計	photoelectrometer
光电效应	光電效應	photoelectric effect
光电旋光计	光電旋光計	photoelectric polarimeter
光电学(＝光电)		
光碟	光碟	optical disc
光度测定法	光測定法	photometric method
光度分析	光測定分析[法]	photometric analysis
光度计	光度計	photometer
光度学	光測定法	photometry
光反应	光反應	photoreaction
光[分]解[反应]	光分解[反應]	photodecomposition, photolysis
光感受体	感光器	photoreceptor
光合作用	光合作用	photosynthesis
光呼吸	光呼吸	photorespiration
光化学	光化學	photochemistry
光化学产量	光化學產率	photochemical yield
光化学臭氧化	光化學臭氧化[反應]	photochemical ozonization
光化学当量	光化學當量	photochemical equivalent
光化学低限	光化學低限	photochemical threshold
光化学动力学	光化學動力學	photochemical kinetics
光化学反应	光化學反應	photochemical reaction
光化学反应器	光化學反應器	photochemical reactor
光化学反应性	光化學反應性	photochemical reactivity
光化学分解	光化學分解[反應]	photochemical decomposition
光化学过程	光化學程序	photochemical process
光化学活性	光化學活性	photochemical activity
光化学激化	光化學激化	photochemical excitation

祖国大陆名	台湾地区名	英 文 名
光化学降解	光化學降解[反應]	photochemical degradation
光化学氯化	光化學氯化[反應]	photochemical chlorination
光化学吸收律	光化學吸收律	photochemical absorption law
光化学效率	光化學效率	photochemical efficiency
光化学诱导	光化學誘導	photochemical induction
光黄化	光黄化	photoyellowing
光活化	光活化[作用]	photoactivation
光检测器	測光器	photodetector
光降解	光降解[反應]	photodegradation
光刻	照相石印法	photolithography
光刻蚀过程	照相蝕刻法	photoetching process
光卤石	光鹵石	carnallite
光氯化	光氯化[反應]	photochlorination
光密度计	光密度計	photodensitometer
光敏剂	光敏劑	photosensitizer
光敏性	感光性	photosensitivity
光敏氧化	感光氧化[反應]	photosensitized oxidation
光谱测定法	光譜法	spectrometry
光谱分析	光譜化學分析, 光譜分析	spectral analysis, spectroscopic analysis
光谱化学	光譜化學分析	spectrochemistry
光谱图	光譜圖, 譜圖	spectrogram
光谱学	①光譜學②光譜法	spectroscopy
光气	光氣	phosgene
光气化法	光氣法	phosgenation process
光热分析	光熱分析[法]	photothermal analysis
光散射	光散射	scattering of light
光弹性	光彈性	photoelasticity
光桐油(中国三年桐的油)	光桐油(中國三年桐)	aleurites fordii oil
光学玻璃	光學玻璃	optical glass
光学玻璃板	光學玻璃板	slab glass
光学补偿器	光學補償器	optical compensator
光学高温计	光學高溫計	optical pyrometer
光[学]密度	光學密度	optical density
光氧化	光氧化[反應]	photooxidation
光氧化老化	光氧化老化	photooxidative aging
光致发光	光發光	photoluminescence
光[致]核反应	光核子反應	photonuclear reaction

祖国大陆名	台湾地区名	英 文 名
光致聚合	光聚合[反應]	photopolymerization
光[致]聚合物	光聚合物	photopolymer
光致抗蚀剂	光阻劑	photoresist
光致离解	光解離	photodissociation
光中子	光中子	photoneutron
光子	光子	photon
光子材料	光子材料	photonic material
广度性质	外延性質	extensive property
广义流态化	廣義流態化	generalized fluidization
广义牛顿流体	廣義牛頓流體	generalized Newtonian fluid
广义坐标	廣義座標	generalized coordinates
归一化	正規化	normalization
规定浓度	當量濃度	normal concentration
规格	規格	specification
规号	規號	gage number
规划	規劃	programming
硅补强剂	矽土氣凝膠	silica aerogel
硅氮聚合物	矽氮聚合物	silicon-nitrogen polymer
硅氮烷聚合物	矽氯烷聚合物	silazane polymer
硅灰石	矽灰石	wollastonite
硅胶	矽膠凝體	silica gel
硅锰铁合金	矽錳鐵齊	ferro-silico-manganese
硅片	矽晶片	silicon wafer
硅[润滑]脂	矽酮脂, 聚矽氧脂	silicone grease
硅石灰石	矽質石灰石	siliceous limestone
硅酸钍矿	矽酸釷礦	thorite
硅酸盐类黏合剂	矽酸鹽類黏合劑	silicate adhesive
硅铁合金	矽鐵齊	ferrosilicon alloy
硅烷	矽甲烷	silane
硅烷类增黏剂	矽烷類增黏劑	silane adhesion promoter
硅烷偶联剂	矽烷類偶聯劑	silane coupling agent
硅线石	矽線石	sillimanite
硅橡胶	矽橡膠	silicon rubber
硅橡胶混炼胶	矽氧橡膠聚合物	silicone rubber compound
硅橡胶黏合剂	矽膠黏著劑	silicone rubber adhesive
硅锌矿	矽鋅礦	willemite
硅岩	矽岩	silica rock
硅氧烷	矽氧烷	siloxane
硅油	矽油	silicone oil

祖国大陆名	台湾地区名	英 文 名
硅藻土	矽藻土	diatomaceous earth ,diatomite, kieselguhr, siliceous earth
硅藻土过滤器	矽藻土過濾器	diatomaceous-earth filter
硅藻土炸药	矽藻土炸藥	kieselguhr dynamite
硅整流器	矽整流器	silicon rectifier
硅质耐火材料	矽石耐火物	silica refractory
硅质耐火黏土	矽質耐火黏土	siliceous fireclay
硅质耐火黏土砖	矽質耐火黏土磚	siliceous fireclay brick
硅质黏土	矽質黏土	siliceous clay
硅砖	矽磚	silica brick
鲑鱼油	鮭油	salmon oil
轨迹	軌跡	locus
癸二酸	癸二酸	sebacic acid
柜式干燥机	箱形乾燥器	shelf drier, shelf dryer
桂皮油	桂皮油, 肉桂油	cassia oil
辊磨(＝辊式捏合机)		
辊式捏合机, 开炼机, 辊磨	輥磨機	roll mill
辊式破碎机	輥式破碎機	roll crusher
辊压机	輥壓機	roller mill press
滚磨机	滾磨機	tumbling mill
滚筒干燥器	轉桶乾燥器	rotating drum dryer
滚筒印花机	滾筒印花機	roller printing machine
滚珠轴承	球軸承	ball bearing
滚柱	輥, 滾筒	roller
滚子轴承	滾筒軸承, 輥軸承	roller bearing
锅	鍋	kettle
锅炉	鍋爐	boiler
锅炉给水	鍋爐給水	boiler feed water
锅盐	鍋鹽	pan salt
国际实用温标	國際實用溫標	international practical temperature scale
果胶	果膠	pectin
果胶酶	果膠酶	pectinase
果胶纤维素	果膠纖維素	pecto-cellulose
果糖	果糖	fructose, levulose
过饱和	過飽和	oversaturation
过饱和度	過飽和度	degree of super saturation
过饱和溶液	過飽和溶液	supersaturated solution
过程变量	程序變數	process variable

祖国大陆名	台湾地区名	英 文 名
过程变异性	程序變異性	process variability
过程辨识	程序辨識, 程序鑑別	process identification
过程不稳定性	程序不穩定性	process instability
过程超载	程序超載	process overload
过程动态学	程序動態[學]	process dynamics
过程反弹	程序反彈	process resilience
过程反应曲线	程序反應曲線	process reaction curve
过程分解	程序分解	process decomposition
过程分析	程序分析	process analysis
过程工程	程序工程, 方法工程	process engineering
过程工程核查表	程序工程檢查報表, 方法工程檢查報表	process engineering check list
过程工业	程序工業	process industry
过程合成(=过程综合)		
过程集成	整合程序, 程序整合	process integration, process integration
过程进料	程序進料	process feed
过程开发	程序開發	process development
过程可靠性	程序可靠性	process reliability
过程控制	程序控制	process control
过程控制模式	程序控制方式	process control pattern
过程灵敏度	程序靈敏度	process sensitivity
过程模拟	程序模擬	process simulation
过程模型化	程序模式化	process modeling
过程能力	程序容量	process capacity
过程评价	程序評估, 程序評價	process evaluation, process evaluation
过程热力学分析	程序熱力分析	thermodynamic analysis of process
过程失稳	程序失穩	process upset
过程时间滞后	程序時間落後	process time lag
过程特性	程序特性	process characteristics
过程稳定性	程序穩定性	process stability
过程系统分析	程序系統分析	process system analysis
过程系统工程, 化工系统工程	程序系統工程	process system engineering
过程响应	程序應答	process response
过程响应曲线	程序應答曲線	process response curve
过程行为特性	程序行為	process behavior
过程应用	程序應用	process application
过程优化	程序最適化	process optimization
过程有效能	程序可用性	process availability

祖国大陆名	台湾地区名	英文名
过程增益	程序增益	process gain
过程自动化	程序自動化	process automation
过程综合,过程合成	程序合成	process synthesis
过程组成元件	程序元件	process element
过渡长度	過渡長度	transition length
过渡金属	過渡金屬	transition metal
过渡流	過渡流動	transition flow
过渡络合物	過渡錯合物	complex transition
过渡区	過渡區[域],轉變區[域]	transition region
过渡态	過渡狀態	transition state
过渡态理论	過渡狀態理論	transition-state theory
过渡元素	過渡元素	transition element
过冷	過冷	supercooling, subcooling
过冷沸腾	過冷沸騰	subcooled boiling
过冷液	過冷液	supercooled liquid
过冷液体	過冷液體	subcooled liquid
过冷蒸气	過冷蒸氣	supercooled vapor
过冷状态	過冷狀態	subcooled state
过量法	過量法	method of excess
过量空气	過量空氣	excess air
过量水	過量水	excess water
过磷酸钙	重過磷酸鈣	superphosphate
过滤	過濾	filtration
过滤介质	過濾介質	filtration medium
过滤流程	過濾器流程	filter flowsheet
过滤面积	過濾面積	filter area, filtration area
过滤灭菌	過濾除菌	filtration sterilization
过滤器堵塞	過濾器阻塞	filter clogging
过滤器容量	過濾器容量	filter capacity
过滤器压缩机	過濾器壓縮機	filter compressor
过滤速率	過濾速率	filtration rate
过滤网调换装置	過濾網調換裝置	screen changer
过滤网组[合]	過濾網組[合]	screen pack
过滤效率	過濾器效率	filter efficiency
过滤性能	過濾器性能	filter performance
过滤周期	過濾周期	filtration cycle
过膨胀现象	過膨脹現象	overexpansion phenomenon
过热	過熱	superheating, overheat

祖国大陆名	台湾地区名	英 文 名
过热[量]	過熱量	superheat
过热器	過熱器	superheater
过热水	過熱水	superheated water
过热状态	過熱狀態	superheated state
过失误差	總誤差	gross error
过失误差检出	總誤差鑑定	gross error identification
过松弛法	過鬆弛法	overrelaxation method
过调节	超越	overshooting
过氧化苯甲酰	過氧化苯甲醯	benzoyl peroxide
过氧化氢异丙苯	氫過氧化異丙苯	cumene hydroperoxide
过氧化物	過氧化物	peroxide
过氧化物分解剂	過氧化物分解劑	peroxide decomposer
过乙酸	過醋酸	peracetic acid
过载	超載, 超負荷	overload
过载磨石机	上動石磨[機]	over-driven buhrstone mill
过早硫化	過早硫化	scorching
过增益	過增益	over gain
过阻尼	過阻尼	overdamping
过阻尼系统	過阻尼系統	overdamped system
过阻尼应答	過阻尼應答	overdamped response

H

祖国大陆名	台湾地区名	英 文 名
哈斯特镍基合金	赫史特合金	hastelloy
海岸污染	海岸污染	coastal pollution
海豹油	海豹油	seal oil
海岛型棉	海島棉	sea island cotton
海绵	海綿	sponge
海绵铂	海綿鉑	sponge platinum
海绵胶	海綿膠	sponge gum
海绵镍	海綿鎳	sponge nickel
海绵钛	鈦海綿	titanium sponge
海绵铁	海綿鐵	sponge iron
海鸟粪	[海]鳥糞	guano
海洋污染	海洋污染	ocean pollution
海藻纤维	海藻纖維	seaweed fiber

祖国大陆名	台湾地区名	英　文　名
亥姆霍兹自由能, 自由能	亥姆霍兹自由能, 自由能	Helmholtz free energy
含铵过磷酸钙	氨化過磷酸鈣	ammoniated superphosphate
含残油软石蜡	粗蠟	slack wax
含尘量	含塵量	dustiness
含尘气体	含塵氣體	dust-laden gas
含钾过磷酸钙	含鉀過磷酸鈣	potash superphosphate
含蜡废油	含蠟污油	paraffin slop
含蜡分馏	蠟分餾	wax fractionation
含蜡油	蠟油, 含蠟油	wax oil, waxy oil
含蜡原油	含蠟原油	waxy crude
含蜡渣油	蒸餘蠟	wax residuum
含量	含量	content
含硫聚合物	含硫聚合物	sulfur-containing polymer
含铅汽油	加鉛汽油	leaded gasoline
含铅氧化锌	含鉛鋅白	leaded zinc
含湿量	①水分②溼氣	moisture
含水量	含水量	water content
含水皂	含水皂	semi-boiled soap
含碳燃料	含碳燃料	carbonaceous fuel
含氧量	含氧量	oxygen content
含皂量	皂含量	soap content
函数发生器	函數波產生器	function generator
函数关系	函數關係[式]	functional relationship
函数平移	函數平移	translation of function
焓	焓	enthalpy
焓浓图	焓-濃度圖	enthalpy-concentration diagram
焓-浓图, P-S 图	P-S 圖	Ponchon-Savarit diagram
焓-熵图, H-i 图	焓-熵圖	enthalpy-entropy diagram
焓-湿图	焓-濕度圖	enthalpy-humidity chart
焓-组成图	焓-組成圖	enthalpy-composition chart, enthalpy-composition diagram
焊接	熔接	welding
夯锤法	鎚結法	rammer process
夯实密度(＝振实密度)		
行列式	行列式	determinant
航空汽油	航空汽油	aviation gasoline
航空燃料	航空燃料	aviation fuel
毫伏特计	毫伏特計	millivoltmeter

祖国大陆名	台湾地区名	英 文 名
毫克	毫克	milligram
毫米	毫米	millimeter
好气[细]菌(=好氧[细]菌)		
好氧处理	需氣處理	aerobic treatment
好氧发酵	需氣發酵	aerobic fermentation
好氧分解	好氣分解	aerobic decomposition
好氧培养	嗜氧培養	aerobic culture
好氧生物氧化	需氣生物氧化[法]	aerobic biological oxidation
好氧污泥消化	需氣污泥消化	aerobic sludge digestion
好氧[细]菌, 好气[细]菌	好氣菌, 需氣菌	aerobic bacteria, aerophilic bacteria
好氧消化	需氣消化	aerobic digestion
好氧氧化	需氣氧化	aerobic oxidation
耗尽	耗盡	depletion
耗尽容许度	耗盡容許度	depletion allowance
耗散	散逸	dissipation
耗散函数	散逸函數	dissipation function
耗散能	散逸能	dissipated energy
耗氧速率	耗氧速率	oxygen consumption rate
合成	合成	synthesis
合成氨	合成氨	synthetic ammonia
合成宝石	合成寶石	synthetic gem
合成单宁	合成單寧	synthetic tannin
合成氮肥	合成氮肥	synthetic nitrogenous fertilizer
合成靛	合成靛	synthetic indigo
合成多肽	合成多肽	synthetic polypeptide
合成反应	合成反應	synthesis reaction, synthetic reaction
合成反应器	合成反應器	synthesis reactor
合成干性油	合成乾性油	synthetic drying oil
合成甘油	合成甘油	synthetic glycerol
合成高分子	合成高分子	synthetic high polymer, synthetic macromolecule
合成革	合成革	synthetic leather
合成过程	合成程序	synthesis process, synthetic process
合成化学	合成化學	synthetic chemistry
合成胶乳	合成膠乳	synthetic latex
合成界面池	合成界面池	synthetic boundary cell
合成聚合物	合成聚合物	synthetic polymer

祖国大陆名	台湾地区名	英 文 名
合成木材	合成木材	synthetic wood
合成泡沫塑料	合成泡沫塑料	synthetic foam
合成气	合成氣, 合成[煤]氣	syngas, synthesis gas, synthetic gas
合成汽油	合成汽油	synthetic gasoline
合成器	合成器	synthesizer
合成燃料	合成燃料	synfuel, synthetic fuel
合成鞣剂	合成鞣劑	syntan
合成鞣料	合成鞣料	synthetic tanning material
合成色素	合成色素	synthetic coloring agent
合成石膏	合成石膏	synthetic gypsum
合成树脂	合成樹脂	synthetic resin
合成树脂胶泥	合成樹脂膠泥	synthetic resin cement
合成树脂黏合剂	合成樹脂黏合劑	synthetic resin adhesive
合成树脂涂料	合成樹脂塗料	synthetic resin varnish
合成塑料	合成塑膠	synthetic plastic
合成弹性体	合成彈性體	synthetic elastomer
合成天然气	合成天然氣	synthetic natural gas
合成天然橡胶	合成天然橡膠	synthetic natural rubber
合成涂布材料	合成塗布材料	synthetic coating material
合成洗涤剂	合成洗滌劑	synthetic detergent
合成纤维	合成纖維	synthetic fiber
合成纤维纸	合成纖維紙	synthetic fiber paper
合成橡胶	合成橡膠	synthetic rubber
合成橡胶胶乳	合成橡膠膠乳	synthetic rubber latex
合成橡胶黏合剂	合成橡膠黏合劑	synthetic rubber adhesive
合成羊毛	合成羊毛	synthetic wool
合成油	合成油	synthetic oil
合成原油	合成原油	syncrude, synthetic crude oil, synthetic crude
合成增塑剂	合成增塑劑	synthetic plasticizer
合成脂肪	合成脂肪	synthetic fat
合成脂肪酸	合成脂肪酸	synthetic fatty acid
合成纸	合成紙	synthetic paper
合成子	合成纖維	synthon
合格质量标准	合格品質水準	acceptable quality level
合金	合金	alloy
合金钢	合金鋼	alloy steel
合作吸附	合作吸附	cooperative adsorption
核弹	核彈	nuclear bomb

祖国大陆名	台湾地区名	英　文　名
核定资本值	資金成本	capitalized cost
核反应堆	核反應器	nuclear reactor
核辐射	核輻射	nuclear radiation
核苷	核苷	nucleoside
核苷酸	核苷酸	nucleotide
核化学	核化學	nuclear chemistry
核聚变	核熔合, 核聚变	nuclear fusion
核能	核能	nuclear energy
核燃料	核燃料	nuclear fuel
[核燃料]后处理工厂	再處理廠	reprocessing plant
核素	核種	nuclide
核酸	核酸	nucleic acid
核糖	核糖	ribose
核糖核酸	核糖核酸	ribonucleic acid(RNA)
核物理学	核[子]物理學	nuclear physics
荷电膜	荷電膜	charged membrane
褐煤	褐煤	lignite
褐铁矿	褐鐵礦	limonite
褐烟煤	褐煙煤	lignitic bituminite
赫斯定律	赫斯定律, 蓋斯定律	Hess's law
黑灰	黑灰	black ash
黑漆	黑漆	Japan black
黑素原染料	硫化染料	melanogen dye
黑体	黑體	black body
黑体辐射	黑體輻射	black radiation
黑体系数	黑體係數	black body coefficient
黑铜矿	黑銅礦	tenorite
黑箱	黑箱	black box
黑箱法	黑箱法	black box method
黑箱模型	黑箱模型	black-box model
黑液	黑液	black liquor
痕量元素	微量元素	trace element
亨利定律	亨利定律	Henry's law
恒沸物(＝共沸物)		
恒沸蒸馏(＝共沸蒸馏)		
恒化器	恆化器	chemostat
恒剪力黏度计	恆剪力黏度計	constant shear viscometer
恒摩尔溢流	定莫耳溢流	constant molar overflow
恒速干燥[阶]段	恆速乾燥期	constant rate drying period

祖国大陆名	台湾地区名	英　文　名
恒速过滤	恆流率過濾	constant[flow]rate filtration
恒温烘箱	恆溫烘箱	constant temperature oven
恒温阱	恆溫阱	thermostatic trap
恒温器	恆溫器	thermostat
恒温箱	保溫箱	incubator
恒温浴	轉速計槽	constant temperature bath
恒压	恆壓, 等壓, 定壓	constant pressure
恒压过滤	恆壓過濾	constant pressure filtration
恒组成溶液	恒組成溶液	constant composition solution
恒[组]分共聚物	共沸共聚物, 共沸共聚合	azeotropic copolymer, azeotropic copolymerization
横式喷雾室	臥式噴霧室	horizontal spray chamber
横向翅片	橫向鰭片	transverse fin
横向结晶	橫向結晶	transcrystallization
烘道	乾燥隧道	drying tunnel
烘干	烘乾	stoving
烘炉	烘爐	baking furnace
烘箱	焙箱, 乾燥烘箱, 爐	baking oven, drying oven, oven
红霉素	紅黴素	erythromycin
红铅	紅鉛, 紅丹	red lead
红外分光光度计	紅外線分光光度計, 紅外線光譜儀	infrared spectrophotometer
红外分析	紅外線分析	infrared analysis
红外干燥	紅外乾燥	infrared drying
红外光谱	紅外線光譜測定法	infrared spectrophotometry
红外光谱法	紅外線光譜學, 紅外線光譜法	infrared spectroscopy
红外线分析仪	紅外線分析儀	infrared analyzer
红外线干燥器	紅外線乾燥器	infrared drier
红锌矿	紅鋅礦	zincite
红柱石	紅柱石	andalusite
宏观动力学	巨觀動力學	macrokinetics
宏观规模行为	巨標[度]行為	macroscale behavior
宏观混合	巨觀混合	macromixing
宏观可逆性	巨觀可逆性	macroscopic reversibility
宏观流体	巨觀流體	macrofluid
宏观平衡	巨觀均衡	macroscopic balance
虹吸管	虹吸管	siphon pipe
虹吸加油器	虹吸加油器	siphon oiler

祖 国 大 陆 名	台 湾 地 区 名	英 文 名
虹吸气压计	虹吸氣壓計	siphon barometer
虹吸式密封	虹吸式密封	syphon seal
虹吸压力计	虹吸壓力計	siphon gauge
喉道流速	喉道流速	throat velocity
喉管	喉道	throat
后焙, 后烘	後焙	postbaking
后产物	後産物	after product
后处理	後處理	after treatment
后发酵	後發酵	after fermentation
后过滤	後過濾	after filtration
后过滤器	後濾器	after filter
后烘(=后焙)		
后净化	後純化	after purification
后硫化	後硫化	post vulcanization
后硫化作用	後硫化	after vulcanization
后期结晶	後期結晶	secondary crystallization
后期燃烧	後燃	after burning
后燃装置	後燃裝置	after burner
后收缩	後收縮	after contraction, after shrinkage
后硬化	後硬化	after hardening
后置冷却器	後冷卻器	after cooler
后轴承	後軸承	rear bearing
厚漆	厚漆	paste paint
厚油	聚合油	polymerized oil
呼吸	呼吸	respiration
呼吸链	呼吸鏈	respiratory chain
呼吸式油罐	通氣槽	breathing tank
弧鞍填料	弧鞍填料	Berl saddle
胡桃油	胡桃油	pecan oil
糊剂	糊	paste
糊精	糊精	dextrin
糊状树脂	糊狀樹脂	paste resin
琥珀	琥珀	amber
互变异构	①互變異構現象②互變異構性	tautomerism
互变异构体	互變異構物	tautomer
互补反馈	互補回饋	complementary feedback, feedback complementary
互混	互混	intermixing

祖国大陆名	台湾地区名	英 文 名
互混度	互混度	degree of intermixing
互溶度	互溶度	mutual solubility
互溶泛滥	互溶汜流[法]	miscible flooding
互溶系统	互溶系統	miscible system
互溶性	互溶性	miscibility
护床	護床	guard bed
花岗岩	花崗岩	granite
花生油	花生油	peanut oil
滑板式泵	滑葉泵	slide vane pump
滑槽	滑槽	chute
滑动	①滑動②泥漿	slip
滑动角	滑動角	angle of slide
滑动摩擦	滑動摩擦	slip friction
滑动型芯	滑動心	slide core
滑动因子	滑動因子	slippage factor
滑阀	滑閥	slide valve
滑石	滑石	talc
滑线	滑線	slidewire
滑移带	滑動區	slip band
滑移速度	滑動速度	slip velocity
滑移系数	滑動因數	slip factor
化工动力学	化工動力學	chemical engineering kinetics
化工分析	化工分析	chemical engineering analysis
化工过程分析	化工程序分析	chemical engineering process analysis
化工热力学	化工熱力學	chemical engineering thermodynamics
化工系统工程(=过程系统工程)		
化合热	化合熱	heat of combination
化合物	化合物	compound
化石	化石	fossil
化石燃料	化石燃料	fossil fuel
化学变化	化學變化	chemical change
化学参数	化學參數	chemical parameter
化学操作	化學操作	chemical operation
化学沉淀	化學沈澱[法]	chemical precipitation
化学成键	化學鍵結	chemical bonding
化学澄清器	化學澄清器	chemical clarifier
化学处理	化學處理	chemical treatment
化学处理过程	化學處理法	chemical treatment process

祖国大陆名	台湾地区名	英 文 名
化学促进剂	化學促進劑	chemical promoter
化学动力过程	機械化學[製漿]法	mechano-chemical process
化学动力学	化學動力學	chemical kinetics
化学动态学	化學動態學	chemical dynamics
化学发光	化學發光	chemiluminescence
化学反应	化學反應	chemical reaction
化学反应工程	化學反應工程	chemical reaction engineering
化学反应平衡	化學反應平衡	chemical reaction equilibrium
化学反应平衡判据	化學反應平衡準則	criterion for chemical reaction equilibrium
化学反应器	化學反應器	chemical reactor
化学反应器稳定性	化學反應器穩定性	chemical reactor stability
化学反应稳定性	化學反應穩定性	chemical reaction stability
化学方程式	化學方程式	chemical equation
化学分离	化學分離	chemical separation
化[学]工厂	化學工廠	chemical plant
化学工程	化學工程	chemical engineering
化学工程建造成本指数	化學工程營建成本指數	chemical engineering construction cost index
化学工程师	化學工程師	chemical engineer
化学工程学	化工科學	chemical engineering science
化学光度计	光量計, 露光計	actinometer
化学过程	化學程序	chemical process
化学[过程]工业	化學程序工業	chemical process industry
化学极化	化學極化	chemical polarization
化学计量混合物	化學計量混合物	stoichiometric mixture
化学计量计算	化學計量計算	stoichiometric calculation
化学计量兼容性	化學計量相容性	stoichiometric compatibility
化学计量数	化學計量數	stoichiometric number
化学计量系数	化學計量係數	stoichiometric coefficient
化学计量学	①化學計量學②化學計量法	stoichiometry
化学键	化學鍵	chemical bond
化学扩散系数	化學擴散係數	chemical diffusivity
化学流体动力学	化學流體動力學	chemohydrodynamics
化学密封	化學密封	chemical seal
化学能	化學能	chemical energy
化学凝聚	化學凝聚	chemical coagulation
化学品	化學品	chemical
化学平衡	化學平衡	chemical equilibrium

祖国大陆名	台湾地区名	英 文 名
化学气相沉积	化學蒸氣沉積[法]	chemical vapor deposition (CVD)
[化学]式	化學式	formula
化学势, 化学位	化學勢, 化學位能	chemical potential
化学输送	化學輸送	chemical transport
化学调理	化學調理[法]	chemical conditioning
化学位(=化学势)		
化学文摘	化學摘要	Chemical Abstracts
化学稳定性	化學穩定性, 化學安定性	chemical stability
化学物理[学]	化學物理[學]	chemical physics
化学物种	化學物種	chemical species
化学吸附	化學吸附	chemical adsorption, chemisorption
化学吸收	化學吸收	chemical absorption
化学吸收剂	化學吸收劑	chemical absorbent
化学性质	化學性質	chemical property
化学需氧量	化學需氧量	chemical oxygen demand (COD)
化学烟雾	化學煙幕	chemical smoke
化学液泛	化學氾流[法]	chemical flooding
化学原料	化學原料	chemical feedstock
化学振荡	化學振盪	chemical oscillation
化学振荡器	化學振盪器	chemical oscillator
化学蒸气	化學蒸氣	chemical vapor
化学制浆法	化學製漿法	chemical pulping process
槐子油	槐子油	ash-seed oil
还原	①還原[反應]②降低	reduction
还原电位	還原電位	reduction potential
还原反应	還原反應	reduction reaction
还原剂	還原劑	reducing agent, reductant
还原区	還原帶	reduction zone
还原染料	染缸	vat dye
还原性漂白	還原漂白	reductive bleaching
还原性脱卤	還原性脱鹵[反應]	reductive dehalogenation
还原值	還原值	reduction value
环滚粉碎机	環輥粉碎機	ring roll pulverizer
环滚磨机	環輥磨機	ring roll mill
环化	環化	cyclization
环化反应	環化反應	cyclization reaction
环化橡胶	環化橡膠	cyclize rubber
环境	環境	environment

祖国大陆名	台湾地区名	英　文　名
环境保护	環境保護	environmental protection
环境成本	環境成本	environmental cost
环境冲击	環境衝擊	environmental impact
环境工程	環境工程	environmental engineering
环境化学	環境化學	environmental chemistry
环境技术	環境技術	environmental technology
环境科学	環境科學	environmental science
环境态	環境態	environmental state
环境温度	周圍溫度, 外圍溫度	ambient temperature, surrounding temperature
环境污染	環境污染	environmental contamination, environmental pollution
环境压力	周圍壓力	ambient pressure
环境引发衰变	環境引發衰退	environmentally sponsored decay
环境质量	環境品質	environmental quality
环流	循環, 流通	circulation
环流反应器	循環反應器, 環流反應器	circulating reactor, loop reactor
环流空间供暖	空間加熱	space heating
环路, 回路	環路, 圈	loop
环路构型	環路組態	loop configuration
环路应答	環路應答	loop response
环路转移函数	環路轉移函數	loop transfer function
环丝氨酸	環絲胺酸	cycloserine
环烷	環烷	naphthene
环隙	環隙	annular space, annulus
环向应力	環應力	hoop stress
环形接头	環線接頭	ring joint
环形压力计	環形測壓計, 環形壓力計	ring piezometer, ring type manometer
环形研磨机	環形研磨機	ring type grinder
环氧化	環氧化[反應]	epoxidation
环氧氯丙烷	環氧氯丙烷	epichlorohydrin
环氧树脂	環氧樹脂	epoxy resin
环状反应器	環狀反應器	annular reactor
环状管路系统	環狀管路系統	looped pipe system
环状流	環狀流動	annular flow
缓冲板	緩衝板	cushion plate
缓冲层	緩衝層	buffer layer

祖国大陆名	台湾地区名	英 文 名
缓冲罐	緩衝罐	buffer tank, surge tank
缓冲辊	緩衝輪	snubber roll
缓冲剂	緩衝劑	buffer
缓冲能力	緩衝能力	buffer capacity
缓冲区	緩衝區[域]	buffer region, buffer zone
缓冲液	緩衝液	buffer solution
缓冲指数	緩衝指數	buffer index
缓冲作用	緩衝作用	buffer action
缓慢反应	緩慢反應	slow reaction
缓蚀剂	腐蝕抑制劑	corrosion inhibitor
缓释	緩釋	slow release
缓释胶囊	緩釋膠囊	slow release capsule
缓释控制	緩釋控制	slow release control
缓释器件	緩釋裝置	slow release device
缓释药物	緩釋藥物	slow release drug
换热	熱交換	heat exchange
换热器, 热交换器	熱交換器	heat exchanger
换热器网路	熱交換器網路	heat exchanger network
换算表	換算表	conversion table
换算常数	換算常數	conversion constant
换算方程式	換算方程式	conversion equation
换算曲线	轉換曲線	conversion curve
换算因子	換算因數	conversion factor
黄饼	黃餅	yellow cake
黄丹	密陀僧	yellow lead
黄化	黃化, 硫化	sulphidizing
黄金分割法	黃金分割搜尋法	golden section searching method
黄磷	黃磷	white phosphorus
黄麻	黃麻	jute
黄芪胶	紫雲英樹膠	tragacanth gum
黄色炸药	代納邁炸藥	dynamite
黄铁矿	黃鐵礦	iron pyrite, pyrite
黄铜矿	黃銅礦	chalcopyrite
黄原酸化[反应]	黃酸化	xanthation
黄原酸纤维素	黃酸纖維素	cellulose xanthate
磺化	磺酸化[反應]	sulfonation
磺化剂	磺酸化劑	sulfonating agent
磺化器	磺酸化器	sulfonator
磺化油	磺酸化油	sulfonated oil

祖国大陆名	台湾地区名	英 文 名
磺酸型离子交联聚合物	磺酸離聚體, 磺酸型離子交聯聚合物	sulphonic acid ionomer
磺酸盐	磺酸鹽	sulfonate
灰分	灰, 灰分	ash
灰分分析	灰分分析	ash analysis
灰分扩散控制	灰[分]擴散控制	ash diffusion control
灰面	灰面	gray surface
灰区	灰區	ash zone
灰色系统	灰色系統	gray system
灰体	灰體	gray body, greybody
恢复时间	回收時間	recovery time
挥发	揮發[作用]	volatilization
挥发度	揮發度	volatility
挥发性悬浮固体	揮發性懸浮固體	volatile suspended solid
辉铜矿	輝銅礦	chalcolite
回归	回歸	regression
回归方程	回歸方程式	regression equation
回归分析	回歸分析	regression analysis
回归系数	回歸係數	regression coefficient
回归线	回歸線	regression line
回火	回火	tempering
回火空气	回火空氣	tempering air
回流	回流	reflux
回流比	回流比	reflux ratio
回流操作	回流操作	reflux operation
回流槽	回流槽	reflux drum
回流管	回流管	reflux tube
回流罐	回流累積量	reflux accumulator
回流阱	回流阱	return trap
回流冷凝器	回流冷凝器	reflux condenser
回流塔	回流塔	reflux column
回流蒸馏	回流蒸餾	reflux distillation
回路(=环路)		
回收	回收	reclamation, regain
回收槽	回收槽	reclaiming tank
回收[率]	回收	recovery
回收[率]曲线	回收曲線	recovery curve
回收石膏	再生石膏	reclaimed gypsum
回收塔	回收塔	recovery tower

祖国大陆名	台湾地区名	英 文 名
回收系数	回收因數	recovery factor
回收系统	回收系統	recovery system
回收装置	回收裝置	reclaiming unit
回水	回水	back water
回缩	退縮	shrink back
回填机	回填機	backfiller
回弯管	回彎管	return bend
回转焙烧炉	旋轉焙燒爐	rotary roaster
回转泵	旋轉泵	rotary pump
回转鼓风机	旋轉鼓風機	rotary blower
回转流量计	迴轉流量計	gyroscopic flowmeter
回转破碎机	迴轉壓碎機	gyratory crusher
回转筛	迴轉篩	revolving screen
回转压缩机	旋轉壓縮機	rotary compressor
回转窑	旋轉窯	rotary kiln
回转轧碎机	迴轉破碎機	gyratory break
回转振动筛	偏旋篩	gyrating screen
绘图仪	製圖儀, 繪圖儀	drawing apparatus
绘图纸	繪圖紙	drawing paper
桧树胶	檜樹膠	sandarac
混合	混合	mixing
混合杯温	混合杯溫	mixing cup temperature
混合槽	混合槽	mixing tank
混合长度	混合長度	mixing length
混合沉降器(=混合澄清器)		
混合程度	混合度	degree of mixing
混合澄清器, 混合沉降器	混合器-沈降器	mixer-settler
混合催化剂	混合觸媒	mixed catalyst
混合度	混合度	mixedness
混合反应器	混合[流動]反應器	mixed reactor
混合肥料	複合肥料	mixed fertilizer
混合罐	混合桶	mixing drum
混合规则	混合法則	mixing rule
混合桨叶	混合槳葉	mixing paddle
混合搅拌机	混合攪拌機	mixing agitator
混合节点	混合節點	mixed node
混合进料	混合進料	mixed feeding

祖国大陆名	台湾地区名	英 文 名
混合流动反应器	混合流動反應器	mixed flow reactor
混合能力	混合能力	mixing capacity
混合器	混合器, 混合機	mixer
混合器功率	混合器功率	mixer power
混合强度	混合強度	mixing intensity
混合热	混合熱	heat of mixing
混合深度	混合深度	mixing depth
混合生产	混合產物	mixed production
混合时间	混合時間	mixing time
混合室	混合室	mixing chamber
混合顺序	混合階	mixed order
混合速率	混合速率	mixing rate
混合酸	混合酸	mixed acid
混合温度	混合溫度	mixing temperature
混合物	混合物	mixture
混合雾沫	混合霧沫	mixing entrainment
混合型式	混合型式	mixing pattern
混合悬浮	混合懸浮	mixed suspension
混合液	混合液	mixed liquor
混合液挥发性悬浮固体	混合液揮發性懸浮固體	mixed-liquor volatile suspended solid
混合液悬浮固体	混合液懸浮固體	mixed-liquor suspended solid
混合用挡板	混合用擋板	mixing baffle
混合整数非线性规划	混合整數非線性規劃	mixed integer nonlinear programming(MINLP)
混合汁	混合汁	mixed juice
混合指数	混合指數	mixing index
混合柱	混合柱	mixing column
混合装置	混合裝置	mixing device
混晶	混合晶體	mixed crystal
混流	混合流動	mixed flow
混流泵	混流泵	mixed flow pump
混凝土	混凝土	concrete
混糖	黃砂糖	muscovado sugar
活度	活性	activity
活度系数	活性係數	activity coefficient
活[管]接头	管套節	union
活化分析	活化分析	activation analysis
活化分子	活化分子	activated molecule

祖国大陆名	台湾地区名	英　文　名
活化硅石	活化矽石	activated silica
活化焓	活化焓	enthalpy of activation
活化化学吸附	活化化學吸附	activated chemisorption
活化剂	活化劑	activator
活化截面	活化截面	activation cross section
活化络合物	活化錯合物	activated complex
活化能	活化能	activation energy
活化熵	活化熵	entropy of activation
活化体积	活化體積	activation volume
活化[作用]	活化[作用]	activation
活塞	活塞	piston
活塞泵	活塞泵	piston pump
活塞阀	活塞閥	piston valve
活塞流(＝平推流)		
活塞流反应器	塞流反應器	piston flow reactor, plug flow reactor
活塞流量计	活塞流量計	piston meter
活塞式压力计	活塞壓力計	piston gage
活塞式执行机构	活塞引動器	piston actuator
活性	活性	activity 'activity'作为'活性'应单独列出
活性表面积	活性表面積	active surface area
活性部位	活性部位	active site
活性部位理论	活性部位理論	active-site theory
活性催化剂	活性觸媒	active catalyst
活性催化中心	觸媒活性中心	active catalytic center
活性点	活性點	active point
活性矾土	活性礬土	activated alumina, active alumina
活性分布	活性分布	activity distribution
活性干酵母	活性乾酵母	active dry yeast
活性固体	活性固體	active solid
活性剂	活性劑	active agent
活性碱	活性鹼	active alkali
活性校正因子	活性校正因數	activity correction factor
活性硫	活性硫	active sulfur
活性络合物	活性錯合物	active complex
活性黏土	活性土	active clay
活性生物过滤	活性生物過濾	activated biological filtration
活性衰变	活性衰變	activity decay
活性炭	活性[木]炭, 活性焦	activated carbon, activated char, activated charcoal, active carbon

祖国大陆名	台湾地区名	英文名
[活性]炭过滤器	碳濾器	carbon filter
活性炭柱	①活性碳管②活性碳塔	activated carbon column
活性污泥	活性污泥	activated sludge, active sludge
活性污泥处理场	活性污泥處理場	activated sludge plant
活性污泥法	活性污泥法	activated sludge process
活性吸附	活化吸附	activated adsorption
活性氧	活性氧	active oxygen
活性氧化铝	活性氧化鋁	activated aluminum oxide
活性因子	活性因數	activity factor
活性指数	活性指數	activity index
活性中心	活性中心	active center
活性中心络合物	活性中心錯合物	active-center complex
火成石英	火成石英	igneous quartz
火成岩	火成岩	igneous rock
火法冶金过程	高溫冶金法	pyrometallurgical process
火管式反应器	火管式反應器	fired-tubular reactor
火花电蚀	火花電蝕, 電火花腐蝕	spark erosion
火花放电	火花放電	spark discharge
火花放电检测器	火花放電檢測器	spark discharge detector
火箭	火箭	rocket
火箭燃料	火箭燃料	rocket fuel
火箭引擎	火箭引擎	rocket engine
火漆	封蠟, 火漆	sealing wax
火山灰	火山灰	pozzolana
火山灰水泥	火山灰水泥	puzzolana cement
火石玻璃	燧石耐火玻璃	flint glass
火焰	火燄	flame
火焰反应器	火燄反應器	flame reactor
火焰分光仪	火燄分光劑	flame spectrometer
火焰温度	火燄溫度	flame temperature
火焰稳定性	火燄固定性	flame stability
火灾	火災危害	fire hazard
火灾及爆炸指数	火災及爆炸指數	fire and explosion index

J

祖国大陆名	台湾地区名	英　文　名
机壳	護罩	casing
机理结构	機構結構	mechanism structure
机理缺陷	機構缺陷	mechanism deficiency
机器人	機器人	robot
机械成型	機械成型[法]	machine molding
机械镀	機械壓紋	mechanical plating
机械分离	機械分離	mechanical separation
机械分离器	機械分離器	mechanical separator
机械浮选池	機械浮選池	mechanical flotation cell
机械功	機械功	mechanical work
机械过滤器	機械過濾器	mechanical filter
机械化学磨光	機械化學磨光	mechano-chemical polishing
机械搅拌	機械攪拌	mechanical agitation
机械离心分离器	機械離心分離器	mechanical centrifugal separator
机械密封	機械密封	mechanical seal
机械能	機械能	mechanical energy
机械能方程式	機械能方程式	mechanical energy equation
机械能量平衡	機械能均衡	mechanical energy balance
机械破碎	機械散解	mechanical disintegration
机械设计	機械設計	mechanical design
机械式控制器	機械式控制器	mechanical controller
机械损失	機械損失	mechanical loss
机械通风	機械通風	mechanical draft
机械通风凉水塔	機械通風冷卻塔	mechanical draft cooling tower
机械-物理分离过程	機械-物理分離程序	mechanical-physical separation process
机械学方程式	機構方程式	mechanistic equation
机械纸浆	機械紙漿	mechanical pulp
机械制浆	機械製漿	mechanical pulping
机械自动化	機械自動化	mechanical automation
迹线	徑線	path line
积分	積分[式]	integral
积分动量方程	積分能量方程式	integral momentum equation
积分法	積分法	integral method
积分反馈	積分回饋	integral feedback
积分反应器	積分反應器	integral reactor

祖国大陆名	台湾地区名	英 文 名
积分分布函数	累積分布曲線	integral distribution curve
积分检验[法]	積分試驗(法)	integral test
积分控制	積分控制	integral control(＝I control)
积分控制作用	積分控制作用	integral control action
积分能量方程	積分能量方程式	integral energy equation
积分绕紧	積分繞緊	integral windup
积分溶解热	積分溶解熱	integral heat of solution
积分时间	積分時間	integral time
积分时间常数	積分時間常數	integral time constant
积分仪	積分器	integrator
积分作用	積分作用	integral action
基本单位	基本單位	fundamental unit
基本性质关系式	基本性質關係式	fundamental property relation
基本原理	基本原理	fundamental principle, fundamentals
基尔霍夫定律	基爾霍夫定律	Kirchhoff's law
基建费用	資本支出	capital expenditure
基体	基體	matrix
基团活度系数	基團活性係數	group activity coefficient
基团专一性	基團異性	group specificity
基因	基因	gene
基因工程细胞	基因工程細胞	genetically engineered cell
基因工程, 遗传工程	基因工程	genetic engineering
基质(＝底物)		
基质材料	基質材料	substrate material
基质专一性	基質專一性	substrate specificity
基准	基準	datum
基准位超越	基準位超越	base-level override
基准位共振	基準位共振	base-level resonance
基准温度	基準溫度	datum temperature
基准物	標準物, 基準物	standard substance
激发态	激態	excited state
激酶	激酶	kinase
吉布斯-杜安方程	吉布斯-杜安方程	Gibbs Duhem equation
吉布斯自由能, 自由焓	吉布斯自由能, 自由焓	Gibbs free energy
级间加热	階間加熱	interstage heating
级间冷却	階間冷卻	interstage cooling
级间循环	階間循環	interstage recirculation
级间注射进料	階間(注射)進料	interstage feed injection
级联反应器	級聯反應器	cascade reactor

祖 国 大 陆 名	台 湾 地 区 名	英 文 名
级联循环	級聯循環	cascade cycle
级效率	階效率	stage efficiency
极大值原理	極大原理	maximum principle
极化	極化	polarization
极化率	極化率	polarizability
极化因子	極化因子	polarization factor
极谱法	極譜法	polarographic method, polarography
极谱仪	極譜儀	polarograph
极限电流	極限電流	limit current
极限环	①極限環圈②極限循環	limit cycle
极限回流条件	極限回流條件	limiting reflux condition
极限灵敏度	極限靈敏度	ultimate sensitivity
极限黏度	極限黏度	limiting viscosity
极限批式尺寸	極限批式產能	limiting batch size
极限因子	極限因數	limiting factor
极限应力	極限應力	ultimate stress
极限周期	極限周期	limiting cycle time
极性	極性	polarity
极性溶剂	極性溶劑	polar solvent
极坐标	極坐標	polar coordinates
极坐标图	極坐標圖	polar plot
急冷系统	抑止系統	quenched system
急骤干燥器	驟乾器	flash drier
集尘	集塵	dust collection
集尘极	集[板]電擊	collector electrode
集尘器	集塵器	dust collector
集成,整合	整合	integration
集成过程	整合程序	integrated process
集管	集管[箱]	header
集散控制系统	分布控制系統	distributed control system
集雾器	集霧器	mist collector
集总参数模型	塊集參數模式	lumped parameter model
集总参数系统	塊集參數系統	lumped parameter system
集总动力学	塊集動力學	lumping kinetics
集总系统	塊集系統	lumped system
几何表面积	幾何表面積	geometric surface area
几何规划	幾何規劃	geometric programming
几何平均	幾何平均	geometric mean

祖国大陆名	台湾地区名	英 文 名
几何平均面积	幾何平均面積	geometric mean area
几何平均温差	幾何平均溫差	geometric mean temperature difference
几何相似	幾何相似性	geometric similarity
几何异构物	幾何異構物	geometric isomer
几何因数	幾何因數	geometric factor
己内酰胺	己内醯胺	caprolactam
挤出	擠壓	extrusion
挤出(=压榨)		
挤出机	擠壓機	extruder
挤压泵	擠壓泵	squeeze pump
给硫剂	給硫劑	sulfur donor agent
给硫体	給硫體	sulfur donor
给水	給水	feed water
给水处理	給水處理	feed water treatment
给水井	進料井	feed well
给体	給予體, 施體	donor
pH 计	pH 計, 酸度計	pH meter
计划评审法	項目評審技術	program evaluation and review technique
计量	計量	metering
计量泵	計量泵	metering pump
计时器	計時器	timer
计数管	計數管	counter tube
计数器	計數器	counter
计数器范围	計數器範圍	counter range
计算	計算	calculation, computation
计算机程序	計算機程式	computer program
计算机辅助工程	計算機輔助工程	computer- aided engineering (CAE)
计算机辅助教学	計算機輔助教學	computer-aided instruction(CAI)
计算机辅助设计	計算機輔助設計	computer-aided design (CAD)
计算机辅助制造	計算機輔助製造	computer-aided manufacturing(CAM)
计算机控制	計算機控制	computer control
计算机模拟	計算機模擬	computer simulation
计算器	計算器	calculator
计温学	測溫法	thermometry
记号	①記號②記法	notation
记录材料	記錄材料, 錄製材料	recording material
记录控制器	記錄控制器	recorder-controller
记录器	記錄器, 錄音機	recorder
记录设备	記錄裝置	recording equipment

祖国大陆名	台湾地区名	英　文　名
记录式气压计	記錄氣壓計	recording barometer
记录图表	記錄紙	recording chart
记忆函数	記憶函數	memory function
记忆流体	記憶流體	memory fluid
记忆衰退材料	記憶衰退材料	fading memory material
记忆细胞	記憶格	memory cell
技术	技術	technology
技术服务	技術服務	technical service
技术评估	技術評估	technical evaluation
技术转移	技術轉移	technology transfer
剂量	劑量	dosage, dose
继电控制	砰砰控制	bang-bang control
继电器	替續器, 繼電器	relay
继动阀	替續閥	relay valve
寄生物	寄生蟲, 寄生物	parasite
加成定律	加成律	additive law
加成反应	加成反應	addition reaction
加成结晶	加成結晶	adductive crystallization
加成聚合	加成聚合(反應)	addition polymerization
加成脱色	加成脫色	additive decolorization
加成性(＝加合性)		
加法器	相加器	summer
加工	處理, 加工	processing
加工变量	加工變數	processing variable
加工工程	加工工程	processing engineering
加工工业	加工工業	processing industry
加工技术	加工技術	processing technology
加合物	加成物	adduct
加合性, 加成性	加成性	additive property
加和点	和點	summing point
加酵母槽	種母槽	pitching tank
加晶种	①加種晶②加種菌	seeding
加料槽	進料槽	feed tank
加料斗	加料斗	charge hopper
加料机	加料機	charger
加料盘	盤飼器	pan feeder
加料升降机	加料升降機	charge elevator
加料速度	進料速率	feed rate
加料运输机	加料運送機	charge conveyor

祖国大陆名	台湾地区名	英 文 名
加氢	氢化(反應)	hydrogenation
加氢重整	加氢重組[反應]	hydroforming, hydroreforming
加氢重整催化剂	加氢重組觸媒	hydroforming catalyst, hydroreforming catalyst
加氢重整过程	加氢重組程序	hydroforming process
加氢重整生成物	加氢重組物	hydroformate
加氢处理	加氢處理	hydrotreating, hydrotreatment
加氢处理装置	加氢處理器	hydrotreater
加氢甲酰基化	加氢甲醯化[反應]	hydroformylation
加氢裂化	加氢裂解[反應]	hydrocracking
加氢裂化单元	加氢裂解單元	hydrocracking unit
加氢裂化器	加氢裂解爐	hydrocracker
加氢气化	加氢氣化[反應]	hydrogasification
加氢气化器	加氢氣化器	hydrogasifier
加氢热解	加氢高溫分解	hydropyrolysis
加氢脱氮	加氢脫氮[反應]	hydrodenitrogenation
加氢脱金属催化剂	加氢脫金屬觸媒	hydrodemetallation catalyst
加氢脱硫	加氢脫硫[反應]	hydrodesulfurization
加氢脱烷基化	加氢脫烷[反應]	hydrodealkylation
加氢液化	加氢液化[反應]	hydroliquefaction
加权残差	加權剩餘	weighted residual
加权函数	加權函數	weighting function
加权平均[值]	加權平均	weighted mean
加权温差	加權溫差	weighted temperature difference
加权因子	加權因數	weighting factor
加热	加熱	heating
加热成本	加熱成本	heating cost
加热成型	加熱成型	hot molding
加热单元	加熱裝置, 加熱單元	heating unit
加热面积	加熱面積	heating area
加热盘管	加熱旋管	heating coil
加热器	加熱器, 加熱爐	heater
加热时间	加熱時間	heating time
加热再生	熱再生	heat regeneration
加水裂化过程	加水裂化程序	aquolization process
加速度计	加速計	accelerometer
加速度误差系数	加速誤差係數	acceleration error coefficient
加速剂	催速劑	accelerator
加速老化	加速老化	accelerated aging

祖国大陆名	台湾地区名	英 文 名
加速器	加速器	accelerator
加速器效能	催速劑效能	potency of accelerator
加速速率	加熱速率	heating rate
加酸器	飼酸箱	acid feed box
加压分馏	加壓分餾	pressure fractionation
加压浮选	加壓浮選	pressure flotation
加压过滤	加壓過濾	pressure filtration
加压馏出物	加壓餾出物	pressure distillate
加压灭菌器	熱壓[反應]器, 高壓釜, 高壓鍋	autoclave
加压膨胀试验	熱壓膨脹試驗	autoclave expansion test
加压砂滤器	加壓砂濾器	pressure sand filter
加压筛	壓力篩	pressure screen
加压氧化	加壓壓力	pressure oxidation
加压叶滤机	加壓葉濾機	pressure leaf filter
加压蒸馏	加壓蒸餾	pressure distillation
加压蒸煮器	加壓蒸煮器	pressure cooker
夹层结构	夾層結構	sandwich structure
夹带床反应器	載送床反應器	entrained bed reactor
夹带剂	①進料器②霧沫器	entrainer
夹带剂泵	進料泵	entrainer pump
夹带速度	霧沫速度	entrainment velocity
夹带物分离器	霧沫阱	entrainment trap
夹点	夾點	pinch point
夹点技术	夾點技術	pinch technology
夹角	挾角	angle of nip
夹丝玻璃	夾網玻璃	wire glass
夹套	套, 夾套	jacket
夹套壁	夾套壁	jacketed wall
夹套锅	[夾]套鍋	jacketed kettle
夹套加热	夾套加熱	jacket heating
夹套加压釜	夾套高壓鍋	jacketed autoclave
夹套结晶器	夾套結晶器	jacketed crystallizer
夹套蒸发器	夾套蒸發器	jacketed evaporator
夹套蒸馏器	夾套蒸餾器	jacketed still
夹心板	三夾板	sandwich panel
甲醇	甲醇, 木精	methanol, methyl alcohol
甲酚	甲酚	cresol
甲硅烷醇	矽烷醇	silanol

祖国大陆名	台湾地区名	英 文 名
甲基丙烯酸甲酯-丙烯腈-丁二烯-苯乙烯共聚物	甲基丙烯酸甲酯-丙烯腈-丁二烯-苯乙烯共聚物	methyl methacrylate-acrylonitrile-butadiene-styrene copolymer (MABS)
甲基橙	甲基橙	methyl orange
甲基对硫磷	甲[基]巴拉松	methyl parathion
甲基红	甲基紅	methyl red
甲基化剂	甲基化劑	methylating agent
甲基化作用	甲基化	methylation
甲基黄	甲基黃	methyl yellow
甲基纤维素	甲基纖維素	methyl cellulose
甲基橡胶	甲基橡膠	methyl rubber
甲基转移作用	轉甲基[反應]	transmethylation
甲基紫	甲基紫	methyl violet
甲烷富气	甲烷富氣	methane rich gas
甲烷化	甲烷化	methanation
甲烷化反应	甲烷化反應	methanation reaction
甲烷菌	甲烷菌	methane bacteria, methane-forming bacteria
甲烷馏除器	去甲烷塔	demethanizer
甲烷贫气	甲烷貧氣	methane lean gas
甲烷水合物	甲烷水合物	hydrate of methane
甲烷转化器	甲烷化器	methanator
甲酰胺	甲醯胺	formamide
甲酰胺法	甲醯胺法	formamide process
钾玻璃	鉀玻璃	potash glass
钾长石	鉀長石	potash feldspar
钾肥	鉀肥	potash fertilizer
钾钙玻璃	鉀鈣玻璃	potash-lime glass
钾铬矾	鉻礬	chrome alum
钾铅玻璃	鉀鉛玻璃	potash-lead glass
钾软皂	鉀皂	potash soap
钾盐	鉀鹽	kali salt
钾盐镁矾	鉀鎂礬, 鉀瀉鹽	kainite
假动力学	假動力學	pseudokinetics
假对比常数	假對比常數	pseudo-reduced constant
假临界常数	假臨界常數	pseudocritical constant
假临界点法	假臨界點法	pseudocritical method
假临界温度	假臨界溫度	pseudocritical temperature
假临界压力	假臨界壓力	pseudocritical pressure
假设	假說	hypothesis

祖国大陆名	台湾地区名	英 文 名
假塑性	假塑性	pseudoplasticity
假塑性流体	假塑性流體	pseudoplastic fluid
假稳态	假穩態	pseudo steady state
假盐	假鹽	pseudosalt
间隔	間隔, 距離	spacing
间规聚丙烯	間規聚丙烯	syndiotactic polypropylene
间接成本	間接成本	indirect cost
间接法	間接法	indirect method
间接隔膜阀	間接隔膜閥	indirect diaphragm valve
间接加热	間接加熱	indirect heating
间接氢化	間接氫化[反應]	indirect hydrogenation
间接热交换	間接熱交換	indirect heat exchange
间接式冷凝器	間接[式]冷凝器	indirect condenser
间接液化	間接液化	indirect liquefaction
间接液位测量	間接液位量測	indirect liquid level measurement
间接蒸发	間接蒸發	indirect evaporation
间同立构单元	間規立構單元	syndiotactic unit
间同立构规整度	間規異構性	syndiotacticity
间同立构加成	間規立構加成	syndiotactic addition
间同立构键接	間規立構鍵接	syndiotactic placement
间同立构聚合	間規聚合	syndiotactic polymerization
间同立构聚合物	間規立構聚合物	syndiotactic polymer
间同立构三[单]元组	間規立構三[單]元組	syndiotactic triad
间同立构序列	間規序列	syndiotactic sequence
间隙	間隙	clearance
间隙测量计	間隙計	clearance meter
间隙体积	間隙體積	clearance volume
间歇操作	批式操作	batch operation
间歇操作法	不連續程序	discontinuous process
间歇沉降	批式沉降	batch settling
间歇萃取	批式萃取	batch extraction
间歇萃取器	批式浸取器	batch extractor
间歇反应器	批式反應器	batch reactor
间歇分馏	批式分餾	batch fractionation
间歇釜式蒸馏	批式蒸餾	batch still distillation
间歇干燥器	批式乾燥器	batch drier
间歇干燥室	間歇乾燥室	intermittent drier
间歇过程	批式程序	batch process
间歇过滤	批式過濾	batch filtration

祖国大陆名	台湾地区名	英 文 名
间歇过滤器	批式過濾機	batch filter
间歇混合器	批式混合器	batch mixer
间歇加工	批式加工[法], 不連續加工[法]	batch processing, discontinuous processing
间歇加料机	批式加料機	batch charger
间歇焦化蒸馏器	批式焦化蒸餾器	batch coke still
间歇搅拌器	批式攪拌器	batch agitator
间歇结晶	批式結晶	batch crystallization
间歇结晶器	批式結晶器	batch crystallizer
间歇进料器	批式進料器	batch feeder
间歇浸取器	批式浸取器	batch extractor
间歇精馏	批式精餾	batch rectification
间歇净化	批式純化	batch purification
间歇控制	批式控制	batch control
间歇离心机	批式離心機	batch centrifuge
间歇培养	批式培養	batch culture
间歇汽化	批式汽化	batch vaporization
间歇生产	批式生產	batch production
间歇系统	批式系統	batch system
间歇窑	週期窯, 間歇窯	periodic kiln
间歇蒸馏	批式蒸餾	batch distillation
间歇蒸馏器	批式蒸餾器	batch still
兼性呼吸	兼性呼吸	facultative respiration
兼性污水塘	兼性污水塘	facultative lagoon
兼性细菌	兼性菌	facultative bacteria
兼性需氧细菌	兼性需氧細菌	facultative aerobic bacteria
监测站	監測站	monitoring station
监督	監督	supervision
监督控制	監督控制	supervisory control
监视	監測	monitoring
监视器	監測器	monitor
减活化能	去活化能	deactivation energy
减量	降低	abatement
减黏裂化	減黏	visbreaking
减湿	除濕	dehumidification
减速齿轮	減速齒輪	reducing gear
减速力	減速力	retarding force
减速扭矩	減速扭矩	retarding torque
减压阀	減壓閥	pressure reducing valve, reducing valve

祖国大陆名	台湾地区名	英 文 名
减压器	減壓器	decompressor
减压蒸馏(＝真空蒸馏)		
减阻	阻力降低	drag reduction
剪切	剪力	shear, shearing
剪切变形	剪切形變	shear deformation
剪切储能模量	剪切儲能模量	shear storage modulus
剪切刚度	剪切剛度	shearing stiffness
剪切力	剪切力	shear force
剪切模量	切變模量, 剪力係數	shear elasticity, shear modulus
剪切挠曲	剪切撓曲	shear flexure
剪切黏度	切變黏度, 剪切黏性	shear viscosity
剪切黏滞系数	切變黏切係數	shear viscosity coefficient
剪切强度	抗切強度, 抗剪強度, 剪斷強度	shearing strength, shear strength
剪切柔量	剪切柔量	shear compliance
剪切蠕变	剪切蠕變	shear creep
剪切试验	剪切試驗	shear test
剪切松弛	剪切弛豫, 剪切鬆弛	shear relaxation
剪切速度	剪速度	shear velocity
剪切速率	切變速率	shear rate
剪切稀化流体	剪切淺稠流體	shear-thinning fluid
剪切应变	剪切應變	shear strain
剪切圆盘式黏度计	剪切圓盤式黏度計	shearing disc viscometer
剪切增稠	剪力增長	shear thickening
剪切增稠流体	剪切增稠流體	shear-thickening fluid
剪应力	剪應力	shear stress
检测器	偵檢器	detector
检查报表	檢查報表	check list
检错	除錯	debug
检流计	電流計	galvanometer
检漏器	偵漏器	leak detector
F 检验	F 試驗	F-test
简单反应	簡單反應	simple reaction
简单剪切	單純剪力	simple shear
简单剪切流动	簡單剪切流動	simple shear flow
简单剪切形变	簡單剪切形變	simple shear deformation
简单[壳式]蒸馏釜	鍋餾器	shell still
简单立方晶格	簡單立方晶格	simple cubic lattice
简单流体	簡單流體	simple fluid

祖国大陆名	台湾地区名	英 文 名
简单伸长	簡單伸長, 單向伸長	simple elongation
简捷法	簡捷法	shortcut method
简捷设计法	簡捷設計法	shortcut design method
简式黄原酸化机	簡式黃化機	simple xanthating machine
简易胶乳比重计	簡易膠乳比重計	simplexometer
碱	鹼	alkali, base
碱测定法	鹼測定法	alklalimetry
碱催化	鹼催化[作用]	base catalysis
碱度	鹼度	alkalinity
碱法	鹼法	alkali process
碱法制浆	鹼式制漿	alkaline pulping
碱量计	鹼計	alkalimeter
碱木质素	鹼木質素	alkali lignin
碱石灰	鹼石灰	soda lime
碱纤维素	鹼纖維素	alkali cellulose, soda cellulose
碱性	鹼度	basicity
碱性催化剂	鹼性觸媒	basic catalyst
碱性固体	鹼性固體	basic solid
碱性硅酸盐	鹼矽酸鹽	alkali silicate
碱性耐火砖	鹼性耐火磚	basic firebrick
碱性染料	鹼性染料	basic dye
碱液	鹼水	lye
碱液泛	鹼液泛流[法]	alkaline flooding
碱釉	鹼性釉	alkaline glaze
建模	模式化	modeling
建筑材料	營建材料	material of construction
渐减曝气	漸減通氣, 漸減曝氣	tapered aeration
渐近解	漸近解	asymptotic solution
渐近稳定区	漸近穩定[域]	region of asymptotic stability
渐近稳定性	漸近穩定性	asymptotic stability
渐近稳定性联合	漸近穩定聯合域	union of asymptotic stability
渐近线	漸近線	asymptote
渐近线图	漸近線圖	asymptotic plot
渐近线中心	漸進線中心	center of asymptote
渐进转化模型	漸進轉化模型	progressive conversion model
渐缩管	漸縮管	convergent tube
渐缩管件	漸縮管件	fitting reducing
溅射	噴濺	sputtering
溅射镀	噴濺沈積[法]	sputtering deposition

祖国大陆名	台湾地区名	英 文 名
溅射蚀刻	噴濺蝕刻[法]	sputtering etching
键	鍵	bond
键角	鍵角	bond angle
键结力	鍵結力	bonding force
键能	鍵能	bond energy
箭头[流程]图	箭頭[流程]圖	arrow diagram
浆糊	漿糊	starch paste
浆尖速度	尖端速度	tip speed
浆料,淤浆	①漿體,泥漿②紙漿	①slurry②pulp
浆料反应	漿體反應	slurry reaction
浆料反应动力学	漿體反應動力學	slurry reaction kinetics
浆料反应器	漿體反應器	slurry reactor
浆料配管系统	漿體配管系統	slurry piping system
桨式混合器	槳式混合器	paddle mixer
桨式搅拌器	槳式攪拌器	paddle agitator, paddle stirrer
桨式絮凝器	槳式絮凝器	paddle flocculator
桨式叶轮	槳式葉輪	paddle impeller
桨叶搅拌器	葉片攪拌器	blade agitator
降低空气污染	空氣污染減少	air pollution abatement
降阶模型	縮減模式	reduced model
降解	①降解②裂解③劣化	degradation
降膜	[下]降膜	falling film
降膜吸收器	降膜式吸收器	falling film type absorber
降膜蒸发器	降膜蒸發器	falling-film evaporator
降速干燥[阶]段	減速乾燥期	falling rate drying period
降速期	減速期	falling rate period
降液	下向流	downflow
降液管	降流管	downcomer
降液管液柱高度	降液管液柱高度	downcomer backup
交叉式构型	對位交叉型	staggered form
交错管排	交錯管	staggered tubes
交迭	交越	crossover
交迭频率	交越頻率	crossover frequency
交迭增益	交越增益	crossover gain
交换剂	交換劑	exchanger
交换能	交換能	exchange energy
交换器	交換器	exchanger
交换吸附	交換吸附	exchange adsorption
交联	交联	cross linkage, cross-linking

祖 国 大 陆 名	台 湾 地 区 名	英 文 名
交联度	交聯度	degree of cross-linking
交联剂	交聯劑	cross-linking agent
交联聚合物	交聯聚合物	cross-linked polymer
交联密度	交聯密度	cross-linking density
交联葡聚糖[凝胶]	交聯葡凝膠,交聯葡凝糖	sephadex gel
交替操作	交替操作	alternative operation
交替操作塔	交替操作塔	alternate operating column
交替共聚合	交替共聚合	alternating copolymerization
交替共聚物	交替共聚物	alternative copolymer
交替系统	交替系統	alternating system
交替[现象]	交替[現象]	alternation
浇道	澆道	runner, sprue
浇注釉	澆釉	pouring glazing
浇铸	鑄造	casting
胶	膠	glue
胶版印刷纸	平板印紙	offset paper
胶带	膠帶	adhesive tape
胶管	軟管, 橡皮軟管	hose, rubber hose
胶合板	合板	plywood
胶化	膠體化	colloidization
胶粒	膠體粒子	colloidal particle
胶囊	膠囊	capsule
胶囊封装	膠囊封裝	encapsulation
胶黏剂	黏著劑, 黏合劑	adhesive agent, adhesives
胶凝化[作用]	膠化	gelation
胶凝剂	膠化劑	gelatining agent
胶凝[作用]	膠凝	gelatination
胶清橡胶	膠清橡膠	skim rubber
胶溶	解膠[作用]	peptization
胶溶剂	助消化劑	peptizing agent
胶乳	乳膠	latex
胶乳漆	乳膠漆	latex paint
胶束	微胞	micelle
胶束催化	微胞催化[作用]	micellar catalysis
胶束泛滥	微胞氾流[法]	micellar flooding
胶束化	微胞化	micellization
胶态白土	膨土, 漿土	colloidal clay
胶态分散	膠態分散	colloidal dispersion

祖国大陆名	台湾地区名	英 文 名
胶体	膠體	colloid
胶体包埋	膠體包埋	gel entrapment
胶体铂	膠體鉑	colloidal platinum
胶体沉淀	膠體沈澱	colloidal precipitation
胶体化学	膠體化學	colloid chemistry
胶体磨	膠體磨機	colloidal mill
胶体溶胶	膠體溶液	colloid solution
胶体溶液	膠體溶液	colloidal solution
胶体稳定化	膠體穩定化	colloid stabilization
胶体现象	膠體現象	colloidal phenomenon
胶体悬浮液	膠體懸浮液	colloidal suspension
胶体状态	膠體狀態	colloidal state
胶原	膠原	collagen
胶质炸药	膠質代納邁炸藥	gelatin dynamite
焦点	焦點	focus
焦[耳]	焦耳	joule
焦耳-汤姆孙系数	焦耳-湯姆遜係數	Joule Thomson coefficient
焦耳-汤姆孙效应	焦耳-湯姆遜效應	Joule Thomson effect
焦化反应	焦化反應	coking reaction
焦化器	煉焦器	coker
焦化蒸馏器	焦化蒸餾器	coke still
焦炉煤气	焦爐氣	coke oven gas
焦炉油	焦爐溚	coke oven tar
焦烧	焦燒	scorching
焦烧试验	焦燒試驗	scorching test
焦炭	焦炭	coke
焦炭孔隙度	焦炭孔隙度	coke porosity
焦糖	焦糖	caramel
焦油	焦油, 溚油,	tar oil, tar
焦油馏出液	溚溜出物	tar distillate
焦油酸	溚酸	tar acid
焦值	焦值	coke number
角冲量	角衝量	angular impulse
角蛋白	角蛋白	keratin
角动量	角動量	angular momentum
角动量守恒	角動量守恆	conservation law for angular momentum
角阀	角閥	angle valve
角畸变	角畸變	angular distortion
角频率	角頻率	corner frequency

祖国大陆名	台湾地区名	英　文　名
角铅矿	角鉛礦	phosgenite
角散射模式	角散射模式	angular scattering pattern
角速度	角速度	angular velocity
角系数	角度因數	angle factor
角[向]变形	角形變	angular deformation
角锥[沉淀]池	錐形選粒器	spitzkasten
绞丝	絲球	skein
搅拌	攪拌	agitation, agitating, stirring
搅拌槽	攪拌槽	agitated vessel, agitating tank
搅拌槽接触器	攪拌槽接觸器	stirred-tank contactor
搅拌反应器	攪拌反應器	agitated reactor, stirred reactor
搅拌干燥器	攪拌乾燥器	agitated drier, agitator drier
搅拌罐	罐式攪拌	stirred tank
搅拌罐式反应器	罐式攪拌反應器	stirred tank reactor
搅拌结晶器	攪拌結晶器	agitated crystallizer
搅拌流反应器	攪拌流動反應器	stirred flow reactor
搅拌流化床	攪拌流體化床	stirred fluidized bed
[搅拌]流量数	流量數	flow number
搅拌器	①攪拌器②攪拌棒	stirrer, agitator
搅浆机	攪漿機	change-can mixer
搅炼	攪煉, 煉鐵	puddle
校正曲线	校正曲線	calibration curve
校正作用	校正作用	corrective action
校准	校正	calibration
校准表	校正圖	calibration chart
酵母	酵母	yeast
酵母菌属	酵母菌	saccharomyces
阶梯分析	階梯分析	step analysis
阶梯干扰	階梯擾動	step disturbance
阶梯环	階梯環	cascade ring
阶跃函数	階梯函數	step function
阶跃函数响应	階梯函數應答	step function response
阶跃输入	階梯輸入	step input
阶跃响应	階梯應答	step response
接触成核	接觸成核	contact nucleation
接触催化	接觸催化作用	contact catalysis
接触点	接觸點	point of contact
接触电阻	接觸阻力	contact resistance

祖国大陆名	台湾地区名	英　文　名
接触方式	接觸方式	contacting pattern
接触过滤	接觸過濾	contact filtration
接触角	接觸角	contact angle
接触冷凝器	接觸冷凝器	contact condenser
接触器	接觸器	contactor
接触器效率	接觸器效率	contactor efficiency
接触热阻	熱接觸阻力	thermal contact resistance
接触室	接觸室	contact chamber
接触稳定过程	接觸穩定程序	contact stabilization process
接触相	接觸相	contacting phase
接点	接點	junction point
接点电位	接點電位	junction potential
接合	接點, 接面, 接頭	junction
接合效率	接頭效率	joint efficiency
接取点	脫離點	take off point
接枝[法]	接枝[法]	grafting
接枝共聚物	接枝聚合物	graft copolymer
接枝聚合	接枝聚合[反應]	graft polymerization
接枝聚合物	接枝共聚物	graft polymer
接种	接種	inoculation
接种物	接種液	inoculum
节点	節點	node
节流	節流	throttling
节流带	節流帶	throttling band
节流阀	節流閥	throttle valve, throttling valve
节流过程	節流程序	throttling process
节流量热计	節流卡計	throttling calorimeter
节流装置	節流裝置	throttling device
节能器	省熱器	economizer
节涌流,弹状流,团状流	塊狀流動,汽胞流動,塞流	slug flow
节涌流反应器	塞流反應器	slug flow reactor
结构单元	結構單元	structural unit
结构分析	結構分析	structure analysis
结构复合材料	結構複合材料	structural composite
结构化模型	結構化模式	structured model
结构黏度	結構黏度	structural viscosity
结构黏合剂	結構黏合劑	structural adhesive
结构泡沫塑料	結構泡沫塑料	structural foam plastic

祖国大陆名	台湾地区名	英 文 名
结构疲劳	結構疲勞	structural fatigue
结构缺陷	結構缺陷	structural defect
结构双折射	結構雙折射	structural birefringence
结构稳定性	結構穩定性	structural stability
结构无序	結構無序	structural disorder
结垢(=污垢)		
结垢系数	結垢係數	scale coefficient
结垢形成	積垢形成	scale formation
结合力	結合力	binding force
结合能	結合能	binding energy
结合水分	結合水分	bound moisture
结焦, 结炭	①煉焦, 煉焦煤②結焦	coking
结焦槽	結焦槽	coke drum
结晶	結晶	crystallization
结晶度	結晶度	degree of crystallization
结晶聚合物	晶質聚合物	crystalline polymer
结晶器	結晶器	crystallizer
结晶热	結晶熱	heat of crystallization
结晶水	結晶水	crystal water, water of crystallization
结晶温度	結晶點	crystallizing point
结晶学	結晶學	crystallography
结块	成餅, 結塊	caking
结块能力	結塊能力	caking power
结块性	結塊性	caking property
结皮[现象]	結皮現象	skinning
结筛	結篩	knot screen
结炭(=结焦)		
结线, 系线	連結線	tie line
截留, 保留	停留	retention
截面	横切面, 截面	cross section
截面积	截面積	cross sectional area
截止阀	阻塞閥	globe valve
截止频率	截止頻率	cut-off frequency
解毒	去毒	detoxification
解聚	解聚合[作用]	depolymerization
解聚集	去聚集[作用]	deaggregation
解离温度	解離溫度	dissociation temperature
解离吸附	解離吸附	dissociative adsorption
解偶环路	解偶環路, 去偶合環路	decoupled loops

祖国大陆名	台湾地区名	英 文 名
解偶联	非偶合, 解偶	uncoupling
解偶器	解偶器, 去偶合器	decoupler
解偶系统	解偶系统, 去偶合系统	decoupled system
解吸	脱附	desorption
解吸剂	解吸剂	stripping agent
解吸塔		stripper
解吸因子	脱附因子, 汽提因數	desorption factor, stripping factor
解析法	解析法	analytical method
解絮凝, 反絮凝	去絮凝[作剂]	deflocculation
介标行为	介標[度]行為	mesoscale behavior
介电常数	介電常數, 比介電常數	dielectric constant, specific inductive capacity
介电加热	介電加熱	dielectric heating
介电体	介電	dielectric
介电吸收	介電吸收	dielectric absorption
介晶态	液晶狀態	mesomorphic state
介稳态(=亚稳态)		
介质	①介質②培養基	medium
介质过滤器	介質過濾器	medium filter
芥子油	芥子油	mustard oil
界面	界面	interface
界面薄膜	界面薄膜	interfacial film
界面工程	界面工程	interfacial engineering
界面活性剂	界面活性劑	interfacial agent
界面聚合	界面聚合	surface polymerization
界面控制	界面控制	interface control
界面面积	界面面積	interfacial area
界面能	界面能	interfacial energy
界面浓度	界面濃度	interfacial concentration
界面势	界面勢	interfacial potential
界面湍流	界面紊流	interfacial turbulence
界面温度	界面溫度	interfacial temperature
界面现象	界面現象	interfacial phenomenon
界面相(=分界表面)		
界面形状	界面形狀	interfacial shape
界面性质	界面性質	interfacial property
界面张力	界面張力	interfacial tension
界面张力计	界面張力計	interfacial tensimeter
界面阻力	界面阻力	interfacial resistance

祖国大陆名	台湾地区名	英 文 名
界面组成	界面組成	interface composition
界限条件	最終條件	terminal condition
金刚砂	剛砂	emery
金红石	金紅石	rutile
金绿宝石	金綠寶石	chrysoberyl
金霉素	金黴素	aureomycin
金相学	金相學	metallography
金属	金屬	metal
金属表面处理	金屬表面處理	metal finishing
金属玻璃	金屬玻璃	metallic glass
金属催化剂	金屬觸媒	metallic catalyst
金属电镀	金屬電鍍	metal plating
金属间化合物	金屬間化合物	intermetallic compound
金属离子惰化剂	金屬離子惰化劑	metal ion inactivator
金属络合酸性染料	金屬錯鹽酸性染料	metal complex acid dye
金属媒染剂	金屬媒染劑	metallic mordant
金属配位染料	金屬化染料	metallized dye
金属喷镀	金屬噴霧	metal spraying
金属塑料	金屬化塑膠	metallized plastic
金属有机化学蒸气沉积	金屬有機化學蒸氣沈積法	metallorganic chemical vapor deposition
金属皂	金屬皂	metallic soap
紧凑型换热器	緊緻熱交換器	compact heat exchanger
紧密度	緊密度	compactness
紧束效应	緊束效應, 質量效應	effective packing
近程	近距離	short range
近程分子内相互作用	分子內近程相互運動	short-range intramolecular interaction
近程结构	近程結構	short-range structure
近程链内曲柄运动	內近程鏈間曲軸運動	short-range intrachain crankshaft movement
近程相互作用	近程	short-range interaction
近程有序	[鏈分子排列的]近程有序	short-range order
近晶态	層列型液晶態	smectic state
近晶相	層列型液晶相	smectic phase
近晶型结构	層列型液晶結構	smectic structure
近似	①近似②近似法	approximation
近似法	近似法	approximate method
近似解	近似解	approximate solution
近似式	近似式	approximation formula

祖国大陆名	台湾地区名	英 文 名
进口长度	進口長度	entry length
进口区	進口區	entrance region
进口效应	進口效應	entrance effect
进料	①進料②饋入③食用	feed
进料板	進料板	feed tray
进料板位置	進料板位置	feed tray location
进料比	進料比	feed ratio
进料带	進料帶	feed belt
进料阀	進料閥	feed valve
进料管	進料管	feed pipe
进料滚柱	進料輥	feed roller
进料孔	進料孔	feed hole
进料流	進料流	feed stream
进料滤网	進料篩網	intake screen
进料器	①進料器,飼機②取食者	feeder, bell and hopper
进料输送机	進料輸送器	feed conveyor
进料调节器	進料調節器	feed regulator
进料温度	進料溫度	feed temperature
进料系统	進料系統	feed system
进料质量	進料品質	feed quality
进料状态线	q 線, 進料狀態線	q-line
进料组成	進料組成	feed composition
进气行程	吸氣衝程	intake stroke
进气压力	進氣壓[力]	intake pressure
浸管液位计	浸管液位計	dip pipe level gage
浸灰	加灰法, 浸灰法	liming
浸灰法	浸灰法	liming process
浸滤	瀝濾	lixiviation
浸没表面	浸沒表面	immersed surface
浸没燃烧	水中燃燒	submerged combustion
浸没燃烧蒸发器	浸沒燃燒蒸發器	evaporator with submerged combustion
浸没式冷凝器	沉式冷凝器	submerged condenser
浸没误差	浸入誤差	immersion error
浸泡(=浸渍)		
浸泡试验	浸泡試驗	soak test
浸漂	浸漂	dip bleaching
浸取	瀝取	leaching
浸取槽	瀝取槽	leaching tank

祖国大陆名	台湾地区名	英　文　名
浸取法	瀝取法	leaching process
浸取液	瀝取液	leachate
浸染	浸染	padding, dip dyeing, steeping
浸染机	浸染機	padding machine
浸染液	浸染液	padding liquor
浸湿法	浸濕法	soakage
浸蚀	浸蝕	etching
浸提物	析出物	educt
浸渍	浸漬	steeping
浸渍, 浸泡	浸漬	dipping, impregnation, maceration
浸渍槽	浸漬槽	dipping tank
浸渍镀锡	浸漬鍍錫	dip tinning
浸渍树脂	浸漬樹脂	solvent impregnated resin
浸渍压榨机	浸漬壓榨機	steeping press
浸渍釉	浸釉	dip glazing
经典流态化	古典流態化	classical fluidization
经济冲击	經濟衝擊	economic impact
经济回流比	經濟回流比	economic reflux ratio
经济可行性	經濟可行性	economic feasibility
经济评价	經濟評估	economic evaluation
经济寿命	經濟壽命	economic life
经济压力降	經濟壓力降	economic pressure drop
经验法	經驗法	empirical method
经验法则	經驗法則	empirical rule
经验方程式	實驗方程式, 經驗方程式	empirical equation
经验模型	經驗模型	empirical model
经验式	經驗式	empirical formula
晶格	晶格	crystal lattice
晶格(＝点阵)		
晶硅石	晶矽石	crystalline silica
晶核	①核②晶核	nucleus
晶核生成(＝成核)		
晶核形成	晶核形成	nucleus formation
晶间腐蚀	粒間腐蝕	intergranular corrosion
晶粒	粒	grain
晶粒长大	顆粒成長	grain growth
晶粒大小	顆粒大小	grain size
晶粒效应	壓紋效應	grain effect

祖国大陆名	台湾地区名	英 文 名
[晶]粒形[状]	粒形	grain shape
晶面	晶面	crystal face
晶片	晶片	wafer
晶石	晶石	spar
晶体	晶體	crystal
晶体成核	晶體成核	crystal nucleation
晶体大小	晶體大小	crystal size
晶体大小分布	晶體粒度大小	crystal size distribution
晶体管	電晶體	transistor
晶体生长	晶體成長	crystal growth
晶体习性	晶體習性	crystal habit
晶系	晶系	crystal system
晶形	晶形	crystal form, crystal shape
晶形结构	晶形結構	crystalline structure
晶形转变	晶形轉變	crystalline transition
晶种	晶種, 種晶	crystal seed, seed crystal
晶种混合槽	晶種混合槽	seed mixer
晶种结晶	晶種結晶	seeded crystallization
晶种粒	結晶種粒	seed grain
晶种纤维	晶種纖維	seed hair
精炼	精煉	affinage, refinement, refining
精炼度	精煉度	degree of refining
精馏	精餾	rectification
精馏板	精餾板	rectifying tray
精馏段	精餾段, 增濃段	rectifying section, rectification section
精馏釜	精餾器, 整流器	rectifying still
精馏器	精餾器	rectifier
精馏塔	精餾塔	rectifier column, rectifying column, rectifying tower
精馏塔板	精餾板	rectifying plate
精[密]度	精確度, 精密度	precision
精密分馏	精密分餾	precise fractionation
精密分馏分离	精分分離	close cut separation
精密分馏馏出物	精分餾出物	close cut distillate
精密分馏馏分	精分餾分	close cut fraction
精密蒸馏	精密分餾	precision fractional distillation
精确解	精確解	exact solution
精筛	精篩	fine screen
精细分散	精細分散	fine dispersion

祖国大陆名	台湾地区名	英　文　名
精细平衡原理	明細均衡原理	principle of detailed balance
精油	香精油	essential oil
精整	最後加工, 整理	finishing
精制机	精製機	refiner
精制棉短绒	棉絨	linter
鲸蜡	鯨蠟	spermaceti wax
鲸蜡油	抹香鯨油, 鯨蜡油, 鯨油	sperm oil, spermaceti oil
鲸油	鯨油	whale oil
鲸脂	鯨脂	whale tallow
井式喷嘴	井型噴嘴	well type nozzle
阱	阱,閘,分離器	trap
肼	肼, 聯胺	hydrazine
净功	淨功	net work
净化过程	純化程序	purification process
净化剂	淨化劑, 清除劑, 消毒劑	scavenger, decontaminant
净化水	淨化水, 純化水	purified water
净化指数(＝去污指数)		
净利润	淨利潤	net profit
净热值	淨熱值	net heating value
净速率	淨速率	net rate
净通量	淨通量	net flux
净现值	淨現值	net present value, net present worth
净正吸压头(＝汽蚀余量)		
净重	淨重	net weight
径向反应器	徑向反應器	radial flow reactor
径向分布函数	徑向分布函數	radial distribution function
径向分散	徑向分散	radial dispersion
径向分散系数	徑向分散係數	radial dispersion coefficient
径向混合	徑向混合	radial mixing
径向流涡轮机	沿徑流渦輪(機)	radial flow turbine
径向速度	徑向速度	radial velocity
径向应力	徑向應力	radial stress
径向轴承	徑向軸承	radial bearing
竞争反应	競爭反應	competitive reaction
竞争吸附	競爭吸附	competitive adsorption
竞争性	競爭性	competitiveness

祖国大陆名	台湾地区名	英　文　名
竞争性抑制	競爭性抑制	competitive inhibition
静电沉降器	靜電集塵器	electrostatic precipitator
静电分离	靜電分離	electrostatic separation
静电分离器	靜電分離器	electrostatic separator
静电介电常数	靜電介電常數	static dielectric constant
静电吸引	靜電吸引	electrostatic attraction
静电消除器	靜電消除器	static eliminator
静电效应	靜電效應	electrostatic effect
静力学	靜力學	statics
静摩擦	靜摩擦	static friction
静态	靜態	static state
静态混合器	靜態混合器	static mixer
静态模量	靜態模量	static modulus
静态疲劳	靜態疲勞	static fatigue
静态热重分析法	靜態熱重量法	static thermogravimetry
静态柔量	靜態柔量	static compliance
静态试验	靜態試驗	static test
静态误差	靜態誤差	static error
静态误差系数	靜態誤差係數	static error coefficient
静态优化	靜態最適化	static optimization
静压	靜壓[力]	static pressure
静压头	靜力高差	static head
静止边界层	靜止邊界層	stagnant boundary layer
静止床	靜止床	quiescent bed
静止期	靜止相	stationary phase
静滞区	靜滯區	dead space
静重活塞压力计	靜重活塞壓力計	dead weight piston gage
静重仪	靜重儀	dead weight gage
镜面磨光	鏡面磨光	mirror polishing
镜式比色计	鏡式比色計	mirror colorimeter
镜式光谱仪	鏡式光譜計	mirror spectrophotometer
酒精	酒精	spirit
酒精汽油	酒精汽油	gasohol
酒石	酒石	tartar
酒石酸	酒石酸	tartaric acid
酒石酸铁钠	酒石酸鐵鈉	sodium iron tartrate
就地回收	就地回收	in-situ recovery
就地加工	就地加工	in-situ processing
就地燃烧	就地燃燒	in-situ combustion

祖国大陆名	台湾地区名	英　文　名
就地再生	就地再生	in-situ regeneration
局部加速度	局部加速度	local acceleration
局部平衡	局部平衡	local equilibrium
局部稳定性	局部穩定性	local stability
局部组成	局部組成	local composition
局部最优[值]	局部最適[值]	local optimum
矩	矩	moment
矩鞍形填料	矩鞍形填料	intalox saddle
矩形脉冲	矩形脈動	rectangular pulse
矩形堰	矩形堰	rectangular weir
矩阵	矩陣	matrix
矩阵求逆	矩陣反轉	matrix inversion
矩阵算术	矩陣算術	matrix arithmetics
巨正则配分函数	巨正則配分函數	grand canonical partition function
巨正则系综	巨正則係綜	grand canonical ensemble
拒虫剂	驅蟲劑, 排斥劑	repellent
距离因数	距離因數	distance factor
距速滞后	距離速度落後	distance velocity lag
锯齿形	鋸齒狀缺口, 鱗片	serration
聚氨基甲酸酯	聚胺甲酸酯	polyurethane
聚氨酯橡胶	聚胺甲酸酯橡膠, PU 橡膠	polyurethane rubber
聚苯乙烯	聚苯乙烯	polystyrene
聚变		fusion
聚丙烯	聚丙烯	polypropylene
聚丙烯腈	聚丙烯腈	polyacrylonitrile
聚丙烯腈纤维	聚丙烯腈纖維	acrylic fiber
聚丙烯酸钠	聚丙烯酸鈉	sodium polyacrylate
聚丙烯酸酯	聚丙烯酸酯	polyacrylate
聚丙烯酸酯树脂	聚丙烯酸酯樹脂	polyacrylate resin
聚丙烯纤维	聚丙烯纖維	polypropylene fiber
聚并, 凝并	聚結	coalescence
聚并模式	聚結模式	coalescence model
聚并器	聚結器	coalescer
聚砜	聚碸	polysulfone
聚合动力学	聚合[反應]動力學	polymerization kinetics
聚合度	聚合度	degree of polymerization
聚合[反应]	聚合[反應]	polymerization, polymerization reaction
聚合反应工程	聚合反應工程	polymerization reaction engineering

祖国大陆名	台湾地区名	英　文　名
聚合反应器	聚合反應器	polymerization reactor
聚合机理	聚合[反應]機構	polymerization mechanism
聚合物泛滥	聚合物氾流[法]	polymer flooding
聚合物加工	聚合物加工, 高分子加工	polymer processing
聚合物加工工程	聚合物加工工程	polymer processing engineering
聚合物加工技术	聚合物加工技術	polymer processing technology
聚合物降解	聚合物降解[反應]	polymer degradation
聚合物土壤改良剂	聚合物土壤改良劑	soil improver of polymer
聚合物稳定剂	聚合物穩定化	polymer stabilization
聚合油	熟油, 亞麻仁油, 聚合油	stand oil
聚集	聚集[作用]	aggregation
聚集结晶	累積結晶	accumulative crystallization
聚集数	聚集數	aggregation number
聚集体	聚集體	aggregate
聚己内酰胺	聚己内醯胺	polycaprolactam
聚甲醛	聚甲醛	polyformaldehyde
聚硫聚合物	多硫聚合物	thiokol polymer
聚硫橡胶	多硫橡膠	polysulfide rubber, thiokol
聚氯乙烯	聚氯乙烯	polyvinyl chloride
聚氯乙烯管	聚氯乙烯管	vinyl pipe
聚氯乙烯树脂	聚氯乙烯樹脂	polyvinyl chloride resin
聚醚	聚醚	polyether
聚偏氟乙烯	聚氟乙烯, 聚偏二氯乙烯	polyvinylidene fluoride, polyvinylidene chloride
聚偏氯乙烯纤维	紗隆, 聚偏氯乙烯樹脂	saran
聚三氟氯乙烯	聚三氟氯乙烯	polytrifluorochloroethylene
聚式流态化	聚集流體化	aggregative fluidization
聚四氟乙烯	聚四氟乙烯, 鐵氟龍	polytetrafluoroethylene, teflon
聚碳酸酯	聚碳酸酯	polycarbonate
聚碳酸酯树脂	聚碳酸酯樹脂	polycarbonate resin
聚烯烃树脂	聚烯烴樹脂	polyolefin resin
聚酰胺	聚醯胺	polyamide
聚酰胺树脂	聚醯胺樹脂	polyamide resin
聚酰胺纤维	聚醯胺纖維	polyamide fiber
聚乙酸乙烯酯	聚乙烯乙酯	polyvinyl acetate
聚乙烯	聚乙烯	polyethylene
聚乙烯吡咯烷酮	聚乙烯四氫吡咯酮	polyvinyl pyrrolidone

祖国大陆名	台湾地区名	英　文　名
聚乙烯醇	聚乙烯醇	polyvinyl alcohol
聚乙烯基醚	聚乙烯醚	polyvinyl ether
聚乙烯醚黏合剂	聚乙烯醚黏合劑	polyvinyl ether adhesive
聚乙烯酯缩丁醛树脂	聚乙烯丁醛樹脂	polyvinyl butyral resin
聚异丁烯	聚異丁烯	polyisobutylene
聚异戊二烯橡胶	聚異戊二烯橡膠, 聚異平橡膠	polyisoprene rubber
聚酯	聚酯	polyester
聚酯类树脂	聚酯樹脂	polyester resin
聚酯类纤维	聚酯纖維	polyester fiber
聚酯类橡胶	聚酯橡膠	polyester rubber
涓流床(＝滴流床)		
卷积	摺積	convolution
卷积积分	摺積積分	convolution integral
卷绕	繞緊	windup
卷缩	卷縮, 扭結	snarl
绢丝橡胶	絹絲橡膠	silk rubber
决策	決策	decision making
决策变量	決策變數	decision variable
决策树	決策樹	decision tree
绝对反应速率	絕對反應速率	absolute reaction rate
绝对沸点	絕對沸點	absolute boiling point
绝对干纤维	絕對乾纖維	absolute dry fiber
绝对计数	絕對計數	absolute counting
绝对零度	絕對零度	absolute zero
绝对黏度	絕對黏度	absolute viscosity
绝对偏差	絕對偏差	absolute deviation
绝对热力学温标	絕對熱力溫標	absolute thermodynamic temperature scale
绝对色值	絕對色值	absolute color value
绝对熵	絕對熵	absolute entropy
绝对湿度	絕對濕度	absolute humidity
绝对湿含量	絕對水分含量	absolute moisture content
绝对温度	絕對溫度	absolute temperature
绝对稳定性	絕對穩定性	absolute stability
绝对误差积分	絕對誤差積分	integral of absolute error
绝对压力	絕對壓力	absolute pressure
绝对压力计	絕對壓力計	absolute pressure gauge
绝对值	絕對值	absolute value
绝对专一性	絕對特異性	absolute specificity

祖国大陆名	台湾地区名	英　文　名
绝对浊度	絕對濁度	absolute turbidity
绝热饱和	絕熱飽和	adiabatic saturation
绝热饱和温度	絕熱飽和溫度	adiabatic saturation temperature
绝热壁	絕熱壁	adiabatic wall
绝热材料	隔熱材料, 保溫材料	insulating material
绝热操作	絕熱操作	adiabatic operation
绝热操作线	絕熱操作線	adiabatic operating line
绝热反应	絕熱反應	adiabatic reaction
绝热反应器	絕熱反應器	adiabatic reactor
绝热干燥	絕熱乾燥	adiabatic drying
绝热管	隔熱管, 保溫管	insulated pipe
绝热过程	絕熱程序	adiabatic process
绝热化学反应	絕熱化學反應	adiabatic chemical reaction
绝热浇道	隔熱澆道	insulated runner
绝热节流过程	絕熱節流程序	adiabatic throttling process
绝热节流装置	絕熱節流裝置	adiabatic throttling device
绝热精馏	絕熱精餾	adiabatic rectification
绝热冷却	絕熱冷卻	adiabatic cooling
绝热冷却曲线	絕熱冷卻曲線	adiabatic cooling curve
绝热冷却线	絕熱冷卻線	adiabatic cooling line
绝热[沥青]纸	絕熱紙	sheathing paper
绝热流	絕熱流動	adiabatic flow
绝热耐火材料	隔熱耐火物	insulating refractory
绝热耐火砖	隔熱耐火磚	insulating firebrick
绝热膨胀	絕熱膨脹	adiabatic expansion
绝热式量热计	絕熱卡計	adiabatic calorimeter
绝热收缩	絕熱收縮	adiabatic contraction
绝热塔	絕熱塔	adiabatic column
绝热条件[状态]	絕熱狀況	adiabatic condition
绝热图	絕熱圖	adiabatic plot
绝热温升	絕熱溫升	adiabatic temperature rise
绝热系统	絕熱系統	adiabatic system
绝热压缩	絕熱壓縮	adiabatic compression
绝热仪器	絕熱裝置	adiabatic apparatus
绝热因子	絕熱因數	adiabatic factor
绝热增湿	絕熱增濕	adiabatic humidification
绝热蒸发	絕熱蒸發	adiabatic evaporation
绝热砖	隔熱磚	insulating brick
绝热转化	絕熱轉化	adiabatic conversion

祖国大陆名	台湾地区名	英 文 名
绝缘材料	絕緣材料	insulating material
绝缘体	絕緣體	insulator
绝缘体催化剂	絕緣體觸媒	insulator catalyst
绝缘纸	絕緣紙	insulating paper
均镀能力	電鍍能力	throwing power
均方根偏差	均方根偏差	root-mean-square deviation
均方根误差	均方根誤差	root-mean-square error
均方根振荡速度	均方根速度	root-mean-square fluctuating velocity
均方误差	均方誤差	mean square error
均衡肥料	均衡肥料	balance fertilizer
均衡伸缩囊	均衡伸縮囊	balance bellow
均衡生长	均衡生長	balanced growth
均衡值	均衡值	balanced value
均聚反应	同元聚合[反應]	homopolymerization
均聚物	同元聚合物	homopolymer
均热因子	均熱因數	soaking factor
均相	勻相	homogeneous phase
均相成核	勻相成核	homogeneous nucleation
均相催化	勻相催化[作用]	homogeneous catalysis
均相反应	勻相反應	homogeneous reaction
均相反应器	勻相反應器	homogeneous reactor
均相共沸混合物	勻相共沸液	homoazeotrope
均相平衡	勻相平衡	homogeneous equilibrium
均相燃烧	勻相燃燒	homogeneous combustion
均相系统	均勻系統	homogeneous system
均压环	測壓環	piezometer ring
均匀分布	均勻分布	uniform distribution
均匀共沸混合物	勻相共沸液	homogeneous azeotrope
均匀结焦	均勻結焦	uniform coking
均匀晶体	均勻晶體	uniform crystal
均匀流	均勻流動	uniform flow
均匀流动	均勻流	uniform stream
均匀流化床	勻相流體化床	homogeneously fluidized bed
均匀体	均勻體	homogeneous body
均匀调节	平均控制	averaging control
均匀湍流	均勻紊流	homogeneous turbulence
均匀系数	均勻係數	uniformity coefficient
均匀性	均勻性	homogeneity, uniformity
均匀液位控制	平均液位控制	averaging level control

祖国大陆名	台湾地区名	英 文 名
均匀载荷	均匀負荷	uniform loading
均匀中毒	均匀中毒	uniform poisoning
均匀转化	均匀轉化	uniform conversion
均匀转化模型	均匀轉化模型	uniform conversion model
均值	①平均②平均值	mean
菌	菌	aspergillus

K

祖国大陆名	台湾地区名	英 文 名
咔唑	咔唑	carbazole
咖啡碾磨机	咖啡磨粉機	coffee mill
卡必醇	卡必醇	carbitol
卡[路里]	卡洛里, 卡	calorie
开敞喷嘴	開敞噴嘴	open nozzle
开敞叶轮	敞動葉輪	open impeller
开发项目	開發專案	development project
开方器	開方器	square root extractor
开关	開關	switch
开关作用	開關作用	on-off action
开环	開環	open loop
开环测试	開環測試	open-loop test
开环极点	開環極點	open-loop pole
开环控制	開環控制	open-loop control
开环零点	開環零點	open-loop zero
开环系统	開環系統	open-loop system
开环转移函数	開環轉移函數	open-loop transfer function
开角(=释放角)		
开炼机(=辊式捏合机)		
开路	開路	open circuit
开路研磨	開路研磨	open circuit grinding
开始[硫化]效应	開始[硫化]效應	set-up effect
开式边界	開式邊界	open boundary
开式反应器	開敞反應器	open reactor
开式过滤器	開敞過濾器	open filter, open sand filter
开式容器	開放容器	open vessel
开式序列反应	非連鎖反應	open sequence reaction
开斯米纶	開司米龍	cashmilon

祖国大陆名	台湾地区名	英　文　名
抗爆	抗震	antiknock
抗爆化合物	抗震化合物	antiknocking compound
抗爆值	抗震值	antiknock value
抗臭氧剂	抗臭氧[老化]劑	antiozonant
抗冻混合物	抗凍混合物	antifreeze mixture
抗刮性	抗刮性	scratch resistance
抗碱性	抗鹼法	alkali resistance
抗焦处理	防焦處理	scorch-resisting treatment
抗静电剂	抗靜電劑	antistatic agent
抗卷绕	抗繞緊	antiwindup
抗菌素(＝抗生素)		
抗拉强度	抗拉強度	tensile strength
抗磨剂	抗磨劑	attrition resistant
抗磨损	抗磨阻力	attrition resistance
抗凝剂	抗凝劑	anticoagulant agent
抗凝效应	抗凝效應	anticoagulant effect
抗日光龟裂剂	抗日光龜裂劑, 抗晒劑	sun-checking agent
抗散裂强度	抗散裂強度	spalling resistance
抗生素, 抗菌素	抗生素	antibiotic
抗水剂	拒水劑	water repellence agent
抗水[作用]	防水劑	water repellent
抗撕强度	撕裂強度	tearing strength
抗体	抗體	antibody
抗微生物活性	抗微生物活性	antimicrobial activity
抗微生物剂	抗微生物藥物	antimicrobial (drug)
抗微生物作用	抗微生物作用	antimicrobial action
抗絮凝剂	去絮凝劑	deflocculant, deflocculation agent
抗氧[化]剂	抗氧化劑	antioxidant
抗原	抗原	antigen
苛化	苛性化	causticization
科尔莫戈罗夫尺度	科爾莫戈羅夫尺度	Kolmogorov's scale
颗粒材料	粒狀物質	granular material
颗粒层过滤器	顆粒層過濾器	granular bed filter
颗粒床	顆粒床	particulate bed
颗粒簇	顆粒團簇	cluster of particles
颗粒过程	顆粒程序	particulate process
颗粒密度	粒子密度	particle density
颗粒群	粒子群	particle swarm
颗粒特性	粒子特性	particle characteristics

祖国大陆名	台湾地区名	英 文 名
颗粒体	颗粒	granule
颗粒形状	粒形	particle shape
颗粒形状因子	粒形因數	particle shape factor
颗粒学	颗粒學	particuology
壳程传热系数	殼側熱傳係數	shell-side heat transfer coefficient
壳模铸造	殼形鑄造, 殼膜法	shell molding
壳模铸造[用]树脂	殼形鑄造樹脂	shell molding resin
壳平衡	殼均衡	shell balance
可编程计算器	可程式計算器	programmable calculator
可编程逻辑控制	可程式邏輯控制	programmable logic control
可编程调制器	可程式控制器	programmable controller
可变孔板流量计	可變孔口計	variable orifice meter
可剥性涂料	可剝塗料	strippable coating
可操作性	可操作性	operability
可测量性	可量測性	measurability
可测量性质	可測性質	measurable property
可的松	皮質酮	cortisone
可纺性	可紡性	spinnability
可观测性	可觀測性	observability
可恢复应力	可復應力	recoverable stress
可获利程度	獲利能力	profitability
可及矩阵	可及矩陣	reachability matrix
可降解塑料	可降解塑膠	degradable plastic
可接受报酬	可接受(的)報酬	acceptable return
可解性	可溶解性, 溶劑合性	solvability
可靠性	可靠性	reliability
可控性	可控性	controllability
可扩展性	延伸性	extensibility
可裂变物质	可分裂材料	fissionable material
可裂变性	可分裂性	fissionability
可流态化颗粒大小	可流體化粒大小	fluidizable particle size
可磨性	可磨性	grindability
可磨性指数	可磨性指數	grindability index
可逆变化	可逆變化	reversible change
可逆沉淀	可逆沈澱	reversible precipitation
可逆电池	可逆電池	reversible cell
可逆度	可逆度	degree of reversibility, reversibility
可逆反应	可逆反應	reversible reaction
可逆分解电位	可逆分解電位	reversible decomposition potential

祖国大陆名	台湾地区名	英 文 名
可逆功	可逆功	reversible work
可逆过程	可逆過程	reversible process
可逆化学反应	可逆化學反應	reversible chemical reaction
可逆机理	可逆機構	reversible mechanism
可逆胶囊	反微胞	reversed micelle
可逆绝热流动	可逆絕熱流動	reversible adiabatic flow
可逆扩散	可逆擴散	reversible diffusion
可逆性	可逆性	reversibility
可燃性	可燃性, 可燃度	flammability
可燃液体规范	可燃液體規範	combustible liquid code
可染性	可染性	dyeability
可溶性	可溶性	solubility
可溶性淀粉	可溶澱粉	soluble starch
可溶性核糖核酸	可溶性核糖核酸, 可溶性 RNA	soluble RNA (sRNA)
可溶性胶	溶性膠	soluble gum
可溶性聚合物	可溶性聚合物	soluble polymer
可溶性树脂	可溶樹脂	soluble resin
可乳化性	可乳化性	emulsifiability
可湿性助剂	助濕劑	wettable agent
可实现性	可真實化性	realizability
可塑性	塑性	plasticity
可调参数	可調參數	adjustable parameter
可调节权	可調節權	adjustable weight
可调锐孔	可調孔口	adjustable orifice
可调轴承	可調軸承	adjustable bearing
可涂期	使用期	spreadable life
可稳性	①穩定性②穩定化度	stabilizability
可洗软膏基	可洗軟膏基	washable ointment base
可行路径法	可行路徑法	feasible path method
可行性	可行性	feasibility
可行性调查	可行性調查	feasibility survey
可行性分析	可行性分析	feasibility analysis
可行性研究	可行性研究	feasibility study
可行域	可行區	feasible region
可压缩流动	可壓縮流動	compressible flow
可压缩流体	可壓縮流體	compressible fluid
可压缩滤饼	可壓縮濾餅	compressible filter cake
可压缩性	壓縮度	compressibility

祖 国 大 陆 名	台 湾 地 区 名	英 文 名
可再生能源	再生能源	renewable energy source
可再生资源	可再生資源	renewable resources
克拉珀龙-克劳修斯方程	克来匹隆-克勞修斯方程	Clapeyron-Clausius equation
克劳修斯不等式	克勞修斯不等式	Clausius inequality
克隆	克隆	clone
克伦舍尔图	克倫舍爾圖	Kremser's diagram
克努森扩散	努生擴散	Knudsen diffusion
克努森数	努生數	Knudsen number
刻度盘指示器	針盤指示計	dial indicator
空管反应器	空管反應器	unpacked-tube reactor
空间点阵	空間晶格	space lattice
空间电荷	空間電荷	space charge
空间电荷限制电流	空間電荷限制電流	SCLC (＝space charge limited current)
空间电荷限制电流	空間電荷限制電流	space charge limited current
空间空调	空間空調	space conditioning
空间冷却	空間冷卻	space cooling
空间群	空間群	space group
空间速度	空間速度	spatial velocity
空间速率	空間速度	space velocity
空间系数	空間因數	space factor
空间效应	位阻效應, 空間效應	steric effect
空间因子	位阻因素	steric factor
空间张力	立構張力, 立構應變	steric strain
空间障碍	空間障礙	steric restriction
空间振荡	空間振盪	spatial oscillation
空气饱和器	空氣飽和器	air saturator
空气吹扫	空氣沖洗	air purge
空气动力学	空氣動力學	aerodynamics
空气动力直径	空氣動力直徑	aerodynamic diameter
空气分离	空氣分離	air separation
空气分离槽	空氣分離槽	air separation tank
空气分离塔	空氣分離塔	air separation tower
空气干燥器	空氣乾燥器	atmospheric drier, air dryer
空气干燥塔	空氣乾燥塔	air drying tower
空气供应	空氣供應	air supply
空气鼓风	鼓風	air blasting
空气鼓泡器	空氣噴佈器	air sparger
空气过滤	空氣過濾	air filtration

祖国大陆名	台湾地区名	英　文　名
空气过滤器	空氣過濾器	air filter
空气混合深度	空氣混合深度	air mixing depth
空气加热器	空氣加熱器	air heater
空气扩散器	空氣擴散器	air diffuser
空气冷凝器	空氣冷凝器	aerial condenser, atmospheric condenser
空气冷却	空氣冷卻	air cooling
空气冷却环	空氣冷卻環	air ring
空气冷却器	氣冷式熱交換器, 空氣冷卻器	air-cooled heat exchanger, air cooler
空气冷却塔	空氣冷卻塔, 風冷塔	atmospheric cooling tower
空气离析器	空氣分離器	air separator
空气溜槽	空氣溜槽	air slide
空气流动	空氣流動	air flow
空气流量计	空氣流量計	air flow meter
空气炉	空氣爐	air furnace
空气灭菌器	空氣滅菌器	air sterilizer
空气膜	空氣膜	air film
空气黏合	氣結	air binding
空气喷射器	空氣噴射器	air ejector
空气喷头	空氣噴嘴	air nozzle
空气膨胀器	空氣膨脹機	air expander
空气-燃料比	空氣-燃料比	air-fuel ratio
空气软管	空氣軟管	air hose
空气调节器	空氣調節器	air regulator
空气污染	空氣污染	aerial contamination, air pollution
空气污染工程	空氣污染工程	air pollution engineering
空气污染控制	空氣污染控制	air pollution control
空气污染物	空氣污染物	air pollutant
空气雾沫剂	空氣霧沫劑	air-entraining agent
空气吸滤器	空氣吸濾器	air suction filter
空气洗涤	空氣洗滌	air washing
空气洗涤器	空氣洗滌器	air scrubber
空气压力	空氣壓力	air pressure
空气压力降	空氣壓力降	air pressure drop
空气压缩机	空氣壓縮機	air compressor
空气扬析	空氣淘析	air elutriation
空气预热器	空氣預熱器	air preheater
空气蒸发	空氣蒸發	air evaporation
空蚀	空洞腐蝕	cavitation corrosion

祖国大陆名	台湾地区名	英　文　名
空调	空氣調節	air conditioning
空调机组	空氣調節單元	air conditioning unit
空调器	空氣調節器	air conditioner
空吸泵	吸取泵	suction pump
空心叶轮搅拌器(＝自吸搅拌器)		
空隙	空隙	void
空隙分数	空隙分率	void fraction
空隙率	空隙度	voidage
空隙速度	間隙速度	interstitial velocity
空隙体积	空隙體積	void volume
空穴辐射	空洞輻射	cavity radiation
孔板	孔口板	orifice plate
孔板计接头	孔口計接頭	orifice tap
孔板计系数	孔口計係數	orifice meter coefficient
孔板流量计	孔口[流量]計	orifice meter, orifice flowmeter
孔板塔	多孔層板塔	perforated plate column, perforated plate tower
孔板蒸馏塔	多孔板蒸餾塔	perforated plate distillation column
孔半径	細孔半徑	pore radius
孔径	孔徑, 細孔大小	pore size
孔径分布	孔徑分布	pore size distribution
孔口	孔口, 小孔	orifice
孔口法兰	孔口法蘭, 孔口凸緣	orifice flange
孔口环	孔口環	orifice ring
孔口排放	孔口排放	orifice discharge
孔口系数	孔口係數	orifice coefficient
孔口中毒	孔口中毒	pore-mouth poisoning
孔流系数	排放係數	discharge coefficient
孔膜隔板	多孔[性]障壁	porous barrier
孔雀绿	孔雀綠	malachite green
孔雀石	孔雀石	malachite
孔容(＝孔体积)		
孔体积, 孔容	細孔體積, 孔隙體積	pore volume
孔隙, 细孔	孔隙, 細孔	pore
孔隙计	孔隙計	porosimeter
孔隙结构	細孔結構	pore structure
孔隙率	孔隙度	porosity
孔隙仪	孔隙儀	porosity apparatus

祖 国 大 陆 名	台 湾 地 区 名	英 文 名
孔效应	孔隙度效應	porosity effect
控制	控制	control
pH 控制	pH 值控制	pH control
控制变量	受控變數, 控制變數	controlled variable, control variable
控制表面	限制表面	control surface
控制策略	控制策略	control strategy
控制单环路	單環路控制	control single-loop
控制点	控制點	control point
控制阀	控制閥	control valve
控制阀范围	控制閥範圍性	rangeability of control valve
控制反应器	反應器控制	control-reactor
控制构型	控制組態	control configuration
控制函数	控制函數	control function
控制回路	控制環路	control loop
控制界限	管制界限	control limit
控制律	控制律	control law
控制盘仪表	控制儀表板	control panel
控制器参数	控制器參數	controller parameter
控制器机理	控制器機構	controller mechanism
控制器校正	控制器校正	controller calibration
控制器匹配	控制器調適	controller adaptation
控制器设定	控制器設定	controller setting
控制器增益	控制器增益	controller gain
控制器振铃	控制器振鈴	controller ringing
控制区间	控制區間	control interval
控[制]释[放]	控制釋放	controlled release
控制算法	控制算則	control algorithm
控制体积	限制體積	control volume
控制图	管制圖	control chart
控制系统	控制系統	control system
控制系统相互作用	控制系統相互作用	interaction of control systems
控制现场	控制區域	controlled field
控制线路	控制線路	control scheme
控制作用	控制作用	control action
枯草杆菌	枯草桿菌	bacillus subtilis
枯草过滤器	乾草過濾器	hay filter
苦艾	苦艾	absinthium
库存控制	存貨控制	inventory control
[库]存量	存貨, 盤存	inventory

祖国大陆名	台湾地区名	英 文 名
库尔特粒度仪	庫爾特粒度儀	Coulter counter
库仑计	電量計	voltameter
跨度	寬度, 跨距	span
快速反应	快速反應	fast reaction
快速过滤器	快速濾器	rapid filter
快速流化床	快速流體化床	fast fluidized bed
快速流态化	快速流體化	fast fluidization
快速砂滤速率	快速砂濾[法]	rapid sand filtration rate
宽床反应器	寬床反應器	massive bed reactor
矿浮选	礦浮選	ore flotation
矿蜡	礦蠟	mineral wax
矿物肥料	礦物肥料	mineral fertilizer
矿物学	礦物學	mineralogy
矿物油	礦油	mineral oil
矿物资源	礦物資源	mineral resources
矿用炸药	礦用炸藥	mining explosive
矿脂	石蠟脂	petrolatum
框架	框, 架	frame
框式搅拌器	框式攪拌器	grid agitator
框式压滤机	框式壓濾機	frame filter press
框图	方塊圖, 塊解圖	block diagram
窥箱	窺箱	sight box
葵花油	葵花油	heliotrope oil
扩大损失	擴大損失	enlargement loss
扩散	擴散	diffusion
扩散泵	擴散泵	diffusion pump
扩散薄膜	擴散薄膜	diffusion film
扩散层	擴散層	diffuse layer
扩散常数	擴散常數	diffusion constant
扩散传质	擴散質傳	diffusion mass transfer
扩散催化	擴散催化[作用]	diffusion catalysis
扩散电流	擴散電流	diffusion current
扩散电位	擴散勢	diffusion potential
扩散方程	擴散方程式	diffusion equation
扩散过程	擴散程序	diffusion process
扩散环	擴散環	diffusion ring
扩散控制	擴散控制	diffusion control
扩散控制反应	擴散控制反應	diffusion-controlled reaction
扩散膜	擴散障壁	diffusion barrier

祖国大陆名	台湾地区名	英 文 名
扩散器	擴散器	diffuser
扩散器护罩	擴散器護罩	diffuser casing
扩散热效应	擴散熱效應	diffusion-thermo effect
扩散时间	擴散時間	diffusion time
扩散速率	擴散速率	diffusion rate
扩散通量	擴散通量	diffusion flux
扩散系数	擴散係數	diffusion coefficient, diffusivity
扩散压力	擴散壓	diffusion pressure
扩散真空泵	擴散真空泵	diffusion vacuum pump
扩散阻力	擴散阻力	diffusion resistance
扩展式表面管	延伸表面管	extended surface tube

L

祖国大陆名	台湾地区名	英 文 名
拉发线	引線	trip wire
拉杆	拉引棒	draw bar
拉坯	拉坯	throwing
拉伸	拉伸	stretching
拉伸比	拉伸比	stretch ratio
拉伸成型	拉伸成型	stretch forming
拉伸纺丝	拉伸紡絲	stretch spinning
拉伸流动	伸長流動	elongational flow
拉伸膜	拉伸膜	stretched membrane
拉伸黏度	伸長黏度, 延伸黏度	elongational viscosity, extensional viscosity
拉伸取向	拉伸取向	stretch orientation
拉伸速率	伸長速率	elongation rate, stretch rate
拉伸性	拉伸性	stretchability
拉乌尔定律	拉午耳定律	Raoult's law
拉西环	拉西環	Rasching ring
蜡	蠟	wax
蜡馏出物	蠟餾出物	wax distillate
蜡纸	蠟紙	wax paper
赖氨酸	離胺酸	lysine
兰金循环	蘭金循環	Rankine cycle
蓝宝石	藍寶石	sapphire
蓝矾	藍礬	blue vitriol
蓝铜矿	藍銅礦	azurite

祖国大陆名	台湾地区名	英 文 名
篮式过滤器	籃式(過)濾器	basket filter
篮式离心机	籃式離心機	basket centrifuge
篮式升降机	籃式升降機	basket elevator
篮式提取器	籃式萃取器	basket extractor
浪动的	波動	rippling
老化试验	老化試驗	aging test
老化试验机	老化試驗機	aging test machine
老化箱	老化箱	aging oven
老化装置	老化裝置	aging equipment
酪蛋白	酪蛋白	casein
酪蛋白黏合剂	酪蛋白黏合劑	casein adhesive
雷蒙磨	雷蒙磨	Raymond mill
雷诺数	雷諾數	Reynolds number
类比	類似	analogy
类比理论	類似理論	analogy theory
类比系统	類比系統	analogy system
类黑素	類黑素	melanoid
类化学近似(=准化学 近似)		
类化学溶液模型(=准 化学溶液模型)		
类聚效应		symbiosis
累积产率	累積產率	integral yield
累积分布	累積分布	cumulative distribution
累积概率	累積機率	cumulative probability
累积过细重量分率	累積過細重量分率	cumulative weight undersizer fraction
累积筛析	累積篩析	cumulative screen analysis
累积速率	累積速率	accumulation rate
累积误差	累積誤差	integrated error
累积重量百分率过粗 部分	累積重量百分率過粗 部分	cumulative weight percent oversize
累积重量百分率过细 部分	累積重量百分率過細 部分	cumulative weight percent undersize
累积重量分布	累積重量分布	cumulative weight distribution
[累计]总量表	總量表	quantity meter
冷冻	冷凍	refrigeration
冷冻吨	冷凍噸	refrigerating ton
冷冻干燥	冷凍乾燥	freeze drying, lyophilization
冷冻干燥机	冷凍乾燥機	freeze drier

祖 国 大 陆 名	台 湾 地 区 名	英 文 名
冷冻过程	冷凍法, 冷凍程序	freezing process
冷冻剂	冷媒, 冷凍劑	refrigerant
冷冻能力	冷凍能力	refrigeration capacity
冷冻凝器	冷凍凝聚	freeze coagulation
冷冻盘管	冷凍旋管	refrigerating coil
冷冻器	冷凍器	freezer
冷冻系数	性能係數	COP(＝coefficient of performance)
冷冻液	冷凍流體	refrigerating fluid
冷光	冷光	luminescence
冷回流	冷回流	cold reflux
冷碱法	冷鹼法	cold-soda process
冷浸灰法	冷汁加灰法	cool liming
冷硫化	冷硫化法	cold vulcanization
冷模合	低溫接點	cold junction
冷凝点	冷凝點	condensation point
冷凝管	冷凝管	condensation tube, condensing tube
冷凝器	冷凝器	condenser
冷凝热	冷凝熱	heat of condensation
冷凝室	冷凝室	condensing chamber
冷凝温度	冷凝溫度	condensing temperature
冷凝液	冷凝液	condensate
冷却段	冷卻區	cooling zone
冷却剂	冷卻劑	coolant, cooling agent
冷却介质	冷卻界質	cooling medium
冷却面	冷卻表面	cooling surface
冷却面积	冷卻面積	cooling area
冷却器	冷卻器	cooler
冷却曲线	冷卻曲線	cooling curve
冷却蛇管	冷卻旋管	cooling coil
冷却时间	冷卻時間	cooling time
冷却水	冷卻水	cooling water
冷却速率	冷卻速率	cooling rate
冷却塔	冷卻塔	cooling tower
冷却系统	冷卻系統	cooling system
冷制橡胶	冷製橡膠	cold rubber
厘泊	厘泊(黏度單位)	centipoise(cp)
厘泡	厘拖(重力黏度單位)	centistokes
离模膨胀	模頭膨脹	die swell
离析	離析	segregation

祖国大陆名	台湾地区名	英 文 名
离析度	隔離度	degree of segregation
离线	離線	off-line
离心	離心鼓風機, 離心送風機	centrifugal blower
离心泵	離心泵	centrifugal pump
离心沉降	離心沉降	centrifugal settling
离心成型	離心成型	centrifugal molding
离心澄清	離心澄清	centrifugal clarification
离心纯化	離心純化	centrifugal purification
离心萃取器	離心萃取器	centrifugal extractor
离心涤气机	離心洗氣器	centrifugal scrubber
离心[分离]	離心[分離], 離心化	centrifugation, centrifugalization
离心分离机	離心分離器	centrifugal separator
离心干燥器	離心乾燥機	centrifugal dryer
离心过滤机	離心過濾機	centrifugal filter
离心机	離心機	centrifugal, centrifuge
离心[机]管	離心管	centrifuge tube
离心机盘	離心機籃	centrifuge basket
离心机叶片	離心機葉片	centrifuge blade
离心机转头	離心機高差	centrifuge head
[离心]精炼法	精煉法	affination
离心力	離心力	centrifugal force
离心碾磨机	離心磨[機]	centrifugal mill
离心喷淋塔	離心噴霧塔	centrifugal spray tower
离心喷雾	離心噴霧	centrifugal spray
离心筛	離心篩	centrifugal screen
离心式固体分离器	離心式固體分離器	centrifugal solid separator
离心式螺旋桨	離心式螺旋槳	centrifugal propeller
离心式液体分离器	離心式液體分離器	centrifugal liquid separator
离心收集器	離心收集器, 離心集塵器	centrifugal collector
离心脱水器	離心脫水機	centrifugal dehydrator
离心旋风器	離心旋風器	centrifugal cyclone
离心压缩机	離心壓縮機	centrifugal compressor
离子电位	離子電位	ionic potential
离子反应	離子反應	ionic reaction
离子浮选	離子浮選[法]	ion flotation
离子交换	離子交換	ion exchange
离子交换分离	離子交換分離[法]	ion exchange separation

祖 国 大 陆 名	台 湾 地 区 名	英 文 名
离子交换过程	離子交換法	ion exchange process
离子交换剂	離子交換器, 離子交換劑	ion exchanger, ion exchange agent
离子交换膜	離子交換薄膜	ion exchange membrane
离子交换平衡	離子交換平衡	ion exchange equilibrium
离子交换容量	離子交換容量	ion exchange capacity
离子交换色谱[法]	離子交換層析法	ion exchange chromatography
离子交换树脂	離子交換樹脂	ion exchange resin
离子交联聚合物	多離子聚合物	ionomer
离子聚合	離子聚合[反應]	ionic polymerization
离子扩散	離子擴散	ionic diffusion
离子排斥	離子排斥	ion exclusion
离子迁移	離子遷移	ionic migration
离子迁移率	離子移動率	ionic mobility
离子强度	離子強度	ionic strength
离子束沉积法	離子束沉積法	ion-beam deposition
离子束蚀刻法	離子束蝕刻法	ion-beam etching
离子型表面活性剂	離子界面活性劑	ionic surfactant
离子注入	離子植入[法]	ion implantation
里迪尔机理	里迪爾機構	Rideal mechanism
理论	理論	theory
理论板数	理論板數	theoretical plate number, theoretical tray number
理论级	理論階	theoretical stage
理论空气量	理論空氣[需要]量	theoretical air
理论模型	理論模式	theoretical model
理论[塔]板	理論板	theoretical plate
理想板	理想板	ideal plate, ideal tray
理想反应器	理想反應器	ideal reactor
理想功	理想功	ideal work
理想管式反应器	理想管式反應器	ideal tubular reactor
理想混合物	理想混合物	ideal mixture
理想级	理想階	ideal stage
理想接触级	理想接觸階	ideal contact stage
理想流动	理想流動	ideal flow
理想流体	理想流體	ideal fluid , perfect fluid
理想气体	理想氣體	ideal gas, perfect gas
理想气体常数	理想氣體常數	ideal gas constant
理想气体定律	理想氣體定律	ideal gas law

祖国大陆名	台湾地区名	英　文　名
理想气体焓	理想氣體焓	ideal gas enthalpy
理想气体温标	理想氣體溫標	ideal gas temperature scale
理想气体状态	理想氣體狀態	ideal gas state
理想溶液	理想溶液	ideal solution, perfect solution
理想周期	理想周期	ideal cycle
锂基润滑脂	鋰皂基潤滑脂	lithium-base grease
力-电流类比	力-電流類比, 力-伏特類比	force-current analogy, force-voltage analogy
力矩	力矩	moment of force
力矩法	矩法	method of moments
力平衡	力均衡	force balance
力势	力勢	force potential
力学	力學	mechanics
力质换算常数	力質量換算常數	conversion constant force-mass
力-质量换算常数	力-質量換算常數	force-mass conversion constant
立方膨胀	體膨脹	cubic(al) expansion
立方膨胀系数	體脹係數	cubical expansion coefficient
立方平均沸点	立方平均沸點	cubic average boiling point
立构重复单元	立構重復單元	stereorepeating unit
立构规整度	立構規整度	stereotacticity
立构规整聚合	定向聚合, 立構規整聚合	stereoregular polymerization, stereospecific polymerization, stereotactic polymerization
立构规整聚合物	立構規整聚合物	stereospecific polymer, stereotactic polymer
立构规整橡胶	立構規整橡膠	stereo rubber
立构规整性	立構規整性	steric regularity
立构嵌段共聚物	立構嵌段共聚物	stereoblock copolymer
立构无规共聚物	無規立構共聚物	stereorandom copolymer
立构显微照相法	立構顯微照相法	stereomicrography
立构有规性	立構規整性	stereoregularity
立构有择聚合	立構選擇聚合	stereoselective polymerization
立构杂化作用	立構雜化作用	stereohybridization
立构中心	立構中心	stereocenter
立式蒸煮器	直立式蒸煮器	vertical digester
立体定向反应	立體特定反應	stereospecific reaction
立体对称均聚物	立體對稱均聚物	stereosymmetrical homopolymer
立体化学	立體化學	stereochemistry
立体结构	立體結構	spatial structure
立体聚合物	立體聚合物	space polymer

祖国大陆名	台湾地区名	英 文 名
立体排列	立體排列	spatial configuration
立体网状聚合物	立體網狀聚合物	space network polymer
立体序列	立體序列	stereosequence
立体序列分布	立體序列分布, 立構序列分布	stereosequence distribution
立体选择性	立體選擇性	stereoselectivity
立体异构体	立體異構體	stereoisomer
立体有规构型	有規立構組態	stereospecific configuration
立体有择催化剂	有規立構催化劑	stereospecific catalyst
立体专一性	立構規整性	stereospecificity
利润	利潤	profit
利润函数	利潤函數	profit function
利润限度	利潤邊際	profit margin
利润因子	利潤因數	profitability factor
利润指数	利潤指數	profitability index
利息	利息	interest
利息回收期	利息回收期	interest recovery period
沥青	瀝青, 柏油	pitch, asphalt
沥青漆	柏油漆	asphalt paint
沥青砂	瀝砂	tar sand
砾磨机	卵石磨	pebble mill
粒度	粒度, 粒子大小	particle size
粒度分布	粒子大小分布, 粒度分布	particle size distribution, size distribution
粒度分离	粒度分離	size separation
粒度分离器	粒度分離器	particle size separator
粒度分析	粒度分析, 粒徑分析	grainsize analysis, granulometry, particle size analysis, size analysis
粒度分析仪	粒度分析儀, 粒徑分析儀	grainsize analyzer, granulometer, particle size analyzer
粒度增大	粒子增大	size enlargement
粒间扩散	粒間擴散	interparticle diffusion
粒径	粒徑	particle diameter
粒内扩散	粒內擴散	intraparticle diffusion
粒内扩散系数	粒內擴散係數	intraparticle diffusivity
粒状催化剂	成粒觸媒	pelleted catalyst
粒状固体过滤器	粒狀固體過濾機	granular solid filter
粒子	粒子	particle
粒子动力学	粒子動力學	particle kinetics

祖 国 大 陆 名	台 湾 地 区 名	英　文　名
粒子旋风分离器	粒子旋風分離器	particle cyclone separator
粒子转化	粒子轉化	particle conversion
连串泵组	串聯泵組	pump series
连串补偿	串聯補償	series compensation
连串反应	串行反應, 連串反應	series reactions, consecutive reaction
连串反应器	串聯反應器	series reactors
连串过程	串聯程序	serial processes
连串连续搅拌反应釜组	串聯連續攪拌槽反應器組	CSTR's in series
连串流动	串流	series flow
连续操作	連續操作	continuous operation
连续萃取	連續萃取	continuous extraction
连续萃取器	連續萃取器	continuous extractor
连续方程	連續方程式	continuity equation
连续分馏	連續分餾	continuous fractionation
连续干燥器	連續乾燥器	continuous drier
连续过程	連續程序	continuous process, continuous type process
连续过滤	連續過濾	continuous filtration
连续过滤器	連續過濾器	continuous filter
连续环圈法	連續環圈法	continuous-cycling method
连续加工	連續加工法	continuous processing
连续搅拌反应釜串联组	串聯連續攪拌槽反應器組	battery of CSTR's
连续搅拌反应釜设计	連續流動攪拌槽反應器設計	CSTR design
连续搅拌反应器	連續攪拌反應器	continuous stirred reactor
连续结晶	連續結晶	continuous crystallization
连续结晶器	連續結晶器	continuous crystallizer
连续介质	連體	continuum
连续聚合	連續聚合	continuous polymerization
连续离心机	連續離心機	continuous centrifuge
连续利息	連續利率	continuous interest
连续[流动]反应釜	連續流動式反應器	continuous flow reactor
连续流动系统	連續流動系統	continuous flow system
连续培养	連續式培養菌	continuous culture
连续皮带鼓式过滤器	連續輸送鼓式過濾器	continuous belt drum filter
连续汽提器	連續汽提器	continuous stripping still
连续倾析	連續傾析	continuous decantation
连续式反应器	連續式反應器	continuous reactor

祖国大陆名	台湾地区名	英 文 名
连续式碳滤器	連續式碳濾器	continuous carbon filter
连续隧道干燥器	連續隧式乾燥器	continuous tunnel drier
连续隧道窑	連續隧式窯	continuous tunnel kiln
连续碳滤法	連續碳濾法	continuous carbon filtration
连续细丝	連續單絲	continuous filament
连续现金流动	連續現金流動	continuous cash flow
连续相	連續相	continuous phase
连续性定律	連續定律	continuity law
连续性方程	連續性方程式	equation of continuity
连续研磨机	連續研磨機	continuous grinder
连续窑	連續窯	continuous kiln
连续蒸馏	連續蒸餾	continuous distillation
连续蒸煮器	連續蒸煮器	continuous digester
联产品	連產物	coproduct
联管节	接頭偶合	coupling
联立模块法(＝双层法)		
敛集效应	緊束效應, 質量效應	packing effect
炼厂气	煉油氣	refinery gas
炼焦过程	煉焦法	coking process
[炼]焦炉	煉焦爐	coke oven, coking coal
炼油厂	煉油廠, 精煉廠	refinery
链段布朗运动	鏈段布朗運動	segmental Brownian motion
链段各向异性	鏈段各向異性	segment anisotropy
链段密度分布	鏈段密度分布	segment-density distribution
链段摩擦因子	鏈段摩擦因子	segmental friction factor
链段相互作用参数	鏈段交互作用參數	segment-interaction parameter
链段旋转	鏈段旋轉	segment rotation
链段跃迁频率	鏈段躍遷頻率	segmental jump frequency
链段运动	鏈段運動	segmental motion, subchain motion
链反应	連鎖反應	chain reaction
链反应引发	連鎖反應[的]起始	initiation of chain reaction
链节, 线段	鏈段	segment
链解聚[作用]	鏈解聚合[反應]	chain depolymerization
链聚合	連鎖聚合[反應]	chain polymerization
链霉素	鏈黴素	streptomycin
链式输送机	鏈式運送機, 鏈運機	chain conveyor
链引发	連鎖反應引發	chain initiation
链增长	連鎖反應增長	chain propagation

祖国大陆名	台湾地区名	英 文 名
链增长反应	增長反應,	propagation reaction, propagation of chain reaction
链终止	連鎖反應終止	chain termination
链终止反应	連鎖終止反應	chain termination reaction
链转移	連鎖反應轉移	chain transfer
两段串级系统	二段串級系統	two-stage cascade system
两段蒸馏	二段蒸餾	two-stage distillation
两级导向阀	二級導引閥	two-stage pilot valve
两流体理论	兩流體理論	two-fluid theory
两相流	兩相流動	two-phase flow
两相流动型式	兩相流動型式	two-phase flow pattern
两相模型	兩相模式	two-phase model
两相区	兩相區[域]	two-phase region
两相系统	兩相系統	two-phase system
两性霉素	抗黴酮	amphotericin
两液体理论	兩液體理論	two-liquid theory
亮度	亮度	brightness
量程可调范围	範圍性	rangeability
量瓶	量瓶	measuring flask
量气管	量氣管	gas measuring tube
量热法	熱量測定法	calorimetry
量热计	卡計, 熱量計	calorimeter
量酸槽	量酸槽	acid measuring tank
量纲不变性	因次不變性	dimensional invariance
量纲分析, 因次分析	因次分析	dimensional analysis
量纲均匀性	因次均勻性	dimensional homogeneity
量纲一致性	因次一致性	dimensional consistency
量子效应	量子效應	quantum effect
料仓松动器	料倉鬆動器	bin activator
料仓卸料器	倉式卸料器	bin discharger
料槽	槽	trough
料封	料封	material seal
料腿	料腿	dipleg
列管换热器	殼管熱交換器	shell-and-tube heat exchanger
列管式反应器	殼管反應器	shell-and-tube reactor
列线图	列線圖	alignment chart
裂变	分裂	fission
裂变产物	分裂產物	fission product
裂变能	分裂能	fission energy

祖国大陆名	台湾地区名	英 文 名
裂缝	裂縫	fissure
裂化	裂解, 裂煉	cracking
裂化催化剂	裂解觸媒	cracking catalyst
裂化度	裂解度	cracking severity
裂化反应	裂解反應	cracking reaction
裂化分馏塔	裂解分餾塔	cracking fractionator
裂化活性	裂解活性	cracking activity
裂化炉	裂解爐	cracking furnace
裂化气	裂解氣	cracked gas
裂化汽油	裂解汽油	cracked gasoline
裂化室	裂解室	cracking chamber
裂化蒸馏	裂解蒸餾	cracking distillation
裂解活性	裂解活性	activity cracking
裂膜纱	裂膜紗	splitting yarn
林产化学品	林材化學品	silvichemical
临界比表面	臨界比表面	critical specific surface
临界比容	臨界比容	critical specific volume
临界常数	臨界常數	critical constant
临界尺寸	臨界尺寸	critical size
临界等温线	臨界等溫線	critical isotherm
临界点	臨界點	critical point
临界点轨迹	臨界點軌跡	locus of critical point
临界反压比	臨界反壓比	critical back pressure ratio
临界共溶温度	臨界共溶溫度, 臨界溶解溫度	consolute temperature, critical solution temperature
临界喉道流速	臨界喉道流速	critical throat velocity
临界混合温度	臨界混合溫度	critical mixing temperature
临界胶束浓度	臨界微胞濃度	critical micelle concentration
临界晶核尺寸	臨界晶核大小	critical nucleus size
临界流	臨界流動	critical flow
临界流动量	臨界流動量	critical flow capacity
临界密度	臨界密度	critical density
临界摩尔体积	臨界莫耳體積	critical molar volume
临界凝析温度	臨界凝固溫度	cricondentherm
临界凝析压力	臨界凝固壓	cricondenbar
临界浓度	臨界濃度	critical concentration
临界频率	臨界頻率	critical frequency
临界溶度	臨界溶解度	critical solubility
临界溶解点	臨界溶解點	critical solution point

祖国大陆名	台湾地区名	英 文 名
临界深度	臨界深度	critical depth
临界湿度	臨界濕度	critical humidity
临界湿含量	臨界含水量	critical moisture content
临界时间	臨界時間	critical time
临界势	臨界勢	critical potential
临界速度	臨界速度	critical velocity
临界速率	臨界速率	critical speed
临界体积	臨界體積	critical volume
临界填料大小	臨界填充大小	critical packing size
临界途径	臨界途徑	critical path
临界温差	臨界溫差	critical temperature difference
临界温度	臨界溫度	critical temperature
临界现象	臨界途徑法	critical phenomenon
临界斜率	臨界斜率	critical slope
临界泄压器件	臨界洩壓性質	critical pressure relief device
临界性质	臨界性質	critical property
临界压力	臨界壓力	critical pressure
临界压力比	臨界壓力比	critical pressure ratio
临界压缩比	臨界壓縮比	critical compressibility ratio
临界压缩因子	臨界壓縮因數	critical compressibility factor
临界音速性质	臨界音速性質	critical sonic property
临界增益	臨界增益	critical gain
临界直径	臨界直徑	critical diameter
临界值	臨界值	critical value
临界质量	臨界質量	critical mass
临界转速	臨界轉速	critical revolution
临界状态	臨界條件	critical condition
临界阻尼	臨界阻尼	critical damping
淋粒反应器	淋粒反應器	raining solid reactor
磷肥	磷肥	phosphate fertilizer
磷光	磷光	phosphorescence
磷硅酸盐玻璃	磷矽酸鹽玻璃	phosphosilicate glass
磷灰石	磷灰石	apatite
磷酸化作用	磷酸化[反應]	phosphorylation
磷酸己糖旁路途径	磷酸己糖旁路途徑	hexose phosphate shunt pathway
磷酸盐玻璃	磷酸鹽玻璃	phosphate glass
磷酸盐缓冲剂	磷酸鹽緩衝劑	phosphate buffer
磷酸盐岩	磷岩	phosphate rock
磷酸盐增效助剂	磷酸鹽補助劑	phosphate builder

祖国大陆名	台湾地区名	英 文 名
磷铁	磷鐵齊	ferrophophorus
磷脂	磷脂	phospholipid
灵敏度	靈敏度	sensitivity
灵敏度分析	靈敏度分析	sensitivity analysis
灵敏度估计	靈敏度估計	sensitivity estimate
菱镁矿	菱鎂礦	magnesite
菱铁矿	菱鐵礦	siderite
菱锌矿	菱鋅礦	zinc spar
零点平衡	零點均衡	null balance
零电位计	零點電位計	null potentiometer
零级反应	零階反應	zero-order reaction
零记忆系统	零記憶系統	zero memory system
零假设	零假設	null hypothesis
零频率增益	零頻率增益	zero frequency gain
零通量表面	零流通量表面	zero-flux surface
零压状态	零壓狀態	zero-pressure state
流变测定法	流變測定法	rheometry
流变破坏	流變破壞	rheodestruction
流变性质	流變性質	rheological property
流变学	流變學	rheology
流变仪	流變儀	rheometer
流程模拟	流程模擬	flowsheeting
流程图	流程圖	flowsheet, flow diagram
流程图符号	流程圖符號	flow diagram symbol, flowsheet symbol
流程图设计	流程圖設計	flowsheet design
流程综合	流程綜合	flowsheet synthesis
流出	流出	effluence
流出管	①放流管②出料管	effluent pipe
流出式黏度计	射流黏度計	efflux viscometer
流出物	流出物	effluent
流[动]	流動	flow
流动比	流動比例	current ratio
流动法	流動法	flow method
流动反应器	流動反應器, 移動反應器	flow reactor, mobile reactor
流动功	流動功	flow work
流动混合器	流動混合器	flow mixer
流动量热计	流動卡計	flow calorimeter
流动落后	流動落後	flow lag

祖国大陆名	台湾地区名	英 文 名
流动双折射	流動雙折射	streaming birefringence
流动网	流動網	flow net
流动系统	流動系統	flow system
流动显示	流動觀測	flow visualization
流动线	流動線	flow line
流动相	移動相	mobile phase
流动行为指数	流動行為指數	flow behavior index
流动资产	流動資產	current asset
流动资金	營運資金	working capital
流动阻力	流動阻力	flow resistance
流股匹配	流股匹配	matching of streams
流函数	流線函數	stream function
流化焙烧炉	流體化焙燒爐	fluidized roaster
流化床	流體床, 流體化床	fluid bed, fluidized bed
流化床反应器	流體床反應器, 流體化床反应器	fluid-bed reactor, fluidized bed reactor
流化床干燥器	流體化床乾燥器	fluidized bed dryer
流化床锅炉	流體化床鍋爐	fluidized bed boiler
流化床气化器	流體化床氣化器	fluidized bed gasifier
流化床燃烧	流體化床燃燒	fluidized bed combustion
流化床燃烧器	流體化床燃燒器	fluidized bed combustor
流化床吸附器	流體化床吸附器	fluidized bed adsorber
流化床转化	流體床轉化	fluid-bed conversion
流化催化剂床	流體化觸媒床	fluidized catalyst bed
流化反应器	流體化反應器	fluidized reactor
流化混合	流體化混合	fluidized mixing
流化数	流體化數	fluidization number
流化速度	流體化速度	fluidizing velocity
流化蒸馏	流體化蒸餾	fluidized distillation
流加培养	流加培養	fed batch culture
流浸膏剂	液體萃取物	liquid extract
流颈, 缩脉	收縮口	vena contracta
流颈接头	收縮口壓力接頭	vena contracta tap
流控技术	流控學	fluidics
流量比例调节	比值流量控制	ratio-flow control
流量测量设备	流量測量裝置	equipment flow measuring
流量测量装置	流量量測裝置, 流血測量裝置	flow measuring equipment
流量积分	流量積分	flow integration

祖 国 大 陆 名	台 湾 地 区 名	英 文 名
流量积分器	流量積分器	flow integrator
流量计	流量計	flow gage, flowmeter
流量控制	流量控制	flow control
流量喷嘴	流動噴嘴	flow nozzle
流量系数	排放係數	discharge coefficient
流量指示计	流量指示計	flow indicator
流率	①流[動]率②流量	flow rate
流能磨	流體能量研磨機	fluid energy mill
流入物	流入物	influent
流水作业	流動程序	flow process
流速计	流速計	current meter
流态化	流體化	fluidization
流体	流體	fluid
流体动力边界层	流體動力邊界層	hydrodynamic boundary layer
流体动力稳定性	流體動力穩定性	hydrodynamic stability
流体动力相互作用	流體動力相互作用	hydrodynamic interaction
流体动力效应	流體動力效應	hydrodynamic effect
流体动力学	流體動力學	fluid dynamics, hydrodynamics
流体高差	流體高差	fluid head
流体-固体反应	流體-固體反應	fluid-solid reaction
流体静力	液體靜力	hydrostatic force
流体静力计	液體靜力計	hydrostatic force meter
流体静压条件	液體靜壓狀態	hydrostatic condition
流体力学	流體力學	fluid mechanics, hydromechanics
流体力学平滑表面	流體動力平滑表面	hydrodynamically smooth surface
流体流动	流體流動	fluid flow
流体流动水力半径	流體流動水力半徑	hydraulic radius for fluid flow
流体流动阻力	流體流動阻力	fluid flow resistance
流体-流体反应	流體-流體反應	fluid-fluid reaction
流体摩擦	流體摩擦	fluid friction
流体速度	流體速度	fluid velocity
流体相	流體相	fluid phase
流体运动学	流體運動學	fluid kinematics, hydrokinematics
流线	流線	streamline
流线截取	流線截取	flow-line interception
流线式过滤器	流線式濾器	streamlined filter
流线型阀	流線形閥	streamlined valve
流线型化	流線化	streamlining
流线型物体	流線形物體	streamlined body

祖国大陆名	台湾地区名	英 文 名
流线运动	流線運動	streamline motion
流线坐标	流線坐標	streamline coordinates
流型	流動型式	flow pattern
硫	硫	sulfur
硫醇	硫醇	mercaptan, thiol
硫醇改性橡胶	硫醇改質橡膠	thiolmodified rubber
硫代水解	硫化氫解[反應]	thiohydrolysis
硫靛蓝	硫靛藍	thioindigo
硫化	硫化[反應], 橡膠硫化反應	sulfurization, vulcanization
硫化度	硫化度	sulfidity
硫化锅	硫化鍋	vulcanizing pan
硫化过程	加硫反應	sulfidation
硫化剂	硫化劑	vulcanizing agent
硫化加速剂	硫化催速劑	accelerator of vulcanization
硫化器	硫化器	vulcanizer
硫化染料	硫化染料	sulfur dye
硫化室	硫化室	vulcanizing chamber
硫化系数	硫化係數	vulcanization coefficient
硫化纤维纸	硬化紙	vulcanized fiber paper
硫化橡胶	硫化橡膠	vulcanizate
硫化油	硫化油	vulcanized oil
硫化状态	熟化狀態	state of cure
硫[黄]硫化	硫[黃]硫化	sulfur vulcanization
硫交联	硫交聯	sulfur crosslinking
硫链丝菌肽	硫鏈絲菌	thiostrepton
硫脲树脂	硫脲樹脂	thiourea resin
硫漂白	硫[黃]漂白	sulfur bleach
硫氢解反应	硫氫解[反應]	thiohydrogenolysis
硫砷铜矿	硫砷銅礦	enargite
硫酸法	硫酸法	sulfuric acid process
硫酸化	硫酸化	sulfating
硫酸化油	硫酸化油	sulfated oil
硫酸盐法	牛皮紙漿法, 硫酸鹽法	kraft process
硫酸盐化	硫酸化[反應]	sulfation
硫酸盐浆	牛皮紙漿	kraft pulp
硫酸盐纸浆	硫酸鹽紙漿, 牛皮紙漿	sulfate pulp
硫酸盐制纸浆法	硫酸鹽法, 牛皮紙漿法	sulfate process
硫酸酯	硫酸酯	sulfuric acid ester

祖国大陆名	台湾地区名	英　文　名
馏程	蒸餾範圍	distillation range, distilling range
馏出液	餾出物	distillate
六方最密堆积晶格	六方最密晶格	hexagonal closest packed lattice
六聚物	六聚物	sexamer
笼合物	晶籠化合物	clathrate
笼合[作用]	晶籠[作用]	clathration
笼式水压机	籠式水壓機	cage hydraulic press
笼效应	籠子效應	cage effect
漏气	漏氣	air false
漏液	滴流	weeping
漏液孔	滴流孔	weeping hole
露点	露點	dew point
露点计	露點計	dew point meter
露点记录器	露點記錄器	dew point recorder
露点温度	露點溫度	dew point temperature
露点压力	露點壓力	dew point pressure
炉缸	爐	hearth
炉黑	爐黑	furnace black
炉气	焦爐氣	oven gas
炉渣	熔渣	slag
炉渣水泥	熔渣水泥	slag cement
卤化	鹵化[反應]	halogenation
鲁棒过程控制	韌性程序控制	robust process control
鲁棒稳定化	韌性穩定化	robust stabilization
鲁棒稳定化度	韌性穩定化度	robust stabilizability
陆地废水处理	陸地[廢水]處理	landfarming
滤板	濾板	filter plate
滤饼	濾餅	cake, filter cake, mud cake
滤饼运输机	濾泥輸送機	cake conveyor
滤布	濾布	filter cloth
滤池深度	濾池深度	filter depth
滤袋	濾袋	filter bag
滤镜反光计	濾鏡反光計	mirror filter reflectometer
滤菌器	濾菌器	bacteria filter
滤框	濾框	filter frame
滤膜	濾膜	filtration membrane
滤泥机	濾泥機	mud press
滤砂	濾砂	filter sand
滤桶	濾桶	filter drum

祖国大陆名	台湾地区名	英 文 名
滤箱	濾箱	filter box
滤液	濾液	filter liquor
滤纸	濾紙	filter paper
路径函数	路徑函數	path function
路径追踪	路徑追蹤	path tracing
辘轳	拉坯盤車	jigger
铝箔	鋁箔	aluminum foil
铝管	鋁管	aluminum pipe
铝媒染剂	鋁媒染劑	aluminum mordant
铝热剂	鋁熱劑	thermite
铝砂	鋁砂	aloxite
铝土矿	鋁礬土	bauxite
铝皂	鋁皂	aluminum soap
铝质黏土	鋁氧黏土	bauxitic clay
绿脱石	矽鐵石	nontronite
绿柱石	鈹土	beryl
氯胺	氯胺	chloramine
氯丁橡胶	新平[橡膠], 氯平橡膠	neoprene rubber, neoprene, chloroprene rubber
氯丁橡胶黏合剂	新平黏合劑	neoprene adhesive
氯化	氯化[反應]	chlorination
氯化苯醌	氯化[苯]醌	chlorinated quinone
氯化酚	氯化酚	chlorinated phenol
氯化钾	氯化鉀	muriate of potash
氯化硫溶液硫化	氯化硫溶液硫化	sulfur chloride vulcanization
氯化石蜡	氯[化]蠟	chlorinated paraffin
氯化烃	氯[化]烴	chlorinated hydrocarbon
氯化物接触室	氯化物接觸室	chloride contact chamber
氯化橡胶	氯化橡膠	chlorinated rubber
氯磺化聚乙烯	氯磺化聚乙烯	chlorosulfonated polyethylene, chlorosulfonic polyethylene
氯磺酰化	氯磺酸化[反應]	chlorosulfonation
氯霉素	氯黴素	chloromycetin
氯乙烯单体	氯乙烯單體	vinyl chloride monomer
卵磷脂	蛋黃態	lecithin
卵石	卵石	pebble
卵石层	卵石床	pebble bed
卵石加热器	卵石加热器	pebble heater
轮虫	輪虫	rotifer

祖国大陆名	台湾地区名	英 文 名
轮碾机	輪輾機	edge runner
轮胎	輪胎	tire
轮胎帘线	輪胎簾布	tire cord
轮胎砂	輪胎砂	tire yarn
罗茨鼓风机	羅次鼓風機	Roots blower
螺带混合机	螺帶混合機	ribbon mixer
螺带搅拌器	螺帶攪拌器	helical ribbon agitator
螺杆	螺桿	screw
螺杆泵	螺旋泵	screw pump
螺杆出料机	螺桿出料機	screw extractor
螺杆挤出反应器	螺桿擠出反應器	screw extruder reactor
螺杆式预塑化注压成型机	螺桿式預塑化成型機	screw preplasticating type injection molding machine
螺杆式注塑成型机	螺桿式注塑成型機	screw type injection molding machine
螺杆效率	螺桿效率	screw efficiency
螺杆芯孔销	螺桿蕊栓	screw core pin
螺杆压出机	螺桿擠出機	screw extruder
螺栓	螺栓	bolt
螺栓和螺帽	螺栓及螺帽	bolt and nut
螺线圈填料, 芬斯克填料	芬斯基填料	Fenske packing
螺型位错	螺旋型差排	screw dislocation
螺旋	螺旋	spiral
螺旋板换热器	螺旋板熱交換器	spiral plate heat exchanger
螺旋板式冷凝器	螺旋冷凝器	spiral condenser
螺旋齿轮泵	螺旋齒輪泵	spiral gear pump
螺旋翅片	螺旋鰭片	helical fin
螺旋管	螺旋管	helical coil
螺旋管加热器	旋管加熱器	spiral coil heater
螺旋滑线	螺旋滑線	helical slidewire
螺旋环	螺旋環	spiral ring
螺旋挤出	螺旋擠出	screw extrusion
螺旋挤出机	螺旋擠製機	auger machine
螺旋桨	螺旋槳	propeller
螺旋桨风扇	螺旋槳風扇	propeller type fan
螺旋桨容量	推進器容量	propeller capacity
螺旋桨式搅拌器	螺槳攪拌器, 螺旋槳混合器	propeller agitator, propeller mixer
螺旋桨式叶轮	螺旋式葉輪	propeller impeller

祖国大陆名	台湾地区名	英　文　名
螺旋搅拌器	螺旋攪拌器	helical agitator
螺旋结构	螺旋結構, 螺桿結構	screw structure
螺旋聚合物	螺旋聚合物	spiropolymer
螺旋流	螺旋流動	spiral flow
螺旋形分离器	螺旋分離器	spiral separator
螺旋形高分子	螺旋形高分子	spiral polymer
螺旋形结构	螺旋形結構	spiral structure
螺旋压力弹簧	螺旋壓力彈簧	helical pressure spring, spiral pressure spring
螺旋运输机	螺旋運送機	spiral conveyor
螺状相	膽固醇型液晶相	cholesteric phase
裸管	裸管	bare pipe, bare tube
裸面	裸面	bare surface
裸热电偶	裸熱電偶	bare thermocouple
络合催化剂	配位觸媒	coordination catalyst
络合滴定	錯離子滴定法	complexometric titration
络合物平衡	錯合物均衡	complex balancing
落后元件	落後元件	lag element
落角	落角	angle of fall
落球法	落球法	falling ball method
落球黏度计	落球黏度計	falling ball viscometer, falling sphere viscometer
落水管	下向流管	downcomer
落体黏度计		falling body viscometer
落针黏度计	落針黏度計	falling needle viscometer

M

祖国大陆名	台湾地区名	英　文　名
麻点	麻點, 凹痕	pitting
麻油	木麻油	hemp seed oil
麻醉剂	麻醉劑	anesthetic
马丁-侯[虞钧]方程	馬丁-侯[虞鈞]方程	Martin Hou equation[of state]
马赫数	馬赫數	Mach number
马居尔方程	馬居爾方程	Margules equation
马拉硫磷	馬拉松[農藥]	malathion
马力	馬力	horsepower (hp)
麦克斯韦关系	馬克斯威爾關係	Maxwell relation

祖国大陆名	台湾地区名	英 文 名
麦芽	麥芽	malt
麦芽糖	麥芽糖	maltose
麦芽糖酶	麥芽酶	maltase
麦芽汁	麥芽汁, 醪	wort, mash
脉冲	脈動	pulse, impulse
脉冲法	脈動法	pulse method
脉冲反应器	脈動反應器	pulse reactor
脉冲函数	脈衝函數	impulse function, pulse function
脉冲搅动	脈衝擾動	impulse disturbance
脉冲流	脈動流動法	pulsing flow
脉冲流技术	脈動流動法	pulse flow technique
脉冲扰动	脈動擾動	pulse disturbance
脉冲筛板塔	脈沖篩板塔	pulsed sieve plate column
脉冲响应	脈衝應答, 脈沖響應	impulse response, pulse response
脉冲型	轉速計	impulse type
脉冲转速计	脈衝轉速計	impulse tachometer
脉动	脈動	pulsation
脉动流化床	脈動流化床	pulsating fluidized bed
脉动输入	脈動輸入	pulse input
脉动速度	脈動速度	pulsation velocity
脉动阻尼器	脈動阻尼器	pulsation damper
莽草酸	莽草酸	shikimic acid
毛刺	毛邊	spew
毛法湿度计	毛法濕度計	hair hygrometer
毛管力	毛細力	capillary force
毛利成本	毛利成本	gross earnings cost
毛细测液器	毛細計	capillarimeter
毛细穿透	毛細穿透	capillary penetration
毛细管	毛細管	capillary, capillary tube
毛细管法	毛細管法	capillary tube method
毛细管膜常数	毛細常數	capillary constant
毛细管膜组件	毛細管模組	capillary module
毛细管黏度计	毛細[管]黏度計	capillary viscometer
毛细管型球	毛細管式球	capillary-type bulb
毛细管准数	毛細數	capillary number
毛细冷凝	毛細冷凝	capillary condensation
毛细瓶	毛細瓶	capillary flask
毛细上升	毛細上升	capillary ascent, capillary elevation
毛细吸附	毛細吸附	capillary adsorption

祖国大陆名	台湾地区名	英 文 名
毛细吸收	毛細吸收	capillary absorption
毛细吸引	毛細吸引	capillary attraction
毛细下降	毛細下降	capillary depression
毛细效应	毛細效應	capillary effect
毛细压力	毛細壓力	capillary pressure
毛细张力	毛細張力	capillary tension
毛纤维	毛纖維	wool fiber
锚式搅拌器	錨式攪拌器	anchor agitator
枚举法	列舉法	enumeration algorithm
梅仁油	梅仁油	plum kernel oil
媒染法	媒染法	mordanting process
媒染剂	媒染劑	mordant
媒染偶氮染料	媒染偶氮染料	mordant azo dye
媒染染料	媒染染料	mordant dye
媒染色料	媒染色料	mordant color
媒染助剂	媒染助劑	mordant assistant
煤床	煤床	coal bed
煤脆性	煤脆性	coal friability
煤的脱挥发作用	煤之去揮發作用	coal devolatilization
煤矸石	礦渣, 礦石	gangue
煤加工技术	煤處理技術	coal processing technology
煤焦油	煤潜, 煤焦油	coal tar
煤孔隙度	煤孔隙度	coal porosity
煤气	煤礦氣	coal gas
煤气表	氣量計	gas meter
煤气发生炉	發生爐	gas producer, producer
煤气化	煤氣化[法]	coal gasification
煤气化过程	煤氣化程序	coal gasification process
煤气化炉	煤氣化器	coal gasifier
煤气喷灯	煤氣爐, 煤氣燈	gas burner
煤热解	煤之高溫分解	coal pyrolysis
煤田	煤田	coal field
煤屑	碎煤	slack coal
煤衍生气	煤衍生氣	coal-derived gas
煤衍生液[体]	煤衍生液[體]	coal-derived liquid
煤液化	煤液化[法]	coal liquefaction
煤液化过程	煤液化程序	coal liquefaction process
煤油	煤油	kerosene, kerosine
煤油裂解	煤油裂解	kerosene cracking

祖国大陆名	台湾地区名	英　文　名
煤油燃料	煤油燃料	kerosene fuel
煤油乳剂	煤油乳劑	kerosene emulsion
煤油脱脂	煤油脱脂	kerosene degreasing
煤蒸馏	煤蒸餾	coal distillation
煤转化	煤轉化	coal conversion
酶	酶, 酵素	enzyme
酶半衰期	酶半衰期	half life of enzyme
酶产物复合物	酶產物複合物	enzyme product complex
酶促电催化	酶電催化	enzymatic electrocatalysis
酶促反应	酶反應	enzymatic reaction
酶催化	酶催化	enzyme catalysis
酶催化反应	酶催化反應	enzyme catalyzed reaction
酶底物反应	酶受質複合物反應	enzyme-substrate reaction
酶-底物复合物	酶-底物複合物	enzyme-substrate complex
酶电极	酶電極	enzyme electrode
酶发酵	酶發酵	enzyme fermentation
酶法	酶試驗法	enzymatic method
酶法分析	酶分析法	enzymatic analysis
酶法水解	酶法水解	enzymatic hydrolysis
酶反应动力学	酶動力學	enzymatic reaction kinetics
酶活力	酶活性	enzyme activity
酶联免疫吸附测定	酶聯免疫吸附測定	enzyme-linked immunosorbent assay
酶免疫分析法	酶免疫分析法	enzyme immunoassay
酶膜	酶膜	enzyme membrane
酶失活	酶去活化	enzyme deactivation
酶稳定	酶穩定化	enzyme stabilization
酶选择性	酶選擇性	enzyme selectivity
[酶]诱导契合学说	[酶]誘導契合學說	induced fit theory
酶专一性	酶專一性	enzyme specificity
霉菌	霉菌	mold, mould
霉菌淀粉酶	霉菌澱粉酶	mold amylase
每分钟转数	每分轉數	revolution per minute (rpm)
美国国家标准协会	美國國家標準協會	American Standards Association (ASA)
美国化学工程师学会	美國化學工程師學會	American Institute of Chemical Engineers (AIChE)
美国石油学会	美國石油協會	American Petroleum Institute (API)
镁橄榄石	矽酸鎂石	forsterite
镁橄榄石瓷	矽酸鎂石瓷	forsterite porcelain
镁橄榄石砖	矽酸鎂石磚	forsterite brick

祖国大陆名	台湾地区名	英 文 名
镁铬砖	鎂鉻磚	magnesite-chrome brick
镁质耐火材料	鎂氧耐火物	magnesite refractory
镁砖	鎂磚	magnesia brick, magnesite brick
蒙特卡罗模拟	蒙地卡羅模擬	Monte Carlo simulation
蒙脱石	微晶高嶺石	montmorillonite
弥散流(=分散流)		
弥散模型	分散模型	dispersion model
弥散系数	分散係數	dispersion coefficient
醚	醚	ether
醚化	醚化[反應]	etherification
醚数	醚值	ether number
醚值	醚值	ether value
米糠油	米糠油	rice oil
米氏常数	米氏常數	Michaelis Menton constant
米氏动力学	米氏動力學	Michaelis Menton kinetics
米氏方程	米氏方程	Michaelis Menton equation
密闭反应器	密閉反應器	closed reactor
密度	密度	density
密度计	密度計	densimeter
密度梯度离心	密度梯度離心	density gradient centrifugation
密度指数	密度指數	density index
密封	密封	seal
密封革	墊皮	packing leather
密封管	密閉管	sealed tube
密封罐	密封罐	seal pot
密封胶	密封膠	seal gum, sealant, sealing compound
密封胶合剂	密封膠合劑	seal cement
密封圈	墊圈	packing ring
密封腿[料腿]	封柱	seal leg
密封性能	密閉性	sealing property
密封液	密封液	sealing liquid
密封油	密封油	sealing oil
密孔板	多孔板	porous plate
密炼机	密閉混合器	internal mixer
密媒分离	濃媒分離	dense medium separation
密相, 浓相	稠相	dense-phase
密相流动	稠相流動	dense-phase flow
密相流化床	稠相流體化床	dense-phase fluidized bed
幂函数型方程	冪函數型方程	power function type equation

祖国大陆名	台湾地区名	英 文 名
幂律流体	幂次律流體	power-law fluid
蜜胺树脂	三聚氰胺樹脂	melamine resin
棉纸	棉紙	cotton paper
棉状纱	棉狀紗	staple yarn
棉子油	棉子油	cotton-seed oil
免疫电泳	免疫電泳	immune eletrophoresis
免疫吸附	免疫吸附	immunoadsorption
面积计	面積計	area meter
面积流量计	面積流量計	area flowmeter, area type flowmeter
面积平均方程式	面積平均方程式	area-averaged equation
面筋	麩質, 麵筋	gluten
面团混合机	調麵機	dough mixer
面心晶格	面心晶格	free-centered lattice
描述函数	描述函數	describing function
描图纸	描圖紙	tracing paper
灭火器	滅火器	fire extinguisher
灭菌	滅菌	sterilization
灭菌器	滅菌器	sterilizer
pH 敏感电极	pH 敏感電極	pH-sensitive electrode
敏感元件	感測元件	sensing element
敏化剂	敏化劑	sensitizer
名义应力	標稱應力	nominal stress
明槽	明渠	open channel
明槽流	明渠流	open channel flow
明矾	明礬, 礬	alum
[明]矾鞣[革]	明礬鞣制	alum tanning
明火加热炉	加熱爐	fired furnace
明火加热器	加熱器	fired heater
明胶	明膠	gelatin
明胶蛋白	明膠蛋白	glutin
明渠	明渠	channel open
命名法	命名[法]	nomenclature
模	模	die
模块	模組, 組件	module
模量	模數	modulus
模量比	模數比	modulus ratio
模拟, 仿真	模擬, 模型化	simulation
模拟仿真	類比模擬	analog simulation
模拟计算	類比計算[法]	analog computation

祖国大陆名	台湾地区名	英　文　名
模拟计算机	類比計算機	analog computer
模拟模型	模擬模式,模擬模型	simulation model
模拟器	模擬器	simulator
模式	型式,方式	pattern
模式识别	模式識別	pattern recognition
模式搜索	模式搜索	pattern search
模塑	①模製②成型	molding
模型	①模型②模式	model
模型比例	模型比例	model scale
模型辨识	模型辨識	model identification
模型参数	模型參數	model parameter
模型建立	模式建立	model building
模型试验	模型試驗	model test
模型试验(=台架试验)		
模造纸	模造紙	simile paper
膜	薄膜	membrane
膜泵	薄膜泵	membrane pump
膜萃取	薄膜萃取	membrane extraction
膜反应器	薄膜反應器	membrane reactor
膜簧式驱动器	膜簧式引動器	diaphragm-spring actuator
膜技术	薄膜技術	membrane technology
膜酵母	膜酵母	film yeast
膜扩散	薄膜擴散	film diffusion
膜扩散控制	薄膜擴散控制	film diffusion control
膜理论	薄膜理論	film theory
膜裂纤维	裂散纖維	split fibre
膜滤器	薄膜過濾器	membrane filter
膜囊	薄膜囊	membrane vesicle
膜强度	薄膜強度	film strength, membrane strength
膜筛	隔膜篩	diaphragm screen
膜渗透	薄膜穿透	membrane permeation
膜生物反应器	薄膜生物反應器	membrane bioreactor
膜式冷凝	薄膜冷凝	film type condensation
膜式洗涤器	膜式洗滌器	film scrubber
膜式压力计	隔膜壓力計	diaphragm pressure gage
膜式蒸发器	薄膜蒸發器	film type evaporator
膜试验	薄膜試驗	film test
膜通透性	薄膜透過性	membrane permeability
膜温度	薄膜溫度	film temperature

祖国大陆名	台湾地区名	英　文　名
膜系数	薄膜係數	film coefficient
膜相	膜相	film phase
膜蒸馏	薄膜蒸餾	membrane distillation
膜支撑物	薄膜支撑物	membrane support
膜状沸腾	薄膜沸騰	film boiling
膜状冷凝	薄膜冷凝	film condensation, filmwise condensation
膜阻力	薄膜阻力	film resistance
膜组件	薄膜組件	membrane module
摩擦	摩擦	friction
摩擦分离器	摩擦分離器	friction separator
摩擦腐蚀	摩擦腐蝕	friction corrosion
摩擦接头	摩擦接頭	friction coupling
摩擦试验	摩擦試驗	friction test
摩擦损失	摩擦損失	friction loss
摩擦损失因子	摩擦損失因數	friction loss factor
摩擦系数	摩擦係數	friction coefficient
摩擦压力降	摩擦壓力降	friction drop
摩擦压头	摩擦高差	friction head
摩擦因子	摩擦因數	friction factor
摩擦阻力	摩擦阻力	frictional resistance, friction drag
摩尔衡算	莫耳均衡	mole balance
摩尔浓度	莫耳濃度	molar concentration
摩尔平均沸点	重量莫耳平均沸點, 莫耳平均沸點	molal average boiling point, molar average boiling point
摩尔平均扩散系数	莫耳平均擴散係數	molar average diffusivity
摩尔平均速度	莫耳平均速度	molar average velocity
摩尔热容	莫耳熱容量	molar heat capacity
摩尔溶液	重量莫耳溶液, 莫耳溶液	molal solution, molar solution
摩尔湿度	莫耳溼度	molar humidity
摩尔体积	莫耳體積	molar volume
摩尔通量	莫耳通量	molar flux
磨床	磨床	grinder, grinding machine
磨光	磨光	polishing
磨耗	磨耗	detrition
磨耗度	抗磨性	abrasiveness
磨耗试验	磨耗試驗	abrasion test
磨耗指数	磨耗指數	abrasion index
磨料	磨[擦]料, 磨擦物	abrasive

祖国大陆名	台湾地区名	英　文　名
磨木浆	磨木漿	groundwood pulp
磨球	磨球	grinding ball
磨砂玻璃	磨砂玻璃	frosted glass
磨蚀	浸蝕	erosion
末端效应	端點效應	end effect
莫来石	富鋁紅柱石	mullite
莫来石瓷	富鋁紅柱石瓷	mullite porcelain
莫来石耐火材料	富鋁紅柱石耐火物	mullite refractory
莫利尔图	莫利爾圖	Mollier diagram
莫诺生长动力学	莫諾生長動力學	Monod growth kinetics
默弗里效率	默弗里效率	Murphree efficiency
模压	模[製]壓[縮]	molding compression
模压成分	模製配料	molding composition
母体核素	母核種	parent nuclide
母体元素	母元素	parent element
母细胞	母細胞	mother cell, parent cell
母液	母液	mother liquor
木材防腐剂	木材防腐劑	wood preservative
木材化学品	木材化学品	wood-derived chemical
木材热解	木材熱解	wood pyrolysis
木材糖化	木材糖化	wood saccharification
木材提取物	木材萃取物	wood extractive
木材蒸馏	木材蒸餾	wood distillation
木瓜蛋白酶	木瓜酶	papain
木浆除滓机	木漿除滓機	slab grating
木焦油	木溚	wood tar
木精	木精, 甲醇	wood spirit, wood alcohol
木煤气	木煤氣	wood gas
木棉	木棉子油	kapok oil
木片干燥器	木片乾燥器	chip drier
木薯淀粉	樹薯[澱粉]	cassava starch
木松香	松香	wood rosin
木素	木質素	lignin
木炭	木炭	charcoal
木糖	木糖	wood sugar, xylose
木酮糖	木酮糖	xylulose
木杂酚油	木雜酚油	wood creosote
木质化	木質化	lignification
木质磺酸	木質磺酸	lignin sulphonic acid

祖国大陆名	台湾地区名	英 文 名
木质酶	木質酶	ligninase
木质纤维素	木質纖維素	lignocellulose
木质纤维素无烟煤	木質纖維素無煙煤	lignocellulosic anthracite
目标管理	目標管理	objective management
目标函数	目標函數	objective function
目标控制	目標控制	objective control
钼催化剂	鉬觸媒	molybdenum catalyst

N

祖国大陆名	台湾地区名	英 文 名
纳米	奈米	nanometer
纳米材料	奈米材料	nanomaterials
纳米过滤	奈米過濾	nanofiltration
纳米技术[的]	奈米技術[的]	nanotechnical
纳米结构	奈米結構	nanostructure
纳米聚合材料	奈米聚合[物]材料	nano-plymer materials
纳米科学[与]技术	奈米科技	nanoscale science and technology
纳米粒子	奈米粒子	nanoparticle
纳米微晶	奈米晶體	nanosized crystals
纳米液滴	奈米液滴	nanodroplet
纳氏泵	納氏泵	Nash pump
钠玻璃	鈉玻璃	soda glass
钠长石	鈉長石	albite, soda feldspar
钠灯	鈉燈	sodium lamp
钠钙玻璃	鈉鈣玻璃	soda-lime glass
钠钙长石	鈉鈣長石	soda-lime feldspar
钠[引发]聚合作用	鈉[引發]聚合作用	sodium polymerization
钠[硬]皂	鈉皂	soda soap
耐电弧性	耐電弧性	arc resistance
耐光作用	光穩定化	photostabilization
耐候性试验	風化試驗, 耐候性試驗	weathering test
耐火表面	耐火表面	refractory surface
耐火材料	耐火物	refractory
耐火瓷	耐火瓷	refractory porcelain
耐火黏土	耐火黏土	fireclay
耐火页岩	耐火頁岩	refractory shale
耐火砖	耐火磚, 耐火黏土磚	firebrick, fireclay brick, refractory brick

祖国大陆名	台湾地区名	英　文　名
[耐火砖]格子散裂试验	屏列剝落試驗	panel spalling test
耐硫酸水泥	抗硫酸鹽水泥	sulfate resistant cement
耐磨圈	耐磨圈	wearing ring
耐磨性	磨耗阻力	abrasion resistance
耐热性	耐熱性	heat resistance, thermal endurance
耐溶剂性	耐溶劑, 耐溶性	solvent resistance
耐水度	①耐水性②耐水度	water tolerance
耐酸板	耐酸板	acid-proof slab
耐酸合金	抗酸合金	acid resisting alloy
耐酸水泥	耐酸水泥	acid-proof cement
耐酸搪瓷	耐酸搪瓷	acid-proof enamel
耐酸物	抗酸劑	acid resistant
耐酸性	抗酸性	acid resistance
耐酸纸	耐酸紙	acid-proof paper
耐酸砖	耐酸磚	acid-proof brick
耐皂洗[色]牢度	耐臭性	soap fastness
耐皂性	耐皂性	soap resistance
萘酚染料	萘酚染料	naphthol dye
难流动[性]	難流動[性]	stiff flow
挠性盘	活動盤	flexitray
脑磷脂	腦磷脂	cephalin, kephalin
内表面	内表面	internal surface
内[部]构件	内[部]構件	internals
内部加热	内部加熱	internal heating
内部收益率	内部收益率	internal rate of return
内插	插值法, 内插法	interpolation
内阀	内閥	inner valve
内含性质(＝强度性质)		
内耗	内摩擦	internal friction
内回流	内回流	internal reflux
内回流控制	内回流控制	internal reflux control
内径	内徑	inside diameter
内聚功	内聚功	cohesion work
内聚力	内聚力	cohesion
内聚能密度	内聚能密度	cohesive density
内扩散	内擴散	internal diffusion
内流	内部流動	internal flow
内酶	内酶	endoenzyme
内磨擦角	内磨擦角	angle of internal friction

祖国大陆名	台湾地区名	英 文 名
内能	内能	internal energy
内燃	内燃	internal combustion
内燃机	内燃機	internal combustion engine
内扰	内在擾動	internal disturbance
内胎	内胎	inner tube
内吸附	内吸附	internal adsorption
内酰胺	内醯胺	lactam
内循环	内循環	internal recycle
内循环反应器	内循環反應器	internal recirculation reactor
内压力	内壓[力]	internal pressure
内在容量	内在容量	internal capacity
内酯	内酯	lactone
内坐标	内部座標	internal coordinates
能汇	能量匯座	energy sink
能级	能階	energy level
能量	能量	energy
能量本质方程式	能量本質方程式	energetic constitutive equation
能量储存	能量儲存	energy storage
能量传递	能量傳送	energy transfer
能量当量	能量當量	energy equivalent
能量方程	能量方程式	energy equation
能量分布	能量分布	energy distribution
能量分离剂	能量分離劑	energy separating agent
能量含量	能量含量	energy content
能量耗散	能量散逸	energy dissipation
能量衡算, 能量平衡	能量均衡	energy balance
能量级线	能量級線	energy grade line
能量集成	能量集成	energy integration
能量交换器	能量交換器	energy exchanger
能量流动	能量流	energy flow
能量密度	能量密度	energy density
能量面	表面能	energy surface
能量平衡(＝能量衡算)		
能量平衡方程式	能量均衡方程式	energy balance equation
能量生成曲线	能量生成曲線	energy generation curve
能量守恒	能量守恆	energy conservation
能量守恒原理	能量守恆原理	principle of conservation of energy
能量损耗	能量消耗	energy consumption
能量损失	能量損失	energy loss

祖国大陆名	台湾地区名	英 文 名
能量损失曲线	能量損失曲線	energy loss curve
能量通量	能量通量	energy flux
能量效率	能量效率	energy efficiency
能量学	能量學	energetics
能量障壁	能量障壁	energy barrier
能量转化器	①能量轉化器②能量轉換物	energy converter
能量转换	能量轉換	energy conversion, energy transformation
能量转换工程	能量轉換工程	energy conversion engineering
能量状态方程	能量狀態方程式	energetic equation of state
能谱	能譜	energy spectrum
能态	能量狀態	energy state
能源	能源	energy source
能源工程	能源工程	energy engineering
能源技术	能量技術, 能源技術	energy technology
能源科学	能量科學	energy science
尼龙	耐綸, 耐隆, 尼龍	nylon
泥浆沉降器	泥漿沈降器	mud settler
泥炭	泥煤	peat
泥炭沼地	泥煤田	peat bog
泥釉	泥釉	slip glaze
拟均相模型	假均相模型	pseudohomogeneous model
拟线性化	似線性化	quasi-linearization
逆变换	逆變換	inverse transformation
逆反冷凝	降壓冷凝	retrograde condensation
逆反应	逆反應	reverse reaction
逆反作用	退減[作用]	retrogradation
逆分馏法	逆分餾[法]	reverse fractionation
逆扩散	逆向擴散	counter diffusion
逆流	逆流, 逆向流動	counterflow, countercurrent flow
逆流操作	逆流操作	countercurrent operation
逆流萃取	逆流萃取	countercurrent extraction
逆流反应器	逆流反應器	countercurrent flow reactor
逆流过程	逆流程序	countercurrent process
逆流喷射式冷凝器	逆流噴凝器	countercurrent jet condenser
逆流倾析	逆流傾析	countercurrent decantation
逆流塔过程	逆流塔程序	countercurrent column process
逆流网络	逆流網路	countercurrent network

祖 国 大 陆 名	台 湾 地 区 名	英 文 名
逆流吸收	逆流吸收	absorption countercurrent, countercurrent absorption
逆流洗涤	逆流洗滌	countercurrent washing
逆溶度曲级	反溶解度曲線	inversed solubility curve
逆向反应	逆反應	backward reaction
逆向进料	逆向進料	backward feed
逆压力梯度	反壓力梯度	adverse pressure gradient
年金表	年金表	annuity table
年龄分布	年齡分布	age distribution
年龄分布函数	年齡分布函數	age distribution function
黏度	黏度	viscosity
黏度测定法	黏度測定法	viscometry, viscosimetry
黏度掺和图	黏度摻配圖	viscosity blending chart
黏度计	黏度計	viscometer, viscosimeter
黏度平均分子量	黏度平均分子量	viscosity average molecular weight
黏度试验	黏度試驗	viscosity test
黏度数	黏度值	viscosity number
黏度系数	黏度係數	viscosity coefficient
黏度指数	黏度指數	viscosity index
黏附	附著, 黏著	adhesion
黏附功	附著功	adhesion work
黏附力	附著力	adhesion force
黏附能	附著能	adhesion energy
黏附强度	附著強度, 黏著強度	adhesion strength, adhesive strength
黏附润湿	附著性潤濕	adhesional wetting
黏附试验	黏著試驗	adhesion test, adhesive test
黏附张力	黏著張力	adhesive tension
黏合	結合, 鍵結, 黏結	bonding
黏合剂	黏結劑	binder, binding agent
黏合剂配方	黏著劑配方	adhesive formulation
黏合能力	結合能力	binder power
黏合强度	鏈結強度, 結合強度	bonding strength
黏合吸引	黏著吸引	adhesive attraction
黏胶人造短纤维	黏液嫘縈棉	viscose staple
黏胶人造丝	黏液嫘縈	viscose rayon
黏菌素	黏菌素	colistin
黏流态	可塑狀態	plastic state
黏泥	黏土	slime
黏塑性	黏塑性	viscoplasticity

祖国大陆名	台湾地区名	英 文 名
黏塑性流体	黏塑性流體	viscoplastic fluid
黏态	黏態, 黏性階段	sticky stage
黏弹性	黏彈性	viscoelasticity
黏弹性流体	黏彈性流體	viscoelastic fluid
黏土	黏土	clay
黏土砖		chamotte
黏性	黏性	stickiness
黏性次层	黏性次層	viscous sublayer
黏性耗散	黏性散逸	viscous dissipation
黏性减震器	黏性制振器	viscous damper
黏性力	黏性力	viscous force
黏性流动	黏性流動	viscous flow
黏性应力	黏性應力	viscous stress
黏性运动	黏性運動	viscous motion
黏性制振器	黏性制振器	dashpot
黏性阻力	黏性阻力	viscous resistance
黏滞发酵	黏液發酵	slimy fermentation
黏着空洞	附著空洞	clinging cavity
碾磨机	銼磨機, 磨碾機	attrition mill, mill
酿酶	解醣酶	zymase
酿造	釀造	brewing
酿造厂	釀造業	brewery
鸟粪磷酸盐	[海]鳥糞磷礦	guano phosphate
鸟嘌呤	鳥嘌呤	guanine
鸟嘌呤酶	鳥嘌呤酶	guanase
尿素	尿素	urea
尿烷	胺甲酸乙酯	urethane
脲酶	尿素酶	urease
捏合	捏揉	kneading
捏合机	捏合機, 捏揉機	incorporating machine, kneader
镍电阻球	鎳電阻球	nickel resistance bulb
镍坩埚	鎳坩堝	nickel crucible
镍钢	鎳鋼	nickel steel
镍铬合金	鎳鉻合金	nichrome
镍铝氧催化剂	鎳鋁氧觸媒	nickel alumina catalyst
镍蓄电池	鎳蓄電池	nickel storage battery
柠檬草油	檸檬草油	lemon-grass oil
凝并(＝聚并)		
凝固	凝固	solidification

祖国大陆名	台湾地区名	英　文　名
凝固点	凝固點, 冰點, 固化點	freezing point, solidification point, solidifying point
凝固点降低	凝固點下降, 冰點下降	freezing point depression, freezing point lowering
凝固过程	固化程序	solidification process
凝固时间	凝固時間, 固化時間,	set time
凝固温度	凝固溫度	setting temperature
凝华作用	反昇華	desublimation
凝胶	凝膠	gel
凝胶过滤	凝膠過濾	gel filtration
凝胶过滤色谱[法]	凝膠過濾層析法	gel filtration chromatography
凝胶免疫电泳	凝膠免疫電泳	gel immunoelectrophoresis
凝胶色谱[法]	凝膠色譜分析[法]	gel chromatography
凝结热	凝聚熱	heat of coagulation
凝结相	凝結相	condensed phase
凝聚槽	凝聚槽	coagulation tank
凝聚剂	凝聚劑	coagulant, coagulating agent
牛顿-拉弗森法	牛頓-拉弗森法	Newton Raphson method
牛顿流体	牛頓流體	Newtonian fluid
牛顿收敛法	牛頓收斂法	Newton method for convergence
牛皮纸	牛皮紙	kraft paper
牛脂	牛脂	tallow
扭辫分析	扭編分析	torsional braid analysis
扭力黏度计	扭力黏度計	torsion viscometer
扭力天平	扭力天平	torsion balance
扭曲度	偏斜度	skewness
扭转	扭轉, 扭力	torsion
扭转流动	扭轉流動	torsional flow
扭转蠕变	扭轉潛變	torsional creep
扭转应力	扭轉應力	torsional stress
农业废物	農業廢棄物	agricultural waste
浓差电池	濃度差電池	concentration cell
浓差极化	濃度極化	concentration polarization
浓度	濃度	concentration
浓度边界层	濃度邊界層	concentration boundary layer
浓度差	濃度差	concentration difference
浓度[分布]剖面[图]	濃度剖面圖	concentration profile
浓度极限	低限濃度	threshold concentration
浓度扩散	濃度擴散	concentration diffusion

祖国大陆名	台湾地区名	英 文 名
浓度速率曲线	濃度速率曲線	concentration-rate curve
浓度梯度	濃度梯度	concentration gradient
浓浆法	濃漿法	thick slurry process
浓硫酸	礬油, 硫酸	oil of vitriol
浓密机(=增稠器)		
浓溶液	濃溶液	rich solution
浓缩器	濃縮器	concentrator
浓缩塔	蒸發塔	evaporating column
浓缩液	濃縮物	concentrate
浓相(=密相)		
努塞特数	努塞特數	Nusselt number

O

祖国大陆名	台湾地区名	英 文 名
欧姆计	歐姆計	ohmmeter
偶氮成分	偶氮成分	azo component
偶氮染料	偶氮染料	azo dye, azoic dye
偶氮染色	偶氮染色	azoic dyeing
偶合	力偶, 偶	couple
偶合方程式	偶合方程式	coupled equations
偶合系统	偶合系統	coupled system
偶极	偶極	dipole
偶极矩	偶極矩	dipole moment
偶图	偶圖	bipartite graph

P

祖国大陆名	台湾地区名	英 文 名
帕(压力单位)	帕斯卡(壓力單位)	pascal
排程计算机控制	排程計算機控制	scheduling computer control
排出阀	排放閥	discharge valve
排出栓	排出栓	drawoff cock
排出压头	排放高差	discharge head
排代(=置换)		
排代泵(=容积式泵)		
排放标准	放流[水]標準, 排放標準	effluent standard, emission standard

祖 国 大 陆 名	台 湾 地 区 名	英 文 名
排放集管	排放集管[箱]	discharge header
排放物清单	排放物清單	emission inventory
排放压力	排放壓力	discharge pressure
排风机	風扇	fan
排管	排管	calandria
排管式蒸发罐	排管式蒸發罐	calandria pan
排管蒸发器(＝中央循环管蒸发器)		
排空		vent
排料	泄料, 排放	blowdown
排气泵	排氣泵	air displacement pump
排气阀	排氣閥, 釋氣閥	exhalation valve, air release valve, air relief valve
排气管	排氣管, 通風管	exhaust pipe, exit tube, vent pipe
排气管线	排收管線	vent line
排气机	排氣機	exhauster
排气及排泄口密封	通氣及排洩口密封	vent and drain seal
排气孔	通氣孔, 排氣孔	vent
排气锐孔	空氣排放孔口	air discharge orifice
排气栓	排氣栓	blow cock
排气系统	排氣系統	air displacement system, gas venting system
排汽	排放蒸汽	exhaust steam
排水管道	排洩管線	drain line
排水阱	排洩閘	drain trap
排水台	排水檯	draining table
排水系统	排水系統	underdrainage system
排泄管	排洩管	drain pipe
排序	排序	precedence ordering
派热克司硬质玻璃	派熱司玻璃	pyrex glass
盘滤机	盤濾機	disc filter
盘式过滤机	盤濾機	disk filter
盘式进料器	進料盤	feed disk
盘式离心机	盤式離心機	disc centrifuge
盘式流量计	盤式流量計	disc meter
盘式压碎机	盤碎機	disc crusher
盘式蒸发器	盤式蒸發器	disc evaporator
盘形膨胀管	盤形膨脹管	coil expansion pipe
盘柱吸收器	盤柱吸收器	disk column absorber
判据	準則	criterion

祖国大陆名	台湾地区名	英 文 名
旁路, 侧流	旁路, 分路	bypass
旁容量	旁容量	side capacitance
旁通比	旁通比	bypass ratio
旁通阀	旁通閥	bypass valve
旁通管	旁通管	bypass pipe
旁通过滤器	旁通過濾器	bypass filter
旁通控制	旁路控制	bypass control
旁通效应	旁通效應	bypassing effect
抛光	擦光	burnish
抛物线阀	抛物線閥	parabolic valve
抛物线溜槽	抛物線槽	parabolic flume
抛物线体聚光器	抛物線體聚光器	paraboloid condenser
泡点	起泡點, 初沸點	bubble point
泡点温度	起泡溫度	bubble point temperature
泡点压力	起泡壓力	bubble point pressure
泡核沸腾	成核沸騰	nucleate boiling
泡沫	泡體, 泡沫	foam, froth
泡沫分离	泡沫分離	foam separation
泡沫分馏	泡沫分餾	foam fractionation
泡沫浮选	泡沫浮選	froth flotation
泡沫硅橡胶	矽氧橡膠泡沫	silicone rubber foam
泡沫混凝土	泡混凝土	foam concrete
泡沫深度	泡沫深度	froth depth
泡沫塑料	泡沫塑膠	foam plastic
泡沫橡胶	泡沫橡膠, 海綿狀橡膠	foam rubber, sponge rubber
泡罩	泡罩	bubble cap , bubbling hood
泡罩板	泡罩板	bubble cap plate, bubble cap tray, bubble plate
泡罩板蒸馏塔	泡罩板蒸餾塔	bubble cap plate distillation tower
泡罩接触器	泡罩接觸器	bubble cap contactor
泡罩塔	泡罩塔	bubble cap column, bubble cap tower, bubble plate column, bubble tower
泡罩塔底	泡罩塔底	bubble tower bottom
泡罩塔顶馏分	泡罩塔頂餾分	bubble tower overhead
泡罩塔盘的反截留	泡罩盤之逆陷	backtrapping of bubble-cap tray
泡罩吸收塔	泡罩吸收塔	bubble cap absorption column
胚胎	胚胎	embryo
培养	培養	culture, cultivation
培养期	培養期	incubation period

祖国大陆名	台湾地区名	英文名
培养器	培養器	cultivator
佩克莱数	佩克萊數	Peclet number
配电盘		panel board
配电器	分配器	distributor
配管,管路	配管,管路	piping
配管标准	配管標準	piping standard
配件当量长度	配件相當長度	fitting equivalent length
配件当量管长	配件相當管長	fitting equivalent pipe length
配件流程图	配件流程圖	fitting flowsheet
配件阻力	配件阻力	fitting resistance
配矩法	配矩法	method of moments matching
配位反应	配位反應	coordination reaction
配位共价	配位共價	coordinates covalence
配位化合物	配位化合物	coordination compound
配位聚合	配位聚合反應	coordination polymerization
喷灯	噴燈	blast burner
喷动床	噴流床	spouted bed
喷镀	①電鍍②壓紋	plating
喷镀金属	金屬化	metallizing
喷发	噴發	eruption
喷粉塔	噴霧塔	spray tower
喷流速度	噴流速度	spouting velocity
喷漆	噴漆	spray paint
喷气燃料	噴射機燃油	jet fuel
喷气式发动机	噴射引擎	jet engine
喷砂	噴砂	sand-blast
喷射泵	噴射泵	jet pump
喷射出口压	噴射出口壓[力]	jet exit pressure
喷射干燥器	噴射乾燥器	jet drier
喷射冷凝器	噴凝器	jet condenser
喷射器	噴射器	ejector, eductor
喷射器排列	噴射器排列	ejector arrangement
喷射器容量	噴射器容量	ejector capacity
喷射器油箱	射出器機油箱	ejector sump
喷射式洗涤器	噴射式洗滌器	jet type washer
喷射压缩	噴射壓縮	jet compression
喷射液泛	噴射氾流	jet flooding
喷水池	噴水池	spray pond
喷丝头	噴絲頭	spinning die, spinning head, spinning jet

祖国大陆名	台湾地区名	英 文 名
喷丝嘴	噴絲嘴	spinneret
喷涂	噴霧	spraying
喷涂法	噴塗法	spray coating process
喷雾	①噴霧②噴霧劑	spray
喷雾弹	噴霧彈	aerosol bomb
喷雾阀	噴霧閥	aerosol valve
喷雾分离器	噴霧分離器	spray separator
喷雾干燥	噴霧乾燥[法]	spray drying
喷雾干燥法	噴霧乾燥法	spray drying process
喷雾干燥器	噴霧乾燥器	spray drier, spraying drier
喷雾冷凝器	噴霧冷凝器	spray condenser
喷雾冷却	噴霧冷卻結晶器	spray cooling
喷雾冷却结晶器	噴霧冷卻結晶器	spray cooled crystallizer
喷雾器	①噴霧器②噴漆器	sprayer
喷雾室	噴霧室	spray chamber
喷雾室接触器	噴室接觸器	spray chamber contactor
喷雾塔	噴霧塔	spray column
喷雾推进剂	推噴劑	aerosol propellant
喷雾雾化器	噴霧[霧化]器	spray atomizer
喷雾[制]橡胶	噴霧[法]橡膠	sprayed rubber
喷洗器	噴洗器	jetter
喷釉	噴釉	spraying glazing
喷嘴	噴嘴	nozzle
喷嘴压力	噴嘴壓力	nozzle pressure
硼玻璃	硼玻璃	borax glass
硼硅玻璃	硼矽酸玻璃	borosilicate glass
硼砂	硼砂, 四硼酸鈉	borax
硼酸盐(或酯)	硼酸鹽	borate
膨胀	擴大, 膨脹	expansion, dilatation, dilation
膨胀测定法	膨脹測定法	dilatometry
膨胀床	膨脹床	expanded bed
膨胀床接触器	膨脹床接觸器	expanded bed contactor
膨胀度	膨脹度	degree of expansion
膨胀功	膨脹功	expansion work
膨胀计	膨脹計	dilatometer
膨胀节	脹縮接頭	expansion joint
膨胀器	膨脹器, 膨脹機	expander
膨胀损失	膨脹損失	expansion loss
膨胀弯管	伸縮圈	expansion loop

祖国大陆名	台湾地区名	英　文　名
膨胀系数	膨脹係數	expansion coefficient
膨胀压力[模具]	膨脹壓力[模具]	swelling pressure
膨胀因子	膨脹因子	expansion factor
膨胀应力	膨脹應力	dilatational stress
碰撞	碰撞	impingement
碰撞理论	碰撞理論	hard sphere theory
批号	批號	batch number
批量大小	批式产能	batch size
皮	表皮	skin
皮带输送机	帶式運送機, 帶運機	belt conveyor
皮带运输机	帶運機	band conveyor
皮革	皮革	leather
皮革态	皮革態	leathery state
皮托管	皮托管	Pitot tube
皮心结构	皮蕊結構	skin-core structure
皮心效应	皮蕊效應	skin and core effect
疲劳强度	疲勞強度	fatigue strength
疲劳试验	疲勞試驗	fatigue test
啤酒	啤酒	beer
啤酒酵母	釀造酵母	brewer yeast
片剂	錠, 片劑	tablet
片型聚合物	片狀聚合物	sheet polymer
片状成型	模壓成型	sheet molding
片状成型料	模壓成型板材	sheet molding compound
偏差	偏差, 偏離	deviation
偏差形式	偏差形式	deviation form
偏光镜	旋光計	polariscope
偏离	偏轉, 偏向, 偏差	deflection, offset
偏离电位计	偏轉電位計	deflection potentiometer
偏离函数	偏離函數	departure function
偏离角	偏向角	angle of deflection
偏摩尔超额物性函数	部分莫耳過量性質	partial molar excess property
偏摩尔焓	偏莫耳焓	partial molar enthalpy
偏摩尔吉布斯自由能	偏莫耳吉布斯自由能	partial molar Gibbs free energy
偏摩尔量	偏莫耳量	partial molar quantity
偏摩尔内能	部分莫耳內能	partial molar internal energy
偏摩尔体积	部分莫耳體積	partial molar volume
偏摩尔物性函数	部分莫耳性質	partial molar property
偏微分方程	偏微分方程式	partial differential equation

祖国大陆名	台湾地区名	英文名
偏相关	部分相關	partial correlation
偏心秤	偏心儀	eccentric scale
偏心度	偏心率	eccentricity
偏心环	偏心環	eccentric ring
偏心渐缩管	偏心漸縮管	eccentric reducer
偏心孔口板	偏心孔口板	eccentric orifice plate
偏心因子	偏離因數	accentric factor
偏压	偏, 偏向, 偏流, 偏壓	bias
偏应力	軸差應力, 偏向應力	deviatoric stress
偏振	偏光	polarization
偏振化因子		polarization factor
偏振片	偏光玻璃	polaroid glass
漂白	漂白	bleaching
漂白粉	漂白粉	bleaching powder, bleach powder
漂白剂	漂白劑	bleaching agent
漂白土	漂白土	fuller's earth
漂白液	漂白液	bleaching liquor
漂白作用	漂白作用	bleaching action
漂移	漂移	drift
撇去浮渣	浮渣	skim
贫气	貧氣	lean gas
贫溶液	貧溶液	lean solution
贫油	貧油	lean oil
频带	頻率帶	frequency band
频率	頻率	frequency
频率测试	頻率試驗法	frequency test
频率示踪器	頻率示蹤器	frequency tracer
频率响应	頻率應答	frequency response
频率响应分析	頻率應答分析	frequency response analysis
频率因子	①頻率因數②指數前因數	frequency factor
频闪仪	頻閃轉速計	stroboscope
频移	頻率移動, 頻率移位	frequency shift
频域	頻率領域	frequency domain
平板	平板	flat plate
平板架	托板	pallet
平板桨	平板漿	flat paddle
平板近似法	平板近似法	flat-plate approximation
平板流动	平板流動	flat-plate flow

祖国大陆名	台湾地区名	英 文 名
平板式换热器	板式熱交換器	plate type heat exchanger
平版印刷术	石板印刷法, 石印法, 印型法	lithography
平槽滤板	平槽濾板	flush plate
平槽压滤板	平槽壓濾板	flush filter plate
平动配分函数	移動配分函數	translational partition function
平管排	平管排	horizontal tube bank
平衡	平衡	equilibrium
平衡槽	均化槽	equalization tank
平衡常数	平衡常數	equilibrium constant
平衡点	平衡點	equilibrium point
平衡电动机	均衡電動機	balancing motor
平衡阀	均衡閥	equalizing valve
平衡法	平衡法	equilibrium approach
平衡方程式	均衡方程式	balance equation
平衡分布	平衡分布	equilibrium distribution
平衡分离过程	平衡分離程序	equilibrium separation process
平衡釜	平衡釜	equilibrium still
平衡含水率	平衡含水量	equilibrium moisture content
平衡化	平衡化	equilibration
平衡环	均衡環	balancing ring
平衡混合物	平衡混合物	equilibrium mixture
平衡级	平衡階	equilibrium stage
平衡近似	平衡近似[法]	equilibrium approximation
平衡冷凝	平衡冷凝	equilibrium condensation
平衡冷凝曲线	平衡冷凍曲線	equilibrium freezing curve
平衡浓度	平衡濃度	equilibrium concentration
平衡判据	平衡準則	equilibrium criterion
平衡汽化	平衡汽化	equilibrium vaporization
平衡曲线	平衡曲線	equilibrium curve
平衡闪蒸	平衡驟汽化	equilibrium flash vaporization
平衡闪蒸曲线	平衡驟汽化曲線	equilibrium flash vaporization curve
平衡收率	平衡產率	equilibrium yield
平衡水	均衡水	balance water
平衡水分	平衡水分	equilibrium moisture
平衡速率	平衡速率	equilibration rate
平衡图	平衡圖	equilibrium diagram, equilibrium plot
平衡系统	平衡系統	equilibrium system
平衡线	平衡線	equilibrium line

祖国大陆名	台湾地区名	英 文 名
平衡蒸馏	平衡蒸餾	equilibrium distillation
平衡转化[率]	平衡轉化率	equilibrium conversion
平衡组成	平衡組成	equilibrium composition
平滑	平滑化, 修匀	smoothing
平滑表面	平滑表面	smooth surface
平桨	槳	paddle
平均半径	平均半徑	mean radius
平均沸点	平均沸點	average boiling point
平均分子量	平均分子量	average molecular weight
平均过滤比阻	平均過濾比阻	average specific filtration resistance
平均偏差	平均偏差	average deviation, mean deviation
平均速度	平均速度	average velocity
平均停留时间	平均滯留時間	average residence time, mean residence time
平均温差	平均溫差	average temperature difference, mean temperature difference
平均误差	平均誤差	average error, mean error
平均自由程	平均自由徑	mean free path
平面应力	平面應力	plane stress
平皿培养	平碟培養[菌]	plate culture
平切薄膜	切片薄膜	sliced film
平切片材	切片	sliced sheet
平台	平臺	platform
平坦效应	平線區效應	plateau effect
平推流返混	塞流逆混	plug flow backmixing
平推流, 活塞流	塞流, 栓流	plug flow
平行处理	平行處理	parallel processing
平行反应	平行反應	parallel reaction
平行进料	平行進料	parallel feed
平行失活	平行失活	parallel deactivation
平移位能	平移能	translational energy
平移因子	平移因子, 移位因子	shift factor
评估	評估	assessment, evaluation
苹果酸	蘋果酸	malic acid
苹果子油	蘋果子油	apple seed oil
屏蔽	屏蔽	shielding, shadowing, screen
屏蔽泵	屏蔽泵	canned-motion pump
屏蔽长度	屏蔽長度	shielding length
屏蔽有效度	篩有效度	screen effectiveness
瓶颈	瓶頸	bottleneck

祖国大陆名	台湾地区名	英 文 名
坡印亭校正	波印亭校正	Poynting correction
坡印亭因子	波印亭因子	Poynting factor
破裂	破裂, 破碎	breakage, break
破乳	乳化液之分解	breakup of emulsion
破碎	壓碎	crush
破碎比	磨碎比	reduction ratio
破碎机	壓碎機, 散解機	crusher, disintegrator
剖面[图]	剖面[圖]	profile
葡聚糖	葡聚糖	dextran, glucan
葡萄糖	葡萄糖	dextrose, glucose
普朗特数	普朗特數	Prandtl number
普鲁士蓝	普魯士藍	Prussian blue
普适方程	通式	generalized equation
普适化	通式化, 一般化	generalization
普适气体常量	通用氣體常數	universal gas constant
普通原料	普通原料, 共用原料	common stock
谱	光譜, 譜	spectrum
暴晒蒸发	太陽能蒸發	solar evaporation
曝光时间	曝露時間	exposure time
曝露界限	曝露界限	exposure limit
曝露试验	曝露試驗	exposure test
曝露中心	曝露中心	exposed center
曝气	曝氣	aeration
曝气浮选	分散空氣浮選[法]	dispersed-air flotation
曝气塘	通氣污水塘, 曝氣污水塘	aerated lagoon
曝气稳定化池	通氣穩定化池, 曝氣穩定化池	aerated stabilization pond
曝气系统	通氣系統, 曝氣系統	aerated system

Q

祖国大陆名	台湾地区名	英 文 名
期望产物	期望產物	desired product
期望值	期望值	desired value
期望值判据	期望值判斷	expected value criterion
漆	噴漆	lacquer
漆革	黑漆皮, 漆革	patent leather

祖国大陆名	台湾地区名	英 文 名
齐白值(100 份纸浆消程的有效氯)	西伯值	Sieber number
奇点	奇點	singular point
奇偶性	奇偶性	parity
歧点	歧點	bifurcation point
歧管	歧管	manifold
企业自动化	商業自動化, 企業自動化	business automation
启发式方法(＝直观推断法)		
启发式规则(＝直观推断法则)		
起爆	起爆	detonation
起动	引動, 發火	priming
起动器	起動機, 起動器	starter
起黏丝性	黏絲性, 黏稠性	stringiness
起燃	點火, 燃燒	ignition
起始流化速度	起始流化速度	incipient fluidizing velocity
起始流化态	起始流化態	incipient fluidization
起酥油	酥脆油	shortening oil
起重机	起重機	hoist
气泵	空氣泵	air pump
气吹磨	風掃磨	air-swept mill
气动测力仪	氣動測力計	pneumatic force meter
气动传动	氣動傳送	pneumatic transmission
气动传动接受器	氣動傳送接受器	pneumatic transmission receiver
气动传动器	氣動傳送器	pneumatic transmitter
气动传动线	氣動傳送線路	pneumatic transmission line
气动传感器	氣動感測器	pneumatic sensor
气动电流继电器	氣動電流替續器	pneumatic electric relay
气动阀	氣動閥	pneumatic valve, air-operated valve
气动机理	氣動機構	pneumatic mechanism
气动搅拌	氣動攪拌	pneumatic stirring
气动控制	氣動控制	pneumatic control
气动模拟	氣動類比	pneumatic analog
气动设定控制器	氣動設定控制器	pneumatic set controller
气动式流动传送器	氣動式流動傳送器	pneumatic flow transmitter
气动式液面计	氣動液面計	pneumatic liquid level gage
气动式液体密度计	氣動液體密度計	pneumatic liquid density gage

祖国大陆名	台湾地区名	英 文 名
气动输送	氣動輸送	pneumatic transport, pneumatic conveying
气动调节阀	氣動控制閥	pneumatic control valve
气动调节器	氣動控制器	pneumatic controller
气动系统	氣動系統	pneumatic system
气动信号	氣動信號	pneumatic signal
气动运输机	空氣運送機	air conveyor
气动执行装置	氣動引動器	pneumatic actuator
气动装置	氣動裝置	pneumatic device
气固反应	氣固反應	gas-solid reaction
气固反应器	氣固反應器	gas-solid reactor
气固平衡	氣固平衡	gas-solid equilibrium
气化	氣化	gasification
气化反应	氣化反應	gasification reaction, gasifying reaction
气化过程	氣化程序	gasification process
气化技术	氣化技術	gasification technology
气化炉	氣化爐, 氣化器	gasifier
气阱	氣阱	gas trap
气冷夹套	氣冷套	air cooled jacket
气力分级器	氣動類析器	pneumatic classifier
气流	氣流	air stream
气流干燥器	氣流乾燥器	pneumatic dryer
气流混合管	混合流通管	mixing draft tube
气流输送器	氣動運送機, 氣運機	pneumatic conveyor
气帽	氣罩	air cap
气膜	氣膜	gas film
气膜接触器	氣膜接觸器	gas film contactor
气膜控制	氣膜控制	gas film control
气囊	安全氣囊	air bag
气泡	氣泡	bubble, air bell
气泡反应器	氣泡反應器	bubble reactor
气泡分馏	氣泡分餾[法]	bubble fractionation
气泡计数器	氣泡計數器	bubble counter
气泡剂	氣泡劑	bubble forming agent
气泡聚并	氣泡聚集	bubble coalescence
气泡流	氣泡流動	bubble flow
[气]泡膜	氣泡膜	bubble film
气泡曝气	氣泡通氣	bubble aeration
气泡系统	氣泡指示系統	air bubble system
气泡型黏度计	泡式黏度計	bubble type viscometer

祖国大陆名	台湾地区名	英 文 名
气泡云, 气泡晕	氣泡雲	bubble cloud
气泡晕(=气泡云)		
气泡指示计	氣泡指示計	bubble gage
气溶胶	①氣溶膠②霧劑	aerosol
气升	①氣提②氣升器	air-lift
气升泵	氣動泵	air-lift pump
气升搅拌	氣升攪拌	air-lift agitation
气升搅拌器	氣升攪拌器	air-lift agitator
气提	氣提	gas stripping
气[体]	①氣②瓦斯	gas
气体常数	氣體常數	gas constant
气体重整	氣重組	gas reforming
气体定律	氣體定律	gas law
气体发射标准	氣體排放標準	gaseous emission standard
气体分布器	氣體分布器	gas distributor
气体分析仪	氣體分析儀	gas analyzer
气体粉碎机	氣流粉碎機	jet mill
气体浮选	氣體浮選	gaseous floatation
气体量热计	煤氣卡計	gas calorimeter
气体燃料	氣體燃料	gaseous fuel
气体溶液	氣體溶液	gaseous solution
气体渗透	氣體滲透	gas permeation
气体收集器	收氣器	gas receiver
气体调节器	氣體調節器	gas regulator
气体温度计	氣體溫度計	gas thermometer
气体洗涤器	洗氣器	gas scrubber
气体总管	氣體總管	gas main
气田	天然氣田	gas field
气味	氣味	odor
气隙	空氣隙	air gap
气相	氣相	gas phase
气相沉积	蒸氣沉積	vapor deposition
气相动能因子	氣相動能因子	gas phase loading factor
气相反应	氣相反應	gas phase reaction
气相固体	氣態固體	gas phase solid
气相控制	氣相控制	gas phase control
气相色谱法	氣相層析法	gas chromatography
气相色谱图	氣相層析圖	gas chromatogram
气相色谱仪	氣相層析儀	gas chromatograph

祖国大陆名	台湾地区名	英　文　名
气压动态学	氣壓動態[學]	air pressure dynamics
气压方程式	氣壓方程式	barometric equation
气压高度	氣壓高度	barometric height
气压管	氣壓管	barometer tube
气压计	氣壓計	barometer, air pressure gage, air gage
气压记录器	氣壓溫度記錄器	barometrograph
气压效应	氣壓效應	barometric effect
气压柱泵	氣壓式泵	barometric leg pump
气液反应	氣液反應	gas-liquid reaction
气液反应器	氣液反應器	gas-liquid reactor
气液分离	氣液分離	gas-liquid separation
气液固反应	氣液固反應	gas-liquid-solid reaction
气液固反应器	氣液固反應器	gas-liquid-solid reactor
气液接触	氣液接觸	gas-liquid contacting
气液接触器	氣液接觸器	gas-liquid contactor
气液平衡	氣液平衡	gas-liquid equilibrium
气液平衡过程	氣液平衡程序	gas-liquid equilibrium process
气液吸收	氣液吸收	gas-liquid absorption
气油比	氣-油比	gas-oil ratio
气闸	氣閘	air lock
汽包	蒸汽鼓	steam drum
汽锤	蒸汽鎚	steam hammer
汽化	汽化	vaporization
汽化器	汽化器	vaporizer
汽化潜热	汽化潛熱	latent heat of vaporization
汽化曲线	汽化曲線	vaporization curve
汽化热	汽化熱	heat of vaporization
汽蚀	空洞現象	cavitation
汽蚀余量, 净正吸压头	淨正吸高差	net positive suction head (NPSH)
汽水	碳酸水	aerated water
汽提	蒸汽汽提	steam stripping
汽提剂	汽提劑	stripping agent
汽提塔	汽提塔, 脱除塔, 汽提器	stripping tower, stripping still, stripping column, stripper
汽提[用]蒸汽	汽提蒸汽	stripping steam
汽相缔合	汽相結合	vapor phase association
汽液分离器	汽液分離器	vapor-liquid separator
汽液平衡	汽液平衡	vapor-liquid equilibrium
汽液平衡比	汽液平衡比	vapor-liquid equilibrium ratio

祖国大陆名	台湾地区名	英　文　名
汽油	汽油, 石油	gasoline, petrol
汽油机	汽油引擎	gasoline engine
汽油阱	汽油阱	gasoline trap
汽蒸	汽蒸, 蒸煮	steaming
迁移	泳動	migration
迁移电位	遷移電位	migration potential
迁移率	①移動率②機動性	mobility
迁移坐标	對流座標	convected coordinates
钎焊	軟焊	soldering
钎焊膏	焊膏	soldering paste
铅	鉛	lead
铅白	鉛白	lead white
铅玻璃	鉛玻璃	lead glass
铅锤式浮标	錘式浮標	plumb bob float
铅丹	鉛丹, 鉛紅	minium
铅管	鉛管	lead pipe
铅基润滑脂	鉛基(潤)滑脂	lead-base grease
铅碱玻璃	鉛鹼玻璃	lead alkali glass
铅室	鉛室	lead chamber
铅室法	鉛室法	chamber process
铅蓄电池	鉛蓄電池	lead storage battery
铅皂	鉛皂	lead soap
前馈	前饋	feedforward
前馈控制	前饋控制	feedforward control
前馈系统	前饋系統	feedforward system
前体	前體	precursor
前沿	前緣	leading edge
潜能	潛能	latent energy
潜热	潛熱	latent heat
潜望镜	潛望鏡	periscope
浅床反应器	淺床反應器	shallow-bed reactor
欠硫	欠硫, 硫化不足, 低硫化	soggy, under cure
欠硫化橡胶	低硫化橡膠	under cured rubber
欠阻尼	欠阻尼	underdamping
欠阻尼系统	欠阻尼系統	underdamped system
欠阻尼响应	欠阻尼應答	underdamped response
茜草	茜草, 茜草染料	madder
茜草根	茜草根	madder root

祖国大陆名	台湾地区名	英 文 名
茜素染料	茜素染料	alizarine dyestuff
茜素色淀	茜素色澱	alizarine lake
嵌段共聚合	多段共聚合	segment copolymerization
嵌段共聚物	段連共聚物, 嵌段共聚物	block copolymer
嵌段聚合	段連聚合[反應], 嵌段聚合[反應]	block polymerization
嵌接	嵌接接頭	scarf joint
强度	強度	intensity, strength
强度性质, 内含性质	內含性質	intensive property
强度状态变量	內含狀態變數	intensive state variable
强力玻璃	強力玻璃	steel glass
强力黏胶纤维	強力黏膠人造絲	strong viscose rayon
强制对流	強制對流	forced convection
强制对流传热	強制對流熱傳送	forced convection heat transfer
强制对流蒸发	強制對流蒸發	forced convection evaporation
强制函数	強制函數	forcing function
强制扩散	強制擴散	forced diffusion
强制通风	強制通風, 鼓風	forced draft
强制通风凉水塔	強制通風冷卻塔	forced draft cooling tower
强制涡旋	強制漩渦	forced vortex
强制循环蒸发	強制循環蒸發	forced circulation evaporation
强制循环蒸发器	強制循環蒸發器	forced circulation evaporator
强制振荡	強制振盪	force oscillation
羟基	羥基	hydroxyl group
羟醛	醛醇	aldol
羟值	羥值	hydroxyl value
翘曲	①翹曲, 翹曲[變形]②[織物]整經	warping, warpage
切段[定长]纤维	定長短纖維	staple fiber
切工组	切工組	draw gang
切换线	切換線	switching line
切口堰	V 形堰	notched weir
切块机	切塊機	slabber
切面直径	切面直徑	cut diameter
切膜扁丝	切膜絲	slit yarn
切膜纤维	切膜纖維	slit fiber
切片	切片	slicing, sectioning
切片机	切片機	slicer

祖国大陆名	台湾地区名	英　文　名
切速稀化	切速稀化	shear-rate thinning
切条机	切條機	slitter
切向力	切線力	tangential force
切向应力	切線應力	tangential stress
切削油	切削油	cutting oil
亲和标记	親和標記	affinity labeling
亲和超滤	親和過濾	affinity ultrafiltration
亲和沉淀	親和沈澱	affinity precipitation
亲和膜	親和膜	affinity membrane
亲和色谱[法]	親和色譜法	affinity chromatography
亲和势	親和性, 親和力	affinity
亲和吸附	親和吸附	affinity adsorption
亲和作用	親和作用	affinity interaction
亲水胶体	親水膠體	hydrophilic colloid
亲水亲油平衡	親水性-親油性均衡	hydrophile-lyophile balance
亲水物	親水物	hydrophile
亲水纤维	親水纖維	hydrophilic fiber
亲水性颗粒	親水性顆粒	hydrophilic particle
亲液胶体	親液膠體	lyophilic colloid
青霉素	青黴素, 盤尼西林	penicillin
青霉素酶	青黴素酶	penicillinase
青铜	青銅	bronze
氢弹	氫彈	hydrogen bomb
氢电极	氫電極	hydrogen electrode
氢放电管	氫放電管	hydrogen discharge tube
氢化催化剂	氫化觸媒	hydrogenation catalyst
氢化反应	氫化反應	hydrogenation reaction
氢化酶	氫酶	hydrogenase
氢化油	氫化油	hydrogenated oil
氢化脂	氫化脂	hydrogenated fat
氢解	氫解(反應)	hydrogenolysis
氢离子比长仪	氫離子比色計	hydrogen ion comparator
氢氯化橡胶	氫氯化橡膠, 鹽酸化橡膠	hydrochloride rubber
氢能化学	氫能化學	chemistry of hydrogen energy
氢气爆炸	氫氣爆炸	hydrogen explosion
氢气泡法	氫氣泡法	hydrogen bubble technique
氢燃料	氫燃料	hydrogen fuel
氢燃料电池	氫燃料電池	hydrogen fuel cell

祖国大陆名	台湾地区名	英 文 名
氢氧根离子	氫氧離子	hydroxide ion
氢氧化钠, 烧碱	氫氧化鈉, 燒鹼	caustic soda
轻垢洗涤剂	輕污清潔劑	light duty detergent
轻灰	輕純鹼	light ash
轻馏分	輕餾分	light ends
轻燃料油	輕燃料油	light fuel oil
轻石脑油	輕石油腦	light naphtha
轻烃	輕烴	light hydrocarbon
轻直馏油	輕直餾油	light straight run
轻质油	輕油	light oil
轻主要组分	輕主成分	light key component
倾角	傾角	angle of inclination
倾析	傾析[法]	decantation
倾析器	傾析器	decanter
倾斜管	傾斜管	dip tube
倾卸装置		tripper
清烤漆	烤清漆	baking varnish
清漆	清漆	varnish
清晰分离	清晰分離	sharp separation
清洗过程	清洗程序	cleaning process
氰硅橡胶	矽腈橡膠	silicone nitrile rubber
琼脂	洋菜	agar
琼脂电泳	瓊脂電泳	agar electrophoresis
琼脂过滤	瓊脂過濾	agar filtration
琼脂扩散	瓊脂擴散	agar diffusion
琼脂扩散技术	瓊脂擴散技術	agar diffusion technology
琼脂培养基	洋菜培基	agar medium
琼脂糖	瓊脂糖	agarose
琼脂糖胶	瓊脂糖膠	agarose gel
求积仪	測面計	planimeter
球晶	球晶	spherulite
球晶结构	球晶結構	spherulite structure
球磨反应器	球磨反應器	ball mill reactor
球磨机	球磨機	ball mill
球磨精制机	球磨精製機	ball mill refiner
球体	球體	sphere
球土	球狀黏土	ball clay
球形度	球形度	sphericity
球形极坐标	球極坐標	spherical polar coordinates

祖国大陆名	台湾地区名	英　文　名
球形油罐	球形槽	spherical tank
球形坐标	球形坐標	spherical coordinates
球状团簇	球狀團簇	spherical clusters
区带电泳	帶域電泳	zone electrophoresis
区[域]	區[域]	region
区域沉降	帶域沉降	zone settling
区域精制	帶域純化法	zone refining method
区域冷冻	帶域冷凍	zone freezing
区域熔炼	帶域熔化	zone melting
区域熔炼法	帶域熔化法	zone melting method
曲率	曲率	curvature
曲线	曲線	curve
E 曲线	E 曲線	E curve
F 曲线	F 曲線	F curve
曲线平直部分	平線區	plateau
曲线坐标	曲線座標	curvilinear coordinates
曲折因子	扭曲度	tortuosity
驱动功率	驅動功率	driver horsepower
驱动函数	驅動函數	driving function
驱动器	推進器	driver
屈伏极限	降伏極限	yield limit
屈服点	降伏點	yield point
屈服强度	降伏強度	yield strength
屈服应力	降伏應力	yield stress
屈尼萃取塔	屈尼萃取塔	Kühni extractor
渠道流	渠流	channel flow
取代法则	代換法則	substitution rule
取代反应	取代反應	substitution reaction
取代基	取代基	substituent
取代基常数	取代基常數	substituent constant
取代基均匀度	取代基均匀度	substituent uniformity
取样	取樣, 抽樣	sampling
取样管	取樣管	sampling tube
取样误差	取樣誤差	sampling error
取样系统	取樣系統, 抽樣系統	sampling system
取样滞后	取樣延遲	sampling delay
去饱和作用	去飽和作用	desaturation
去垢力	清潔力	detergency
去光	去光	delusting

祖 国 大 陆 名	台 湾 地 区 名	英 文 名
去光剂	去光劑	delusterant
去过热	去過熱	desuperheat
去过热器	去過熱器	desuperheater
去灰分	去灰分	deashing
去灰分过程	去灰分程序	deashing process
去灰煤	去灰煤	deashed coal
去活化级	去活化階	order of deactivation
去矿化	去礦[物]質	demineralization
去离子水	去離子水	deionized water
去离子作用	去離子[作用]	deionization
去木素作用	去木質素	delignification
去溶剂化	去溶劑[作用]	desolvation
去湿器	溼氣分離器	moisture separator
去稳定作用	去安定[作用]	destabilization
去污剂	去污劑	decontaminant
去污指数, 净化指数	去污指數	decontamination factor
圈转电流计	圈轉電流計	moving coil galvanometer
全部循环过程	全部循環程序	total recycle process
全导数	全導數	total derivative
全分析	整體分析	bulk analysis
全回流	全回流	total reflux
全混	全混, 完全混合	complete mixing, perfect mixing
全混流	全混流	complete mixing flow
全同立构聚丙烯	同規立構聚丙烯	isotactic polypropylene
全同立构聚合物, 等规 聚合物	同規立構聚合物	isotactic polymer
全纤维素	全纖維素	holocellulose
权函数	權函數, 權數	weight function
权因子	權因數	weight factor
权重分布	重量分布	weight distribution
醛缩醇	縮醛類	acetal
缺胶层	缺膠層	starved line
缺胶接头	缺膠接頭	starved joint
缺陷固体	缺陷固體	defect solid
缺陷晶体	缺陷晶體	defect crystal
缺陷值	隱含值	default value
确定性模型	決定性模型	deterministic model
裙式给料机	裙飼機	apron feeder
裙式输送机	裙式輸送機	apron carrier

祖国大陆名	台湾地区名	英 文 名
裙式运输机	裙式運送機	apron conveyor
群青	群青	ultramarine
群青黄	群黄	ultramarine yellow
群青蓝	群青	ultramarine blue
群青绿	群綠	ultramarine green
群青紫	群紫	ultramarine violet
群青棕	群棕	ultramarine brown

R

祖国大陆名	台湾地区名	英 文 名
燃点	燃點, 燃燒點, 火點	ignition point, burning point
燃尽	燃盡	burn-out
燃料	燃料	fuel
燃料比	燃料比	fuel ratio
燃料成本	燃料成本	fuel cost
燃料床	燃料床	fuel bed
燃料电池	燃料電池	fuel cell
燃料工业	燃料工業	fuel industry
燃料酒精	燃料酒精	fuel alcohol
燃料空气比	燃料空氣比	fuel-air ratio
燃料油	燃料油	fuel oil
燃气轮发动机	氣渦輪引擎	gas turbine engine
燃气轮机	氣渦輪機	gas turbine
燃烧	燃燒	combustion
燃烧反应	燃燒反應	combustion reaction
燃烧反应器	燃燒反應器	combustion reactor
燃烧过程	燃燒程序	combustion process
燃烧机理	燃燒機構	combustion mechanism
燃烧率	燃燒速率	burning rate
燃烧面	燃燒峰面	combustion front
燃烧器	燃燒器	burner
燃烧区	燃燒區	combustion zone
燃烧曲线	燃燒曲線	combustion curve
燃烧热	燃燒熱	heat of combustion
燃烧室	燃燒室	combustion chamber
燃烧收缩	燃燒收縮	burning shrinkage
燃烧速率	燃燒速率	combustion rate

祖国大陆名	台湾地区名	英　文　名
燃烧特性	燃燒特性	combustion characteristics
燃烧效率	燃燒效率	combustion efficiency
染料	染料	dye
染料亲和色谱[法]	染料親和色層法	dye affinity chromatography
染料摄入法	染料攝入法	dye uptake method
染色	染色	dyeing
染色机	染色機	dyeing machine
扰动	擾動	disturbance
绕成绞	成絞	skeining
绕射(=衍射)		
热	熱	heat
热安培计	熱偶安培計	thermoammeter
热本构方程	熱本質方程式	thermal constitutive equation
热泵	熱泵	heat pump
热波	熱波	thermal wave
热沉降	熱沉降	thermal precipitation
热成型	熱重組, 熱壓成型	thermoforming
热重整	熱重組	thermal reforming
热冲击试验	熱震實驗	thermal shock test
热处理	熱處理	heat treatment
热传导	熱傳導	conduction heat transfer, conductive heat transfer
热传导率测定槽	熱傳導係數測定槽	thermo-conductivity cell
热传递	熱輸送	heat transport
热磁分析	熱磁分析[法]	thermomagnetic analysis, thermomagnetism
热单位	熱單位	thermal unit
热导检测器	熱傳導係數偵檢器	thermal conductivity detector
热滴定[法]	溫度滴定[法]	thermometric titration
热点	熱點	hot spot
热电	熱電	thermoelectricity
热电池	熱電池	thermobattery
热电堆	熱電堆	thermopile
热电法	熱電法	thermoelectrometry
热电高温计	熱電高溫計	thermoelectric pyrometer
热电计	熱電計	thermoelectrometer
热电流	熱電流	thermocurrent
热电流计	熱電流計	thermogalvanometer
热电偶	熱電偶	thermocouple
热电偶参考接点	熱電偶參考接點	thermocouple reference junction

祖国大陆名	台湾地区名	英 文 名
热电偶高温计	熱電偶高溫計	thermocouple pyrometer
热电偶接点	熱電[偶]接點	thermoelectric junction
热电偶式流速计	熱電偶式流速計	thermocouple type anemometer
热电式电位计	熱電[式]電位計	thermoelectric potentiometer
热电温度计	熱電溫度計	thermoelectric thermometer
热电泳	熱泳法	thermophoresis
热对流	熱對流	heat convection
热发射	熱發射	heat emission
热反射法	熱反應[分析]法	thermoreflectometry
热方法	熱分析法	thermal method, thermal technique
热分解	熱分解	thermal decomposition
热分析	熱分析[法]	thermal analysis, thermoanalysis
热分析法	熱分析法	thermoanalytical method
热分析器	熱分析儀	thermoanalyzer
热分析显微镜	熱分析顯微鏡	thermoanalytical microscopy
热风炉	熱風器	hot blast heater
热封黏合剂	熱封黏合劑	heat sealing adhesive
热辐射	熱輻射	radiation of heat, thermal radiation
热负载	熱負載	heat duty
热功当量	熱功當量	mechanical equivalent of heat
热功转化率	熱功轉化率	heat rate
热固定	熱固定	thermofixation
热固化	熱固性	thermosetting
热固性树脂	熱固性樹脂	thermosetting resin
热固性塑料	熱固物	thermoset
热管	熱管	heat-pipe
热管换热器	熱管交換器	heat-pipe exchanger
热光分析法	熱光[分析]法	thermophotometry
热光学法	熱光學法	thermooptometry
热含量	熱含量	heat content
热耗散	熱散逸	heat dissipation, thermal dissipation
热合	熱合	sealing
热合成	熱合成	thermal synthesis
热合机	熱合機, 封焊機	sealer
热合接头	熱合接頭	sealing joint
热核反应	熱核反應	thermonuclear reaction
热虹吸管	熱虹吸	thermo siphon
热虹吸式再沸器	熱虹吸再沸器	thermo siphon reboiler
热化学	熱化學	thermochemistry

祖国大陆名	台湾地区名	英 文 名
热回流	熱回流	hot reflux
热机	熱機	heat engine
热机械法	熱機械法	thermomechanometry
热机械分析	熱機械法	thermomechanical analysis
热集成	熱整合	heat integration
热降解	熱降解	thermal degradation
热交换器	熱交換器	heat interchanger
热交换器(＝换热器)		
热胶凝	熱膠凝	thermogelling
热接点	熱接點	hot junction
热解	熱解	pyrolysis
热解过程	熱解程序, 裂解程序	pyrolysis process
热解聚	熱解聚合	thermal depolymerization
热解炉	裂解爐, 熱解爐	pyrolysis furnace
热解油	熱解油	pyrolysis oil
热浸	熱浸漬	hot dipping
热经济学	熱經濟學	thermo-economics
热阱	熱匯座, 熱壑	heat sink
热静力学	靜熱力學	thermostatics
热聚合	熱聚合[反應]	thermal polymerization, thermopolymerization
热颗粒分析	熱顆粒分析[法]	thermoparticulate analysis
热坑干燥室	熱坑乾燥室	hot floor drier
热扩散	熱擴散	thermal diffusion, thermodiffusion
热扩散系数, 导温系数	熱擴散係數	thermal diffusivity, thermal diffusion coefficient
热离解	熱解離	thermal dissociation
热力学	熱力學	thermodynamics
热力学本构方程	熱力本質方程式	thermodynamic constitutive equation
热力学第二定律	熱力學第二定律	second law of thermodynamics
热力学第零定律	熱力學第零定律	zeroth law of thermodynamics
热力学第三定律	熱力學第三定律	third law of thermodynamics
热力学第一定律	熱力學第一定律	first law of thermodynamics
热力学分析	熱力分析	thermodynamic analysis
热力学概率	熱力擴散	thermodynamic probability
热力学关系	熱力關係	thermodynamic relation
热力学函数	熱力學函數	thermodynamic function
[热力学]环境	外圍	surroundings
热力学力	熱力學力	thermodynamic force

祖国大陆名	台湾地区名	英　文　名
热力学平衡	熱力學平衡	thermodynamic equilibrium
热力学特性函数	熱力學特性函數	thermodynamic characteristic function
热力学通量	熱力學通量	thermodynamic flux
热力学图	熱力圖	thermodynamic diagram
热力学温标	熱力溫標	thermodynamic temperature scale
热力学温度	熱力學溫度	thermodynamic temperature
热力学系统	熱力學系統	thermodynamic system
热力学效率	熱力效率	thermodynamic efficiency
热力学性质	熱力學性質	thermodynamic property
热力学性质表	熱力性質表	thermodynamic property table
热力学一致性	熱力一致性	thermodynamic consistency
热力学一致性检验	熱力一致性試驗[法]	thermodynamic consistency test
热量传递(＝传热)		
热量衡算	熱量均衡	heat balance
热量回收	熱回收	heat recovery
热[量]平衡	熱量均衡	heat balance, thermal balance
热量去除曲线	熱量去除曲線	heat removal curve, heat removal line
热裂变	熱裂变	thermal fission
热裂化	熱裂解	thermal cracking
热流	熱流動	heat flow
热流道	熱澆道	hot runner
热流[量]	熱流動	heat flow
热流通量	熱通量	heat flux
热硫化	熱硫化	hot vulcanization
热敏电阻器	熱阻體, 熱阻器, 電熱調節器	thermistor
热能	熱能	heat energy , thermal energy
热能方程式	熱能方程式	thermal energy equation
热膨胀	熱膨脹	thermal expansion
热膨胀系数	熱脹係數	thermal expansion coefficient
热平衡	熱平衡	thermal equilibrium
热气再循环过程	熱氣循環程序	hot gas recycle process
热汽	熱汽	hot vapor
热容	熱容量	heat capacity
热容量	熱容量	thermal capacity
热容流率	熱容流率	heat capacity flow rate
热渗透	熱滲透[作用]	thermoosmosis
热生成	熱生成	heat generation
热生成曲线	熱生成曲線	heat generation curve

祖国大陆名	台湾地区名	英 文 名
热生成线	熱生成線	heat generation line
热声法	熱聲[分析]法	thermosonimetry
热声强测量法	熱測音法	thermoacoustimetry
热失活	熱衰化	thermal deactivation
热石灰法	熱石灰法	hot lime process
热释发光	熱發光	thermoluminescence
[热]收缩机	[熱]收縮機	shrinking machine
热塑性	熱塑性	thermoplasticity
热塑性聚合物	熱塑性聚合物	thermoplastic polymer
热塑性树脂	熱塑性樹脂	thermoplastic resin
热塑性塑料	熱塑性塑膠	thermoplastic
热损失	熱損失	heat loss
热炭黑	熱碳黑	thermal black
热套管	熱套管	thermal well
热特性	熱示性[法]	thermal characterization
热天平	熱天平	thermobalance
热通量	熱通量	heat flux
热脱挥发分	熱去揮發物作用	thermal devolatilization
热烷化	熱烷化[反應]	thermal alkylation
热稳定性	熱安定性	thermal stability
热稳性	熱安定性	thermostability
热污染	熱污染	heat pollution, thermal pollution
热物理性质	熱物理性質	thermophysical property
热效率	熱效率	heat efficiency, thermal efficiency
热效应	熱效應	heat effect
热行为	熱行為	thermal behavior
热循环	熱循環	thermal cycle
热压	熱壓	hot pressing
热压机	熱壓機	hot press
热氧化降解	熱氧化降解	thermooxidative degradation
热氧化老化	熱氧化老化	thermooxidative aging
热应变	熱應變	thermal strain
热应力	熱應力	thermal stress
热原子法	熱原子法	thermoatomic process
热源	熱源	heat source, thermal source
热运动学	熱運動學	thermokinematics
热震	熱震	thermal shock
热汁加灰法	熱汁加灰法	hot liming
热值	熱值	heating value, heat value, thermal value

祖国大陆名	台湾地区名	英 文 名
热值试验	熱值試驗	thermal value test
热致液晶	熱致性液晶	thermotropic liquid crystal
热重法	熱重量分析法	thermogravimetry
热重量分析	熱重量分析[法]	thermogravimetric analysis
热重量分析器	熱重量分析儀	thermogravimetric analyzer
热自氧化	熱自氧化	thermal autoxidation
热阻	熱阻力	thermal resistance
人工操作	人工操作	manual operation
人工成本	人工成本	labor cost
人工肥料	人造肥料	artificial fertilizer
人工干燥室	人工乾燥室	artificial drier
人工老化	人工老化	artificial aging
人工器官	人造器官	artificial organ
[人工]神经网络	[人工]神經網絡	artificial neural network(ANN)
人工肾脏	人工腎臟	artificial kidney
人工甜味剂	人工甜味料	artificial sweetener
人工智能	人工智能	artificial intelligence(AI)
人孔	人孔	manhole
人血清清蛋白	人血清清蛋白	human serum albumin
人造短纤维	①人造短纖維②纖維長度	staple
人造沸石	人造沸石	permutit
人造革	人造皮	artificial leather
人造黄油	人造奶油	margarine
人造棉	嫘縈棉	staple rayon
人造燃料	人造燃料	artificial fuel
人造肉	人造肉	simulated meat
人造丝	人造絲, 嫘縈	artificial silk, rayon
人造丝短纤维	嫘縈棉	rayon staple
人造丝浆	嫘縈紙漿	rayon pulp
人造丝束	嫘縈束	rayon tow
人造纤维	人造纖維	artificial fiber, man-made fiber
认可概率	驗收機率	probability of acceptance
韧度	韌度	tenacity, toughness
韧皮纤维	韌皮纖維	stem fiber
韧性聚苯乙烯	韌化聚苯乙烯	toughened polystyrene
日本[天然]漆	高清漆, 黑漆	Japan
日晒变色	日晒變色	sun-discoloration
容差	公差, 容許度, 容許量	tolerance, allowance

祖国大陆名	台湾地区名	英　文　名
容积传氧系数, 体积传氧系数	體積傳氧係數	volumetric oxygen transfer coefficient
容积式泵, 排代泵	正位移泵	positive displacement pump
容积式流量计	正位移流量計	positive displacement flowmeter
容量	容量	capacity
容量滴定法	容量滴定[法]	volumetric titration
容量分析法	容積分析法	volumetry
容量估计	容量估計	capacity estimate
容量因子	容量因數	capacity factor
容器	容器	container
容许粗糙度	容許粗糙度	admissible roughness
容许间隔	容許間隔	tolerance interval
溶度参数	溶度參數	solubility parameter
溶度分级	溶度分級	solubility fractionation
溶度分析	溶度分析	solubility analysis
溶度积	溶度積	solubility product
溶度极限	溶度極限	solubility limit
溶纺纤维	溶紡纖維	solvent spun fiber
溶剂	溶劑	solvent
溶剂萃取	溶劑萃取	solvent extraction
溶剂法	溶劑法	solvent process
溶剂纺丝	溶劑紡絲	solvent spinning
溶剂分解作用	溶劑分解	solvolysis
溶剂分离器	溶劑分離器	solvent separator
溶剂锋面	溶劑鋒面	solvent front
溶剂复合	溶劑複合	solvent lamination
溶剂感敏性黏合剂	易溶膠黏性	solvent sensitive adhesive
溶剂合物	溶劑合物	solvate
溶剂化	溶合作用	solvation
溶剂化反应	溶劑分解反應	solvolysis reaction
溶剂化能	溶劑合能	solvation energy
溶剂化能力	溶劑化能力	solvating power
溶剂化效应	溶劑化效應, 溶劑合效應	solvating effect, solvation effect
溶剂化增塑剂	溶劑化塑化劑	solvating plasticizer
溶剂基黏合剂	溶劑基黏合劑	solvent based adhesive
溶剂基涂料	溶劑基涂料	solvent based coating
溶剂胶浆	液狀凝固劑	solvent cement
溶剂精炼煤	溶劑精煉煤	solvent-refined coal

祖国大陆名	台湾地区名	英 文 名
溶剂精制	溶劑精煉[法]	solvent refining
溶剂-链节相互作用	溶劑敏性黏合劑	solvent-segment interaction
溶剂裂化	溶劑裂解	solvent cracking
溶剂裂纹法	溶劑銀紋	solvent crazing
溶剂流铸法	溶劑流鑄法	solvent cast process
溶剂滤油	溶劑油	solvent naphtha
溶剂浓度	溶劑濃度	solvent strength
溶剂去灰分法	溶劑去灰分[法]	solvent deashing
溶剂去灰分过程	溶劑去灰分程序	solvent deashing process
溶剂染色	溶劑染色	solvent dyeing
溶剂蚀刻	溶劑腐蝕	solvent etching
溶剂试验	溶劑試驗	solvent test
溶剂梯度洗脱	溶劑梯度洗脱	solvent-gradient elution
溶剂相互作用参数	溶劑相互作用參數	solvent interaction parameter
溶剂效应	溶劑效應	solvent effect
溶剂型黏合剂	溶劑型黏合劑	solvent type adhesive
溶剂型增塑剂	溶劑型增塑劑	solvent type plasticizer
溶剂粘接	溶劑黏接	solvent welding
溶剂滞留量	溶劑滯留	solvent hold-up
溶剂助染	溶劑助染劑	solvent assisted dyeing
溶剂最大容限	溶劑容許溶解力	solvent tolerance
溶胶	膠溶體	sol
溶胶部分	溶膠部份	sol fraction
溶胶化	溶膠	solation
溶胶-凝胶转换	溶膠-凝膠轉化	sol-gel transformation
溶胶橡胶	溶橡膠	sol rubber
溶解	溶解	dissolution
溶解度	溶解度	solubility
溶解度隔板	溶解度障壁	solubility barrier
溶[解]度积常数	溶[解]度積常數	solubility product constant
溶解度试验	溶度試驗	solubility test
溶解度-温度曲线	溶解度-溫度曲線	solubility-temperature curve
溶解固体	溶解固體	dissolved solid
溶解空气	溶解空氣	dissolved air
溶解器	溶解器	dissolver
溶解热	溶解熱	heat of solution
溶解压	溶解壓力	solution pressure
溶解氧	溶氧	dissolved oxygen
溶解张力	溶解張力	solution tension

祖国大陆名	台湾地区名	英文名
溶菌酶	溶菌酶	lysozyme
溶气浮选	溶解空气浮选[法]	dissolved-air flotation
溶线	溶度曲线	solubility curve
溶氧分析仪	溶氧分析仪	dissolved oxygen analyzer
溶氧探头	溶氧探针	dissolved oxygen probe
溶液	溶液	solution
溶液的依数性	溶液的依数性	colligative property of solution
溶液丁苯橡胶	溶液丁苯橡胶	solution styrene-butadiene rubber
溶液纺丝	溶液纺丝	solution spinning
溶液化学	溶液化学	solution chemistry
溶液浇铸	溶液浇铸	solution casting
溶液聚合	溶液聚合	solution polymerization
溶液双重折射	溶液双折射	solution birefringence
溶液中基团分率	溶液中基团分率	group fraction in solution
溶胀	膨胀, 润胀	swell
溶胀比	膨润比	swelling ratio
溶胀度	膨润度	degree of swelling
溶胀剂	溶胀剂	swelling agent
溶胀量	溶胀量	swelling capacity
溶胀能力	溶胀能力	swelling power
溶胀压	①溶胀压	swelling pressure
溶胀值	溶胀值	swelling value
溶质	溶质	solute
熔点	熔点	fusion point, melting point
熔点下降	熔点下降	melting point depression
熔纺	熔纺[丝]	melt spinning
熔化	熔态	melt
熔化曲线	熔化曲线	fusion curve
熔化热	熔化热	heat of fusion
熔化温度	熔化温度	melting temperature
熔化物	熔体	melt
熔矩	熔点范围	melting range
熔炼	熔炼	smelting
熔炼坩埚	熔炼坩埚	smelting pot
熔炼炉	熔炼炉	smelting furnace
熔融	熔化, 熔合	fusion
熔融黏度	熔体黏度	melt viscosity
熔蚀作用	熔蚀作用	slag action
熔体流动指数	熔化指数	melt flow index

祖国大陆名	台湾地区名	英 文 名
熔体指数	熔化指數	melt index
熔浴气化器	熔浴氯化器	molten-bath gasifier
融化	融化	thawing
冗余[度]	冗餘	redundancy
冗余方程	冗餘方程	redundant equation
揉捏机	捏泥機	pug mill
鞣革厂	製革廠	tannery
鞣制	鞣製	tannage
肉桂油	桂皮油, 肉桂油	cinnamon oil
蠕变	①潛變②蠕變	creep
蠕变试验	潛變試驗	creep test
蠕变运动	蠕流運動	creeping motion
蠕流	蠕流	creeping flow
乳白玻璃	乳白玻璃	milky glass, opalescent glass
乳二糖	乳糖	lactobiose
乳化剂	乳化劑	emulsifying agent
乳化剂	乳化劑	emulsifier
乳化胶体	乳化膠體	emulsifying colloid
乳化漆	乳化漆	emulsion paint
乳化器	乳化器	emulsifier
乳化软膏	乳化軟膏	emulsifying ointment
乳化稳定	乳態穩定性	emulsion stability
乳化液膜	乳液薄膜	emulsion liquid membrane
乳化作用	乳化	emulsification
乳胶	乳化膠體, 類乳化液	emulsoid
乳清蛋白	乳白蛋白	lactalbumin
乳酸	乳酸	lactic acid
乳酸钙	乳酸鈣	calcium lactate
乳糖	乳糖	lactose
乳糖酶	乳糖酶	lactase
乳相	乳化相	emulsion phase
乳液, 乳浊液	乳液, 乳膠, 乳態	emulsion
乳液聚合	乳化聚合	emulsion polymerization
乳油分离机	乳油分離機	skimming machine
乳浊玻璃	不透明玻璃	opaque glass
乳浊剂	乳白劑, 失透劑	opacifier, opacifying agent
乳浊液(=乳液)		
入口	進口	inlet
软玻璃	軟玻璃	soft glass

祖国大陆名	台湾地区名	英　文　名
软段	軟段	soft segment
软钢	軟鋼	mild steel
软膏基质	軟膏基	ointment base
软化	軟化	softening
软化点	軟化點	softening point
软化剂	軟化劑	soft agent, softener, softening agent
软化器	去礦[物]質劑	demineralizer
软化水	軟化水, 去礦[物]質水	softened water, demineralized water, demineral water
软化温度	軟化溫度	softening temperature
软钾镁矾	硫酸鉀鎂礦	schonite
软件	軟體	software
软件包	套裝軟體	package software
软[链]段	柔性鏈段	soft segment
软锰矿	軟錳礦	pyrolusite
软木	軟木	softwood
软木粉	軟木粉	softwood flour
软泥法	軟泥法	soft-mud process
软润滑脂	軟潤滑脂	soft grease
软石蜡	軟石蠟	soft paraffin
软水	軟水	soft water
软水剂	①水軟化劑②水軟化器	water softener
软水铝石	水鋁土	boehmite
软酸	軟酸	soft acid
软纤维	軟纖維	soft fiber
软性 X 射线	弱 X 射線	soft X-ray
软性洗涤剂	軟性清潔劑	soft detergent
软质聚合物	軟質聚合物	soft polymer
软质蜡	軟蠟	soft wax
软质橡胶	軟橡膠	soft rubber
锐孔管	孔口管	orifice tube
锐缘孔口	銳緣孔口	sharp edged orifice
瑞利数	瑞利數	Rayleigh number
瑞斯托菌素	利黴素	ristocetin
润滑	潤滑	lubrication
润滑剂	潤滑劑	lubricant
润滑力学	潤滑力學	lubrication mechanics
润滑性	潤滑性	lubricity

祖国大陆名	台湾地区名	英 文 名
润滑油	潤滑油	lubricating oil
润滑脂	[潤]滑脂	grease, lubricating grease
润湿表面积	潤濕表面積	wetted surface area
润湿剂	保濕劑, 潤濕劑	humectant, wetting agent
润湿率	潤濕率	irrigation rate
润湿效应	潤濕效應	wetting effect
润湿周边	潤濕周邊	wetted perimeter
弱电解质	弱電解質	weak electrolyte
弱碱	弱鹼	weak base
弱可逆性	弱可逆性	weak reversibility
弱酸	弱酸	weak acid

S

祖国大陆名	台湾地区名	英 文 名
洒水器	灑水器	sprinkler
萨纶管	紗隆管	saran pipe
赛璐玢	賽珞凡	cellophane
赛璐珞	賽珞珞	celluloid
三次采油	三次石油回收	tertiary oil recovery
三对角矩阵	三對角矩陣	tridiagonal matrix
三分子反应	三分子反應	termolecular reaction, trimolecular reaction
三合盐	三合鹽	triple salt
三级处理	三級處理	tertiary treatment
三级化学澄清器	三級化學澄清器	tertiary chemical clarifier
三角缺口	三角凹槽	triangular notch
三角图	三角圖	triangular diagram
三角坐标	三角座標	triangular coordinates
三聚丙烯	三聚丙烯	propylene trimer
三聚氰胺-甲醛树脂	三聚氰胺甲醛樹脂	melamine formaldehyde resin
三聚氰胺黏合剂	三聚氰胺黏合劑, 美耐皿黏合劑	melamine adhesive
三聚体	三聚物	trimer
三氯乙醛	三氯乙醛	chloral
三式控制	三式控制	three-mode control
三式控制器	三式控制器	three-mode controller
三水铝石	水礬土, 水鋁氣	gibbsite
三水铝石耐火土	水礬土耐火土	gibbsite fireclay

祖国大陆名	台湾地区名	英　文　名
三羧酸循环	三羧酸循環	tricarboxylic acid cycle
三通	三通, T 型管	T-piece, tee
三相点	三相點	triple point
三相反应器	三相反應器	three-phase reactor
三相流化床	三相流化床	three-phase fluidized bed
三相流态化	三相流態化	three-phase fluidization
三元共沸物	三元共沸液	ternary azeotrope
三元系[统], 三组分系统	三成分系統, 三元系	ternary system
三轴应力	三軸應力	triaxial stress
三组分系统(＝三元系[统])		
散堆填料	隨機填充	random packing
散裂	散裂	spall, spallation, spalling
散裂试验	散裂試驗	spalling test
散裂试验板台	散裂試驗板台	spalling test panel
散热片	散熱片	cooling fin
散热器	散熱器	radiator
散射	散射	scattering
散射池	散射池	scattering cell
散射角	散射角	scattering angle
散射体积	散射體積	scattering volume
散射系数	散射係數	scattering coefficient
散射形式	散射形式	scattering pattern
散式流态化	顆粒式流體化, 散式流體化	particulate fluidization
散纤维染色	散纖維染色	stock dyeing
散装储存	散裝儲存	bulk storage
散装填料	堆積填充	dumped packing
扫描电子显微镜	掃描電子顯微鏡	scanning electron microscope (SEM)
扫描电子显微镜法	掃描電子顯微鏡法	scanning electron microscopy
扫描量热法	掃描量熱法	scanning calorimetry
扫描速度	掃描速度	scanning rate
色淀	色澱, 沈澱色料	lake
色谱电泳	電泳層析	chromatoelectrophoresis
色谱法, 层析法	層析法	chromatography, chromatographic method
色谱分离	層析分離	chromatographic separation
色谱聚焦	聚焦層析	chromatofocusing
色谱图	層析圖	chromatogram

祖国大陆名	台湾地区名	英 文 名
色谱仪	層析儀	chromatograph
色谱柱	層析柱	chromatographic column
色谱柱式脉冲反应堆	層析柱式脈動反應器	chromatographic column pulse reactor
色散力	分散力	dispersion force
色散曲线	分散曲線	dispersion curve
杀虫剂	殺蟲劑	insecticide, pesticide
杀稻瘟菌素	殺稻瘟菌素	blasticidin
杀菌剂	殺菌劑	fungicide, germicide
杀螨剂	殺壁虱劑, 殺蟎劑	acaricide, miticide
杀鼠剂	殺鼠劑	rodenticide
沙丁鱼油	沙丁魚油	sardine oil
纱	紗, 線	yarn
纱染机	紗染機	yarn dyeing machine
砂布	砂布	emery cloth
砂浆	灰泥	mortar
砂滤法	砂濾法	sand filtration
砂滤器	砂濾器	sand-bed filter
砂轮	砂輪	emery wheel
砂磨	砂磨	sand mill
砂纸	砂紙, 磨擦紙	sand paper, abrasive paper
鲨鱼皮	鯊魚皮	sharkskin
筛	篩	sieve, sifter
筛板	篩板, 篩盤	sieve plate, sieve tray, screen plate
筛板塔	篩板塔	sieve plate column, sieve plate tower
筛布	篩布	bolting cloth
筛分	篩[分]	sieving
筛分试验	篩分試驗	sieve test
筛孔	篩孔	screen aperture, screen opening
筛滤器	篩濾機	screen filter
筛磨	篩曆機	screen mill
筛目	網目	mesh
筛目效率	篩孔效率	mesh efficiency
筛上物分布曲线	篩上物分布曲線	oversize distribution curve
筛网	篩子	screen
筛网分离器	篩孔分離器	mesh separator
筛网过滤器	篩網過濾器	mesh filter
筛网输送机	篩式運送機, 篩運機	screen conveyor
筛析	篩析	mesh analysis, screen analysis, sieve analysis

祖国大陆名	台湾地区名	英 文 名
筛选	篩選	screening
筛选测验	篩轉試驗	screening test
筛选机	篩選機, 篩粉機	bolting mill, sifting machine
筛选效率	篩選效率	screening efficiency
筛选装置	篩選裝置	screening device
筛用丝	篩用絲	bolting silk
筛组	篩組	screen bank, bank screen
晒裂	晒裂	suncrack, sun-cracking, sun-crazing
山梨糖醇酐单油酸酯	去水山梨糖醇三硬脂酸酯	sorbitan monooleate
闪点	閃[火]點	flash point
闪点测定器	閃[火]點測定器	flash point apparatus
闪烁火焰	閃爍火燄	flickering flame
闪速焙烧炉	急驟焙燒爐	flash roaster
闪速干燥	驟沸乾燥	flash drying
闪速蒸馏	驟餾	flash distillation
闪锌矿	閃鋅礦	zinc blende
闪蒸	驟蒸發, 驟汽化	flash evaporation, flash vaporization
闪蒸槽	驟沸桶	flash drum
闪蒸计算	驟沸計算	flash calculation
闪蒸浓缩器	急驟濃縮器	flash concentrator
闪蒸气	驟蒸氣	flash vapor
闪蒸曲线	驟汽化曲線	flash vaporization curve
闪蒸室	驟沸室	flash chamber
闪蒸塔	驟餾塔	flash distillation column, flash tower, flashing liquid
闪蒸液	驟沸液	flashing liquid
扇形排风机	通風扇	fan blower
商业发展	商業發展	commercial development
熵	熵	entropy
熵变化	熵變化	entropy change
熵产生	熵生成	entropy generation, entropy production
熵方程式	熵方程式	entropy equation
熵衡算	熵均衡	entropy balance
熵流, 熵通量	熵流, 熵通量	entropy flow
熵通量(＝熵流)		
熵增原理	熵增原理	principle of entropy increase
上光剂	發光劑	lustering agent
上光盘	上光機	polishing pan

祖国大陆名	台湾地区名	英 文 名
上极限	上限	upper limit
上浆机	上漿機	sizing machine
上胶剂	上膠劑	sizing agent
上控制限	管制上限	upper control limit
上流塔	上流塔, 上流柱	upflow column
上清液	上澄液	supernatant
上升	上升	lift
上升系数	上升係數	lift coefficient
上限	上限	upper bound
上相(=顶相)		
上行管	上升管	ascending pipe
上行色谱法	上升層析法	ascending chromatography
上行窑	階級窯	ascending kiln
上溢(=溢流)		
上游	上游	upstream
上游压力	上游壓力	upstream pressure
上釉	浸釉[法]	glazing
上脂	上脂	stuffing
烧碱(=氢氧化钠)		
烧碱法浆	苛性鈉紙漿	soda pulp
烧碱法(制浆)	鹼法	soda process
烧结	燒結, 燒結法	sinter, sintering
烧结玻璃	燒結玻璃	sintered glass
烧结玻璃过滤器	燒結玻璃過濾器	sintered glass filter
烧结成型	燒結成型	sinter forming
烧结带	燒結區	sintering zone
烧结料	燒結物料	sintered material
烧结膜	燒結膜	sinter membrane
烧瓶	燒瓶	flask
烧蚀	融磨	ablation
舌形板	舌形板	jet tray
蛇管	旋管, 蛇管, 盤管	coil
蛇管冷凝器	旋管冷凝器	coil condenser
蛇管散热器	旋管散熱器	coiled radiator
蛇管式蒸发器	旋管蒸發器	coil evaporator
蛇笼型聚电解质	蛇籠型聚電解質	snake-cage polyelectrolyte
蛇笼型两性离子交换 树脂	蛇籠型兩性離子交換 樹脂	snake-cage amphoteric ion exchange resin
蛇形管	旋管	coiled pipe

祖国大陆名	台湾地区名	英　文　名
设备安全	裝置安全性	equipment safety
设备成本	裝置成本	equipment cost
设备规格表	設備規格	specification of equipment
设备可靠性	裝置可靠性	equipment reliability
设备流程图	裝置流程圖	equipment flowsheet
设定值	設定點	set point
设定值变动	設定點變動	set point variation
设定值跟踪	設定點追蹤	set point tracking
设定值控制	設定點控制	set point control
设定值扰动	設定點擾動	set point disturbance
设定值响应	設定點應答	set point response
设计	設計	design
设计变量	設計變數	design variable
设计标准	設計標準	design standard
设计参数	設計參數	design parameter
设计方程式	設計方程式	design equation
设计工程师	設計工程師	design engineer
设计工具	設計工具	design tool
设计模型	設計模型	design mode
设计数据	設計數據	design data
设计图	設計圖	design chart
设计项目	設計專案	design project
设计主管工程师	專案工程師	project engineer
设计准则	設計準則	design criterion
社会成本	社會成本	social cost
射流	①射流, 流出②流出量	efflux
射流穿透长度	射流穿透長度	jet penetration length
射流反应器	射流反應器	jet reactor
射流时间	射流時間	efflux time
射流速度	射流速度	efflux velocity
射流系数	射流係數	efflux coefficient
射气热分析	放射性熱分析[法]	emanation thermal analysis
X 射线谱	X 射線譜	X-ray spectrum
X 射线衍射	X 射線繞射	X-ray diffraction
X 射线荧光	X 射線螢光	X-ray fluorescence
摄动理论(=微扰理论)		
摄谱仪	攝譜儀	spectrograph
摄取	攝取, 升道	uptake
摄取速率	攝取速率	uptake rate

祖国大陆名	台湾地区名	英　文　名
摄氧	攝氧[量]	oxygen uptake
摄氧速率	攝氧速率	oxygen uptake rate
伸长	延伸	extension
砷杀虫剂	砷殺蟲劑	arsenical insecticide
砷杀鼠剂	砷殺鼠劑	arsenical rodenticide
深度冷冻	低溫程序	cryogenic process
深水井注入	深井注入[法]	deep-well injection
神经网络训练	神經網絡訓練	neural network training
神经元	神經元	neuron
渗出	滲出, 泌出	exudation
渗滤	滲濾	diafiltration, infiltration, percolation
渗滤器	滲濾器	percolation extractor
渗滤液	滲濾液	percolate
渗入测试	浸透試驗, 穿透試驗	penetration test
渗入度	針入度	degree of penetration
渗碳	①膠接②滲碳③硬化	cementation
渗碳法	①膠接程序②滲碳程序	cementation process
渗透萃取	透[過]萃[取]法	pertraction
渗透电位	穿透電位	penetration potential
渗透剂	浸透劑	penetrant, penetrating agent
渗透扩散	滲透擴散	osmotic diffusion
渗透试验	透過性試驗	permeability test
渗透试验器	滲透試驗器	osmoscope
渗透通量	滲透通量	permeation flux
渗透物	滲透物	permeate
渗透系数	滲透係數	osmotic coefficient
渗透性	透過性	permeability
渗透压	滲透壓	osmotic pressure
渗透压计	滲[透]壓[力]計	osmometer
渗透仪	透過性測定儀	permeability apparatus
渗透蒸发	滲透蒸發	pervaporation
渗透[作用]	滲透	osmosis
渗析, 透析	透析[作用]	dialysis
渗析槽	透析槽	dialysis cell
渗析培养, 透析培养	透析培養	dialysis culture
渗析器, 透析器	透析儀	dialyzator, dialyzer
渗析液, 透析液	透析液	dialyzate
渗余物	滲餘物	retentate

祖国大陆名	台湾地区名	英 文 名
升华	昇華	sublimation
升华干燥法	昇華乾燥[法]	sublimate drying
升华器	①昇華器②昇華材料	sublimer
升华热	昇華熱	heat of sublimation
升华物	昇華物	sublimate
升华装置	昇華裝置	sublimation apparatus
升降机	升降機, 升液器	lift
升降色谱法	升降層析法	ascending-descending chromatography
升膜蒸发器	升膜蒸發器	climbing-film evaporator
升压泵	增壓泵	booster pump
升压喷射器	助力噴射器	booster ejector
升压器	增壓器	booster
升压压力表	升壓壓力錶	boost gage
生产成本	生產成本	production cost
生产过剩	過度生產	overproduction
生产进度表	生產時程表	production schedule
生产控制	生產管制	production control
生产力	生產力	productivity
生产率	生產速率	production rate
生产能力	生產能力	production capacity
生产能力过剩	超負荷	overcapacity
生产数据	生產數據	production data
生产线	生產線	production line
生成焓	生成焓	enthalpy of formation
生成热	生成熱	heat of formation
生成速率	生成速率	birth rate
生存理论	生存理論	survival theory
生存率	生存率	survival rate
生化反应	生化反應	biochemical reaction
生化反应工程	生化反應工程	biochemical reaction engineering
生化反应器	生化反應器	biochemical reactor
生化分离	生化分離	biochemical separation
生化工程	生化工程	biochemical engineering
生化热力学	生化熱力學	biochemical thermodynamics,
生化需氧量	生物需氧量, 生化需氧量	biochemical oxygen demand(BOD)
生化需氧量负载	生化需氧量負載	BOD load
生活污水	衛生污水, 家庭污水	sanitary wastewater
生能反应	生能反應	energy producing reaction

祖 国 大 陆 名	台 湾 地 区 名	英 文 名
生石灰	生石灰	quick lime
生丝精练	生絲精練, 絲綢精練	silk degumming, silk scouring
生丝强伸力试验计	強力延伸試驗器	serimeter
生态系统	生態系統	ecosystem
生铁	生鐵	ingot iron, pig iron
生物变质	生物腐蝕	biodeterioration
生物材料	生物材料	biological material, biomaterial
生物测定	生物測定	bioassay
生物处理	生物處理[法]	biological treatment
生物处理过程	生物處理法	biological treatment process
生物传感器	生物感測器	biosensor
生物催化	生物催化[作用]	biocatalysis
生物催化反应	生物催化反應	biocatalytic reaction
生物催化剂	生物催化劑	biocatalyst
生物大分子	生物大分子	biomacromolecule
生物电池	生物電池	biocell
生物毒性	生物毒性	biotoxicity
生物毒性试验	生物毒性試驗	biotoxicity test
生物发光	生物發光	bioluminescence
生物反应器	生物反應器	biological reactor, bioreactor
生物分离	生物分離	bioseparation
生物分散	生物分散	biological dispersion
生物工程	生物工程學	bioengineering
生物功能试剂	生物功能試劑	biofunctional reagent
生物固体	生物性固體	biological solid
生物过程	生物程序, 生物過程	biological process, bioprocess
生物合成	生物合成	biosynthesis
生物合成反应	生物合成反應	biosynthetic reaction
生物化学	生物化學	biochemistry
生物技术	生物技術	biotechnology
生物碱	生物鹼	alkaloid
生物降解	生物分解, 生物降解	biodegradation
生物降解塑料	生物可分解塑膠	biodegradable plastic
生物降解洗涤剂	生物降解清潔劑	biodegradable detergent
生物降解性	生物降解性, 生物分解性	biodegradability
生物浸取	生物浸取	bioleaching
生物聚合物	生物聚合物	biological polymer, biopolymer
生物可利用率	生物可利用率	bioavailability

祖国大陆名	台湾地区名	英 文 名
生物流体力学	生物流體力學	biofluid mechanics
生物滤器	生物濾器	biological filter
生物膜	生物[薄]膜	biological membrane, biological film
生物能学	生物能量學	bioenergetics
生物群	生物相	biota
生物燃烧	生化燃燒	biochemical combustion
生物渗透	生物滲透	bio-osmosis
生物塔	生物塔	biological tower
生物特异性连结	生物特異性連結	biospecifically binding
生物需氧量	生物需氧量	biological oxygen demand
生物絮凝	生物絮凝	biological flocculation
生物延时	生物時間落後	biological time lag
生物氧化	生物氧化	biological oxidation
生物医学	生物醫學	biomedicine
生物医学工程	生醫工程	biomedical engineering
生物医学技术	生醫技術	biomedical technology
生物医药器件	生醫裝置	biomedical device
生物制剂	生物製劑	biological agent
生物质	生質	biomass
生物质能	生質能	biomass energy
生物质气化	生質氣化	biomass gasification
生物质燃料	生質燃料	biomass fuel
生物质热解	生質高溫分解	biomass pyrolysis
生物质转化	生質轉化	biomass conversion
生物质转化过程	生質轉化程序	biomass conversion process
生物转化	生物轉化	bioconversion, biotransformation
生物转盘法	[旋轉]生物盤法	rotating-disc process
生锈	生銹	stain
生长参数	生長參數	growth parameter
生长动力学	生長動力學	growth kinetics
生长静止期	生長靜止期	stationary growth phase
生长率	生長速率	growth rate
生长曲线	生長曲線	growth curve
生长收率	生長產量	growth yield
生长收率系数, 增殖收率系数	生長產量係數	growth yield coefficient
生长因子	生長因子	growth factor
生殖周期	繁殖周期	reproductive cycle
声控	聲控	acoustic control

祖国大陆名	台湾地区名	英 文 名
声速	音速	sonic velocity
声速流动	聲速流動, 音速流動	acoustic flow, sonic flow
声学	聲學	acoustics
声致发光	聲發光	sonoluminescence
剩余贡献(=残余贡献)		
剩余焓(=残余焓)		
剩余熵(=残余熵)		
剩余体积(=残余体积)		
剩余项(=残余项)		
剩余性质(=残余性质)		
失活	失活, 去活化	inactivation, deactivation
失活剂	去活化劑	deactivator
失控	失控	runaway
施肥	施肥	fertilization
施工图设计	細部工程設計	detailed engineering design
施密特数	西門諾夫數	Schmidt number
湿壁塔	濕壁管	wetted wall column, wetted-wall tower
湿表面	濕[表]面	wet surface
湿存水	溼水分	hygroscopic water
湿度	溼度	humidity
湿度比	溼度比	humidity ratio
湿度表	溼度表	humid chart, humidity chart
湿度测定法	溼度測定法	psychrometry
湿度测定器	水分測定器	moisture determination apparatus
湿度差	溫度計量差	psychrometric difference
湿度计	溼度計	hygrometer, moisture meter
湿度图	濕度圖	psychrometric chart
湿度线	溼度線	psychrometric line
湿度因数	水分因數	moisture factor
湿法分离	濕法分離	wet separation
湿法过程	濕製程	wet process
湿法冶金[学]	溼法冶金[學]	hydrometallurgy
湿纺	濕紡[絲]	wet spinning
湿含量	含水量	moisture content
湿化	水化	slaking
湿井	濕井	wet well
湿面积	潤濕面積	wetted area
湿磨	濕磨	wet grinding
湿气	濕[天然]氣, 含油氣	wet gas

祖国大陆名	台湾地区名	英 文 名
湿气计	濕氣計	wet gas meter
湿球温度	濕球溫度	wet bulb temperature
湿球温度计	濕球溫度計	wet-bulb thermometer
湿球下降	濕球下降	wet-bulb depression
湿热	溼熱	humid heat
湿筛选	濕篩選	wet screening
湿式密闭容器	濕式密閉容器	wet seal holder
湿式燃烧	濕式燃燒	wet combustion
湿式氧化	濕式氧化[法]	wet oxidation
湿体积	溼體積	humid volume
湿物料	濕物料	moist material
湿辗机	濕輾機	wet pan mill
湿蒸汽	濕蒸汽	wet steam
十六烷值	十六烷值	cetane number
十六烷指数	十六烷指數	cetane index
石膏	石膏	gypsum
石膏模	石膏模	plaster mold
石灰	石灰	lime
石灰乳	石灰乳	lime milk
石灰石	[石]灰石	limestone
石灰苏打法	石灰蘇打法	lime-soda process
石灰窑	石灰窯	lime kiln
石灰皂	鈣皂	lime soap
石蜡	石蠟	paraffin, paraffin wax
石蜡馏分	石蠟餾出物	paraffin distillate
石蜡油	石蠟油	paraffin oil
石蜡皂	石蠟皂	paraffin soap
石蜡纸	石蠟紙	paraffin paper
石蜡转化	石蠟轉化	paraffin conversion
石棉	石綿	asbestos
石棉过滤器	石綿過濾器	asbestos filter
石磨修整	石磨刻鑿	buhrstone mill dressing
石墨	石墨纖維	graphite
石墨电极	石墨電極	graphite electrode
石墨坩埚	石墨坩堝	graphite crucible
石墨化	石墨化	graphitization
石墨耐火材料	石墨耐火物	graphite refractory
石墨润滑脂	石墨滑脂	graphite grease
石墨纤维	石墨纖維	graphitic fiber

祖 国 大 陆 名	台 湾 地 区 名	英 文 名
石墨纤维强化塑料	石墨纖維強化塑膠	graphite fiber reinforced plastics
石脑油	①石油腦②輕油	naphtha, petroleum naphtha
石脑油裂解	輕油裂解	naphtha cracking
石脑油热解	輕油裂解	naphtha pyrolysis
石蕊试纸	石蕊試紙	litmus paper
石盐	岩鹽	halite
石英	石英	quartz
石英玻璃	石英玻璃, 矽石玻璃	quartz glass, silica glass
石英管	石英管	quartz tube
石英细粉	石英粉	silica flour
石英岩	石英石	quartzite
石油	石油	petroleum
石油工程	石油工程	petroleum engineering
石油化工厂	石[油]化[學]工業區	petrochemical complex
石油化学	石油化學	petrochemistry
石油化学工业	石油化學工業, 石化工業	petrochemical industry
石油化学品	石油化學品, 石化品	petrochemical
石油回收	石油回收	oil recovery
石油加工工程	石油煉製工程	petroleum processing engineering
石油加工技术	石油煉製技術	petroleum processing technology
石油焦	石油焦	petroleum coke
[石油]沥青	瀝青	asphaltum
石油炼制	石油煉製	petroleum refining
石油醚	石油醚	petroleum ether
石油气	石油氣	petroleum gas
时变系统	時變系統	time variant system, time varying system
时程表	時程表	time schedule
时间比率法	時間比率法	time ratio method
时间标度	時間標度	time scaling
时间常量	時間常數	time constant
时间导数	時間導數	time derivative
时间关连	時間關連[性]	time dependence
时间控制	時間控制	time control
时间线	時間線	time line
时间序列	時間序列	time series
时间序列模型	時間序列模式	time series model
时间最短控制	時間最短控制	time minimum control
时空	空間時間	space time

祖国大陆名	台湾地区名	英 文 名
时效硬化	經時硬化	age hardening
时延	時間遲延, 時間延遲	time delay
时域	時間領域	time domain
时滞	時間落後	time lag
实报酬率	實報酬率	true rate of return
实沸点蒸馏曲线	真沸點曲線	true boiling point curve
实际阶段	實際階	real stage
实际可实现性	實際可真實化性	physical realizability
实际上升	實際上升	actual lift
实际[塔]板	實際板	actual plate
实际塔板数	實際板數	actual plate number
实际稳定区	實用穩定區[域]	region of practical stability
实际稳定性	實用穩定性	practical stability
实平均	實平均	true mean
实验设计	實驗設計	experimental design
实验室反应器	實驗室反應器	laboratory reactor
实验室手册	實驗室手冊	laboratory manual
实验数据	實驗數據	data experimental, experimental data
实验误差	實驗誤差	experimental error
实验用反应器	實驗用反應器	experimental reactor
炻器	缸器, 陶石器	stoneware
蚀刻	蝕刻	etching
蚀刻剂	蝕刻液	etchant
食品加工	食品加工	food processing
食物-微生物比	食物-微生物比	food-to-microorganism ratio
食盐	食鹽	table salt
食用靛蓝	靛胭脂	indigo carmine
食用牛脂	食用牛脂	edible tallow
食用脂肪	食用脂肪	edible fat
使用期	適用期, 耐用期	serviceable life
市场成本	市場成本	market cost
市场研究	市場研究	market research
示波器	示波器	oscillograph, oscilloscope
示差热膨胀测量术	微差膨脹測定法	differential dilatometry
示范装置	示範裝置	demonstration unit
示意图	示意圖	schematic diagram, schematics
示踪化学	示蹤[劑]化學	tracer chemistry
示踪技术	示蹤劑法	tracer technique
示踪剂	示蹤劑, 示蹤器	tracer

祖国大陆名	台湾地区名	英　文　名
示踪曲线	示蹤劑應答曲線	tracer curve
示踪同位素	示蹤同位素	tracer isotope
示踪响应曲线	示蹤劑應答曲線	tracer response curve
示踪信息	示蹤劑資訊	tracer information
事件矩阵	事件矩陣	occurrence matrix
势	勢	potential
势差	勢差	potential difference
势垒	勢障	potential barrier
势流	勢流	potential flow
势能	位能	potential energy
势能面	位能面	potential energy surface
势偏差	勢偏差	potential deviation
势线	勢線	potential line
势修正	勢修正	potential correction
视镜	窺鏡	sight glass
视孔	視孔	sight hole
视密度, 表观密度	視密度	apparent density
视因数	視因數	view factor
试饼	試餅	pat
试管	試管	test tube
试验	試驗法	testing
适合性	適合性	suitability
适应调谐	適應調諧	adapt tuning
适用性	應用性	applicability
释放角, 开角	釋放角	angle of release
嗜冷菌	嗜冷菌	psychrophile, psychrophilic bacteria
嗜热细菌	嗜熱菌	thermophilic bacteria
噬菌体	噬菌體	phage
收集环	收集環	collecting ring
收集剂	收集器	collector
收敛加速	加速收斂	convergence acceleration
收敛判据	收斂判斷	convergence criterion
收率	①產率, 產量②降伏	yield
收率系数	產率係數	yield coefficient
收缩剂	收縮劑	shrinking agent
收缩裂纹	收縮開裂	shrinkage crack
收缩率	收縮率	shrinkage rate
收缩能力	收縮能力	shrinking power
收缩曲线	收縮曲線	shrinkage curve

祖国大陆名	台湾地区名	英　文　名
收缩损失	收縮損失	contraction loss
收缩稳定性	收縮穩定性	contraction stability
收缩系数	收縮係數	coefficient contraction, contraction coefficient
收缩应力	收縮應力	shrinking stress
收缩张力试验机	收縮張力試驗機	shrinkage tension tester
收益率	收益率	earning rate
手动重整	手動重整	manual reset
手性分离	對掌性分離	chiral separation
手摇干湿度计	搖轉濕度計	sling psychrometer
守恒方程式	守恆方程式	conservation equation
守恒律	守恆律	conservation law
受槽	接受器	receiver
受激分子	受激分子	energized molecule
受激响应技术	刺激應答法	stimulus-response technique
受控变量	受控變數	variable controlled
受体	接受體, 接收器	acceptor
受阻沉降	受阻沉降	hindered sedimentation, hindered settling
枢轴	樞, 樞軸	pivot
疏水泵	排洩泵	drain pump
疏水胶体	疏水膠體	hydrophobic colloid
疏水器	蒸汽阱, 祛水器	steam trap
疏水色谱[法]	疏水層析法	hydrophobic chromatography
疏水物	①疏水性②疏水物	hydrophobe
疏水纤维	疏水纖維	hydrophobic fiber
疏水性颗粒	疏水性顆粒	hydrophobic particle
疏液胶体	疏液膠體	lyophobic colloid
舒尔茨-齐姆分布	舒爾茨-齊姆分布	Schulz-Zimm distribution
输出[变]量	輸出變數	output variable
输出层	輸出層	output layer
输出函数	輸出函數	output function
输出集	輸出集	output set
输出信号	輸出信號	output signal
输出压力	輸出壓力	output pressure
输入	輸入	input
输入变量	輸入變數	input variable
输入函数	輸入函數	input function
输入搅动	輸入擾動	input disturbance
输入节点	輸入節點	input node

祖国大陆名	台湾地区名	英 文 名
输入信号	輸入信號	input signal
输送	運送	conveying
输运现象(=传递现象)		
熟化器	老化器	aging machine
熟化时间	熟化時間	curing time
熟化试验	硬化試驗	curing test
熟料	燒粉	chamotte
熟石膏	熟石膏	plaster of Paris
熟石灰	水合石灰, 熟石灰	hydrated lime
熟铁	熟鐵	wrought iron
熟油	熟[煉]油	boiled oil
曙红染料	曙紅染料	eosin dye
曙红色淀	曙紅色澱	eosin lake
树胶	膠	gum
树脂	樹脂	resin
树脂鞣法	樹脂鞣製	resin tannage
树脂皂	樹脂皂	resin soap
竖管式蒸发器	豎管蒸發管	vertical tube evaporator
竖甑	豎甑	vertical retort
数据	數據	datum
数据	數據	data
数据不准性	數據不準性	data inaccuracy
数据处理	數據處理	data processing
数据分析	數據分析	data analysis
数据校正(=数据调谐)		
数据拟合	數據擬合	data fitting
数据日志	數據日誌	data log
数据筛选	數據篩選	data screening
数据调谐, 数据校正	數據调和	data reconciliation
数均分子量	數量平均分子量	number average molecular weight
数量级	量階	order of magnitude
数式化	①配方②公式化	formulation
数学程序	數學規劃	mathematical programming
数学模拟	數學模式化	mathematical modeling
数学模型	數學模型	mathematical model
数值方法	數值方法	numerical method
数值分析	數值分析	number analysis, numerical analysis
数值解	數值解	numerical solution
数字计算机	數位計算機	digital computer

祖国大陆名	台湾地区名	英 文 名
数字记录器	數位記錄器	digital recorder
数字控制	數位控制, 數值控制	digital control, numerical control
数字模拟计算机	拼合計算機, 複合計算機	hybrid computer
衰变	衰變, 衰退	decay
衰变比	衰退比, 衰變比	decay ratio
衰变常数	衰變常數	decay constant
衰变催化剂	衰退觸媒	decaying catalyst
衰变机理	蛻變機構	disintegration mechanism
衰变影响选择性	衰退影響之選擇性	decay-affected selectivity
衰减	衰減[作用]	attenuation
衰减器	衰減器	attenuator
衰减效应	衰減效應	attenuation effect
衰减因子	衰減因數	attenuation factor
衰亡期	衰減期	decline phase
双倍定率递减折旧法	雙倍定率遞減折舊法	double declining balance method
双臂捏合机	雙臂捏合機	double arm kneading mixer
双部位机理	雙部位機構	dual-site mechanism
双层法, 联立模块法	雙層法	two tier approach
双层隔热玻璃板	雙層隔熱玻璃板	double-glazed insulating pane, thermopane
双层过滤器	雙介質過濾器	dual-medium filter
双层绝热量热器	絕熱雙卡計	adiabatic twin calorimeter
双层中空玻璃	雙層玻璃	pair glass
双重促进催化剂	雙重促進化觸媒	double promoted catalyst
双重上胶	雙重上膠	double sizing
双重塔	雙重塔	double column
双重循环	雙重循環	dual recirculation
双动泵	雙動泵	double acting pump
双动搅拌器	雙動攪拌器	double motion agitator
双分散	雙分散	bidispersion
双分子反应	雙分子反應	bimolecular reaction
双峰的	雙模式	bimodal
双峰孔径分布	雙模式孔徑分布	bimodal pore-size distribution
双缸往复泵	雙缸往復泵	duplex reciprocating pump
双功能催化剂	雙功能觸媒	dual-function catalyst
双辊轧碎机	雙輥軋碎機	double roll crusher
双滑线电桥	雙滑線電橋	double slide wire bridge
双极矩	雙極矩	bipolar moment
双极性电极	雙極電極	bipolar electrode

祖国大陆名	台湾地区名	英 文 名
双桨混合器	雙漿混合器	double paddle mixer
双结点溶度曲线	雙結溶解度曲線	binodal solubility curve
双金属簇	雙金屬團	bimetallic cluster
双金属催化剂	雙金屬觸媒	bimetallic catalyst
双金属调温计	雙金屬調溫計	bimetallic thermoregulator
双金属温度计	雙金屬溫度計	bimetallic thermometer
双口阀	雙口閥	double ported valve
双连续结构	雙連續結構	bicontinuous structure
双螺带混合机	雙臂帶混合機	double helical ribbon mixer
双螺杆挤出机	雙螺桿擠出機	twin screw extruder
双模式系统	雙模式系統	dual-mode system
双膜理论	雙薄膜理論	two-film theory
双曲线型[动力学]方程	雙曲線型[動力學]方程	hyperbolic type[kinetic]equation
双溶剂萃取	雙溶劑萃取	double solvent extraction
双式过滤器	雙式濾器	double filter
双式控制	雙式控制	dual-mode control, two-mode control
双室稳定化池	雙室穩定化池	two-cell stabilization pond
双水相萃取	雙水相萃取	aqueous two-phase extraction
双水相体系(=双水相系统)		
双水相系统, 双水相体系	雙水相系統	aqueous two-phase system
双糖	雙醣	disaccharide
双头螺纹短接头	短絲接管	short nipple
双位控制器	雙位控制器	two-position controller
双位驱动器	雙位引動器	two-position actuator
双稳定装置	雙穩定裝置	bistable device
双吸叶轮机	雙吸取葉輪機	double suction impeller
双向狭缝	雙向狹縫	bilateral slit
双效蒸发	雙效蒸發	double effect evaporation
双效蒸发器	雙效蒸發器	double effect evaporator
双旋光	雙旋光	birotation
双循环	二元循環, 雙循環	binary cycle
双用控制	雙用控制	duplex control
双折射	雙折射	birefraction
双折射流体	雙折射流體	birefringent fluid
双锥掺合机	雙錐摻合機	double cone blender
双锥分级机	雙錐類析器	double cone classifier

祖国大陆名	台湾地区名	英　文　名
双阻力理论	雙阻力理論	two-resistance theory
双组分混合物(＝二元混和物)		
双座阀	雙座閥	double seat valve
爽身粉	滑石粉	talcum powder
水/油乳剂软膏基	水/油乳劑軟膏基	W/O emulsion ointment base
水包油乳状液	水中油乳液	oil in water emulsion
水玻璃	水玻璃	water glass
水簸机	水簸器	hydraulic jig
水处理	水處理	water treatment
水锤	水鎚	water hammer
水淬	水淬	shredding
水萃取物	水相萃取物	aqueous extract
水阀	水閥	water valve
水分离器	水分離器	water separator
水封	水密封	water seal
水合	水合	hydration
水合器	水合器	hydrator
水合热	水合熱	heat of hydration
水合物	水合物	hydrate
水合盐	水合鹽	hydrated salt
水化酶	水合酶	hydrase
水加热器	水加熱器	water heater
水浆涂料	泥漿塗料	slurry coating
水胶体	親水膠體	hydrocolloid
水解	水解(作用)	hydrolysis
水解产物	水解產物	hydrolysate
水解酶	水解酶	hydrolase, hyrolytic enzyme
水解吸附	水解吸附	hydrolytic adsorption
水净化	水淨化, 水純化	water purification
水冷结晶器	水冷結晶器	water cooled crystallizer
水冷却	水冷卻	water cooling
水冷却器	水冷卻器	water cooler
水力半径	水力半徑	hydraulic radius
水力泵	水力泵	hydraulic pump
水力超载量	水力超載量	hydraulic overload
水力传感器	水力感測器	hydraulic sensor
水力阀	水力閥	hydraulic valve
水力分类	水力類析	hydraulic classification

祖国大陆名	台湾地区名	英　文　名
水力分离	水力分離	hydraulic separation
水力分离器	水力分離器	hydraulic separator
水力负载量	水力負載量	hydraulic load
水力共振	水力共振	hydraulic resonance
水力平滑壁	水力平滑壁	hydraulically smooth wall
水力平滑管	水力平滑管	hydraulically smooth pipe
水力平均直径	水力平均直徑	hydraulic mean diameter
水力坡度线	水力級線	hydraulic grade line
水力效率	水力效率	hydraulic efficiency
水力旋流分离器	液體旋風分離器	liquid cyclone separator
水力旋流器	液體旋風器	liquid cyclone
水力学	水力學	hydraulics
水力压头	水力高差	hydraulic head
水力直径	水力直徑	hydraulic diameter
水力转速计	水力[式]轉速計	hydraulic tachometer
水力阻力	水力阻力	hydraulic resistance
水帘	水簾	water curtain
水淋冷凝器	滴水冷凝器	drip condenser
水轮机	水力渦輪機	hydraulic turbine
水煤浆	煤水漿	coal-water slurry
水煤气	水煤氣	water gas, blue gas
水煤气变换	水煤氣轉化	gas shift
[水煤气]变换反应	[水煤氣]轉化[反應]	shift conversion, gas shift reaction, water gas shift reaction
[水煤气]变换反应器	[水煤氣]轉化反應器	shift reactor
水煤气变换过程	水煤氣轉化程序	gas shift process
水镁矾	硫酸鎂石	kieserite
水锰矿	水錳礦	manganite
水磨光机	水高純化器	water polisher
水泥	水泥	cement
水泥管	水泥管	concrete pipe
水泥输送泵	水泥泵	cement pump
[水泥]硬化率	[水泥]硬化指數	cementation index
水喷头	灑水頭	sprinkler head
水平管式蒸发器	平管蒸發器	horizontal tube evaporator
水平列管蒸发器	平管式蒸發器	evaporator with horizontal tubes
水溶胶	水溶膠	hydrosol
水溶性软膏基	水溶性軟膏基	water-soluble ointment base
水溶液	水溶液	aqueous solution

祖国大陆名	台湾地区名	英 文 名
水溶助长剂	增溶劑, 增水溶劑	hydrotropic agent, hydrotrope
水溶助长性	增水溶性	hydrotropy
水软化	水軟化	water softening
水套	水套	water jacket
水调理	水調理	water conditioning
水位压头	水位高差	water head
水污染	水污染	water pollution
水污染工程	水污染工程	water pollution engineering
水污染降低	水污染降低	water pollution abatement
水污染控制	水污染控制	water pollution control
水污染物	水污染物	water pollutant
水性漆	水性漆	water base paint, water paint
水压机	水壓機	hydraulic press
水压用水	水壓用水	hydraulic water
水杨酸	柳酸, 水楊酸	salicylic acid
水银电解槽	汞電池	mercury cell
水银温度计	水銀溫度計	mercury thermometer
水硬石灰	水凝石灰	hydraulic lime
水硬水泥	水凝水泥	hydraulic cement
水浴	水浴	water bath
水跃	水躍	hydraulic jump
水蒸气	蒸汽, 水蒸汽	steam
水蒸气阀容量	水蒸汽閥容量	steam valve capacity
水蒸气图表	蒸汽表	steam table
水蒸气蒸馏	蒸汽蒸餾	steam distillation
水质	水質	water quality
水质分析	水質分析	water analysis
水质判据	水質準則	water quality criterion
水质硬度	水質硬度	water hardness
税	税	tax
税率	税率	tax rate
顺磁体	順磁體	paramagnetic body
顺磁效应	順磁效應	paramagnetic effect
顺磁性	順磁性	paramagnetism
顺磁氧分析仪	順磁氧分析儀	paramagnetic oxygen analyzer
顺式	順式	cis-form
顺[式]丁[二烯]橡胶	丁二烯橡膠	cis-butadiene rubber
顺式异戊二烯橡胶	順異戊二烯橡膠	cis-isoprene rubber
顺向进料	順向進料	forward feed

祖国大陆名	台湾地区名	英 文 名
顺序操作	順序操作	sequential operation
顺序阀	順序閥	sequencing valve
顺序控制	順序控制	sequence control
瞬时产量分率	瞬間產量分率	instantaneous fractional yield
瞬时反应	瞬間反應	instantaneous reaction
瞬时反应速率	瞬間反應速率	instantaneous reaction rate
瞬时杀菌器	瞬間殺菌器	flash pasteurizer
瞬时系统	瞬間系統	instantaneous system
瞬态(=暂态)		
瞬态扩散	暫態擴散	transient diffusion
瞬态现象	暫態現象	transient phenomenon
瞬态响应	暫態應答	transient response
瞬态响应分析	暫態應答分析	transient response analysis
瞬态行为	暫態行為	transient behavior
朔佩尔张力试验机	肖伯張力試驗機	Schopper tensile tester
丝纺	絲紡	silk spinning
丝光处理	絲光處理, 絲光化	mercerizing, mercerization
丝光棉	絲光棉	mercerized cotton
丝光纱	絲光紗	mercerized yarn
丝胶蛋白	絲膠	sericin
丝鸣[绸料摩擦声]	絲鳴	scroop
丝纤维	絲纖維	silk fiber
丝状细菌	絲狀細菌	filamentous bacteria
斯科特挠度计	斯科特曲度計	Scott flexometer
斯潘塞[分级]法	斯彭瑟[分級]法	Spencer method
斯氏黏度计	斯氏黏度計	Stormer's viscometer
斯托丁格黏度定律	斯托丁格黏度定律	Staudinger's viscosity law
斯托克斯(动力黏度单位)	斯托克斯(動力黏度單位)	stokes
斯托克斯近似法	斯托克斯近似法	Stokes approximation
斯托克斯线	斯托克斯線	Stokes line
撕裂应力	撕裂應力	tearing stress
撕碎机	撕碎機	shredder
撕碎装置	撕碎裝置	shredding device
死带	靜滯帶, 死帶	dead band
死区	靜滯區	dead zone
死水区	死水區[域]	deadwater region
死态	靜滯態	dead state
死亡期	死亡期	death phase

祖国大陆名	台湾地区名	英　文　名
死亡速率	死亡速率	death rate
四环素	四環素, 四環黴素, 鉑黴素	tetracycline, achromycin
四聚丙烯	四聚丙烯	propylene tetramer
四聚物	四聚[合]物	tetramer
四氯苯醌	四氯[苯]醌	chloranil
四通阀	四通閥	four-way valve
四乙铅	四乙基鉛	tetraethyl lead
伺服机构	伺服機構	servo mechanism
伺服机构补偿	伺服機構補償	servomechanism compensation
伺服控制	伺服控制	servo control
伺服问题	伺服問題	servomechanism-type problem, servo problem
松弛变量	假擬變數	slack variable
松弛法, 弛豫法	鬆弛法	relaxation method, relaxation technique
松弛试验	鬆弛試驗	relaxation test
松弛丝光	鬆弛絲光	slack mercerization
松弛现象	鬆弛現象	relaxation phenomenon
松弛状态	鬆弛狀態	relaxed state
松节油	松節油	turpentine oil
松香	松香, 松脂	rosin
松香胶	松香膠料	rosin size
松香施胶	松香上膠	rosin sizing
松香油	松香油	rosin oil
松香皂	松香皂	rosined soap
松油	松香油, 松樹油	pine oil, pine tree oil
松子油	松子油	pine seed oil
苏打灰	鹼灰, 純鹼, 碳酸鈉	soda ash
速成胶质	速成膠	accelerated gum
速度边界层	速度邊界層	velocity boundary layer
速度边界层厚度	速度邊界層厚度	velocity boundary layer thickness
速度分布	速度分布	velocity distribution
速度[分布]剖面[图]	速度剖面圖	velocity profile
速度计	速度計	velocity meter
速度距离滞后	速度距離落後	velocity-distance lag
速度势	速度勢	velocity potential
速度梯度	速度梯度	velocity gradient
速度压头	速度高差	velocity head
速率	速率	speed

祖国大陆名	台湾地区名	英　文　名
速凝水泥	快乾水泥, 速凝水泥	accelerated cement
速启阀	速啟閥	quick-opening valve
速止剂	速止劑	shortstopper, shortstopping agent
速止聚合	速止聚合	shortstopped polymerization
塑度计	塑性計	plastimeter, plastometer
塑解剂	解膠劑	peptizing agent
塑料	塑膠	plastic
塑料工程	塑膠工程	plastic engineering
塑料计	塑性試驗計	plasticimeter
塑料加工	塑膠加工	plastic processing
塑料黏合剂	塑膠黏結劑	plastic binder
塑料黏结剂	塑膠膠合劑	plastic cement
塑料泡沫	塑膠泡體, 塑膠泡沫	plastic foam
塑料填料	塑膠填充物	plastic packing
塑性材料	塑性材料, 塑膠材料	plastic material
塑性计	塑性測定器	plastograph
塑性流体	塑性流體	plastic fluid
塑性凝胶	塑性凝膠	plastigel
塑性屈伏	塑性降伏	plastic yield
塑性屈伏点	塑性降伏點	plastic yield point
塑性指数	塑性指數	plastic index
酸败	酸腐	souring
酸败发酵	酸敗發酵	sour fermentation
酸泵	酸泵	acid pump
酸槽	酸液儲槽	acid tank
酸槽车	酸槽車	acid tank car
酸催化	酸催化[作用]	acid catalysis
酸度	酸度	acidity
酸度检定	酸度試驗	acidity test
酸度指数	酸度指數	acidity index
酸法	酸法	acid process
酸化	酸化, 加酸[作用]	acidification, acidulation
酸化槽	加酸槽	acidulating tank
酸化剂	酸化劑	acidulant, acidifier
酸化器		acidifier, acidulator
酸价(＝酸價)		
酸冷却器	酸冷卻器	acid cooler
酸气	酸氣	sour gas
酸受体	酸受體	acid acceptor

祖国大陆名	台湾地区名	英 文 名
酸水	酸水	sour water
酸水汽提	酸水汽提	sour-water stripping
酸雾	酸霧	acid mist
酸洗	浸酸, 酸洗	pickling
酸洗槽	浸酸浴	pickle bath, pickling bath
酸洗法	浸酸法	pickling process
酸洗液	浸酸液	pickle liquor
酸性	酸性	acidity
酸性成分	酸性成分	acid component
酸性催化剂	酸性觸媒	acidic catalyst
酸性催化[作用]	酸性催化[作用]	acidic catalysis
酸性肥料	酸性肥料	acid fertilizer
酸性铬媒染料	酸性鉻媒染料	acid chrome dye
酸性耐火材料	酸性耐火物	acid refractory
酸性耐火砖	酸性耐火磚	acid firebrick , acid refractory brick
酸性气体	酸氣	acid gas
酸性染料	酸性染料	acid color, acid dye
酸性污泥	酸性污泥	acid sludge
酸烟	酸煙	acid fume
酸雨	酸雨	acid rain
酸值, 酸价	①酸值②酸價	acid number
算法	算法	algorithm
算法合成技术	算法合成技術	algorithmic synthesis technique
算术平均	算術平均	arithmetic mean
算术平均温差	算術平均溫差	arithmetic mean temperature difference
算术平均值	算術平均值	arithmetic mean value
随管加热蒸汽	隨管加熱蒸汽	tracing steam
随机变量	隨機變數	random variable
随机表面更新	隨機表面更新	random surface renewal
随机波动	隨機波動	random fluctuation
随机抽样	隨機取樣	random sampling
随机分析	隨機分析	stochastic analysis
随机负荷变化	隨機負載變化	random load change
随机过程	隨機程序, 概率過程	probabilistic process , random process, stochastic process
随机扰动	隨機擾動	random disturbance
随机数	隨機數	random number
随机搜索	隨機搜索	random search
随机特性	隨機特質	stochastic feature

祖国大陆名	台湾地区名	英　文　名
随机稳定性	隨機穩定性	stochastic stability
随机误差	隨機誤差	random error
随机信号	隨機信號	random signal
随机性	隨機性	randomness
碎玻璃	玻璃屑	cullet
碎料板	塑合板, 碎屑膠合板	particle board
碎裂	碎片化[作用]	fragmentation
碎裂	碎片化[作用]	fragmentation
碎形	分形	fractal
隧道干燥器	隧式乾燥器	tunnel drier
隧道窑	隧式窯	tunnel kiln
燧卵石	燧卵石	flint pebble
燧石	燧石	flint
损耗	損失	loss
损耗角正切	損失模數比, 損失正切	loss tangent
损失功	損失功	lost work
羧基化	羧化[作用]	carboxylation
羧基酶	羧基酶	carboxylase
羧甲基纤维素	羧甲纖維素	carboxymethyl cellulose
羧酸盐	羧酸鹽	carboxylate
缩二脲	縮二脲	biuret
缩合反应	縮合反應	condensation reaction
缩合系数	縮合係數	condensation coefficient
缩痕	收縮標誌, 收縮皺紋	shrink mark
缩径管接头	漸縮接頭	reducing coupling
缩聚[反应]	聚縮合[反應]	condensation polymerization, polycondensation
缩扩喷嘴	細腰噴嘴	converging-diverging nozzle
缩脉(＝流颈)		
缩醛树脂	縮醛樹脂	acetal resin
缩小	縮小	scale down
所得税	所得稅	income tax
索亥俄法	索亥俄法	Sohio process

T

祖国大陆名	台湾地区名	英　文　名
塔	塔	column

祖国大陆名	台湾地区名	英 文 名
塔板	塔板, 板, 盤	column plate, column tray, plate, tray
塔板间距	板间距	tray spacing
[塔]板效率	板效率	plate efficiency
塔顶冷凝器	頂部冷凝器	overhead condenser
塔顶蒸气	頂部蒸氣	overhead vapor
塔内件	塔内件	column internals
塔器	塔	tower
塔填充物	塔填充物	column packing, tower packing
塔贮留量	塔貯留量	tower hold-up
台秤	檯秤	platform balance
台架反应器	實驗室規模反應器	bench scale reactor
台架试验, 模型试验	實驗室規模試驗	bench scale test
太古油	茜草油	alizarine oil
太阳常数	太陽常數	solar constant
太阳辐射	太陽輻射	solar radiation
太阳光谱	太陽光譜	solar spectrum
太阳能	太陽能	solar energy
太阳能电池	太陽電池	solar cell
太阳能发动机	太陽引擎	solar engine
太阳能集热器	太陽能收集器	solar collector
太阳能水耕	太陽能水耕	solar aquafarming
太阳能蒸馏	太陽能蒸餾	solar distillation
太阳能贮池	太陽能貯池	solar pond
太阳油	太陽油	solar oil
肽	肽	peptide
肽酶	肽酶	peptidase
钛白	鈦白	titanium white
钛白陶	鈦白陶[器]	titania whiteware
钛瓷	鈦瓷[器]	titania porcelain
钛合金	鈦合金	titanium alloy
钛搪瓷	鈦搪瓷	titanium enamel
钛铁矿	鈦鐵礦	ilmenite
泰乐菌素	泰黴素	tylosin
泰勒标准筛	泰勒標準篩	Tyler standard sieve
酞菁蓝	酞花青藍	phthalocyanine blue
酞菁染料	酞花青染料	phthalocyanine dye
酞菁颜料	酞花青顏料	phthalocyanine pigment
酞青绿	酞花青綠	phthalocyanine green
檀香油	檀香油	sandal oil

祖国大陆名	台湾地区名	英 文 名
弹簧常量	彈簧常數	spring constant
弹簧秤	彈簧秤	spring balance, spring scale
弹簧吊架	彈簧吊架	spring hanger
弹簧管压力计	彈簧管壓力計	Bourdon gauge
弹簧力	彈簧力	spring force
弹簧平衡型钟式压力计	彈簧均衡鐘型壓力計	spring balanced type bell gauge
弹簧式安全阀	彈簧安全閥	spring safety valve
弹簧式硬度试验仪	彈簧硬度試驗儀	spring type hardness tester
弹簧载荷减压阀	彈簧負載減壓閥	spring loaded reducing valve
弹簧载荷面积计	彈簧負載面積計	spring loaded area meter
弹簧载荷调压器	彈簧負載調壓器	spring loaded pressure regulator
弹簧执行机构	彈簧引動器	spring actuator
弹簧-质量-阻尼系统	彈簧-質量-阻尼系統	spring-mass-damper system
弹力丝	彈力絲	stretch yarn
弹力纤维	彈性纖維, 鬆緊纖維	spandexfiber, spandex
弹力性	彈簧性	springness
弹性	①彈性②反彈	elasticity, resilience
弹性胶囊	彈性膠囊	elastic capsule
弹性聚合物	彈性聚合物	elastopolymer
弹性硫	彈性硫, 塑性硫	elastic sulfursulfur, plastic sulfur
弹性模量	彈性模數	elastic modulus
弹性能	彈性能	elastic energy
弹性试验	彈性試驗	elasticity test
弹性数	彈性數	elasticity number
弹性体	彈性體	elastomer
弹性体共混物	彈性體摻合物	elastomer blend
弹性纤维	彈性纖維, 鬆緊纖維	snapback fiber, elastic fiber
弹性液体	彈性液體	elastic liquid
弹状流(＝节涌流)		
炭	炭	char
炭黑	碳黑	carbon black
炭滤	碳濾[法]	carbon filtration
炭刷	碳刷	carbon brush
炭砖	碳磚	carbon brick
探头	探針, 探桿	probe
碳电极	碳電極	carbon electrode
碳化钙	碳化鈣	calcium carbide
碳化硅	碳化矽	carbide silicon, silicon carbide

祖国大陆名	台湾地区名	英 文 名
碳化硅纤维	碳化矽纖維	silicon carbide fiber
碳化硅砖	碳化矽磚	carbide brick
碳化硼	碳化硼	boron carbide
碳平衡	碳均衡	carbon balance
碳氢化合物	烴, 碳氫化合物	hydrocarbon
碳水化合物	碳水化合物	carbohydrate
碳素钢	碳鋼	carbon steel
碳塑性体	彈性塑膠	elastoplastic
碳酸钙	碳酸鈣	calcium carbonate
碳酸气饱充法	碳酸法	carbonation process
碳酸氢钠	①焙用鹼②碳酸氫鈉	baking soda
碳吸附	碳吸附	carbon adsorption
碳纤维	碳纖維	carbon fiber
碳纤维强化塑料	碳纖維強化塑膠	carbon fiber reinforced plastic
碳循环	碳循環	carbon cycle
碳质耐火材料	碳[質]耐火物	carbon refractory
羰基合成	羰氫化法	oxo process
羰基化	羰化[反應]	carbonylation
羰基镍催化剂	鎳羰觸媒	nickel carbonyl catalyst
搪瓷	搪瓷	porcelain enamel
搪瓷钢	搪瓷鋼	porcelain enameled steel
搪塑	冷凝模塑	slush molding
糖	糖	sugar
糖化[反应]	糖化	saccharification
糖化剂	糖化劑	saccharifying agent
糖浆剂	①糖漿②漿	syrup
糖酵解	醣解	glycolysis
糖酵解途径	糖酵解途徑	glycolytic pathway
糖精	糖精	saccharin
糖类	糖類	saccharide
糖量计	糖量計	saccharimeter
糖蜜	糖蜜	molasses
糖蜜晶粒	糖蜜起晶	molasses graining
糖蜜酒精	糖蜜酒精	molasses alcohol
糖蜜塔养基	糖蜜培養基	molasses medium
淌流	淌流	sag flow
淌流板	淌流板	sag flow board
桃红釉	桃红釉, 桃花浪釉	peach bloom glaze
桃仁油	桃仁油	peach-kernel oil

祖国大陆名	台湾地区名	英　文　名
陶瓷	陶業, 陶藝, 陶瓷學	ceramics
陶瓷管	陶瓷管	ceramic tube
陶瓷金属	金屬瓷料	cermet
陶瓷金属电阻器	金屬陶瓷電阻器	cermet resistor
陶瓷耐火材料	陶瓷耐火物	ceramic refractory
陶瓷器皿	陶器	ceramic ware
陶瓷釉	陶瓷釉	ceramic glaze
陶瓷制品	陶瓷, 陶器	ceramic
陶器	陶器, 土器	earthenware
陶土	陶土	earthenware clay
淘矿机	播洗器	vanner
淘析器	淘析器	elutriator
套管反应器	套管反應器	double pipe reactor
套管接头	套筒接頭	sleeve joint
套管冷却结晶器	套管冷卻結晶器	votator apparatus
套管冷却器	套管冷卻器	double pipe cooler
套管热交换器	套管熱交換器	double-pipe heat exchanger
套筒联轴节	套筒連結器	sleeve coupling
特定牌号	特定名稱	specific designation
特级耐火材料	超強耐火物, 特級耐火物	super duty refractory
特殊性能高分子	特殊性能高分子	speciality polymer
特性常数	特性常數	characteristic constant
特性反应时间	特性反應時間	characteristic reaction time
特性方程	特性方程式	characteristic equation
特性根	特性根	characteristic root
特性函数	特性函數	characteristic function
特性化	特性化, 示性	characterization
特性黏度	極限黏度	intrinsic viscosity
特性黏合	特性黏合, 比黏合	specific adhesion
特性判据	性能準則	performance criterion
特性曲线	特性曲線	characteristic curve
特性数	性能值	performance number
特性因数	特性因數, 特性化因數	characteristic factor, characterization factor
特异性	特定性, 特異性	specificity
特征长度	特性長度	characteristic length
特征函数	特徵函數	eigenfunction
特征时间	特性時間	characteristic time
特征值	特徵值, 特性值	eigenvalue, characteristic value

祖国大陆名	台湾地区名	英 文 名
特种黏合剂	特殊膠黏劑	special adhesive
特种涂料	特殊塗佈	special coating
梯度	梯度	gradient
梯度法	梯度法	gradient method
梯形堰	梯形堰	trapezoidal weir
锑白	銻白	antimony white
锑电极	銻電極	antimony electrode
锑黄	銻黃	antimony yellow
锑基硅氧烷聚合物	銻硅氧烷聚合物	stibinosiloxane polymer
锑皂	銻皂	antimonial soap
提纯	純化, 淨化	purification
提斗浸取器	斗式浸取器	bucket-elevator extractor
提高采收率	增強石油回收[法]	enhanced oil recovery
提馏	汽提, 脫除	stripping
提馏段	汽提段	stripping section
提浓段	增濃段	enriching section
提升	[品質]提升, 升級	upgrading
提升管	①上升管②豎板	riser
提升管反应器	上升管反應器	riser reactor
体积	體積, 容積	volume
体积表面直径	體積表面[相當]直徑	volume surface diameter
体积传氧系数(＝容积传氧系数)		
体积分率	體積分率	volume fraction
体积功	體積功	volume work
体积流量(＝体积流率)		
体积流率, 体积流量, 体积流速	體積流量	volumetric flow rate
体积流速(＝体积流率)		
体积模数	體積[彈性]模數	volume modulus
体[积]黏性	整體黏度	bulk viscosity
体积平均沸点	體積平均沸點	volumetric average boiling point
体积平均速度	體積平均速度	volume average velocity
体积平均直径	體積平均直徑	volume mean diameter
体积收缩	體積收縮	volume shrinkage
体积效率	體積效率	volumetric efficiency
体积指数	體積指數	volume index
体扩散	體擴散	bulk diffusion
体膨胀率	體積膨脹係數	volume expansivity

祖国大陆名	台湾地区名	英　文　名
体外	體外	in vitro
体系(＝系统)		
体心点格	體心晶格	body-centered lattice
替代天然气	替代天然氣	substitute natural gas
替代物	替代物	substituent
替换评估	替換評估	evaluation replacement
天然碱	天然鹼	natural soda
天然碱	碳酸鈉石	trona
天然气	天然氣	natural gas
天然气重整	天然氣重組	natural gas reforming
天然气水合物	天然氣水合物	natural gas hydrates
天然气液化	天然氣液化	natural gas liquefaction
天然汽油	天然汽油, 井口汽油	natural gasoline, casing head gasoline
天然清漆	天然清漆	natural varnish
天然橡胶	天然橡膠	natural rubber
添加剂	添加劑	additive
添加酵母	添加酵母, 投種	pitching
甜菜糖	甜菜糖	beet sugar
甜味剂	低硫天然甜味料	sweetener
填充	填料, 裝填	filling
填充床	填充床	packed bed
填充床反应器	填充床反應器	packed bed reactor
填充高度	填充高度	packed height
填充剂	填充劑, 增容填料, 補助劑	builder, bulking filler
填充剂	填充劑	stuffing
填充密度	填充密度	packed density, packing density
填充式萃取塔	填充式萃取塔, 填充式萃取器	packed extractor
填充式喷淋塔	填充噴霧塔	packed spray tower
填充式洗涤器	填充洗氣器	packed scrubber
填充物, 填料材料	填充物	packing material
填充蒸馏塔	填充蒸餾塔	packed distillation column
填充柱	填充柱	packed column
填料	填料	filler
填料	填料	stuffing
填料材料(＝填充物)		
填料函式密封	填料箱密封	stuffing box seal
填料环	填充環	packing ring

祖国大陆名	台湾地区名	英　文　名
填料塔	填充塔	packed tower
填料塔	填料塔	packed column
填料吸收器	填充塔吸收器	packed column absorber
填料箱	填料箱	stuffing box
填料因子	填充因數	packing factor
填料支承板	填充物支架	packing support
条件稳定	有條件的穩定性	conditional stability
条纹	①條紋②加條紋	stripe
条纹线	煙線	streak line
调合漆	調和漆	ready-mixed paint, blended paint
调和汽油	摻配汽油	blended gasoline
调节剂	調理劑, 調節劑	conditioning agent, regulator
调节控制	調節控制	regulatory control
调节器	調節器	regulator
调节器参数整定	控制器調協	controller tuning
调节箱	調節箱	regulating box
调味剂	調味料	flavoring materials
调温旋管	調溫旋管	tempering coil
调谐	調諧	tuning
调谐参数	調諧參數	tuning parameter
调压变压器	可變變壓器	variable transformer
调压阀	壓力調節閥	pressure regulating valve
调压器	壓力調節器	pressure regulator
调优操作	調優操作	evolutionary operation
调优法	調優法	evolutionary method
调整	調整	adjustment
调制	調變	modulation
跳开装置	跳動裝置	tripper
跳汰流化床	跳汰流化床	jigged fluidized bed
贴壁细胞	貼壁細胞	anchorage-dependent cell
萜	萜	terpene
萜油	萜油	terpene oil
铁氨催化剂	鐵氨觸媒	catalyst iron-ammonia, iron-ammonia catalyst
铁催化剂	鐵觸媒	iron catalyst
铁合金	鐵合金	ferro-alloy
铁铝氧催化剂	鐵鋁氧觸媒	catalyst iron-alumina, iron-alumina catalyst
铁媒染剂	鐵媒染劑	iron mordant
铁锰合金	錳鐵齊	ferromanganese

祖国大陆名	台湾地区名	英　文　名
铁鞣	鐵鞣	iron tannage
铁石棉[填充剂]	石綿礦	amosite
铁细菌	鐵細菌	iron bacteria
烃发酵	烴發酵	hydrocarbon fermentation
烃黑	烴黑	hydrocarbon black
停车控制	停工控制	shutdown control
停工检查	停工檢查	shutdown inspection
停工期	停工期	shutdown period
停工日程表	停工時程	shutdown schedule
停工时间	停工時間	shutdown time
停留期间	貯留期間	holding period
停留时间	滯留時間, 貯留時間	holding time, retention time, residence time
停留时间分布	滯留時間分布	residence time distribution(RTD)
停留时间分布密度函数	滯留時間分布密度函數	residence time distribution density function
停修时间	維護停工時間	maintenance downtime
停止流动法	止流法	stopped flow method
停滞期间	停留期間	detention period
停滞时间	停留時間, 遲延時間, 延遲時間	detention time, dead time
通道	通道	channel
通断控制	開關控制	on-off control
通风	通風	ventilation
通风比例	通風比率	vent ratio
通风机	通風器	ventilator
通风计	通風計	draft gage
通风扇	通風扇	draft fan
通风需求	通風需求	ventilation requirement
通过量, 产量	產量	throughput
通量	①通量②助溶劑	flux
通量密度	通量密度	flux density
通量密度矢量	通量向量密度	flux density vector
通气管	通氣管	aeration candle
通式	通式, 一般化關係式	generalized correlation
通用服务检查表	一般服務檢查表	general service check list
通用简约梯度法	通用簡約梯度法	general reduced gradient method, GRG method
通用冷凝器	通用冷凝器	universal condenser
通用清洁剂	通用清潔劑	general purpose detergent

祖国大陆名	台湾地区名	英 文 名
通用指示剂	通用指示計	universal indicator
同步传送器	同步傳送器	synchro-transmitter
同步互穿网络	同步互穿網絡	simultaneous interpenetrating network
同步机	同步儀	synchro
同步加速器	同步加速器	synchrotron
同步控制变压器	同步控制變壓器	synchro-control transformer
同电子排列体	等容線	isostere
同化	同化	assimilation
同上传热单元高度	熱傳單元高度	height of a heat transfer unit (HTU)
同时操作	同時操作	simultaneous operation
同时反应	併發反應	simultaneous reactions
同时聚合	同時聚合	simultaneous polymerization
同时吸收	同時吸收	simultaneous absorption
同位素	同位素	isotope
同位素法	同位素法	isotopic method
同位素分离	同位素分離	isotope separation
同位素丰度	同位素含量, 同位素豐度	isotopic abundance
同位素指示剂	同位素示蹤劑	isotopic tracer
同系化合物	同系物	homologous compound
同系物	同系物	homologue
同向流(=并流)		
同心变径管	同心漸縮管	concentric reducer
同心浮球	同心浮子	concentric float
同心管柱	同心管柱	concentric-tube column
同心环	同心環	concentric ring
同心刻度盘	同心儀	concentric dial
同心孔口板	同心孔口板	concentric orifice plate
同心柱黏度计	同軸圓柱黏度計	concentric cylinder viscometer
同型发酵菌	純種發酵菌	homofermentative bacteria
桐油	桐油	tung oil
铜铵人造纤维	銅氨縲縈	cuprammonium rayon
统计单元	統計單元	statistical unit
统计分析	統計分析	statistical analysis
统计估计	統計估計	statistical estimation
统计[结构]共聚物	統計[結構]共聚物	statistical copolymer
统计均匀性	統計均勻性	statistical homogeneity
统计理论	統計理論	statistical theory
统计力学	統計力學	statistical mechanics

祖国大陆名	台湾地区名	英 文 名
统计链	統計鏈	statistical chain
统计链段	統計鏈段	statistical segment
统计权重矩阵	統計權重矩陣	statistical weight matrix
统计热力学	統計熱力學	statistical thermodynamics
统计设计	統計設計	statistical design
统计误差	統計誤差	statistical error
统计线团	統計線團	statistical coil
统计学	統計學	statistics
桶磨机	桶磨機	barrel mill
桶式锅炉	桶式鍋爐	drum boiler
桶式浸取	槽式浸取	vat leaching
桶式去皮机	桶式去皮機	barking drum, drum barker
筒	桶, 鼓	drum
筒式进料机	進料鼓	feed drum
头馏分	頂部餾出物	overhead distillate
投入产出	輸入投出	input-output
投资回收	投資回收	investment recovery
投资回收期	投資回收期	payback period
投资收益率	投資收益率	rate of return on investment, return on investment
投资资本	投資資本	investment capital
透光度	①透光度②透射係數	transmittance
透光率	分光透射率	spectral transmittance
透明度	透明度	transparency
透明釉	透明釉	transparent glaze
透明皂	透明皂	transparent soap
透气性	透氣性	air permeability, gas permeability
透热壁	無熱阻壁	diathermal wall
透射率	透射係數	transmissivity
透水性	透水性	water permeability
透析(=渗析)		
透析培养(=渗析培养)		
透析器(=渗析器)		
透析液(=渗析液)		
透析蒸馏	透析蒸餾	perdistillation
突变	突變	mutation
突变论	災難理論	catastrophe theory
突变模型	災難模型	catastrophic model
突变频率	突變頻率	mutation frequency

祖国大陆名	台湾地区名	英　文　名
突开式安全阀	急洩閥	pop valve
突然膨胀	突然膨脹	sudden expansion
图	圖	plot
H-i 图(＝焓-熵图)		
P-S 图(＝焓-浓图)		
T-H 图(＝温湿图)		
图解	圖解	graphical solution
图解法	圖解法	graphical method
图解分析	圖解分析法	graphical analysis
图解积分	圖解積分法	graphical integration
图解计算	圖解計算	graphical calculation
图解仪表板	圖解儀表板	graphic panel
图论	圖論	graph theory
图像分析仪	圖像分析儀	quantimet
图形流程图	圖形流程圖	pictorial flowsheet
涂布	塗布	spread
涂布量	塗布量	spread
涂胶压延机	塗膠壓延機, 擦膠壓延機	spreading calender
涂料	塗料	paint
涂料技术	塗裝技術	coating technology
途径	途徑	route
土地污染	土地污染	land pollution
土霉素	土黴素, 地靈黴素	terramycin
土壤	土壤	soil
土壤承受压力	泥土承受壓力	soil bearing pressure
土壤处理	土壤處理	soil treatment
土壤肥力	土壤肥度	soil fertility
土壤腐蚀	土壤腐蝕	soil corrosion
土壤改良剂	土壤改良劑	soil conditioner
土壤酵母	土壤酵母	soil yeast
土壤力学	土壤力學	soil mechanics
土壤溶胶	土壤溶膠	soil sol
土壤酸度	土壤酸度	soil acidity
土壤微生物	土壤微生物	soil organism
土壤稳定剂	土壤穩定劑	soil stabilizer
土壤污染	土壤受污, 土壤污染	soil contamination, soil pollution
土壤消毒剂	土壤消毒劑	soil sterilant
土石膏	土[狀]石膏	gypsite

祖国大陆名	台湾地区名	英　文　名
土水泥	土壤水泥	soil cement
土质颜料	土質颜料	earth pigment
吐酒石	吐酒石	tartar emetic
湍动流化床	湍動流化床	turbulent fluidized bed
湍流边界层	紊流邊界層	turbulent boundary layer
湍流标度	紊流標度	turbulence scale
湍流促进器	紊流促進器	turbulence promoter
湍流度	紊流度	degree of turbulence
湍流反应器	紊流反應器	turbulent flow reactor
湍流核心	紊流核心	turbulent core
湍流剪应力	紊流剪應力	turbulent shear stress
湍流扩散	紊流擴散	turbulent diffusion
湍流扩散系数	紊流擴散係數	turbulent diffusivity
湍流能谱	紊流能譜	turbulent-energy spectrum
湍流强度	紊流強度	intensity of turbulence
湍流松弛现象	紊流鬆弛現象	turbulent relaxation phenomenon
湍流, 紊流	紊流	turbulent flow
湍流应力	紊流應力	turbulent stress
湍流运动	紊流運動	turbulent motion
团聚	团聚	agglomeration
团聚值	团聚值	agglomerating value
团块	团聚	agglomerate
团状流(=节涌流)		
推迟时间	减速時間, 遲延時間	retardation time
推斥势	推斥勢	repulsive potential
推动力	驅動力	driving force
推进剂	推進劑	propellant
推理控制	推算控制	inferential control
退化	退化	degeneracy, degeneration, obsolescence
退化形式	退化形式	degeneration form
退火	退火, 徐冷	annealing
退火点	徐冷點	annealing point
退火炉	退火窯	lehr
褪色剂	褪色劑	stripping agent
托(压力单位)	托(壓力單位)	torr
托板	托板	support plate
拖铲	拖刮機	drag scraper
拖尾	尾料, 尾渣	tailing
拖尾旋涡	拖尾旋渦	trailing vortex

祖国大陆名	台湾地区名	英 文 名
脱氨基	去胺[反應]	deamination
脱氨基酶	去胺酶	deaminase
脱玻化	①反玻化②結晶化③失透明[作用]	devitrification
脱氮	脱氮[反應], 脱氮化[作用], 去氮	denitrogenation
脱丁烷塔	去丁烷塔	debutanizer
脱芳构化	去芳香化[反應]	dearomatization
脱辅[基]酶	酶蛋白	apoenzyme
脱附等温线	脱附等溫線	desorption isotherm
脱附控制	脱附控制	desorption control
脱甲烷	去甲烷[反應]	demethanization
脱胶	脱膠, 脱色, 脱脂	boiling-off
脱蜡	脱蠟	dewaxing
脱硫	脱硫[反應], 去硫化,	sulfur elimination, devulcanization, desulfurization
脱卤	脱鹵[反應]	dehalogenation
脱氯塔	去氯器	dechlorinator
脱模板	脱模板	stripper plate
脱模机		stripper
脱模剂		stripping agent
脱模剂	脱模劑	mold-release agent, parting agent, release agent
脱漆剂	去漆劑	paint remover
脱气	除氣	deaeration, degasification
脱气器	除氣器	degasifier
脱气塔	除氣器	deaerator
脱氢	脱氫[反應]	dehydrogenation
脱氢催化剂	脱氫觸媒	catalyst dehydrogenation, dehydrogenation catalyst
脱氢环化作用	脱氫環化[反應]	dehydrocyclization
脱氢酶	去氫酶	dehydrogenase
脱溶剂器	脱溶劑器	desolventizer
脱色	脱色	decolorization
脱色剂	脱色劑	decolorizer
脱水	脱水	dehydration, dewatering
脱水酒精	脱水酒精	dehydrated alcohol
脱水器	脱水器	dehydrator
脱水收缩	脱水收縮	syneresis

祖国大陆名	台湾地区名	英文名
脱水物	脱水物	dehydrate
脱羧	脱羧[反應]	decarboxylation
脱碳	去碳	decarburization
脱烷基化	脱烷, 去烷[反應]	dealkylation
脱硝	脱硝	denitration
脱硝塔	脱硝塔	denitration tower, denitrator
脱盐	淡化	desalination
脱氧	去氧[反應]	deoxidation, deoxygenation
脱乙烷塔	去乙烷塔	deethanizer
脱脂	脱脂	degrease
脱脂奶	脱脂牛乳	skim milk
妥尔油	松油	tall oil
妥尔油松香	松香	tall-oil rosin
妥尔油皂	松香油皂	tall-oil soap

W

祖国大陆名	台湾地区名	英文名
瓦楞纸	瓦楞紙	corrugating paper
瓦斯油	製氣油, 柴油	gas oil
外表面	外表面	external surface
外表面积	外表面積	external area
外表面浓度	外表面濃度	external surface concentration
外部施胶	外部上膠	external sizing
外部坐标	外部座標	external coordinates
外换热器	外熱交換器	external heat exchanger
外回流	[塔]外回流	outside reflux, external reflux
外回流比	[塔]外回流比	outside reflux ratio
外径	外徑	external diameter, outside diameter
外控制环路	外控制環路	external control loop
外扩散	外擴散	external diffusion
外力	外力	external force
外流	外部流動	external flow
外能	外能	external energy
外扰[动]	外來擾動	external disturbance
外热法	外熱法	external heating
外渗	外滲透	exosmosis
外消旋化	消旋反應	racemization

祖国大陆名	台湾地区名	英 文 名
外消旋混合物	消旋混合物	racemic mixture
外循环	外循環	external recycle
外循环反应器	外循環反應器	external recycle reactor
外压[力]	外壓力	external pressure
外延	磊晶	epitaxy
外延生长	磊晶生長, 晶體同軸生長	epitaxial growth
外在容积	外在容積	external capacity
弯管	彎管	pipe bend, bend
弯管当量管长	彎管相當管長	bend equivalent pipe length
弯管机	彎管機	pipe bending machine
弯曲试验	彎曲試驗	bending test
弯曲因子	扭曲[度]因數	tortuosity factor
弯[曲]应力	抗彎應力, 塑流應力	bending stress, flexural stress
弯头	①彎頭②肘管	elbow
弯头喷嘴	彎頭噴嘴	elbow nozzle
完全混合反应器	完全混合反應器	perfectly mixed reactor
完全混合消化	完全混合消化	complete-mixing digester
完全解偶	完全解偶	perfect decoupling
完全控制	完美控制	perfect control
顽辉石	頑火輝石	enstatite
烷基苯磺酸盐	烷[基]苯磺酸鹽	alkylbenzene sulfonate
烷基化	烷化	alkylation
烷基转移	轉烷化[反應]	transalkylation
晚材	夏材	summer wood
碗形磨	碗形磨	bowl mill
万古霉素	萬古黴素	vancomycin
万能批量工厂	目標批式工廠	multipurpose batch plant
万向接头	通用接頭	universal joint
王水	王水	aqua regia
网鞍填料	網鞍填料	McMahon packing
θ网环	網環	Dixon ring
网孔	篩孔	mesh
网孔塔板	網孔塔板	perform tray
网络	網路	network
网筛	網篩	mesh screening
网状层	網狀層, 網狀組織	stratum reticular
网状催化剂	觸媒綱	gauze catalyst
往复板[式]萃取塔	往復板萃取塔	reciprocating plate extraction column

祖国大陆名	台湾地区名	英　文　名
往复板[式]塔	往復板萃取塔	reciprocating plate column
往复加料器	往復飼機	reciprocating feeder
往复耙	往復耙	reciprocating rake
往复筛	往復篩	reciprocating screen
往复式	往復運動	reciprocating
往复式活塞压缩机	往復式活塞壓縮機	reciprocating piston compressor
往复式压缩机	往復[式]壓縮機	reciprocating compressor
往复式真空泵	往復真空泵	reciprocating vacuum pump
往复循环反应器	循環反應器	reciprocating reactor
危险废物	危害性廢棄物	hazardous waste
危险性评估	危害性評估	hazard evaluation
危险指数	危險指數	hazard index
威尔逊方程	威爾遜方程	Wilson equation
威尔逊方法	威耳生方法	Wilson method
微波	微波	microwave
微波波谱学	微波光譜學	microwave spectroscopy
微波干燥	微波乾燥	microwave drying
微波谱	微波譜	microwave spectrum
微差测温法	微差測溫法	differential thermometry
微差间隙	微差間隙	differential gap
微差接触设备	微差接觸設備	differential contact equipment
微差温度计	微差溫度計	differential thermometer
微滴	小滴	droplet
微电路	微電路	microcircuit
微分表面	微分表面	differential surface
微分产率	微分產率	differential yield
微分沉降	微分沉降	differential settling
微分法	微分法	differentiation, differential method, differential technique
微分反应器	微分反應器	differential reactor
微分方程	微分方程式	differential equation
微分控制	微分控制	derivative control (＝D control), differential control
微分冷凝	微分冷凝	differential condensation
微分能量平衡	微分能量均衡	differential energy balance
微分器	微分器	differentiator
微分溶解热	微分溶解熱	differential heat of solution
微分时间	微分時間	derivative time
微分时间常数	微分時間常數	derivative time constant

祖国大陆名	台湾地区名	英 文 名
微分稀释热	微分稀釋熱	differential heat of dilution
微分蒸馏	微分蒸餾	differential distillation
微分质量平衡	微分質量均衡	differential mass balance
微分重量分布	微差重量分布	differential weight distribution
微粉磨	微磨機	micronizer
微观混合	微觀混合	micromixing
微观可逆性	微觀可逆性	microreversibility, microscopic reversibility
微观可逆性原理	微觀可逆性原理	principle of microscopic reversibility
微观流体	微觀流體	microfluid
微观平衡	微觀均衡	microscopic balance
微观态	微觀狀態	microstate
微积分	微積分	calculus
微胶囊	微膠囊	microcapsule
微结构	微結構	microstructure
微结构材料	微結構化材料	microstructured material
微晶	微晶	crystallite, microcrystal, microcrystalline
微晶结构	微晶結構	microcrystalline structure
微晶蜡	微晶蠟	microcrystalline wax
微孔	微孔	micropore
微孔材料	海綿狀材料	cellular material
微孔分离器	微孔分離器	microporous separator
微孔过滤器	微孔過濾器	microporous filter
微孔扩散	細孔擴散	pore diffusion
微孔扩散阻力	細孔擴散阻力	pore diffusion resistance
微孔滤膜	微孔膜	microporous membrane
微孔[透壁]分气法	細孔氣體分離法	atmolysis
微粒	顆粒, 粒狀物	particulate
微粒学	微粒學	micromeritics
微量分析	微量分析	microanalysis
微量化学	微量化學	microchemistry
微量现象	微標[度]現象	microscale phenomenon
微滤	微孔過濾, 微[過]濾	microfiltration, microstraining
微滤器	濾微器	microstrainer
微米	微米	micron
微囊化	微膠囊化	microencapsulation
微球	微球	microsphere
微扰	微擾	perturbation
微扰变量	微擾變數	perturbation variable
微扰法	微擾法	perturbation method

祖国大陆名	台湾地区名	英　文　名
微扰理论, 摄动理论	微擾理論	perturbation theory
微扰硬链理论	微擾硬鏈理論	perturbed hard chain theory, PHC theory
微乳	微乳液	microemulsion
微生物	微生物	microbe, microorganism
微生物动力学	微生物動態[學]	microbial dynamics
微生物动力学	微生物動力學	microbial kinetics
微生物发酵	微生物發酵	microbial fermentation
微生物法	微生物程序, 微生物法	microbial process
微生物反应	微生物反應	microbial reaction
微生物反应器	微生物反應器	microbial reactor
微生物腐蚀	厭氣腐蝕	anaerobic corrosion
微生物腐蚀	微生物[引起的]腐蝕	microbial corrosion
微生物过程	微生物程序, 微生物法	microbiological process
微生物结垢	微生物積垢	microbial fouling
微生物膜	微生物薄膜	microbial film
微生物学	微生物學	microbiology
微型催化反应器	微觸媒反應器	microcatalytic reactor
微型反应工程	微反應工程	microreaction engineering
微型反应器	微反應器	microreactor
微型计算机	微算機, 微電腦	microcomputer
微型仪器	小型儀器	miniature instrument
[微]元	①元素②元件③要素	element
微载体	微載體	microcarrier
微正则配分函数	微正則配分函數	microcanonical partition function
微正则系综	微正則系綜	microcanonical ensemble
韦[伯](磁通量单位)	韋伯	Weber
韦伯数	韋伯數	Weber number
韦格斯坦法	魏格斯坦法	Wegstein method
韦斯模数	魏斯模數	Weisz modulus
桅杆清漆	晶石清漆	spar varnish
唯象模型	現象模式	phenomenological model
唯象系数	現象係數	phenomenological coefficient
维纶	維尼綸	vinylon
维生素	維生素, 維他命	vitamin
维修	維護, 保養	maintenance
维修费	維護成本	maintenance cost
维修管理	維護管理	maintenance management
伟晶岩	偉晶花崗石	pegmatite
尾部液体	稀溶液	tail liquid

祖国大陆名	台湾地区名	英文名
尾管	尾管	tail pipe
尾流, 尾涡	尾流	wake
尾气	尾氣	end gas, tail gas
尾涡(＝尾流)		
卫星工厂	衛星工廠	satellite plant
未处理废水	原廢水	raw wastewater
未处理污水	原污水	raw sewage
未处理淤泥	原污泥	raw sludge
未反应核模型	未反應核模型	unreacted core model
未活化化学吸附	未活化化學吸附	nonactivated chemisorption
未经净化水	原水	raw water
未控流	未控流	wild stream
位力方程	維里方程	virial equation
位力系数	維里係數	virial coefficient
位势	位能	potential
位头	勢差, 位能差	potential head
位相关系	相關係	phase relation
位形配分函数, 构型配分函数	組態分配函數	configurational partition function
位形性质, 构型性质	組態性質	configurational property
位移计	位移[流量]計	displacement meter
位移流量计	位移流量計	displacement flowmeter
位移压力计	位移壓力計	displacement pressure gage
位置模型理论	位置模型理論	site-model theory
位阻	位阻[現象], 空間障礙	steric hindrance
喂料装置	稱量飼機	weighing feeder
温标	溫標	temperature scale
温差	溫差	temperature difference
温差电效应	熱電效應	thermoelectric effect
温差推动力	溫差驅動力	temperature difference driving force
温度	溫度	temperature
温度边界层	熱邊界層	thermal boundary layer
温度变送器	溫度傳送器	temperature transmitter
温度程序	溫度排程	temperature schedule
温度滴定法	溫度滴定[分析]法	thermometric titrimetry
温度范围	溫度範圍	temperature range
温度分布	溫度分布	temperature distribution
温度[分布]剖面[图]	溫度剖面圖	temperature profile
温度计	溫度計	temperature gage, thermometer

祖国大陆名	台湾地区名	英　文　名
温度计套管	熱套管	thermowell
温度记录器	溫度記錄器	temperature recorder
温度校正	溫度校正	temperature calibration, temperature correction
温度进展	溫度進展, 溫度歷程	temperature progression
温度控制	溫度控制	temperature control
温度控制器	溫度控制器	temperature controller
温度水平	溫階	temperature level
温度梯度	溫度梯度	temperature gradient
温度调节阀	熱調閥	thermo regulating valve
温度调节器	溫度調節器, 調溫器	temperature regulator, thermoregulator
温度指示计	溫度指示計	temperature indicator
温度指示控制器	溫度指示控制器	temperature indicating controller
温度组成图	溫度-組成[關係]圖	temperature-composition diagram
温泡	溫度球莖	temperature bulb
温熵图	溫度-熵[關係]圖	temperature-entropy diagram
温湿图, T-H 图	溫濕圖	temperature-humidity chart
温室效应	溫室效應	greenhouse effect
文丘里管	文氏管, 細腰管	Venturi tube
文丘里流量计	文氏流量計, 文氏計, 細腰流量計, 細腰計	Venturi flowmeter, Venturi meter
文丘里喷嘴	文氏噴嘴, 細腰噴嘴	Venturi nozzle
文丘里洗涤器	文式洗滌器	Venturi scrubber
文献	文獻	literature
文献调查	文獻調查	literature survey
紊流(=湍流)		
稳定操作条件	穩定操作條件	stable operating condition
稳定成分	穩定成份	stabilizing ingredient
稳定重整油	穩定重組油	stabilized reformate
稳定度	穩定度, 安定度	degree of stability
稳定化槽	穩定化槽	stabilization tank
稳定化池	穩定化槽, 穩定化池	stabilization basin, stabilization pond
稳定极限	穩定極限	stability limit
稳定剂	穩定劑	stabiliser, stabilizing agent, stabilizator
稳定界限	穩定邊限	margin of stability
稳定近似	穩態近似法	steady state approximation
稳定流	穩流	steady flow
稳定汽油	穩定汽油	stabilized gasoline
稳定器	穩定器	stabilizator

祖国大陆名	台湾地区名	英　文　名
稳定区域	穩定區[域]	stability region, stable region
稳定溶液	①穩定解②安定溶液	stable solution
稳定乳液	穩定乳液, 穩定乳態	stable emulsion
稳定条件	穩定條件	stability condition
稳定同位素	穩定同位素	stable isotope
稳定稳态	穩定穩態	stable steady state
稳定响应	穩定應答	stable response
稳定性	穩定性	stability
稳定性分析	穩定性分析	stability analysis
稳定性判据	穩定準則	stability criterion
稳定性试验	穩定性試驗	stability test
稳定性受损	持久中毒	stability poisoning
稳定[作用]	穩定[作用]	stabilization
稳泡剂	泡沫穩定劑	foam stabilizer
稳态	穩定狀態	stable state
稳态操作	穩態操作	steady state operation
稳态分析	穩態分析	steady state analysis
稳态过程	穩態程序	steady state process
稳态近似	穩態近似法	stationary state approximation
稳态扩散速率	穩態擴散速率	steady-state diffusion rate
稳态流动	穩態流動	steady state flow
稳态模型	穩態模式	steady-state model
稳态柔量	穩態柔量	steady-state compliance
稳态蠕变	穩態蠕變	steady-state creep
稳态唯一性	穩態唯一性	uniqueness of steady state
稳态稳定性	穩態穩定性	stability of steady state
稳态误差	穩態誤差	steady state error
稳态响应	穩態應答	steady state response
稳态性能	穩態性能	steady-state performance
稳态增益	穩態增益	steady state gain
稳态周期操作	平穩週期操作	steady periodic operation
涡度	漩渦度	vorticity
涡壳泵	渦卷泵	volute pump
涡流	渦流	eddy flow
涡流长度	渦流長度	eddy length
涡流传递	渦流輸送	eddy transport
涡流扩散	渦流擴散	eddy diffusion
涡流扩散系数	渦流擴散係數	eddy diffusivity
涡流黏度	渦流黏度	eddy viscosity

祖国大陆名	台湾地区名	英　文　名
涡流损耗	渦流損失	eddy loss
涡流停留时间	渦流滯留時間	eddy residence time
涡流吸收	渦流吸收	eddy absorption
涡流消除器	碎漩渦器	vortex breaker
涡流卸掉	渦流分離	eddy shedding
涡流应力	渦流應力	eddy stress
涡流运动	渦流運動	eddy motion
涡流运动黏度	渦流動黏度	eddy kinematic viscosity
涡流转速计	渦流轉速計	eddy current tachometer
涡流阻力	渦流阻力	eddy resistance
涡轮	渦輪機	turbine
涡轮泵	渦輪泵	turbine pump, turbo-pump
涡轮干燥机	渦輪乾燥機	turbo-dryer
涡轮鼓风机	渦輪鼓風機	turbo-blower
涡轮机	渦輪機	turbo-machine
涡轮搅拌器	渦輪攪拌器	turbine agitator
涡轮离心泵	渦輪離心泵	turbine centrifugal pump
涡轮流量计	渦輪[式]流量計	turbine flowmeter
涡轮压缩机	渦輪壓縮機	turbo-compressor
涡轮叶轮	渦輪[式]葉輪	turbine impeller
涡片	漩渦面	vortex sheet
涡线	漩渦線	vortex line
涡旋	漩渦	vortex
涡旋搅拌器	漩渦攪拌器	vortex agitator
涡旋势	漩渦勢	vortex potential
涡旋脱落	漩渦分離	vortex shedding
涡旋尾迹	漩渦尾跡	vortex trail
涡旋运动	漩渦運動	vortex motion
蜗结	渦輪, 渦形板	snail
蜗壳	渦卷形	volute
沃尔展开式	沃爾展開式	Wohl expansion
卧式混合器	臥式混合器	horizontal mixer
卧式离心筛	臥式離心篩	horizontal centrifugal screen
卧式炉	臥式爐	horizontal furnace
卧式蒸煮器	臥式蒸煮器	horizontal digester
乌木蜡	烏木蠟	ebonite wax, ebony wax
污点	污點	speck
污垢, 结垢	積垢	fouling, scale

祖国大陆名	台湾地区名	英　文　名
污垢系数	結垢係數, 積垢因數, 積垢係數	scale coefficient, fouling factor, fouling coefficient
污泥层	污泥層	sludge blanket
污泥处理	污泥處理	sludge disposal
污泥处理过程	污泥處理法	sludge disposal process
污泥焚烧炉	污泥焚燒爐	sludge furnace, sludge incinerator
污泥过滤	污泥過濾	sludge filtration
污泥密度指数	活泥密度指數	sludge density index
污泥浓缩	污泥增稠	sludge thickening
污泥浓缩器	污泥增稠器	sludge thickener
污泥蓬松现象	污泥蓬鬆[現象]	sludge bulking
污泥体积指数	污泥體積指數	sludge volume index
污泥消化	污泥消化	sludge digestion
污泥增稠	污泥增稠	thickening sludge
污染	污染	pollution, contamination
污染工程	污染工程	pollution engineering
污染减少	污染減量	pollution abatement
污染控制	污染控制, 污染防治	pollution control
污染水	受污染水	polluted water
污染土壤	受污土壤	contaminated soil
污染物	污染物	contaminant, pollutant
污染物标准指数	污染物標準指數	pollutant standards index
污染源	污染源	pollution source
污水处理	污水處理	sewage disposal
污水管	污水管	sewer pipe
污油	污油	slop
无电镀	無電鍍	electroless plating
无电涂	無電塗, 無電塗裝	electroless coating
无定形聚合物	無定形聚合物, 非晶質聚合物	amorphous polymer
无定形碳	非晶[形]碳	amorphous carbon
无动混合器	無動混合器	motionless mixer
无反应性	無反應性	nonreactivity
无反应性组分	無反應性成分	nonreactive component
无纺布	不織布	nonwoven fabric
无缝管	無縫管	seamless pipe
无缝夹套锅	無縫[夾]套鍋	jacketed seamless kettle
无光油漆	無光油漆	matt paint
无光釉	無光釉	matt glaze

祖国大陆名	台湾地区名	英　文　名
无规分布	隨機分布	random distribution
无规共聚物	隨機共聚物	random copolymer
无规降解	無規降解[反應]	random degradation
无规扩散	隨機擴散	random diffusion
无规立构聚合物	無聚合物	atactic polymer
无滑动条件	無滑動條件	no slip condition
无灰滤纸	無灰濾紙	ashless filter paper
无混合流动反应器	無混合流動反應器	unmixed flow reactor
无机染料	礦物性染料	mineral dye
无机酸	礦酸	mineral acid
无菌操作	無菌操作	aseptic operation
无菌密封	無菌密封	aseptic seal
无孔碳	卡珀(炭精製品)	karbate
无量纲参数	無因次參數	dimensionless parameter
无量纲数	無因次數	dimensionless number
无量纲数群	無因次群	dimensionless group
无硫硫化	無硫硫化	sulfurless cure
无摩擦流动	無摩擦流動	frictionless flow
无铅汽油	無鉛汽油	unleaded gasoline
无热溶液	恒焓溶液	athermal solution
无溶剂基准	無溶劑基準	solvent-free basis
无溶剂黏合剂	無溶劑黏合劑	solventless adhesive
无溶剂漆	無溶劑漆	solventless paint
无溶剂涂料	無溶劑塗料	solventless coating
无水乙醇	無水酒精, 絕對酒精	absolute alcohol
无弹簧执行机构	無彈簧引動器	springless actuator
无碳复写	無碳複本	carbonless copy
无梯度反应器	無梯度反應器	gradientless reactor
无梯度接触器	無梯度接觸槽	gradientless contactor
无限回流	無限回流	infinite reflux
无限稀释	無限稀釋	infinite dilution
无相互作用系统	無相互作用系統	noninteracting system
无效能(=㶲)		
无旋流	不旋轉流動	irrotational flow
无旋运动	不旋轉運動	irrotational motion
无烟火药	無煙火藥, 無煙葯	smokeless powder
无烟煤	無煙煤	anthracite, anthracite coal
无烟燃料	無煙燃料	smokeless fuel
无烟推进剂	無煙推進劑	smokeless propellant

祖国大陆名	台湾地区名	英　文　名
无盐聚电解质溶液	無鹽聚電解質溶液	salt-free polyelectrolyte solution
无液流量计	無液流量計	aneroid flowmeter
无液气压计	無液氣壓計	aneroid barometer
无液压力计	無液壓力計	aneroid manometer
无应力皱纹	無應力皺紋	stress wrinkless
无约束优化	無約束最佳化	unconstrained optimization
无载体催化剂	非受載觸媒	unsupported catalyst
无阻尼系统	無阻尼系統	undamped system
无阻尼响应	無阻巴應答	undamped response
无阻尼自然频率	無阻尼自然頻率	undamped natural frequency
炕, 无效能	無效能	anergy
物理参数	物理參數	physical parameter
物理常数	物理常數	physical constant
物理处理	物理處理	physical treatment
物理处理过程	物理處理法	physical treatment process
物理传递步骤	物理輸送步驟	physical transport step
物理促进剂	物理促進劑	physical promoter
物理定律	物理定律	physical law
物理分离	物理分離	physical separation
物理过程	物理程序	physical process
物理化学	物理化學	physical chemistry
物理化学吸收	物理化學吸收	physicochemical absorption
物理模型	物理模型	physical model
物理气相沉积	物理蒸氣沈積法	physical vapor deposition
物理示踪剂	物理示蹤劑	physical tracer
物理试验	物理試驗	physical test
物理吸附	物理吸附	physical adsorption, physisorption
物理吸收	物理吸收	physical absorption
物理吸收剂	物理吸收劑	physical absorbent
物理系统	物理系統	physical system
物理性质	物理性質, 物性	physical property
物理约束	物理限制	physical constraint
物量流程图	計量流程圖	quantitative flow diagram
物料	材料	material
物料衡算流程图	物質均衡流程圖	material balance flowsheet
物料衡算, 物料平衡	物質均衡	material balance
物料平衡(＝物料衡算)		
物料与能量平衡	質能均衡	material and energy balance
物态参量	狀態性質	state property

祖国大陆名	台湾地区名	英 文 名
物质	物質	matter
物质导数	物質導數	material derivative
物质分离剂	質量分離劑	mass separating agent
物质函数	物質函數	material function
物质客观性原理	物質客觀性原理	principle of material objectivity
物种	物種	species
物种形成	物種形成, 物種演變	speciation
误差	誤差	error
误差比	誤差比	error ratio
误差常数	誤差常數	error constant
误差传播	誤差傳遞	error propagation
误差传递	誤差傳遞	propagation of error
误差积分	誤差積分	error integral
误差判据	誤差準則	error criterion
误差平方积分	誤差平方積分	integral of square error
误差平方控制	誤差平方控制	error-squared control
误差信号	誤差信號	error signal
误校正	誤校正	miscalibration
雾	霧	mist
雾化	霧化	atomization
雾化喷管	霧化噴嘴	atomizing spray nozzle
雾化喷嘴	霧化噴嘴, 噴嘴	atomizing nozzle, spray nozzle
雾化器	霧化器, 噴槍	atomizer, spray gun
雾化燃烧器	霧化燃油器	atomizing burner
雾化[用]空气	霧化空氣	atomizing air
雾化油	霧化油	atomized oil
雾化蒸汽	霧化噴蒸汽	atomizing steam
雾粒	霧粒	mist particle
雾沫混合器	霧沫混合器	entrainment mixer
[雾沫]夹带	霧沫	entrainment
雾沫蒸发器	霧沫蒸發器	entrainment evaporator
雾喷嘴	霧噴嘴	fog nozzle
雾状流	霧沫流動	mist flow

X

祖国大陆名	台湾地区名	英 文 名
吸附	吸附, 吸著	sorption, adsorption

祖国大陆名	台湾地区名	英　文　名
吸附比	吸附比	sorption ratio
吸附表面	吸附表面	adsorption surface
吸附波	吸附波	adsorption wave
吸附部位	吸附部位	adsorption site
吸附等容线	吸附等容線	adsorption isostere
吸附等温线	吸附等溫線	adsorption isotherm, sorption isotherm
吸附等压线	吸附等壓線	adsorption isobar
吸附法	吸附法	adsorption method
吸附管	吸附管	adsorption tube
吸附过滤	吸附過濾	adsorption filtration
吸附剂	吸附劑	adsorbent, adsorptive agent
吸附控制	吸附控制	adsorption control
吸附量	吸附容量	adsorptive capacity
吸附络合物	被吸附[之]錯合物	adsorbed complex
吸附能力	①吸附性②吸附度	adsorptivity, adsorbability
吸附平衡	吸附平衡	adsorption equilibrium
吸附器	吸附器, 吸附塔	adsorber
吸附热	吸附熱	heat of adsorption
吸附容量	吸附容量	adsorption capacity
吸附色谱法	吸附層析法	adsorption chromatography
吸附势	吸附勢	adsorption potential
吸附速率	吸附速率	adsorption rate
吸附图表	吸附圖表	adsorption chart
吸附-脱附	吸附-脱附	adsorption-desorption
吸附系数	吸附係數	adsorption coefficient
吸附质	吸附質	adsorbate
吸附滞后现象	吸附滯後現象	sorption hysteresis
吸附中心	吸附中心	adsorption center
吸附柱	吸附塔	adsorption column
吸光度	吸收度, 吸光度	absorbance
吸滤器	吸濾器	suction filter
吸墨纸	吸墨紙	blotting paper
吸气器	[噴水]抽氣器	aspirator
吸气式替续器	吸引式替續器	aspirating relay
吸气速率	吸氣速率	air suction rate
吸取高差	吸取高差	suction head
吸热反应	吸熱反應	endothermic reaction
吸热过程	吸熱程序	endothermic process
吸热性	吸熱度	endothermicity

祖国大陆名	台湾地区名	英　文　名
吸入速度	吸入速度	suction velocity
吸入压力	吸入壓[力]	suction pressure
吸湿	吸濕, 水吸著	sorption of water
吸湿性	吸溼能力, 吸溼性	hygroscopicity
吸收	吸收	absorption
吸收绷带	吸收繃帶	absorbent bandage
吸收不连续性	吸收不連續性	absorption discontinuity
吸收层	吸收床	absorbent bed
吸收场	吸收場	absorption field
吸收池	吸收槽	absorption cell
吸收等温线	吸收等溫線	absorption isotherm
吸收度曲线	吸收率曲線	absorbency curve
吸收反应	吸收反應	absorption reaction
吸收光谱	吸收譜	absorption spectrum
吸收过程	吸收程序	absorption process
吸收过滤器	吸收過濾器	absorbent filter
吸收基质	吸收基質	absorption base
吸收剂	吸收劑	absorbent
吸收剂量	吸收劑量	absorbed dose
吸收校正	吸收修正	absorption correction
吸收截面	吸收截面	absorption cross-section
吸收率	吸收率, 吸收比	absorptivity, absorption ratio
吸收能力	吸收能力	absorption power
吸收谱带	吸收帶, 吸光帶	absorption band
吸收器	吸收塔, 吸收槽	absorber
吸收曲线	吸收曲線	absorption curve
吸收热	吸收熱	absorption heat, heat of absorption
吸收色谱	吸收層析	absorption chromatography
吸收式过滤器	吸收式濾器	absorptive-type filter
吸收试验	吸收試驗	absorption test
吸收数	吸收數	absorption number
吸收速率	吸收速率	absorption rate
吸收塔	吸收塔	absorption tower
吸收系数	吸收係數	absorption coefficient
吸收系统	吸收系統	absorption system
吸收性软膏基	吸收性軟膏基	absorption ointment base
吸收性纤维素	吸收性纖維素	absorbent cellulose
吸收因子	吸收因數, 吸收因子	absorption factor
吸收油	吸收油	absorption oil

祖国大陆名	台湾地区名	英 文 名
吸收指示剂	吸收指示劑	absorption indicator
吸收指数	吸收指數	absorption index
吸收制冷	吸收製冷	absorption refrigeration
吸收柱	吸收管, 吸收塔	absorption column
吸收装置	吸收裝置	absorbing apparatus
吸水率	吸水率	water absorbency
吸引管线	吸入管線	suction line
吸引力	吸引力	attractive force
吸引力区[域]	吸引區[域], 漸近穩定區[域]	region of attraction
吸引升液器	吸引上升	suction lift
吸引势	吸入勢, 相吸勢	suction potential, attractive potential
吸着剂	吸著劑	sorbent
吸着树脂	包藏樹脂	occluded resin
析因设计	因子設計	factorial design
析因实验	因子實驗	factorial experiment
烯类聚合物	乙烯系聚合物	vinyl polymer
硒聚合物	硒聚合物	selenium polymer
稀糊化淀粉	稀糊化澱粉	thin-boiling starch
稀胶法	稀膠法	broth dilution
稀释	稀釋	dilution
稀释剂	稀釋劑, 減黏劑	diluent, thinner
稀释热	稀釋熱	heat of dilution
稀释效应	稀釋效應	dilution effect
稀疏矩阵	零散矩陣	sparse matrix
稀相	稀相	dilute phase
稀相流化床	稀相流體化床	dilute-phase fluidized bed
锡箔	錫箔	tinfoil
锡媒染剂	錫媒染劑	tin mordant
锡皂	錫皂	tin soap
熄灭	熄滅	extinction
熄灭温度	熄滅溫度	extinction temperature
席夫碱	希夫鹼	Schiff base
席夫碱聚合物	希夫鹼聚合物	Schiff base polymer
洗出液	洗出液	eluate
洗涤剂	清潔劑	detergent
洗涤碱	洗滌, 碳酸鈉	washing soda
洗涤器	洗氣器, 洗滌器	scrubber
洗涤速率	洗滌速率	washing rate

祖国大陆名	台湾地区名	英 文 名
洗涤塔	洗滌塔, 洗氣塔	column scrubber, scrubbing tower, wash column, washing tower
洗涤液	洗滌水	washings
洗涤皂	洗衣皂	laundry soap
洗毛	洗毛	scouring
洗糖法	洗糖	affination
洗糖数	洗糖數	affination number
洗糖值	洗糖值	affination value
洗提液	洗提液, 洗滌液, 洗析液	eluant
洗脱	洗提, 洗析	elution
洗脱色谱法	洗析層析[法]	elution chromatography
洗衣碱	洗衣鹼	laundry soda
洗衣洗涤剂	洗衣清潔劑	laundry detergent
系统, 体系	系統	system
系统边界	系統邊界	system boundary
系统参数	系統參數	system parameter
系统动力学	系統動態	system dynamics
系统分析	系統分析	system analysis
系统工程	系統工程	system engineering
系统功能		system function
系统函数	系統函數	system function
系统稳定性	系統穩定性	system stability
系统性能	系統性能	system performance
系统优化	系統最適化	system optimization
系线(=结线)		
细胞抽取物	細胞抽取物	cell extract
细胞分离	細胞分離	cell separation
细胞负载	細胞負載	cell loading
细胞密度	細胞密度	cell density
细胞培养	細胞培養	cell culture
细胞破碎	細胞破碎	cell disruption
细胞溶解	細胞分解	cell lysis, cytolysis
细胞融合	細胞融合	cell fusion
细胞收集	細胞收集	cell harvesting
细胞碎片	細胞碎片	cell debris
细胞悬浮培养	細胞懸浮培養	cell suspension culture
细胞匀浆	細胞勻漿	cell homogenate
细胞周期	細胞周期	cell cycle

祖国大陆名	台湾地区名	英　文　名
细长度	細長, 微小	slenderness
细菌	細菌	bacteria
细菌床	細菌床	bacteria bed
细菌过滤	除菌過濾	bacteriologic filtration
细菌含量	細菌含量	bacteria content
细菌数	菌數	bacteria count
细菌[性]发酵	細菌[性]發酵	bacterial fermentation
细菌氧化	細菌氧化	bacterial oxidation
细颗粒	細顆粒	fine particle
细孔	細孔	mesopore
细孔(＝孔隙)		
细孔方向分布	細孔方向分布	pore orientation distribution
细孔入口	細孔入口	pore entrance
细孔形状	細孔[幾何]形狀	pore geometry
细孔形状因子	細孔形狀因素	pore shape factor
细碎机	細級壓碎機	fine crusher
细致平衡	明細均衡	detailed balancing
霞石	霞石	nephelite
下传动式石磨机	底動石磨[機]	under-driven buhrstone mill
下降性共沸混合物	負共沸液	negative azeotrope
下限	下限	lower bound, lower limit
下相(＝底相)		
下行式反应器	下向流反應器	downflow reactor
下游	下游	downstream
下游处理	下游處理	downstream processing
下游压力	下游壓力	downstream pressure
夏季油	夏油	summer oil
纤维	纖維	fibre
纤维编织	纖維編織	satin weave
纤维长度	纖維長度	staple length
[纤维]长度分布图	[纖維]長度分布圖	staple diagram
[纤维长度]分析器	[纖維長度]分析器	sorter
[纤维]分级	[纖維]分級	sorting
纤维纲	支架	spider
纤维素	纖維素	cellulose
纤维素黄酸钠	黃酸纖維素鈉	sodium cellulose xanthate
纤维素胶体	纖維素膠體	cellulose colloid
纤维素酶	纖維素酶	cellulase
纤维素黏合剂	纖維素黏合劑	cellulose adhesive

祖国大陆名	台湾地区名	英　文　名
纤维素喷漆	纖維素噴漆	cellulose lacquer
纤维素衍生物	纖維素衍生物	cellulose derivative
纤维增强塑料	纖維強化塑膠	fiber reinforced plastics
弦线	弦線	chord line
显函数	顯函數	explicit function
显热	顯熱	sensible heat
显色剂	顯色劑	developer
显式法	顯式法	explicit method
显微光度计	微光度計	microphotometer
显微术	顯微術, 顯微法	microscopy
显影剂	顯影劑	developer
现金比率	現金比率	cash ratio
现金回收期	現金回收期	cash recovery period
现金流通图	現金流通圖	cash-flow diagram
现值	現值	present value, present worth
现值法	現值法	present worth method
现值因子	現值因數	present worth factor
线段(=链节)		
线汇座	線匯座, 線壑	line sink
线路损失	線路損失	line loss
线上计算机控制	線上計算機控制	on-line computer control
线上调谐	線上調諧	on-line tuning
线速度	線速度	linear velocity
线型低密度聚乙烯	線性低密度聚乙烯	linear low density polyethylene
线性	線性	linearity
线性凹槽	線性凹槽	linear notch
线性超前	線性超前	linear lead
线性动量原理	線性動量原理	linear momentum principle
线性阀	線性閥	linear valve
线性分析	線性分析	linear analysis
线性共沸液	線性共沸液	linear azeotrope
线性规划	線性規劃	linear programming (LP)
线性化	線性化	linearization
线性化系统	線性化系統	linearized system
线性回归分析	線性迴歸分析	linear regression analysis
线性控制	線性控制	linear control
线性控制阀	線性控制閥	linear control valve
线性模型	線性模式	linear model
线性黏弹性	線性黏彈性	linear viscoelasticity

祖国大陆名	台湾地区名	英　文　名
线性吸收系数	線性吸收係數	linear absorption coefficient
线性系统	線性系統	linear system
线源	線源	line source
限度控制	限度控制	limit control
限量成分	限量成分	limiting component
限制器	限制器	limiter
限制因素	限制因素	limiting factor
相等	均化	equalization
相对饱和度	相對飽和	relative saturation
相对粗糙度	相對粗糙度	relative roughness
相对丰度	相對含量, 相對豐度	relative abundance
相对过饱和度	相對過飽和	relative supersaturation
相对黑度	相對黑度	relative blackness
相对挥发度	相對揮發度	relative volatility
相对黏度	相對黏度	relative viscosity
相对偏差	相對偏差	relative deviation
相对湿度	相對濕度	relative humidity
相对速度	相對速度	relative velocity
相对速度因子	相對速度因數	relative velocity factor
相对速率常数	相對速率常數	relative rate constant
相对稳定性	相對穩定性	relative stability
相对误差	相對誤差	relative error
相对吸附能力	相對吸附度	relative adsorptivity
相对吸收率	相對吸收率	relative absorptivity
相对增益	相對增益	relative gain
相对蒸气压	相對蒸氣壓	relative vaporvapor pressure
相关函数	相關函數	correlation function
相关数据	相關數據	correlating data
相互扩散	相互擴散	interdiffusion
相互凝聚	相互凝聚	intercoagulation
相互作用	相互作用	interaction
相互作用度	相互作用度	degree of interaction
相互作用环路	相互作用環路	interacting loops
相互作用力	相互作用力	interaction force
相互作用能	相互作用能	interaction energy
相互作用势	相互作用勢	interaction potential
相互作用系数	交互作用係數	interaction coefficient
相互作用系统	相互作用系統	interacting system
相互作用因数	相互作用因數	interaction factor

祖国大陆名	台湾地区名	英 文 名
相互作用指数	相互作用指數	interaction index
相邻矩阵	相鄰矩陣	adjacency matrix
相容性	互適性, 相容性	compatibility
相似变换	相似變換	similarity transformation
相似定律	相似定律	similar law
相似理论	相似性理論	similarity theory
相似性	相似性, 模擬	similarity
相似性解	相似解	similarity solution
相	相	phase
相变	相變化, 相變換	phase change, phase transformation
相际传质	相間質傳	interphase mass transfer
相间反应	相間反應	interphase reaction
相间交换系数	相間交換係數	interphase exchange coefficient
相间扩散	相間擴散	interphase diffusion
相间温度	相間溫度	interphase temperature
相间相内扩散	相間相內擴散	interphase-intraphase diffusion
相间相内有效性	相間相內有效度	interphase-intraphase effectiveness
相角	相位, 相角	phase angle
相角轨迹	相角軌跡	phase angle locus
相空间	相空間	phase space
相律	相律	phase rule
相内产率	相內產率	intraphase yield
相内温度	相內溫度	intraphase temperature
相内有效性	相內有效度	intraphase effectiveness
相平衡	相平衡	phase equilibrium
相平面	相平面	phase plane
相平面标绘	相平面圖	phase plane plot
相平面分析	相平面分析	phase plane analysis
相平面图	相平面圖	phase plane portrait
相速度	相速度	phase velocity
相特性	相行為	phase behavior
相图	相圖	phase diagram
相位	相位	phase
相[位]差	相位差	phase difference
相位超前	相位超前, 相位領先	phase advance, phase lead
相位交叉	相位交越	phase crossover
相位交叉点	相位交越點	phase crossover point
相位交叉频率	相位交越頻率	phase crossover frequency
相位裕量	相位邊限	phase margin

祖 国 大 陆 名	台 湾 地 区 名	英 文 名
相位滞后	相位落後	phase lag
相移	相位位移	phase shift
相转移	相轉移	phase transfer
相转移催化	相[間]轉移催化[作用]	phase transfer catalysis
相转移催化反应	相[間]轉移催化反應	phase transfer catalyzed reaction
相转移催化剂	相[間]轉移觸媒	phase transfer catalyst
香草	香草	vanilla
香草醛	香草精	vanillin
香精回收	香精回收	essence recovery
香料	香料	perfume
香料固定剂	香料固定劑	perfume fixative, perfume fixing agent
香茅醇	香茅醇	citronellol
香茅醛	香茅醛	citronellal
香茅油	香茅油	citronella oil
香柠檬油	香柑油	bergamot oil
香皂	香皂	toilet soap
箱式干燥器	廂式乾燥計, 盤乾燥器	compartment drier, tray drier
箱式压滤机	廂式壓濾機	chamber filter press
箱式研磨机	箱式研磨機	magazine grinder
响应面	應答面	response surface
响应曲线	應答曲線	response curve
响应时间	應答時間	response time
响应速度	應答速率	response speed
向列态	向列液晶態	nematic state
向列型液晶	向列型液晶	nematic liquid crystal
向热性	向熱性	thermotropism
向日葵油	向日葵油	sunflower oil
向心加速度	向心加速度	centripetal acceleration
向心力	向心力	centripetal force
项目	專案, 計劃	project
项目工程	專案工程	project engineering
橡胶	橡膠	rubber
橡胶衬里管	橡膠襯管	rubber-lined pipe
橡胶胶乳	橡膠乳膠	rubber latex
橡胶磨	橡膠磨碾機	rubber mill
橡胶态液体	彈性液體	rubber-like liquid
橡胶状态	橡膠狀態	rubbery state
橡皮管	橡皮管	rubber tube, rubber tubing
橡实油	櫟實油	acorn oil

祖国大陆名	台湾地区名	英　文　名
削皮机	削皮機	skiving machine
削匀	修裹	shaving
消除反应	脱離反應	elimination reaction
消除器	去污劑, 除草劑	eradicator
消毒	消毒	disinfection
消毒剂	消毒劑	disinfectant
消防水	消防水	hydrant water
消光系数	消光係數	extinction coefficient
消化	消化	digestion
消化槽	消化槽	digester
消泡剂	消泡劑	defoaming agent
消泡器	除泡機	foam breaker
硝饼	硝餅	niter cake
硝饼炉	硝餅爐	niter cake furnace
硝锅	硝鍋	niter pot
硝化法	硝化程序, 硝化法	nitration process
硝化甘油炸药	硝化甘油炸藥	nitroglycerine explosive
硝化菌	氮硝化菌	nitrifying bacteria
硝化器	硝化器	nitrator
硝化[作用]	硝化, 氮硝化	nitration, nitrification
硝基火药	硝基火藥	nitroexplosive
硝基染料	硝基染料	nitro dye
硝炉	硝爐	niter oven
硝棉	強[硝化]棉	gun cotton
硝石	硝石	niter, saltpeter
硝酸	硝酸	aqua fortis
硝酸纤维素	硝化纖維素, 硝酸纖維素	cellulose nitrate, nitrocellulose
硝酸纤维素火药	硝化纖維素無煙火藥	nitrocellulose powder
硝酸纤维素塑料	硝化纖維素塑膠, 硝酸纖維素塑膠	cellulose nitrate plastic
小角光散射	小角光散射	small angle light scattering (SALS)
小角散射	小角散射	small angle scattering(SAS)
小角 X 射线散射	小角 X 射線散射	small angle X-ray scattering (SAXS)
小角 X 射线衍射	小角 X 射線繞射	small angle X-ray diffraction (SAXD)
[小时]体积空[间]速[度]	單位體積空間時速	volumetric hourly space velocity
小型超级计算机	小型超級計算機	minisupercomputer
小型计算机	小型電腦, 小型計算機	minicomputer

祖 国 大 陆 名	台 湾 地 区 名	英 文 名
小中取大效用判据	小中取大效用判據	maximin utility criterion
肖伯回弹性	肖伯回彈性	Schob resilience
肖伯弹力试验机	肖伯彈力試驗機	Schob elastometer
肖氏弹性计	邵氏彈性計	Shore elastometer
肖氏硬度	邵氏硬度	Shore hardness
肖氏硬度计	邵氏硬度計	Shore durometer
效率	效率	efficiency
楔	楔形, 楔形體	wedge
协同效应	增效效應	synergistic effect
斜板	斜板	inclined plate
斜管压力计	斜管壓力計	inclined manometer
斜管蒸发器	斜管蒸發器	inclined tube evaporator
斜钠钙石	單斜鈉石灰	gaylussite
斜坡函数	斜坡函數	ramp function
斜坡扰动	斜坡擾動	ramp disturbance
斜坡响应	斜坡應答	ramp response
斜弯头	斜彎頭	miter elbow
斜栅冷却器	斜柵冷卻器	inclined grate cooler
谐波分析	諧波分析	harmonic analysis
谐波应答	諧波應答	harmonic response
谐和平均	諧和平均	harmonic mean
谐和循环	諧和循環	harmonic cycling
谐振子	諧波振盪器	harmonic oscillator
泄放器	泄放器	bleeder
泄放装置	釋放裝置	relief device
泄料池	泄料池	blow pit
泄料箱	排放槽	blowdown tank
泄漏	洩漏	leakage
泄漏试验	漏洩試驗	leakage test
卸出损失	洩流損失, 排放損失	discharge loss
卸料器	傾卸車	dumper
卸料时间	排放時間	discharge time
卸煤加速器	卸煤加速器	coal accelerator
卸载	卸載	unloading
辛二醛聚合物	辛二醛聚合物	suberaldehyde polymer
辛烷值	辛烷值	octane number
辛烷值测试	辛烷值測試	octane number test
辛烷值增进剂	辛烷值增進劑	octane enhancer
锌白	鋅白	zinc white

祖国大陆名	台湾地区名	英 文 名
锌钡白	鋅鋇白	lithopone
锌铬黄	鋅鉻黃, 鋅黃	zinc chrome, zinc yellow
锌海绵	鋅綿	zinc sponge
锌合金	鋅合金	zinc alloy
锌华	鋅華	zinc flower
锌绿	鋅綠	zinc green
锌冕玻璃	鋅冕玻璃	zinc-crown glass
锌青铜	鋅青銅	zinc bronze
锌铁尖晶石	鋅鐵礦	franklinite
锌蓄电池	鋅蓄電池	zinc storage battery
新霉素	新黴素	neomycin
新生霉素	新生黴素	novobiocin
新闻纸	新聞紙	news-printing paper
信号传递滞后	信號傳送落後	signal transfer lag
信号流程图	信號流程圖	signal flow diagram, signal flow graph
信号烟	信號煙	signal smoke
信息	資訊	information
信息板	資訊板	information board
信息储存	資訊儲存	information storage
信息处理	資訊處理	information handling
信息聚合物	資訊聚合物	informational polymer
信息流图	資訊流圖	information flow diagram
星形聚合物	星形聚合物	star polymer
行星式搅拌器	迴繞式攪拌器	planet agitator
行政成本	行政成本	administrative cost
U 形管换热器	U 形管熱交換器	U-tube heat exchanger
S 形塔板	長條泡罩板	uniflux tray
形状数	形狀數	aspect number
形状双折射	形狀雙折射	shape birefringence
形状稳定性	形狀穩定性	shape stability
形状系数	形種因數	shape factor
形状阻力	形狀阻力	form drag
U 型管	U 形管	U tube
U 型管换热器	髮夾管換熱器	hairpin tube heat exchanger
T 型管件当量长度	T 型管件相關管長	tee equivalent pipe length
U 型结晶器	U 型結晶器	U type crystallizer
型坯	型坯	parison
V 型缺口	V 形凹槽	V notch
杏仁油	杏仁油	almond oil, apricot kernel oil

祖国大陆名	台湾地区名	英 文 名
性能	性能	performance
性能方程	性能方程式	performance equation
性能控制	性能控制	performance control
性能曲线	性能曲線	performance curve
性能试验	性能試驗	performance test
性能试验报告	性能試驗報告	performance test report
性能系数	性能係數	coefficient of performance (COP)
性能指标	性能指數	performance index
性质变化	性質變化	property change
性质关联式	性質關係[式]	property relationship
休止角	休止角	angle of respose
修理	修理	repair
修匀常数	修匀常數	smoothing constant
修匀信号	修匀信號	smoothed signal
修正密度(=有效密度)		
锈	銹	rust
溴化	溴化	bromination
虚拟变量	虚擬變量	pseudovariable
虚拟反应	假反應	pseudoreaction
虚拟活性	虚擬活性	dummy activity
虚拟机理	假機構	pseudomechanism
虚拟系数	虚擬係數	pseudoparameter
需氧量	需氧量	oxygen demand
序贯抽样	順序取樣	sequential sampling
序贯分析	順序分析	sequential analysis
序列	序列	sequence
序列长度	序列長度	sequence length
序列长度分布	序列長度分布	sequence-length distribution
序列共聚物	序列共聚物	sequential copolymer
序列聚合	序列聚合	sequential polymerization
絮凝	絮凝作用	flocculation
絮凝槽	絮凝槽	flocculation tank
絮凝测验	絮凝測驗	floc test
絮凝沉降	絮凝沉降	flocculent settling
絮凝反应	絮凝反應	flocculation reaction
絮凝机理	絮凝機構	flocculating mechanism
絮凝剂	絮凝劑	flocculant, flocculent
絮凝粒子	絮凝粒子	flocculent particle, floc particle
絮凝器	絮凝器	flocculator

祖 国 大 陆 名	台 湾 地 区 名	英 文 名
絮凝物	絮凝物, 絮凝體	flocculate, floc
絮凝物成长	絮凝體成長	floc growth
絮凝物大小	絮凝體大小	floc size
絮凝悬浮液	絮凝懸浮液	flocculent suspension
蓄电池	蓄電池	storage battery, storage cell
蓄热器	①蓄熱器②熱媒	heat accumulator, heat regenerator
蓄热式换热器	再生熱交換器	regenerative heat exchanger
悬滴试验	懸滴試驗	pendant drop test
悬浮	懸浮	suspension
悬浮催化剂	懸浮觸媒	suspended catalyst
悬浮胶体	懸膠體	suspension colloid
悬浮聚合	懸浮聚合	suspension polymerization
悬浮黏合剂	懸浮黏合劑	suspension adhesive
悬浮区法	浮動帶域[純化]法	floating zone method
悬浮生长	懸浮生長	suspended growth
悬浮生长系统	懸浮生長系統	suspended-growth system
悬浮体	懸溶膠體	suspensoid
悬浮稳定剂	懸浮液安定劑	suspension stabilizer
悬浮物负载	懸浮物負載	suspension load
悬浮系统	懸浮系統	suspension system
悬浮线	懸線	suspension wire
悬浮液		suspension
悬筐蒸发器	籃式蒸發器	basket-type evaporator
旋风除尘器	粉塵旋風器	dust cyclone
旋风分离	旋風分離	cyclone separation
旋风分离器	旋風分離器, 分離旋風器	cyclone separator, separation cyclone
旋风排气机	旋風排氣機	cyclone exhauster
旋风器	旋風器	cyclone
旋风器效率	旋風器效率	cyclone efficiency
旋风洗涤器	旋風洗氣器	cyclone scrubber
旋风蒸发器	旋風蒸發器	cyclone evaporator
旋光计	旋光計	polarimeter
旋光率	比旋光度	specific rotation
旋光能力	旋光能力	rotatory power
旋光色散	旋光色散	optical rotatory dispersion
旋光异构体	光學異構物	optical isomer
旋筐反应器	旋籃反應器	rotating-basket reactor
旋塞	活栓, 旋塞	cock

祖国大陆名	台湾地区名	英 文 名
旋塞阀	塞閥	plug valve
旋涂	旋轉塗佈	spin coating
[旋]涡	渦流	eddy
旋涡空穴	漩渦空洞	vortex cavity
旋液分离器	流體旋風器	hydrocyclone
旋转萃取器	旋轉萃取器	rotating extractor
旋转度	旋轉度	rotationality
旋转阀	旋轉閥	rotary valve
旋转分配器	旋轉分配器	rotary distributor
旋转干燥器	旋轉乾燥器	rotary dryer
旋转过滤机	旋濾機	rotary filter
旋转焊接	旋轉焊接	spin welding
旋转结晶器	旋轉結晶器	rotary crystallizer
旋转进料器	旋轉飼機	rotary feeder
旋转冷却器	旋轉冷卻器	rotary cooler
旋转连续过滤机	旋轉續濾器	rotary continuous filter
旋转能级	旋轉能階	rotational energy level
旋转黏度计	旋轉黏度計	rotary viscometer, rotation viscometer
旋转排代泵	旋轉排量泵	rotary displacement pump
旋转筛	旋轉篩	rotary screen, rotary sieve
旋转筛摇床	旋轉搖篩器	rotary sieve shaker
旋转式喷头	旋轉噴灑器	rotating sprinkler
旋转洗涤机	旋轉洗滌機	rotary washer
旋转效率	旋轉效率	rotary efficiency
旋转压滤机	旋轉壓濾機	rotary filter press, rotary crusher
旋转延伸	旋轉延伸	spin-drawing
旋转摇床	旋轉搖動器	rotary shaker
旋转真空泵	旋轉真空泵	rotary vacuum pump
选矿	選礦	ore dressing
选区衍射	選擇繞射	selected diffraction
选择定则	選擇法則	selection rule
选择加氢	選擇性氫化	selective hydrogenation
选择精馏	選擇[性]精餾	selective rectification
选择聚合	選擇聚合	selective polymerization
选择裂化	選擇性裂解	selective cracking
选择器	選擇器	selector
选择吸附	優先吸附	preferential adsorption
选择吸收	選擇性吸收	selective absorption
选择吸收剂	選擇性吸收劑	selective absorbent

祖国大陆名	台湾地区名	英 文 名
选择性比	選擇性比	selectivity ratio
选择性萃取	選擇性萃取	selective extraction
选择性发酵	選擇性發酵	selective fermentation
选择性分散曲线	選擇性分散曲線	selectivity dispersion curve
选择性浮选	選擇性浮選	selective flotation
选择性共沸蒸馏	選擇性共沸蒸餾	selective azeotropic distillation
选择性浸取	選擇性瀝取	selective leaching
选择性控制	選擇性控制	selective control
选择性控制器	選擇性控制器	selective controller
选择性溶剂	選擇性溶劑	selective solvent
选择性系数	選擇性係數	selectivity coefficient
选择性中毒	選擇[性]中毒	selective poisoning
学习控制系统	學習控制系統	learning control system
血清白蛋白黏合剂	血清蛋白黏合劑	serum albumin adhesive
血清球蛋白	血清球蛋白質	serum globulin
熏蒸剂	燻蒸劑	fumigant
熏制	燻燒	smoking
驯化	環境適應	acclimation
驯化污泥	環境污泥	acclimation sludge
循环	循環	cycling, recycle, recycling
循环泵混合器	循環泵混合器	circulating pump mixer
循环比	循環比	recirculating ratio, recirculation ratio, recycle ratio
循环层	循環層	circulation layer
循环滴	循環流動滴	circulating drop
循环法	循環法	circulation method
循环反应器	循環反應器	recirculation reactor
循环管	循環管	circulation pipe, circulation tube
循环过程	循環程序	cyclic process, recycle process
循环合成气	循環合成氣	recycle synthesis gas
循环回流	循環回流	circulating reflux
循环加料	循環原料	recycle feedstock
循环流动	循環流動	circulation flow, recirculation flow
循环流化床	循環流化床	circulating fluidized bed
循环模式	循環型式	recirculation pattern
循环判据	循環準則	circulating criterion
循环式混合器	循環式混合器	circulation mixer
循环物流	循環流動	recycle stream
循环系统	循環系統	recycle system

祖国大陆名	台湾地区名	英 文 名
循环因子	循環因數	recirculation factor
循环油	循環油	recycle oil
蕈状蒸馏塔	菇形蒸餾塔	mushroom distilling column

Y

祖国大陆名	台湾地区名	英 文 名
压电晶体	壓電晶體	piezoelectric crystal
压焓图	壓力-焓圖	pressure-enthalpy diagram
压花	壓紋	embossing
压花机	壓紋機	embossing machine
压花压光机	壓紋機	embossing calender
压降	壓[力]降	pressure drop
压紧	壓緊	compaction
压块	壓塊	briquetting
压力变送器	壓力傳送器	pressure transmitter
压力补偿	壓力補償	pressure compensation
压力测量	壓力量測	pressure measurement
压[力]差	壓力差, 微差壓力, 差壓	pressure difference , differential pressure
压力传感器	壓力轉換器	pressure transducer
压力范围	壓力範圍	pressure range
压力分布	壓力分布	pressure distribution
压力计, 压力表	壓力計	pressure gage, pressure meter
压力纪录器	壓力記錄器	pressure recorder
压力校正	壓力修正	pressure correction
压力校准	壓力校正	pressure calibration
压力控制	壓力控制	pressure control
压力控制器	壓力控制器	pressure controller
压力扩散	壓力擴散	pressure diffusion
压力能	壓[力]能	pressure energy
压力平衡阀	壓力均衡閥	pressure balanced valve
压力平衡器	均壓器	pressure equalizer
压力-容积功	壓力-體積功	pressure-volume work
压力容器	壓力容器, 壓力槽	pressure vessel
压力式温度计	壓力式溫度計	pressure type thermometer
压力损失	壓力損失	pressure loss
压力梯度	壓力梯度	pressure gradient

祖国大陆名	台湾地区名	英 文 名
压力-温度图	壓力-溫度圖	pressure-temperature diagram
压力系数	壓力係數	pressure coefficient
压力系统	壓力系統	pressure system
压力效率	壓力效率	pressure efficiency
压力效应	壓力效應	pressure effect
压力信号	壓力信號	pressure signal
压力-真空阀	壓力真空閥	pressure-vacuum valve
压力指示计	壓力指示計	pressure indicator
压力-组成图	壓力-組成圖	pressure-composition diagram
压滤饼	壓濾餅	filter press cake, press cake
压滤布	壓濾布	filter press cloth
压滤机	壓濾機	filter press, pressure filter
压敏胶	壓感黏合劑	pressure sensitive adhesive
压模	壓模	press mold
压模机	模壓機	molding press
压片	壓片, 乾壓製錠, 薄層	tabletting, slugging tablet making, sheeting
压容图	壓力-體積圖	pressure-volume diagram
压实度	壓密度	degree of consolidation
压缩	壓縮	compression
压缩比	壓縮比	compression ratio
压缩波	壓縮波	compression wave
压缩功	壓縮功	compression work
压缩过程	壓縮程序	compression process
压缩机	壓縮機	compressor
压缩机规格	壓縮機規格	compressor specification
压缩空气	壓縮空氣	compressed air
压缩空气喷雾器	氣壓噴霧器	compressed air sprayer
压缩力	壓縮力	compressive force
压缩率判据	壓縮度準則	compressibility criterion
压缩强度	抗壓強度	compressive strength
压缩区	壓縮區	compression zone
压缩热	壓縮熱	heat of compression
压缩渗透技巧	壓縮滲透技術	compression-permeability technique
压缩系数	壓縮係數	compressibility coefficient
压缩压力	壓縮壓力	compressive pressure
压缩因子	壓縮因數	compressibility factor
压缩因子通用图	Z 值通用圖, 壓縮比通用圖	generalized Z chart
压头	壓力高差	pressure head

祖国大陆名	台湾地区名	英 文 名
压头损失	高差損失, 落差損失	head loss
压延机	壓延機, 研光機, 光機軋	calender
压延孔板波纹填料	壓延孔板波紋填料	protruded corrugated sheet packing
压延效应	壓延效應	calender effect
压榨常数	壓榨常數	expression constant
压榨机	擠壓機	squeezer
压榨, 挤出	壓榨	expression
压铸	壓鑄, 模鑄	die casting
压铸成型	壓鑄成型	die casting molding
压铸硫化	壓鑄硫化	spue
芽殖	出芽	budding
芽殖细胞	出芽細胞	budding cell
芽殖真菌	出芽真菌	budding fungi
哑变数	啞變數	dummy variable
雅可比矩阵	雅可比矩陣	Jacobian matrix
亚单元	亞單元	subunit
亚单元结构	亞單元結構	subunit structure
亚分子	亞分子	submolecule
亚甲蓝	亞甲藍	methylene blue
亚晶胞	亞晶胞	subcell
亚硫酸氢盐	二亞硫酸鹽	hydrosulfite
亚硫酸氢盐漂白	二亞硫酸鹽漂白	hydrosulfite bleaching
亚硫酸盐法	亞硫酸鹽法	sulfite process
亚硫酸盐浆	亞硫酸鹽紙漿	sulfite pulp
亚麻	亞麻	flax
亚麻布	亞麻布	linen
亚麻纤维	亞麻纖維	flax fiber
亚麻子油	亞麻仁油	flax-seed oil, linseed oil
亚声速喷嘴	次音速噴嘴	subsonic nozzle
亚微观的	亞微觀的	submicroscopic
亚微观断裂	亞微觀斷裂	submicrofracture
亚微观胶束	亞微觀膠束	submicroscopic micelle
亚微观结构	亞微觀結構	submicroscopic structure
亚微裂纹	亞微裂紋	submicrocrack
亚稳单相极限线	亞穩均相極限線	spinodal
亚稳定性	介穩定性	metastability
亚稳界限	介穩定界限	metastable limit
亚稳区	介穩定區[域]	metastable region

祖国大陆名	台湾地区名	英文名
亚稳态, 介稳态	介穩定狀態	metastable state
亚硝胺红	亞硝胺紅	nitrosamine red
亚硝基染料	亞硝基染料	nitroso dye
烟道	煙道	flue
烟道气	煙道氣	flue gas, stack gas
烟点	發煙點	smoke point
烟煤	煙煤	bituminite, soft coal
烟幕弹	煙幕彈	smoke bomb
烟片[橡胶]	煙片[橡膠]	smoked sheet
烟曲霉素	煙黴素	fumagillin
烟鞣制	煙鞣製	smoke tanning
烟雾发生器	發煙器	smoke generator
烟雾剂	發煙劑	smoke agent
烟雾手榴弹	煙幕手榴彈	smoke grenade
烟雾烛[缸]	發煙罐	smoke candle
腌制	醃製	salting
延迟	延遲, 遲延	delay
延迟时间	遲延時間, 延遲時間	delay time
延迟作用	遲延作用, 延遲作用	delay action
延期年金	延期年金	deferred annuity
延伸颈	延伸頸	extension neck
延伸流动	延伸流動	extensional flow
延时曝气	延時曝氣	extended aeration
延性物料	延展性材料	ductile material
严格法	嚴格法	rigorous method
研究	研究	research
研磨	研磨, 粉碎	grind, grinding
研磨辅料	助磨劑, 研磨添加劑	grinding aids, grinding additive
研磨滚柱	壓榨輥	mill roller
研磨介质	研磨介質	grinding medium
研磨能力	研磨能力	mill capacity, milling capacity
研磨室	粉碎室	mill room
研磨效率	研磨效率	mill efficiency
盐	鹽	salt
盐捕集器	受鹽器	salt catcher
盐皮	鹽皮, 腌皮	salted hide
盐溶	鹽溶	salting in
盐水	鹽水	brine
盐酸	鹽酸	muriatic acid

祖国大陆名	台湾地区名	英 文 名
盐析	鹽析	salting out
盐析剂	鹽析劑	salting out agent
盐洗	鹽洗	brine wash
盐纤维素	鹽纖維素	salt cellulose
盐效应	鹽效應	salt effect
盐釉	鹽釉	salt glaze
颜料	顏料	pigment
衍射, 绕射	繞射	diffraction
衍生物	衍生物	derivative
掩蔽能力	遮蓋能力	masking power
厌气[细]菌(=厌氧[细]菌)		
厌氧处理	厭氣處理	anaerobic treatment
厌氧发酵	厭氣發酵	anaerobic fermentation
厌氧分解	厭氣分解	anaerobic breakdown, anaerobic decomposition
厌氧培养	厭養培養	anaerobic culture
厌氧污泥	厭氣污泥	anaerobic sludge
厌氧污泥消化	厭氣污泥消化	anaerobic sludge digestion
厌氧[细]菌, 厌气[细]菌	厭氣[細]菌	anaerobic bacteria, anaerobe
厌氧消化	厭氣消化	anaerobic digestion
厌氧消化槽	厭氣消化槽	anaerobic digester
厌氧氧化	厭氣氧化	anaerobic oxidation
验收抽样	驗收取樣	acceptance sampling
验收抽样计划	驗收取樣計畫	acceptance sampling plan
验收试验	驗收試驗	acceptance test
验丝计	驗絲計	serimeter
堰	堰	weir
堰高	堰高	weir height
扬程	高差, 揚程	head
扬析	淘析	elutriation
扬析常数	淘析常數	elutriation constant
羊毛蜡	羊毛蠟	wool wax
羊毛油	羊毛油	wool oil
羊毛脂	[粗]羊毛脂	wool grease, lanolin, wool fat
羊毛脂醇	羊毛醇	wool alcohol
羊皮纸	羊皮紙	parchment paper
阳极	陽極	anode

祖国大陆名	台湾地区名	英　文　名
阳极电位	陽極電位	anode potential
阳极反应	陽極反應	anodic reaction
阳极腐蚀	陽極腐蝕	anodic corrosion
阳极泥	陽極泥	slime
阳极氧化	陽極氧化	anode oxidation
阳离子	陽離子	cation
阳离子淀粉	陽離子澱粉	cationic starch
阳离子交换	陽離子交換	cation exchange
阳离子交换剂	陽離子交換劑	cation exchanger
阳离子交换器	陽離子交換器	cation exchanger
阳离子交换树脂	陽離子交換樹脂	cation exchange resin
阳离子聚合	陽離子聚合[反應]	cationic polymerization
阳离子型表面活性剂	陽離子界面活性劑	cationic surface active agent, cationic surfactant
阳碳离子	碳離子	carbonium ion
阳碳离子机理	碳離子機構	carbonium mechanism
阳碳离子染料	碳離子型染料	carbonium dyes
氧传递	氧傳遞	oxygen transfer
氧分析仪	氧分析儀	oxygen analyzer
氧合作用	加氧[反應], 充氧	oxygenation
氧化	氧化[反應]	oxidation
氧化苯乙烯聚合物	氧化苯乙烯聚合物	styrene oxide polymer
氧化铂催化剂	氧化鉑觸媒	platinum oxide catalyst
氧化催化剂	氧化觸媒	oxidation catalyst
氧化电位	氧化電位	oxidation potential
氧化淀粉	氧化澱粉	oxidized starch
氧化发光	氧發光	oxyluminescence
氧化反应	氧化反應	oxidation reaction
氧化锆玻璃	鋯玻璃	zirconia glass
氧化锆耐火材料	氧化鋯耐火物	zirconium oxide refractory
氧化铬-氧化铝催化剂	鉻鋁氧觸媒	chromia-alumina catalyst
氧化过程	氧化程序	oxidation process
氧化还原滴定	氧化還原滴定[法]	oxidation reduction titration
氧化还原电池	氧化還原電池	oxidation reduction cell
氧化还原电位	氧化還原電位	oxidation reduction potential
氧化还原反应	氧化還原反應	oxidation reduction reaction, redox reaction
氧化还原酶	氧化還原酶	oxydo-reductase
氧化剂	氧化劑	oxidant
氧化降解	氧化降解[反應]	oxidative degradation

祖国大陆名	台湾地区名	英 文 名
氧化酶	氧化酶	oxidase, oxydase
氧化镁	氧化鎂, 鎂氧, 鎂礬土	magnesia
氧化期	氧化期	oxidation period
氧化渠	氧化渠	oxidation ditch
氧化热裂化	氧化熱解[反應]	oxidative pyrolysis
氧化塘	氧化池, 污水塘	oxidation pond, lagoon
氧化抑制剂	氧化抑制劑	oxidation inhibitor
氧化油	氧化油	oxidized oil
氧量耗尽	氧之耗盡	oxygen depletion
氧收率系数	氧收率係數	oxygen yield coefficient
样品	樣品	sample
样品系统	樣品系統	sample system
窑	窯	kiln
[窑]炉	爐	furnace
摇动筛	搖動篩	shaking screen
摇饲机	搖飼機	cradle feeder
遥测	遙測	telemetering
遥控	遙控	remote control
药典	藥典	pharmacopoeia
药动学	藥物動力學	pharmacokinetics
药剂学	藥劑學	pharmaceutics
药理学	藥物學, 藥理學	pharmacology
药物化学	藥物化學	pharmaceutical chemistry
药物制剂	藥劑	pharmaceutical preparation
椰子油	椰子油	coconut oil
噎塞	噎塞, 阻塞	choking, choke
耶特算法	耶特算法	Yate's algorithm
冶金过程	冶金程序	metallurgical process
冶金焦	冶金焦	metallurgical coke
冶金学	冶金學	metallurgy
冶炼厂	①冶煉廠, 熔煉廠②熔煉爐	smelter, smeltery
冶炼过程	熔煉法	smelting process
冶炼炉	熔煉爐	smelter hearth
冶炼炉气	熔煉爐氣	smelter gas
叶红素	葉紅素	erythrophyll
叶蜡石	葉蠟石	pyrophyllite
叶滤机	葉濾器	leaf filter
叶绿素	葉綠素	chlorophyll

祖国大陆名	台湾地区名	英 文 名
叶轮	槳輪, 槳	paddle wheel, impeller
叶轮混合器	葉輪混合器	impeller mixer
叶轮搅拌器	葉輪攪拌器	impeller agitator
叶轮式流量计	葉輪流量計	impeller type flowmeter
叶轮直径	葉輪直徑	impeller diameter
叶轮轴	葉輪軸	impeller shaft
叶片	葉片	blade
叶片式鼓风机	葉輪式鼓風機	vane type blower
叶片式流动	葉輪流動	vane flow
叶片转子	葉片轉子	bladed rotor
曳力	①阻力②拉力	drag force
曳力矩式黏度计	拖扭矩式黏度計	drag torque type viscosimeter
曳引效应	阻力效應	drag effect
页岩	頁岩	shale
页岩油	頁岩油	shale oil
液氨	液氨	liquid ammonia
液氮	液態氮	liquid nitrogen
液滴瓦解	液滴瓦解	drop breakup
液泛	氾流	flooding
液封	液體密封	liquid seal
液固反应	液固反應	liquid-solid reaction
液固旋风器	液固旋風器	liquid-solid cyclone
液化	液化	liquefaction
液化点	液化點	liquefaction point
液化反应器	液化反應器	liquefaction reactor
液化过程	液化程序	liquefaction process
液化气体	液化氣体	liquefied gas
液化器	液化器	liquefier
液化燃料	液化燃料	liquefied fuel
液化热	液化熱	heat of liquefaction
液化石油气	液化石油氣	liquefied petroleum gas
液化天然气	液化天然氣	liquefied natural gas
液环压缩机	液環壓縮機	liquid ring compressor
液解	液解[作用]	lyolysis
液晶	液晶	liquid crystal
液晶聚合物	液晶聚合物	liquid crystal polymer
液晶显示	液晶顯示	liquid crystal display
液晶转变	液晶轉變	liquid crystal transition
液面计	液位計	liquid level gage

祖国大陆名	台湾地区名	英 文 名
液面控制	液位控制	liquid level control
液膜	液[態薄]膜	liquid membrane, liquid film
液膜接触器	液膜接觸器	liquid film contactor
液膜系数	液膜係數	liquid film coefficient
液膜阻力	液膜阻力	liquid film resistance
液气反应	液氣反應	liquid-gas reaction
液汽平衡	液汽平衡	liquid-vapor equilibrium
液态反应	液態反應	liquid reaction
液态空气	液態空氣	liquid air
液态空速	每小時之液體空間速度	liquid hourly space velocity
液体	液體, 液相	liquid
[液体]比重计	液體比重計	araeometer
液体浮选	液體浮選[法]	liquid flotation
液体火箭燃料	液體推進劑	liquid propellant
液体界面控制	液體界面控制	liquid interface control
液体静力学	液體靜力學	hydrostatics
液体静压[强]	液體靜壓[力]	hydrostatic pressure
液体流化床	液體流體化床	liquid fluidized bed
液体密度测定法	液體比重測定法	hydrometry
液体密度计	比重計	hydrometer
液体平衡	液液平衡	liquid-liquid equilibrium
液体燃料	液體燃料	liquid fuel
液体溶液	液體溶液	liquid solution
液体石蜡	液態石蠟	liquid paraffin
液体温度计	液體溫度計	liquid thermometer
液体压头	液體高差	liquid head
液位	液位, 液面	liquid level
液位传送器	液位傳送器	level transmitter
液位计	液位計	level gauge
液位控制	液位控制	level control
液位控制器	液位控制器	level controller
液位指示器	液位指示計	liquid level indicator
液下泵	沉式泵	submerged pump
液相	液相	liquid phase
液相反应	液相反應	liquid phase reaction
液相反应器	液相反應器	liquid phase reactor
液相裂解	液相裂解	liquid phase cracking
液相色谱法	液相層析[法]	liquid chromatography

祖国大陆名	台湾地区名	英 文 名
液相色谱图	液相層析圖	liquid chromatogram
液相色谱仪	液相層析儀	liquid chromatograph
液压	水壓	hydraulic pressure
液压控制	液壓控制	hydraulic control
液压控制器	液壓控制器	hydraulic controller
液压离心机	液壓離心機	hydraulic centrifuge
液压流体	液壓流體	hydraulic fluid
液压破裂法	液壓破裂[法]	hydraulic fracturing
液压筒	液壓筒	hydraulic cylinder
液压运送	液壓運送	hydraulic conveying
液压制动器	液壓引動器	hydraulic actuator
液氧	液態氧	liquid oxygen
液液萃取	液液萃取	liquid-liquid extraction
液液反应	液液反應	liquid-liquid reaction
液液分离器	液液分離器	liquid-liquid separator
液柱静压头	液體靜力高差	hydrostatic head
液柱压力计	液柱壓力計, [流體]壓力計	liquid column gage, manometer
一步法[酚醛]树脂	一步法[酚醛]樹脂	single-stage resin
一次测量	主量測[值]	primary measurement
一次处理	初級處理	primary treatment
一级标准	一級標準	primary standard
一级动力学	一階動力學	first order kinetics
一级反应	一階反應	first order reaction
一级过程	一階程序	first order process
一级控制系统	一階控制系統	first order control system
一级落后	一階落後	first order lag
一级系统	一階系統	first order system
一级相变	一階相變	first order phase transition
一级应答	一階應答	first order response
一级正应力系数	主正應力係數	primary normal stress coefficient
一级转变温度	一階轉移溫度	first order transition temperature
一维模型	一維模型	one-dimensional model
一致性	一致性	consistency
一致性试验	一致性試驗	consistency test
伊利石	白雲母石	illite
医用高分子	生醫聚合物	biomedical polymer
仪表板	儀表板	panel board
仪表流程图	儀器流程圖	instrument flowsheet

祖国大陆名	台湾地区名	英　文　名
仪表配置	儀器配置, 儀器規劃	instrumentation
仪表配置图	儀器配置圖	instrumentation diagram
仪器	儀器, 裝置	apparatus, instrument
仪器分析	儀器分析	instrumental analysis
仪器伺服机构	儀器伺服機構	instrument servomechanism
胰岛素	胰島素	insulin
移标式液位计	移標式液位計	displacement float level gage
移动床	移動床	moving bed
移动床法	移動床法	moving bed process
移动床反应器	移動床反應器	moving bed reactor
移动床气化器	移動床氣化器	moving bed gasifier
移动床吸附器	移動床吸附器	moving bed adsorber
移动催化剂床	移動觸媒床	moving catalyst bed
移动界面	移動界面	moving boundary
移动筛	移動篩	traveling screen
移动相	移動相	moving phase
遗传工程(＝基因工程)		
乙丙橡胶	乙烯-丙烯三共聚物	ethylene-propylene rubber
乙二醇	乙二醇	glycol
乙二磺酸钠聚合物	乙烯基磺酸鈉聚合物	sodium ethylene sulfonate polymer
乙基纤维素	乙基纖維素	ethyl cellulose
乙腈	乙腈	acetonitrile
乙炔黑	乙炔黑	acetylene black
乙酸-丁酸纤维素	乙酸-丁酸纖維素	cellulose acetate butyrate
乙烯基化合物	乙烯系化合物	vinyl compound
乙烯系树脂	乙烯系樹脂	vinyl resin
乙酰化	乙醯化	acetylation
乙酰值	乙醯值	acetyl value
艺术印刷纸	美術印刷銅版紙	art-printing paper
异丙苯	異丙苯	isopropylbenzene, cumene
异构重整	異構重組	isoforming
异构化	異構化	isomerization
异构体	異構物	isomer
异化	異化	dissimilation
异径管	漸縮管	pipe reducer, reducer
异径管件	漸縮管件	reducing fitting
异径管阻力	漸縮管阻力	reducer resistance
异径三通管	漸縮 T 型管, 漸縮三通管	reducing tee

祖国大陆名	台湾地区名	英 文 名
异径弯头	漸縮彎頭	reducing elbow
异氰酸酯聚合物	異氰酸酯聚合物	isocyanate polymer
异氰酸酯黏合剂	異氰酸酯黏合劑	isocyanate adhesive
异戊二烯橡胶	異戊二烯橡膠, 異平橡膠	isoprene rubber
异相成核	不勻相成核	heterogeneous nucleation
异型发酵菌	雜發酵菌	heterofermentative bacteria
抑制	抑制[作用]	inhibition
抑制剂	抑制劑, 制止器, 壓制劑	inhibitor, restrainer
易燃性	可燃性	inflammability
易碎性	易碎性	fragility
疫苗	疫苗	vaccine
逸出气体分析法	釋出氣體分析法	evolved gas analysis
逸出气体检测	釋出氣體偵測法	evolved gas detection
逸度	逸壓	fugacity
逸度系数	逸壓係數	fugacity coefficient
意外费用因数	經常開銷費用因數	contingency factor
溢料槽	溢料槽	spew groove
溢料缝	溢料	spew
溢料面	溢料面	spew area
溢料线	溢料線	spew line
溢流槽	半成品槽	run-down tank
溢流管	溢流管, 下向流管	overflow pipe, return tube, downcomer, down pipe
溢流警报	溢流警報	overflow alarm
溢流, 上溢	溢流	overflow
溢流速度	溢流速度	overflow velocity
溢流速率	溢流速率	overflow rate
溢流吸入管线	氾流吸入管線	flooded suction line
溢流堰	溢流堰	overflow weir
因变量	因變數	dependent variable
因次分析(=量纲分析)		
因次均一性原理	因次均勻原理	principle of dimensional homogeneity
因果性	因果性	causality
因子	因子	factor
F 因子	F 因子	F factor
j 因子	j 因數	j-factor
j_D 因子, 传质 j 因子	j_D-因數	j_D-factor

祖国大陆名	台湾地区名	英　文　名
j_H因子,传热 j 因子	j_H-因數	j_H-factor
阴丹士林蓝	陰丹士林藍	indanthrene blue
阴丹士林染料	陰丹士林染料	indanthrene dye
阴极	陰極	cathode
阴极反应	陰極反應	cathodic reaction
阴离子	陰離子	anion
阴离子[催化]聚合	陰離子聚合[反應]	anionic polymerization
阴离子淀粉	陰離子澱粉	anionic starch
阴离子交换剂	①陰離子交換器②陰離子交換劑	anion exchanger
阴离子交换树脂	陰離子交換樹脂	anion exchange resin, anionic exchange resin
阴离子型表面活性剂	陰離子界面活性劑	anionic surface-active agent, anionic surfactant
阴离子型表面活性剂	陰離子界面活性劑	
音速	音速	acoustic velocity
银颈	銀頸	silver necking
银值	銀值	silver number
引发剂	啟發劑, 引發劑	initiator
引风	抽氣通風	induced draft
引风式凉水塔	抽氣通風冷卻塔	induced draft cooling tower
引力	重力	gravitation
引力常量	重力常數	gravitational constant
引力场	重力場	gravitational field
引燃剂	引燃劑	combustion initiator
引诱剂	誘引劑	attractant
饮用水	飲用水	potable water
隐蔽层	隱含層	hidden layer
隐函数	隱函數	implicit function
隐枚举法	隱枚舉法	implicit enumeration method
隐色基	無色基	leuco base
隐色体	無色化合物	leuco compound
隐式法	隱式法	implicit method
英国药典	英國藥典	British pharmacopoeia
英热单位	英熱單位	British thermal unit (BTU)
罂子桐油	子桐油	aleurites ardata oil
荧光	螢光	fluorescence
荧光分光镜	螢光譜儀	fluorescence spectroscope
荧光光谱	螢光光譜	fluorescent spectrum

祖国大陆名	台湾地区名	英 文 名
荧光黄	螢光黃	fluorescein
荧光晶体	螢光晶體	fluorescent crystal
荧光染料	螢光染料	fluorescent dye
荧光增白剂	螢光增白劑	fluorescent whitening agent
荧石	氟石, 螢石	fluorspar
影式模型	影式模型	cinematic model
映射	映射	mapping
应变光学系数	應變光學係數	strain optical coefficient
应变能	應變能	strain energy
应变能函数	應變能函數	strain energy function
应变黏合剂	應變黏合劑	strain adhesive
应变软化	應變軟化	strain softening
应变双折射	應變雙折射	strain birefringence
应变椭球	應變橢圓	strain ellipsoid
应变仪	應變儀	strain gauge
应变硬化	應變硬化	strain hardening
应变张量	應變張量	strain tensor
应急操作	緊急操作	emergency operation
应力	應力	stress
应力弛豫	應力弛豫, 應力松弛	stress relaxation
应力弛豫模量	應力弛豫模量, 應力鬆弛模量	stress relaxation modulus
应力弛豫曲线	應力弛豫曲線, 應力鬆弛曲線	stress relaxation curve
应力弛豫试验机	應力弛豫試驗機, 應力鬆弛試驗機	stress relaxometer
应力传递机理	應力傳遞機構	stress-transfer mechanism
应力分布	應力分布	stress distribution
应力分析	應力分析	stress analysis
应力腐蚀	應力腐蝕	stress corrosion
应力功率	應力功率	stress power
应力光学系数	應力光學係數	stress optical coefficient(SOC)
应力集中	應力集中	stress concentration
应力集中系数	應力集中因子	stress concentration factor
应力结晶性	應力結晶性	stress crystallinity
应力开裂	應力開裂	stress cracking
应力强度因子	應力強度因子	stress intensity factor
应力软化	應力軟化	stress softening
应力石墨化	應力石墨化	stress graphitization

祖国大陆名	台湾地区名	英 文 名
应力史	應力史	stress history
应力双折射	應力雙折射	stress birefringence
应力椭球体	應力椭球	stress ellipsoid
应力消除试验	應力消除試驗	stress relief test
应力-形变曲线	應力-形變曲線	stress-deformation curve
应力银纹	應力銀紋	stress crazing
应力-应变曲线	應力-應變曲線	stress-strain curve
应力-应变曲线关系式	應力-應變曲線關係式	stress-strain curve relation
应力-应变响应	應力-應變響應	stress-strain response
应力-应变行为	應力-應變行為	stress-strain behavior
应力-应变滞后圈	應力-應變滞後圈	stress-strain loop
应力诱导极化	應力誘導極化	stress-induced polarization
应力诱导结晶	應力誘導結晶	stress-induced crystallization
应力诱导取向	應力誘導取向	stress-induced orientation
应力诱导生长	應力誘導生長	stress-induced growth
应力张量	應力張量	stress tensor
应力致白	應力致白	stress whitening
应用	應用	application
应用化学	應用化學	applied chemistry
应用热力学	應用熱力學	applied thermodynamics
应用微生物学	應用微生物學	applied microbiology
硬度	硬度	degree of hardness, hardness
硬度标度	硬度尺標	hardness scale
硬度计	硬度計, 邵氏硬度計	sclerometer, durometer
硬度试验	硬度試驗	hardness test
硬化	硬化	hardening
硬化剂	硬化劑, 硬挺[整理]劑	stiffening agent, stiffener, hardener
硬化油	硬化油	hardened oil
硬件	硬體	hardware
硬蜡	硬蠟	hard wax
硬泥法	硬泥法	stiff-mud process
硬球	硬球	hard sphere
硬球理论	硬球理論	hard sphere theory
硬石膏	石膏, 硬膏	plaster
硬水	硬水	hard water
硬水铝石	水鋁石	diaspore
硬橡胶	硬橡膠	hard rubber, vulcanite
硬性链	硬性鏈, 剛性鏈	stiff chain
硬性洗涤剂	硬性清潔劑	hard detergent

祖国大陆名	台湾地区名	英　文　名
硬质胶	硬橡膠	ebonite
硬质耐火黏土	燧石耐火黏土	flint fireclay
硬质黏土	燧石黏土	flint clay
㶲, 有效能	可用性	availability, exergy
㶲衡算, 有效能衡算	有效能均衡	exergy balance
㶲损失	有效能损失	exergy loss
永久变形	永久變形, 永久定形	permanent deformation, permanent set
永久定形	永久定形	permanent press
永久清除	永久處理	permanent disposal
永久扰动	永久擾動	permanent disturbance
永久损失	永久損失	permanent loss
永久硬度	永久硬度	permanent hardness
泳动电势	流動電位	streaming potential
涌波	湧波	surge wave
涌料	冲流	flushing
用剂	用劑	dosing
优化, 最优化	最適化	optimization
油	油	oil
油包水乳状液	油中水乳液	water in oil emulsion
油饼	油餅	oil cake
油船	油輪	oil tanker
油冬化	①油冬化②油低溫脱脂	oil winterizing
油浮选	油浮選	oil flotation
油脚	油腳, 油渣	oil foot
油井气	井口天然氣	casing head gas
油轮	油輪	tanker
油媒染剂	油媒染劑	oil mordant
油喷雾器	噴油器	oil atomizer
油漆	油漆	paint
油气	油氣	oil gas
油润滑脂	棕櫚油[潤]滑脂	palm oil grease
油砂	油砂	oil sand
油水分离器	油水分離器	oil-water separator
油田	油田	oil field
油雾分离器	分霧器	mist separator
油性树脂	油性樹脂	oleo-resin
油页岩	油頁岩	oil shale
油浴	油浴	oil bath

祖国大陆名	台湾地区名	英　文　名
油毡	油毡, 油布	linoleum
油脂	油脂	fats and oils
油脂涂料	油性漆	oil paint
油纸	油纸	oiled paper, saturating paper
游标尺	游標尺	Venturi scale
游标卡尺	游標卡尺	Venturi caliper
游离碱	自由鹼	free alkali
游离酶	游離酶	free enzyme
游离水分, 自由水分	自由水分	free moisture
游离酸	自由酸	free acid
有毒材料	有毒材料, 毒物	toxic material
有规结构流体	結構化流體	structured fluid
有规立构共聚	有規立構共聚	stereospecific copolymerization
有规立构聚合物	有規立構聚合物	stereoregular polymer
有规立构橡胶	有規立構橡膠	stereospecific rubber
有害气体	有害氣體	noxious gas
有核结晶	成核結晶	nucleate crystallization
有机补助剂	有機補助劑	organic builder
有机超负载	有機物超負載	organic overload
有机肥料	有機肥料	organic fertilizer
有机负荷量	有機物負載	organic load
有机汞	有機汞	organo mercury
有机固体	有機固體	organic solid
有机硅瓷漆	聚矽氧瓷漆	silicone enamel
有机硅聚合物	矽氧聚合物	silicone polymer
有机硅黏合剂	矽氧樹脂黏合劑	silicone adhesive
有机硅树脂	矽氧樹脂	silicone resin
有机硅塑料	矽[酮]塑膠	silicone plastic
有机硅弹性体	矽酮彈性體	silicone elastomer
有机硅脱模剂	矽[酮]脱膜劑	silicone release
有机金属化合物	有機金屬化合物	organo metallic compound
有机黏合剂	有機黏合劑	organic binder
有机黏土	有機黏土	organoclay
有机酸	有機酸	organic acid
有机碳	有機碳	organic carbon
有机体	微生物	organisms
有机污泥	有機污泥	organic sludge
有机增塑糊	有機溶膠	organosol
有理函数	有理函數	rational function

祖国大陆名	台湾地区名	英 文 名
有理数	有理數	rational number
有鳞[片]纤维[黏胶]	有鱗[片]纖維[黏膠]	scale fiber
有限边界元素	有限邊界原素	finite boundary element
有限差分	有限差分	finite difference
有限差分法	有限差分法	method of finite difference
有限扰动	有限擾動	finite disturbance
有限时间稳定性	有限時間穩定性	finite-time stability
有限元法	有限原素法	finite element method
有向图	有向圖	digraph
有效表面	有效表面	effective surface
有效导热系数	有效導熱係數	effective thermal conductivity
有效功率	有效功率	effective power
有效挥发度	有效揮發度	effective volatility
有效颗粒直径	顆粒直徑	effective particle diameter
有效孔半径	有效細孔半徑	effective pore radius
有效扩散系数	有效擴散係數	effective diffusivity
有效利息	有效利息	effective interest
有效粒度	有效顆粒大小	effective grain size
有效氯	有效氯	available chlorine
有效密度, 修正密度	有效密度	effective density
有效能(=㶲)		
有效能分析(=㶲分析)		
有效能衡算(=㶲衡算)		
有效浓度	有效濃度	effective concentration
有效曝气	有效通氣	effective aeration
有效系数	有效係數	effectiveness coefficient
有效性	有效度, 有效性	effectiveness
有效因子	有效因子	effectiveness factor
有效值	有效值	effective value
有旋流	旋轉流動	rotational flow
有载体催化剂	受載觸媒	supported catalyst
有载体金属催化剂	受載金屬觸媒	supported metal catalyst
有载体双金属催化剂	受載雙金屬觸媒	supported bimetallic catalyst
有择溶解度	優先溶解度	preferential solubility
诱变剂	突變劑	mutagen
诱导偶极	感應偶極	induced dipole
诱导期	誘導期	induction period
诱发反应	誘導反應	induced reaction
诱发阻力	感應阻力	induced drag

祖国大陆名	台湾地区名	英 文 名
釉	釉	glaze
釉烧窑	釉窑	glost kiln
淤浆(=浆料)		
淤浆法	淤漿法	slurry process
淤浆聚合	漿狀聚合	slurry polymerization
淤泥法	污泥法	sludge process
淤渣生成促进剂	污泥促進劑	sludge promoter
淤渣再循环	污泥循環	sludge recycle
鱼肝油	鱈魚肝油	cod-liver oil
鱼藤酮	魚藤酮	rotenone
玉米淀粉	玉米澱粉	corn starch
玉米糖	玉米糖	corn sugar
玉米糖浆	玉米糖漿	corn syrup
玉米油	玉米油	corn oil
预焙	預焙	prebaking
预测	預測	prediction
预测控制	預測控制	predictive control
预处理	預處理, 初步處理	pretreatment, preliminary treatment
预发酵期	預發酵期	prefermentation period
预防	預防處理	prevention
预防处理	預防處理	preventative treatment
预分布	預分布	predistribution
预分馏	初分餾	prefractionation
预估值	預估值	prior estimate
预混模制物	預混模製[物]	premix molding
预冷器	前冷卻器	precooler
预滤	預濾	prefiltration
预热	預熱	preheating
预热炉	預熱爐	preheat furnace
预热期	預熱期	preheating period
预热器	預熱器	preheater
预热区	預熱區	preheating zone
预热蒸发器	預熱蒸發器	preheating evaporator
预闪蒸塔	初餾塔	preflash tower
预设计成本估算	設計前成本估計	predesign cost estimate
预先控制	預先控制	anticipatory control
预应力外壳	預應力外殼	stressed shell
预载荷	預負載	preload
预蒸发器	預蒸發器	preevaporator

祖国大陆名	台湾地区名	英文名
阈能	低限能量	threshold energy
阈频[率]	低限频率	threshold frequency
阈限值	低限值	threshold limit value
阈[值]	低限	threshold
元反应	基本反應	elementary reaction
元素周期表	週期表	periodic table
原电池	電池	galvanic cell
原核成核	初級成核	primary nucleation
原理	原理	principle
原料	原料	feed stock, raw material, stock
原料成本	原料成本	raw material cost
原料储罐	原料儲存	raw material storage
原料检查报表	原料檢查報表	raw material check list
原料气	原料氣體	raw gas
原生动物	原生動物	protozoa
原生质	原生質	protoplasm
原生质体	原生體	protoplast
原污水	原污水	sewage raw
原型	雛型	prototype
原型试验	雛型實驗	prototype experiment
原液	原液	primary liquid
原油	原油	crude oil, crude
原油拔顶(蒸出轻馏分)过程	粗餾程序	skimming process
原油蒸馏	原油蒸餾	crude distillation
原油资源	原油資源	crude resources
原汁	原汁	absolute juice
原装置	初級工廠	primary plant
原子弹	分裂彈	fission bomb
原子间距	原子間距	atomic spacing
原子量	原子量	atomic weight
原子量表	原子量表	atomic weight table
原子热容	原子熱容量	atomic heat capacity
原子[相]互作用	原子相互作用	atomic interaction
原子序[数]	原子序	atomic number
原子质量	原子質量	atomic mass
圆孔口	圆孔口	rounded orifice
圆盘	圆盤	disk
圆盘干燥器	盤式乾燥器	disk dryer

祖国大陆名	台湾地区名	英 文 名
圆盘给料机	盤飼機	disk feeder
圆盘浇口	圓盤澆口	disk gate
圆盘流动式冷凝器	圓盤流動式冷凝器	disc flow condenser
圆盘磨	盤式研磨	disc attrition mill
圆盘塔	盤式塔	disc column
圆桶喷雾器	圓桶噴霧器	barrel sprayer
圆筒干燥器	圓筒乾燥器	cylinder drier
圆筒筛	礦石篩, 轉筒篩	trommel
圆形标度指示计	圓形標[度指]示計	circular scale indicator
圆形度	圓形度	circularity
圆形图	圓狀圖	circular chart
圆周速度	圓周速度, 周邊速度	circumferential velocity, peripheral speed, peripheral velocity
圆锥破碎机	錐碎機	cone crusher
源	源	source
约化质量	折合質量	reduced mass
约束[条件]	限制	constraint
[月]桂叶油	香葉油	bay oil
月桂油	月桂油	laurel oil
月桂子油	月桂子油	bayberry oil
云母	雲母	mica
云母铀矿	瀝青鈾礦	uranite
匀化	均質化	homogenization
匀化器	均質機	homogenizer
匀浆	均質	homogenate
运动	運動	motion
运动方程	運動方程式	equation of motion
运动黏度	動黏度	kinematic viscosity
运动黏度计	動黏度計	kinematic viscometer
运动相似	運動相似性	kinematic similarity
运动性质	運動性質	kinematic property
运动学	運動學	kinematics
运输	運輸, 輸送	transportation
运输带	運送機帶	conveyor belt
运输机	運送機	conveyer
运输滞后	輸送落後	transportation lag
运转	運轉	run
蕴压	蘊壓[力]	intrinsic pressure

Z

祖国大陆名	台湾地区名	英文名
杂醇油	雜醇油	fusel oil
杂多酸	異聚酸觸媒	heteropolyacid catalyst
杂酚油	雜酚油	creosote
杂交	雜交	hybridization
杂交瘤	雜交瘤	hybrid tumor
杂胶(＝废橡胶)		
杂菌感染	微生物感染	microbial contamination
杂菌群	雜菌群	heterogeneous population
杂散光	雜散光	stray light
杂质	不純物, 雜質	impurity
甾醇, 固醇	固醇	sterol
载点	負荷點	loading point
载气	載體氣體	carrier gas
载热体	載熱體, 熱媒, 加熱介質, 熱傳送介質	heat carrier, heating medium, heat transfer medium, heating medium
载体	①輸送機②載體	carrier
载体分馏	載體蒸餾	carrier distillation
再沉积	再沉積	redeposition
再充氧作用	再充氧	reoxygenation
再处理	再處理	reprocessing
再锻烧	再燃燒[反應]	recalcination
再锻烧过程	再燃燒程序	recalcining process
再沸冷凝器	再沸冷凝器	reboiler condenser
再沸器动力学	再沸器動能	reboiler dynamics
再沸器, 重沸器	再沸器	reboiler
再归一化	再正規化	renormalization
再活化	再活化	reactivation
再加热	再熱	reheating
再结晶	再結晶	recrystallization
再聚合	再聚合[反應]	repolymerization
再聚集	再聚集[作用]	reaggregation
再曝气	再曝氣	reaeration
再热器	再熱器	reheater
再生	再生	regeneration
再生材料	再生材料	reclaimed material

祖国大陆名	台湾地区名	英 文 名
再生胶	再生橡膠	reclaimed rubber
再生炉	再生爐	regeneration furnace
再生器	再生器	regenerator
再生热	再生熱	regenerative heat
再生速率	繁殖速率	reproduction rate
再生系统	再生系統	regenerative system
再生纤维	再生纖維	regenerated fiber
再生纤维素	再生纖維素	regenerated cellulose
再生橡胶	再生橡膠	regenerated rubber
再生循环[周期]	再生循環	regenerative cycle
再生周期运动	再生振盪	regenerative oscillation
再现度	再現度	degree of reproducibility
再现性	重複性, 再現性	reproducibility
再现因素	繁殖因數	reproduction factor
再悬浮	再懸浮	resuspension
再循环	循環	recirculation
再压	再壓	repress
再蒸馏收率	重餾產率	rerun yield
再蒸馏水	再蒸餾水	double distilled water
再蒸馏塔	重餾塔	rerun column, rerunning tower
再装满	再填充物	refill
在线	線上	on-line
在线适应	線上調適	on-line adaptation
暂态, 瞬态	暫態	transient state
藻胶	褐藻素	algin
藻类	藻	algae
藻酸钠	薄酸鈉	sodium alginate
藻酸纤维	褐藻素纖維	alginate fiber, algin fiber
皂糊	皂糊	soap paste
皂化	皂化反應	saponification
皂化剂	皂化劑	saponifying agent
皂化聚合物	皂化聚合物	saponified polymer
皂化纤维素乙酸酯	皂化纖維素乙酸酯	saponified cellulose acetate
皂化乙酸人造丝	皂化乙酸人造絲	saponified acetate rayon
皂化值	皂化值	saponification number
皂胶束	皂膠粒	soap micelle
皂脚	皂腳	niger
皂脚油	皂腳油	niger oil
皂片	皂片	soap flake

祖国大陆名	台湾地区名	英 文 名
皂液萃取	皂液萃取	soap extraction
造粒	造粒	granulation, pelletizing
造粒机	製粒機, 製丸機	pelletizer, nodulizer
造粒塔	製粒塔	prilling tower
造粒转窑	粒化轉窯	nodulizing kiln
噪声	①雜訊, 噪音②干擾	noise
噪声水平	噪音水平	noise level
噪声温度计	雜訊溫度計	noise thermometer
噪声污染	噪音污染	noise pollution
择形催化剂	形狀選擇觸媒	shape-selective catalyst
增稠	增稠	thickening
增稠过程	增稠程序	thickening process
增稠剂	增稠劑	thickening agent
增稠能力	增稠能力	thickening capacity
增稠器, 浓密机	增稠器, 稠厚器, 增稠器	thickener
增感染料	感光染料	sensitizing dye
增强材料	增強材料	reinforcing material
增强塑料	強化塑膠	reinforced plastics
增强填充料	增強填料	reinforcing filler
增强因子	增進因數	enhancement factor
增强子	①增味劑②增進劑	enhancer
增溶反应	增溶反應	solubilizing reaction
增溶剂	助溶劑, 增溶劑	solubilizing agent, solubilizer, solubilizing agent
增溶色谱法	增溶色譜法	solubilization chromatography
增溶作用	增溶作用	solubilization
增升装置	舉升裝備	high lift device
增湿	增溼	humidification, humidifying
增湿过程	增溼程序	humidification process
增湿剂	增溼器, 增溼劑	humidizer
增湿器	增溼器	humidifier
增湿图	增溼圖	humidifying chart
增塑剂	增塑劑, 塑化劑, 可塑劑	plasticizer
增塑溶胶	塑料溶膠	plastisol
增塑作用	塑化	plasticization
增碳燃气	增碳燃氣	carburetted gas
增碳水煤气	增碳水煤氣	carburetted water gas

祖国大陆名	台湾地区名	英　文　名
增效混合剂	增效混合劑	synergistic mixture
增效机理	增效機構	synergistic mechanism
增效剂	增效劑	synergist
增效添加剂	增效填加劑	synergistic additive
增效性的	增效性的	synergistic
增效性稳定剂	增效穩定劑	synergistic stabilizer
增效性阻燃剂	增效燃劑	synergistic flame retardant
增效作用	增效, 綜效	synergism
增压压缩机	增壓壓縮機	booster compressor
增益	增益	gain
增益补偿	增益補償	gain compensation
增益分隔频率	增益交越頻率	gain crossover frequency
增益交点	增益交越點	gain crossover point
增益交迭	交越增益	gain crossover
增益矩阵	增益矩陣	gain matrix
增益相图	增益-相位圖	gain-phase plot
增益裕量	增益邊限	gain margin
增值	附加價值	added value
增殖动力学(＝生长动力学)		
增殖收率系数(＝生长收率系数)		
增资	增資	incremental investment
甑, 干馏釜	甑	retort
甑馏法	甑[蒸]餾法	retort process
渣棉	礦渣, 溶渣	slag wool
渣油	蒸餘物, 殘渣油	residuum
轧机	輥軋機	rolling mill
轧制机	輥軋機	roller mill
闸阀	閘閥	gate valve
闸函数	障壁函數	barrier function
闸喉喷嘴	閘喉噴嘴	choked nozzle
闸门	①閘②澆口	gate
栅格	柵格	hurdle
炸药	炸藥	explosive
债务基金	償債基金, 債務基金	sinking funding
展幅机	展幅機, 拉幅機	stenter
张力	張力	tension
张力计	張力針	tensiometer

祖国大陆名	台湾地区名	英文名
张力试验机	張力試驗機	tension testing machine
张量	張量	tensor
樟脑油	樟腦油	camphor oil
胀塑性	膨胀性	dilatancy
胀塑性流体	膨胀性流體	dilatant fluid
账面价值	帳面價值	book value
沼气	沼氣, [微]生物生成氣	marsh gas, biogas
赵[广绪]-西得方法	趙-西德方法	Chao-Seader's method
照相增感剂	感光敏化劑	photographic sensitizer
照相纸	感光紙	photographic paper
折旧	折舊, 貶值	depreciation
折旧备用金	折舊準備金	depreciation reserve
折旧费	折舊成本	depreciation cost
折流板	擋板器	baffler
折流板(＝档板)		
折流分离	擋板分離	baffle separation
折流分离器	折流分離器	deflection separator
折流蒸发器	擋板蒸發器	baffled evaporator
折射计	折射計	refractometer
折射率	折射率	refractive index
折射温度系数	折射溫度係數	refraction temperature coefficient
折射系数	比折射率	specific refractivity
折现收益率	折現收益率	discounted cash flow rate of return
折现因子	折現因數	discount factor
折现盈亏平衡点	折現損益均衡點	discount breakeven point
赭石	赭石	ocher
褶	褶, 疊	plait
褶点(＝共溶点)		
蔗糖	蔗糖	cane sugar, saccharose, sucrose
蔗屑	蔗屑	bagassillo
蔗渣	蔗渣	bagasse
蔗[渣]板	蔗[渣]板	bagasse board
蔗渣蒸煮罐	蔗渣蒸煮器	bagasse digester
针点浇口	針點澆口	pin-point gate
针阀	針形閥	needle valve
针孔数	針孔數	needle number
针入度测定计	穿透計	penetrometer
针入度指数	穿透指數	penetration index
珍珠岩	珠岩	perlite

祖国大陆名	台湾地区名	英 文 名
真方差	實變異	true variance
真沸点	真沸點	true boiling point
真菌	真菌	fungi
真空	真空	vacuum
真空泵	真空泵	vacuum pump
真空成型	真空成形	vacuum forming, vacuum molding
真空阀	真空閥	vacuum valve
真空放电	真空放電	vacuum discharge
真空分馏器	真空分餾器, 真空分餾塔	vacuum fractionator
真空浮选	真空浮選	vacuum flotation
真空干燥	真空乾燥	vacuum drying
真空干燥器	真空乾燥器	vacuum drying apparatus, vacuum desiccator
真空干燥箱	真空乾燥箱	vacuum drying oven
真空高差	真空高差	vacuum head
真空鼓式过滤器	真空濾桶	vacuum drum filter
真空管	真空管	vacuum tube
真空锅	真空罐	vacuum pan
真空过程	真空程序	vacuum process
真空过滤机	真空過濾機	vacuum filter
真空烘箱	真空烘箱	vacuum oven
真空计	真空計	vacuometer, vacuum gage
真空焦油	真空殘渣油, 真空潛	vacuum tar
真空结晶	真空結晶	vacuum crystallization
真空结晶器	真空結晶器	vacuum crystallizer
真空解除	真空釋放	vacuum relief
真空密封	真空密封	vacuum seal
真空浓缩	真空濃縮	vacuum concentration
真空盘架干燥器	真空廂乾燥器	vacuum shelf drier, vacuum tray drier
真空润滑油	真空潤滑脂	vacuum grease
真空闪蒸	真空驟汽化	vacuum flash vaporization
真空脱水器	真空脱水器	vacuum dehydrator
真空压	真空壓[力]	vacuum pressure
真空压力计	真空壓力計	vacuum manometer
真空压制	真空壓製	vacuum pressing
真空压制机	真空壓製機	vacuum press
真空油	真空油	vacuum oil
真空甑	真空甑	vacuum retort
真空蒸发	真空蒸發	vacuum evaporation

祖国大陆名	台湾地区名	英　文　名
真空蒸发器	真空蒸發器	vacuum evaporator
真空蒸馏釜	真空蒸餾器	vacuum still
真空蒸馏, 减压蒸馏	真空蒸餾	vacuum distillation
真空蒸馏塔	真空蒸餾塔	vacuum column
真密度	真密度	true density
真实气体	真實氣體	real gas
真实组成	真實組成	real composition
振荡	振盪, 振動, 擺動	oscillation
振荡反应	振盪反應	oscillating reaction
振荡频率	振盪頻率	oscillation frequency
振荡元件	振盪元件	oscillatory element
振动给料器	振動飼機	vibratory feeder
振动流化床	振動流體化床	vibrated fluidized bed
振动能	振動能	vibrational energy
振动能级	振動能階	vibrational energy level
振动配分函数	振動分配係數函數	vibration partition function
振动器	振動器	vibrator
振动筛	擺動篩	oscillating screen, vibrating screen
振动输送机	搖運機	shaking conveyor
振幅	振幅	amplitude
振幅比	振幅比	amplitude ration
振铃极点	振鈴極點	ringing pole
振筛器	搖篩器	sieve shaker
振实密度, 夯实密度	振實密度	tap density
镇静钢	全靜碳鋼, 全靜鋼	killed carbon steel, killed steel
震动器	搖動器	shaker
震凝性	觸變性	rheopexy
震凝性流体	搖變增黏流體	rheopectic fluid
蒸氨塔	氨蒸餾器	ammonia still
蒸发	蒸發	evaporation
蒸发管	蒸發管	evaporating pipe
蒸发结晶	蒸發結晶	evaporative crystallization
蒸发结晶器	蒸發結晶器	evaporative crystallizer
蒸发冷冻	蒸發冷凍	evaporative refrigeration
蒸发冷凝器	蒸發冷凝器	evaporative condenser
蒸发冷却	蒸發冷卻	evaporation cooling, evaporative cooling
蒸发率	蒸發效率	evaporation rate, rate of evaporation
蒸发器	蒸發器	evaporator
蒸发器污垢	蒸發器積垢	evaporator scale

祖国大陆名	台湾地区名	英　文　名
蒸发热	蒸發熱	heat of evaporation
蒸发室	蒸發室	evaporator room
蒸发速度测定器	蒸發速率測定器	atometer
蒸发损失	蒸發損失	evaporation loss
蒸发系数	蒸發係數	evaporation coefficient
蒸发效率	蒸發效率	evaporation efficiency
蒸馏	蒸餾	distillation
蒸馏釜	蒸餾鍋, 蒸餾器	distillation still, still
蒸馏阱	蒸餾閘	distillation trap
蒸馏冷凝器	蒸餾冷凝器	distilling condenser
蒸馏瓶	蒸餾瓶	distilling flask
蒸馏器	蒸餾器	distillating apparatus, distiller, distilling apparatus
蒸馏区	蒸餾區	distillation zone
蒸馏试验	蒸餾試驗	distillation test
蒸馏水	蒸餾水	distilled water
蒸馏塔	蒸餾塔	distillation column
蒸馏塔板	蒸餾板	distillation tray
蒸馏值	蒸餾值	distillation value
蒸气	蒸氣	vapor
蒸气扩散泵	蒸氣擴散泵	vapor diffusion pump
蒸气旁路	蒸氣傍路	vapor bypass
蒸气压	蒸氣壓	vapor pressure
蒸气压缩	蒸氣壓縮	vapor compression
蒸汽吹洗	蒸汽清除	steam purge
蒸汽动力发电厂	蒸汽發電廠	steam power plant
蒸汽阀	蒸汽閥	steam valve
蒸汽管	蒸汽管	steam tube
蒸汽锅炉	蒸汽鍋爐	steam boiler
蒸汽机	蒸汽機	steam engine
蒸汽夹套	蒸汽套鍋	steam jacket
蒸汽经济	蒸汽經濟	steam economy
蒸汽冷凝器	蒸汽冷凝器	steam condenser
蒸汽冷凝水	蒸汽冷凝液	steam condensate
蒸汽裂解	蒸汽裂解	steam cracking
蒸汽硫化	蒸汽硫化	steam cure
蒸汽灭菌	蒸汽滅菌	steam sterilization
蒸汽喷射泵	蒸汽抽氣泵, 噴汽抽氣機	steam jet pump, steam jet ejector

祖国大陆名	台湾地区名	英　文　名
蒸汽喷射器	蒸汽射出器, 蒸汽抽氣機	steam ejector
蒸汽品质	蒸汽乾度, 蒸汽品質	steam quality
蒸汽驱动泵	蒸汽驅動泵	steam-driven pump
蒸汽省热器	蒸汽省熱器	steam economizer
蒸汽室	蒸汽室	steam chamber
蒸汽释放阀	蒸汽釋放閥	steam relief valve
蒸汽熟化	蒸汽固化	steam cure
蒸汽水分离器	蒸汽水分離器	steam separator
蒸汽涡轮	蒸汽渦輪	steam turbine
蒸汽消毒器	蒸汽滅菌器	steam sterilizer
蒸汽旋管	蒸汽旋管, 蒸汽盤管	steam coil
蒸汽浴	蒸汽浴	steam bath
蒸汽再生法	蒸汽再生[法]	steam regeneration
蒸汽转化	蒸汽重組	steam reforming
蒸汽总管	蒸汽集管[箱]	steam header
蒸腾	蒸散[作用], 發汗[現象]	transpiration
蒸煮	蒸煮	cook
蒸煮槽	蒸煮槽	cooking tank
蒸煮过程	蒸煮程序	cooking process
蒸煮器	蒸煮器	digester
蒸煮酸	蒸煮酸	cooking acid
蒸煮液	蒸煮液	cooking liquor
整合(＝集成)		
整流器	整流器	rectifier
整数规划	整數規劃	integer programming
整体流	整體流動	bulk flow
整体浓度	整體濃度	bulk concentration, concentration bulk
整体强度	整體強度	bulk intensity
整体通气器	整體通氣器	bulk aerator
整体指数	整體指數	bulk index
整重	整體重量	bulk weight
整装催化剂	整裝催化劑	monolithic catalyst
整装填料	結構填充物	structured packing
正比计数器	比例計數器	proportional counter
正常操作	正常操作	normal operation
正常反应动力学	正常反應動力學	normal reaction kinetics
正常渠深	正常渠深	normal channel depth

祖国大陆名	台湾地区名	英　文　名
正常溶液	當量溶液	normal solution
正常线性化	常態線性化	normal linearization
正常状态	常態	normal condition
正齿轮泵	正齒輪泵	spur gear pump
正反馈	正回饋	feedback positive, positive feedback
正反应	正反應	forward reaction
[正负电子]对湮没	對偶互毀	pair annihilation
正负共沸物	正負共沸液	positive-negative azeotrope
正割模量	正割係數	secant modulus
正共沸混合物	正共沸液	positive azeotrope
正规溶液	正規溶液	regular solution
正规溶液理论	規則溶液理論	regular solution theory
正激波	正震波	normal shock wave
正交流过程	正交流程序	orthoflow process
正交配置	正交配置	orthogonal collocation
正交性	正交性	orthogonality
正偏差	正偏離	positive deviation
正式报告	正式報告	formal report
正态分布	常態分布	normal distribution
正态分布函数	常態分布函數	normal distribution function
正位移	正位移	positive displacement
正位移计	正壓位移計	positive displacement meter
正弦波	正弦波	sine wave, sinusoidal wave
正弦波发生器	正弦波產生器	sine wave generator
正弦分析	正弦分析	sinusoidal analysis
正弦扰动	正弦擾動	sinusoidal disturbance
正弦式变形	正弦式變形	sinusoidal deformation
正弦响应	正弦應答	sine response, sinusoidal response
正弦信号	正弦信號	sinusoidal signal
正则变换	正準變換	canonical transformation
正则方程	正準方程式	canonical equation
正则矩阵	正準矩陣	canonical matrix
正则配分函数	正準分配函數	canonical partition function
正则系综	正則系綜	canonical ensemble
正则形式	正準形式	canonical form
正则坐标	正準座標	canonical coordinates
支撑液膜	支撐液膜	immobilized liquid membrane
支承板	支撐板	support plate
支持膜	支持膜	supporting film

祖国大陆名	台湾地区名	英　文　名
支付	支付	payout
支管	支管	side tube
支链	支鏈	branched chain
支配方程	統御方程式	governing equation
支座	拱台	abutment
芝麻油	芝麻油	sesame oil
织机	織機	weaving machine
织物过滤器	織物過濾器	fabric filter
织物整理	纖物整理	textile finishing
脂肪醇	脂肪醇	fatty alcohol
脂肪酶	脂酶	lipase
脂肪酸	脂肪酸	fatty acid
脂肪族烃	脂肪族烴	aliphatic hydrocarbon
执行机构	引動器	actuator
直边孔口	方緣孔口, 直邊孔口	square edge orifice
直方图	直立圖, 矩形圖	histogram
直观推断法, 启发式方法	啟發式法	heuristic method
直观推断法则, 启发式规则	啟發式法則	heuristic rule
直浇道	注道	sprue
直角坐标	直角坐標	rectangular coordinates
直接成本	直接成本	direct cost
直接代入法	直接代入法	direct substitution
直接法	直接法	direct method
直接还原	直接還原[法]	direct reduction
直接火焰炉	直接火加熱爐	direct-fired furnace
直接加热型蒸发器	直接加熱型蒸發器	evaporator with direct heating
直接浇口	直接澆口	direct gate
直接控制	直接控制	direct control
直接煤液化	直接煤液化[法]	direct coal liquefaction
直接氢化	直接氫化[反應]	direct hydrogenation
直接染料	直接染料	direct dye
直接热解	直接高溫分解	direct pyrolysis
直接生产成本	直接生產成本	direct production cost
直接数字控制	直接數位控制	direct digital control
直接搜索法	直接搜尋法	direct search method
直接液化	直接液化	direct liquefaction
直接作用控制器	直接作用控制器	direct-acting controller

祖国大陆名	台湾地区名	英 文 名
直径	直徑	diameter
直链聚合物	直鏈聚合物	straight-chain polymer
直列管排	直列管排	in-line tube arrangement
直馏法	直餾	straight-run distillation
直馏油	直餾油	straight run
直闪石	斜方角閃石	anthophyllite
直线折旧	直線折舊[法]	straight-line depreciation
K 值	汽液平衡比例, K 值	K value
pH 值	pH 值, 酸度值	pH value
植物激素	植物激素	plant hormone
植物生长激素	植物生長激素	plant growth hormone
植物生长调节剂	植物生長調節劑	plant growth regulator
植物油	植物油	vegetable oil
植物甾醇	植固醇	phytosterol
止回阀	止回閥	back pressure valve
止逆阀, 单向阀	止回閥, 單向閥	check valve
纸电泳	紙電泳法	paper electrophoresis
纸分配色谱	紙分配層析[法]	paper partition chromatography
纸浆[粕]液	漿[粕]液	slush pulp
纸色谱法	紙層析[法]	paper chromatography
指示计	指示計	indicator
指示剂	指示劑	indicator
指数分布	指數分布	exponential distribution
指数函数	指數函數	exponential function
指数律	指數律	exponential law
指数落后	指數落後	exponential lag
指数前因子	指數前因數, 頻率因數	preexponential factor
酯	酯	ester
酯化作用	酯化[反應]	esterification
酯交换	轉酯化[反應]	transesterification
酯值	酯值	ester number, ester value
制革	鞣製	tanning
制管机	製管機	tubing machine
制浆	製漿	pulping
制浆法	製漿法	pulping process
制冷循环	冷凍週期	refrigeration cycle
制霉菌素	奈黴素, 制黴菌素	nystatin
制药工业	製藥工業	pharmaceutical industry
制造	製造	manufacture, fabrication

祖国大陆名	台湾地区名	英 文 名
制造成本	製造成本	manufacturing cost
制造技术	製造技術	manufacturing technology
质量	質量	mass
质量传递(=传质)		
质量分率	質量分率	mass fraction
质量规格	品質規格	quality specification
质量衡算	質量均衡	mass balance
质量控制	品質管制	quality control
质量扩散系数	質量擴散係數	mass diffusivity
质量流	質量流動	mass flow
质量流量(=质量流率)		
质量流量计	質量流量計	mass flowmeter
质量流率, 质量流量	質量流量	mass flow rate
质量流速	質量速度	mass velocity
质量浓度	質量濃度	mass concentration
质量平均速度	質量平均速度	mass average velocity
质量迁移	質量輸送	mass transport
质量去除曲线	質量去除曲線	mass removal curve
质量生成曲线	質量生成曲線	mass generation curve
质量守恒	質量守恆	mass conservation
质量守恒原理	質量守恆原理	principle of conservation of mass
质量通量	質量通量	mass flux
质量吸收系数	質量吸收係數	mass absorption coefficient
质[量中]心	質量中心, 質心	center of mass, centroid
质量作用定律	整體作用定律	law of mass action
质量作用动力学	質量作用動力學	mass action kinetics
质谱	質譜	mass spectrum
质谱法	質譜法	mass spectrometry
质谱图	質譜圖	mass spectrogram
质谱仪	質譜儀	mass spectrometer
质子溶剂	質子溶劑	protic solvent
滞后	落後, 遲滯	lag, hysteresis 滞后补偿
滞后期	遲滯期	lag phase
滞流(=层流)		
滞留槽	貯留槽	holding tank
滞留池	貯留池	holding pond
滞留量	貯留量	hold-up
滞留时间	貯留時間	hold-up time
滞止膜	[停]滯膜	stagnant film

祖国大陆名	台湾地区名	英 文 名
滞止相	[停]滞相	stagnant phase
置换	置换, 替代	replacement
置换, 排代	取代, 置换	displacement
置换反应	取代反應, 置换反應	displacement reaction
置信水平	可信水平	confidence level
置信限	可信度限制	confidence limit
置信域	可信區域	confidence region
中和	中和	neutralization
中和滴定	中和滴定[法]	neutralization titration
中和点	中和點	neutralization point
中和反应	中和反應	neutralization reaction
中和剂	中和劑	neutralizing agent
中和耐火材料	中性耐火物	neutral refractory
中和区	中和區	neutral zone
中和热	中和熱	heat of neutralization
中和值	中和值	neutralization number, neutralization value
中级筛选	中級篩選	medium screening
中级筛选	中級篩選	medium sizing
中级压碎	中級壓碎	medium crushing
中级研磨	中級研磨	medium grinding
中间产物	中間產物	intermediate product
中间产物储存	中間產物儲存	intermediate storage
中间澄清器	中間澄清器, 中間沉降槽	intermediate clarifier
中间冷凝器	中間冷凝器	intercondenser
中间冷却器	中間冷卻器	intercooler
中间试验装置	實驗工廠	pilot plant
中间体	中間產物	intermediate
中间相	液晶相	mesophase
中间压碎机	中級壓碎機	intermediate crusher
中空纤维组件	中空纖維組件	hollow-fiber module
中空铸型法	中空模型	slush molding
中弯头	中彎頭	medium sweep elbow
中位数	中點, 中值	median
中温菌	嗜溫菌	mesophile, mesophilic bacteria
中性玻璃	中性玻璃	neutral glass
中性肥料	中性肥料	neutral fertilizer
中性肥皂	中性肥皂	neutral soap
中性耐火砖	中性耐火磚	neutral firebrick

祖国大陆名	台湾地区名	英 文 名
中性施胶	中性膠料	neutral size
中性亚硫酸盐半化学法	中性亞硫酸[鹽]半化學法	neutral sulfite semichemical process
中性亚硫酸盐法	中性亞硫酸[鹽]法	neutral sulfite process
中央循环管蒸发器, 排管蒸发器	中央循環管蒸發器	calandria type evaporator
中油	中油	middle oil
中值选择器	中點選擇器	median selector
终点	終點	end point
终点控制	終點控制	end-point control
终端	①終端機②[末]端	terminal
终端速度	終端速度	terminal velocity
终端下降速度	最終下降速度	terminal falling velocity
终沸点	終沸點	final boiling point
终值原理	終值定理	final-value theorem
终止反应	終止反應	termination reaction
钟乳石	鐘乳石	stalactite
钟形流量计	鐘流量計	bell flowmeter
种群参数	雜群參數	population parameter
种群动态	族群動態	population dynamics
种群密度函数	族群密度函數	population density function
种群平衡	族群均衡	population balance
种子聚合	晶種聚合	seeding polymerization
中毒	中毒	poisoning
仲氦	仲氦	parahelium
仲氢	仲氫	parahydrogen
重差计(测定比重用)	比重計	gravity meter
重化学品	重化學品	heavy chemical
重晶石	重晶石	barite
重均沸点	重量平均沸點	weight average boiling point
重均分子量	重量平均分子量	weight average molecular weight
重力	重力	gravitational force, gravity
重力沉降	重力沉降	gravitation settling, gravity settling
重力沉降分离器	重力沉降分離器	gravity settler separator
重力沉降器	重力沉降器	gravity settler
重力电池	重力電池	gravity cell
重力分离	重力分離	gravitational separation, gravity separation
重力分离器	重力分離器	gravity separator
重力分凝	重力分凝, 重力離析	gravity segregation

祖国大陆名	台湾地区名	英　文　名
重力过滤	重力過濾	gravity filtration
重力过滤器	重力過濾器	gravitational filter
重力换算因数	重力換算因數	gravitational conversion factor
重力流动	重力流動	gravity flow
重力式过滤器	重力過濾器	gravity filter
重力循环	重力循環	gravity circulation
重力压头	重力高差	gravity head
重力增稠	重力增稠	gravity thickening
重力增稠器	重力增稠器	gravity thickener
重量分率	重量分率	weight fraction
重量分析[法]	重量分析[法]	gravimetric analysis
重量摩尔浓度	重量莫耳濃度	molal concentration
重量摩尔浓度	重量莫耳濃度	molality
重量摩尔热容	重量莫耳熱容	molal heat capacity
重量平均直径	重量平均直徑	weight mean diameter
重量损失	重量損失	weight loss
重氢	重氫	heavy hydrogen
重燃料油	重燃料油	heavy fuel oil
重水	重水	heavy water
重苏打灰	重鹼灰	dense soda ash
重烃	重烴	heavy hydrocarbon
重尾馏分	重餾分	heavy ends
重直馏油	重直餾油	heavy straight-run
重质油	重油	heavy oil
周期	週期, 期間	period
周期操作	周期操作	cyclic operation, periodic operation
周期反应	周期反應	cyclic reaction
周期负荷	週期負載	periodic load
周期解	週期解	periodic solution
周期流	週期流動	periodic flow
周期流反应器	週期流動反應器	periodic flow reactor
周期凝聚	週期凝聚	periodic coagulation
周期时间	周期	cycle length, cycle time
周期式反应器	周期式反應器	cyclic reactor
周期现象	週期現象	periodic phenomenon
周期性	週期性	periodicity
周期性应力	週期性應力	cyclic stress
周期振荡	週期振盪	periodic oscillation
周转率	週轉率	turnover ratio

祖国大陆名	台湾地区名	英 文 名
轴	轴	shaft
轴衬	轴襯, 襯套	bush
轴承	轴承	bearing
轴承衬套	轴承襯套	bearing bushing
轴承滚珠	球轴承	bearing ball
轴承圈	轴承圈	bearing collar
轴承式混合器	轴承式混合器	bearing mixer
轴承托座	轴承架	bearing bracket
轴承箱	轴承箱	bearing box, bearing housing
轴对称流	轴向對稱流動	axial symmetric flow
轴功	轉轴功	shaft work
轴角	轴角	axial angle
轴流泵	轴向流泵	axial flow pump
轴流反应器	轴流反應器	axial-flow reactor
轴流[式]风机	轴向風扇	axial-flow fan
轴流式压缩机	轴向流動壓縮機	axial-flow compressor
轴向对称	轴向對稱	axial symmetry
轴向分散	轴向分散	axial dispersion
轴向分散系数	轴向分散係數	axial dispersion coefficient
轴向混合	轴向混合	axial mixing
轴向流	轴向流動	axial flow
轴向速度	轴向速度	axial velocity
轴制动马力	轴制動馬力	shaft brake horsepower
肘形管	肘形管	elbow pipe
骤混	骤混	flashing mixing
骤混器	骤混器	flashing mixer
骤冷	骤冷, 淬火	quenching, flash chilling
骤冷水	骤冷水	quenching water
骤冷油	骤冷油	quenching oil
骤氯化	骤氯化[反應]	flash chlorination
猪油	豬油	lard oil
竹节丝	竹節絲	slub yarn
竹桃霉素	安徽素, 奥連徽素	oleandomycin
逐步反应	逐步反應	stepwise reaction
逐步共聚合	逐步共聚合	step copolymerization
逐步加成聚合	逐步加成聚合[反應]	stepwise addition polymerization
逐步加成聚合物	逐步加成聚合物	step addition polymer
逐步聚合	逐步聚合	step reaction polymerization, stepwise polymerization

祖国大陆名	台湾地区名	英 文 名
逐步冷冻[法]	漸進冷凍	progressive freezing
逐次反应	逐次反應	successive reaction
逐次聚合	逐次聚合	successive polymerization
逐次开闭操作	逐次開閉操作	sequencing on-off operation
逐次撕裂强度	連續撕裂強度	successive tear strength
逐级反应	逐級反應	staircase reaction
逐级计算	分階計算	stage calculation
逐级接触	分階接觸	stage contacting
逐级接触器	分階接觸器	stage contactor
主变量	主要變數	primary variable
主产品	主產物	primary product
主导极点	主要極點	dominant pole
主发酵	主發酵	principal fermentation
主反应	主反應	primary reaction
主干	主桿	stem
主根	主根	dominant root
主环路	主環路	primary loop
主加料机	主飼機, 初級取食者	primary feeder
主令控制器	主控制器	master controller
主要本征值	主要特性值	dominant eigenvalue
主要尺寸	主因次	primary dimension
主要冲击	主要衝擊	primary impact
主要解偶	主要解偶	major decoupling
主要设计变量	主要設計變數	primary design variable
主要元素	主要元素	element primary
主应力	主應力	principal stress
主增塑剂	主塑化劑	primary plasticizer
主轴	主軸	principal axis
助促进剂	助加速劑	secondary accelerator
助催化剂	助催化劑, 促進劑, 促進體	co-catalyst , promoter
助抗氧剂	助抗氧劑	secondary antioxidant
助滤剂	助濾器	filter aid
助凝剂	助凝劑	coagulant aid
助色团	助色團	auxochrome
助悬剂	懸浮劑	suspending agent
注残料顶杆	注殘料頂桿	sprue ejector
注浆成型过程	注漿法	slip casting process
注射成型	射出成型	injection molding

祖国大陆名	台湾地区名	英 文 名
注射量	射出容量, 注射量	shot capacity, shot size, shot weight
注射器	注射器, 射出器	injector, syringe
注射速率	射出速率	shot rate
注射体积	注射體積	shot volume
注射体积控制器	注射體積控制器	shot volume controller
注射周期	射出週期	shot cycle
贮槽	貯槽, 蓄水池	reservoir
贮存稳定性	貯存穩定性	storage stability
驻点	停滯點, 穩定點	stagnation point, stationary point
驻点温度	停滯溫度	stagnation temperature
驻点性质	停滯性質	stagnation property
驻点压力	停滯壓力	stagnation pressure
柱	柱	column
柱晶白霉素	白黴素	leucomycin
柱面坐标	圓柱座標	cylindrical coordinates
柱容量	塔柱容量	column capacity
柱塞	柱塞	plunger
柱塞泵	柱塞泵	plunger pump
柱色谱法	柱式層析法	column chromatography
柱式蒸馏器	柱式蒸餾器	column still
柱特性因素	塔柱示性因數	column characterization factor
铸钢	鑄鋼	cast steel
铸件	鑄件	casting
[铸件等]气孔	氣孔	blow hole
铸模	鑄模	casting mold
铸塑聚合	鑄造聚合[反應]	cast polymerization
铸塑树脂	鑄模樹脂	cast resin
铸铁	鑄鐵	cast iron
专家系统	專家系統	expert system
专利	專利	patent
专利权使用费	權利金	royalty
专用化学品	特用化學品	speciality chemical
砖	磚	brick
转变点	轉變點	transition point
转变范围	轉變範圍	transition range
转变热	轉移熱, 轉變熱	heat of transition
转变温度	轉移溫度	transition temperature
转带浸取器	轉帶浸取器	belt extractor
转化	轉化, 變換, 轉換	conversion, transformation

祖国大陆名	台湾地区名	英 文 名
转化参量	轉化參數	conversion parameter
转化点	轉化點, 反轉點	inversion point
转化率	轉化率, 轉化分率	conversion rate, fractional conversion
转化率-温度图	轉化率-溫度圖	conversion-temperature chart
转化酶	轉化酶	invertase
转化器	轉化器	converter
转化热	轉化熱	heat of conversion
转化深度	轉化程度	conversion level
转化时间	轉化時間	conversion time
转化数	觸媒變率, 酶變率	turnover number
转化糖	轉化糖	invert sugar
转化温度	轉化溫度	conversion temperature, inversion temperature
转化原理	轉化原理	conversion principle
Z 转换	Z 轉換	Z transformation
转换涂层	變換塗層	conversion coating
转角接头	轉角接頭	corner tap
转位	移位	dislocation
转向阀	轉向閥	diversion valve
转动惯量	慣性力矩	moment of inertia
转动能	旋轉能	rotational energy
转动配分函数	轉動分配函數	rotational partition function
转鼓过滤机	轉桶濾機	rotary drum filter
转鼓混合机	轉鼓混合機	tumbler mixer
转矩	扭矩	torque
转矩管	扭犖管	torque tube
转矩流变仪	扭矩流變儀	torque rheometer
转子	轉子	rotor
转子流量计	浮子流量計	rotameter
转子叶片	轉子葉片	rotor blade
转盘过滤机	轉盤濾機	rotary disc filter
转盘流量计	轉盤流量計	rotary disc meter
转盘塔	轉盤塔	rotating disc contactor(RDC)
转速计	轉速計, 轉數計	tachometer
转速计常数	轉速計常數	tachometer constant
转筒筛	轉筒篩, 圓筒篩選	bolting reel, screening trommel, drum sifting
转筒吸收器	轉桶吸收器	rotating-drum absorber
转筒洗涤	轉筒洗滌	drum washing
转筒真空过滤机	轉桶真空過濾機	rotary vacuum drum filter

祖 国 大 陆 名	台 湾 地 区 名	英 文 名
装配	安装	mounting
装配线	裝配線	assembly line
装卸设备	裝卸設備	handling facility
装置	裝置	device
装置建立成本	配置成本	setup cost
状态	狀態	state
状态变量	狀態變數	state variable
PR[状态]方程	PR 方程	Peng-Robinson equation
状态方程	狀態方程	equation of state
状态函数	狀態函數	state function
状态和	狀態和	sum over state
追赶法	追趕法	chasing method
追踪等值线	追蹤等能線	tracing contour
锥板式黏度计	板錐黏度計	cone-and-plate viscometer
锥顶油罐	錐頂油槽	cone roof tank
锥管螺纹	管接頭	pipe tap
锥形除渣器	淨漿機	centri cleaner
锥形管	錐形管	tapered pipe
锥形给料器	錐形進料器	cone feeder
锥形流动	錐形流動	cone flow
锥形瓶	錐形瓶	erlenmeyer flask
锥形雾化器	錐形霧化器	cone atomizer
准定常态	似穩態	quasi steady state
准化学近似, 类化学近似	似化學近似	quasi-chemical approximation
准化学溶液模型, 类化学溶液模型	似化學溶液模型	quasi-chemical solution model
准静态	似定態	quasi-stationary state
准静态过程	似靜態過程	quasi-static process
准静态热重量分析法	似靜態熱重量法	quasistatic thermograrimetry
准确度	準確度, 準確性	accuracy
灼烧	焚化, 灰化	incineration
浊度	濁度	turbidity
浊度计	濁度計	turbidmeter
浊水	濁水	black water
着色	著色	staining, stain
着色法	著色法	staining method
着色试验	著色試驗	staining test
资本	資本	capital

祖国大陆名	台湾地区名	英　文　名
资本比率	資本比率	capital ratio
资本回收因数	資本回收因數	capital recovery factor
资本所得	資本所得	capital gain
资本投入	資本投資	capital investment
资本投资成本因数	資本投資成本因數	capital investment cost factor
资产负债表	資産負債表, 平衡表	balance sheet
资产会计	資産會計	asset accounting
资产型式	資産型式	asset type
资金的时间价值	貨幣的時間價值	time value of money
资源成本	資源成本	resources cost
资源分配	資源配置	allocation of resource
籽棉	種棉	seed cotton
子程序	次程式	subprogram, subroutine
子细胞	子細胞	daughter cell
子元素	子元素	daughter element
紫胶	蟲漆	lac, shellac
紫胶蜡	蟲膠蠟	shellac wax
紫胶清漆	蟲膠清漆	shellac varnish
紫霉素	群紫	viomycin
紫苏子油	蘇子油	perilla oil
紫外灯	紫外線燈	ultraviolet lamp
紫外光	紫外光	ultraviolet light
紫外光谱仪	紫外線光譜儀	UV spectrophometer
紫外线	紫外線	ultraviolet ray
紫外线比色计	紫外線比色計	ultraviolet colorimeter
紫外线发生器	紫外線產生器	ultraviolet generator
紫外[线]分光光度计	紫外線分光光度計, 紫外線光譜儀	ultraviolet spectrophotometer
紫外线熟化	紫外線熟化	ultraviolet curing
紫外线吸收	紫外線吸收	ultraviolet absorption
紫外线吸收剂	紫外線吸收劑	ultraviolet absorber
紫外线显微法	紫外線顯微法	ultraviolet microscopy
紫外线显微镜	紫外線顯微鏡	ultraviolet microscope
紫外线荧光	紫外線螢光	ultraviolet fluorescence
T 字模	T 字模	T-die
自变数	自變數	independent variable
自持续	自續	self sustaining
自持续反应	自續反應	self-sustained reaction
自持振荡	持續振盪	sustained oscillation

祖 国 大 陆 名	台 湾 地 区 名	英 文 名
自催化	自催化反應	autocatalysis
自动过程控制器	自動程序控制器	automatic process controller
自动过滤器	自動過濾器	automatic filter
自动恒温器	自動恒溫器	automatic thermostat
自动记录器	自動記錄器	auto recorder
自动加料器	自動進料器	automatic feeder
自动加热性	自熱性	self-heating property
自动进料	自動進料	automatic feed
自[动]聚合	自聚合	auto-polymerization
自动控制	自動控制	automatic control
自动控制器	自動控制器	automatic controller
自动控制系统	自動控制系統	automatic control system
自动硫化	自動硫化	self-curing, self-vulcanizing
自动灭火	自熄	self-extinguishing
自动凝结	自凝聚	auto-coagulation
自动皮带秤	稱重計	weightometer
自动手控开关	自動手動開關	auto-manual switch
自动调节器	自動調節器	automatic regulator
自动稳定	自穩定	auto-stabilization
自动选择器	自動選擇器	auto-selector
自动选择器控制	自動選擇器控制	auto-selector control
自发成核[作用]	自發成核[作用]	sporadic nucleation
自发反应	自發反應	spontaneous reaction
自发过程	自發程序	spontaneous process
自发极化	自然極化	spontaneous polarization
自发结晶	自發結晶	spontaneous crystallization
自发乳化	自發乳化	spontaneous emulsification
自发形核	自發成核	spontaneous nucleation
自固化	自塑	self-cure
自固化环氧树脂	自塑化環氧樹脂	self-cure epoxy resin
自固化黏合剂	自塑化黏合劑	self-curing adhesive
自记气压计	氣壓記錄器	barograph
自加速作用	自加速[作用]	autoacceleration
自交联丙烯酸树脂	自交聯丙烯酸樹脂	self-crosslinking acrylic resin
自交联丙烯酸酯橡胶	自交聯丙烯酸酯橡膠	self-crosslinking acrylate rubber
自胶合	自膠合	auto-agglutination
自校正调节器	自調諧調節器	self-tuning regulator, self-tuning controller
自净化	自淨化[作用]	self purification
自聚合	自聚合	self-polymerization

祖国大陆名	台湾地区名	英　文　名
自扩散	自擴散	self-diffusion
自来水	自來水	tap water
自老化	自老化	self ageing
自硫化	自硫化	auto-vulcanization
自磨机	自磨機	autogenous mill
自黏合纤维	自黏合纖維	self-bonded fiber
自黏胶带	自黏膠帶	self-adhesive tape
自黏着剂	自黏著劑	self-adhesive
自平衡电桥	自均衡電橋	self-balancing bridge
自平衡电位计	自均衡電位計	self-balancing potentiometer
自起动泵	自引泵	self-priming pump
自然对流	自然對流	natural convection
自然老化	自然老化	natural aging
自然冷却	自然冷卻	natural cooling
自然凝结	自然凝固	spontaneous coagulation
自然频率	自然頻率	natural frequency
自然通风	自然通風	natural draft, natural draught
自然通风冷却塔	自然通風冷卻塔	natural draft cooling tower
自然循环蒸发器	自然循環蒸發器	natural circulation evaporator
自然蒸发	自然蒸發	natural evaporation, spontaneous evaporation
自然周期	自然週期	natural period
自然资源	天然資源	natural resources
自燃	自燃	self-ignition, spontaneous combustion
自燃温度	燃點, 自燃溫度	autoignition temperature, self-ignition temperature
自热反应	自熱反應	autothermal reaction
自溶	自消化, 自分解	autolysis
自溶脂肪酶	自解脂酶	autolytic lipase
自身低聚化	自身低聚化	self-oligomerization
自适应控制	自適應控制, 適應控制	self-adaptive control, adaptive control
自适应去耦器	適應解偶器	adaptive decoupler
自调整	自調節	self regulation
自吸搅拌器, 空心叶轮搅拌器	空心葉輪攪拌器	hollow agitator
自熄性	自熄性	self extinguish ability
自旋	自轉, 自述	spin
自旋回波	自轉回聲, 自轉回波	spin echo

祖国大陆名	台湾地区名	英　文　名
自旋回波法	自旋回波法, 自轉回聲法	spin echo method
自旋晶格弛豫	自旋晶格弛豫, 自旋晶格鬆弛	spin-lattice relaxation
自旋晶格弛豫时间	自旋-格子鬆弛時間	spin-lattice relaxation time
自旋偶合	自旋偶合	spin coupling
自旋偶合常数	自旋偶合常數	spin coupling constant
自旋去偶	自旋去偶	spin decoupling
自旋-自旋弛豫	旋-旋傳遞	spin-spin relaxation
自旋-自旋弛豫时间	旋-旋傳遞時間	spin-spin relaxation time
自旋-自旋偶合	自旋耦合	spin-spin coupling
自氧化	自氧化	autoxidation
自引发	自發	self-initiation
自由表面	自由表面	free surface
自由部位浓度	自由部位濃度	free site concentration
自由沉降	自由沉降	free settling
自由沉降速度	自由沉降速度	free falling velocity
自由程	自由徑	free path
自由度	自由度	degree of freedom
自由对流	自然對流	free convection
自由焓(＝吉布斯自由能)		
自由基	自由基	free radical
自由基捕获剂	自由基捕捉劑	radical scavenger
自由基反应	自由基反應	free radical reaction
自由基机理	自由基機構	free radical mechanism
自由基聚合	自由基聚合[反應]	free radical polymerization
自由空间, 分离空间	自由空間, 分離空間	freeboard
自由流	自由流	free stream
自由流动	自由流動	free flow
自由流动压头	自由流動高差	free flow head
自由能	自由能	free energy
自由能组成图	自由能組成圖	free energy-composition diagram
自由水	自由水	free water
自由水分(＝游离水分)		
自由水含量	自由水含量	free moisture content
自由涡旋	自由漩渦	free vortex
自由压碎	自然壓碎	free crushing
自诱导反应	自誘發反應	auto-induced reaction

祖国大陆名	台湾地区名	英　文　名
自增强聚合物	自增強聚合物	self-reinforcing polymer
自增长	自增長	self-propagation
自治系统	自發系統	autonomous system
自终止	自終止	self-termination
综合壁	複合壁	compositive wall
棕榈蜡	棕櫚蠟	palm wax
棕榈仁油	棕櫚仁油, 棕櫚實油	palm kernel oil, palm nut oil
棕榈油	棕櫚油	palm oil
棕土	富錳棕土	umber
总产量分率	總產量分率	overall fractional yield
总传递单元	總傳送單位	overall transfer unit
总传递函数	總轉移函數	overall transfer function
总传热系数	總熱傳係數	overall heat transfer coefficient
总传质单元数	總質傳單元數	number of overall transfer units
总传质系数	總質傳係數	overall mass transfer coefficient
总发射系数	總發射係數	total emissivity
总固体	總固體[量]	total solid
总过程	總程序	overall process
总回收	總回收[率]	overall recovery
总能	總能	total energy
总能量平衡	總能量均衡	overall energy balance
总热平衡	總熱量均衡	overall heat balance
总热值	總熱值	gross heating value
总熵平衡	總熵均衡	overall entropy balance
总速率	總速率	overall rate
总体反应	總體反應	global reaction
总体反应速率	總體反應速率	global reaction rate
总体速率	總速率	global rate
总体稳定性	總體穩定性	global stability
总体最优[值]	總體最適[值]	global optimum
总吸收系数	總吸收係數	overall absorption coefficient
总系数	總係數	overall coefficient
总效率	總效率	overall efficiency
总压法	總壓法	total pressure method
总压力	總壓[力]	total pressure
总压头	總高差	total head
总硬度	總硬度	total hardness
总有机碳	總有機碳	total organic carbon
总质量平衡	總質量均衡	overall mass balance

祖国大陆名	台湾地区名	英 文 名
总资本投资额	總資本投資額	total capital investment
总组成	總組成	overall composition
纵向翅片	縱向鰭片	longitudinal fin
纵向分散系数	縱向分散係數	longitudinal dispersion coefficient
纵向应力	縱向應力	longitudinal stress
走旁路	旁流, 側流	bypassing
租金	租金	rent
阻迟运动	阻滯運動	retarded motion
阻化剂	終止劑	stopping agent
阻抗	阻抗	impedance
阻力	阻力	resistance
阻力系数	阻力係數	drag coefficient, resistance coefficient
阻尼	阻尼	damping
阻尼比	阻尼比	damping ratio
阻尼常数	阻尼常數	damping constant
阻尼力	阻尼力, 制振力	damping force
阻尼流体	阻尼流體	damping fluid
阻尼器	①阻尼器, 制振器②氣閘③擋板	damper
阻尼系数	阻尼係數	damping coefficient
阻尼振荡	阻尼振盪	damped oscillation
阻尼振荡器	阻尼振盪器	damped oscillator
阻燃剂	阻燃劑, 防火劑, 耐燃劑	fire retardant, flame resisting agent
阻塞压碎	阻塞壓碎	choke crushing
阻滞沉降	阻滯沉降	impeded settling
阻滞剂	阻滯劑	retarder
阻滞因子	阻滯因數	retardation factor
组成	組成	composition
组成规格	組成規格	composition specification
组成轨道	組成軌跡	composition trajectory
组成控制	組成控制	composition control
组分	組成, 成分	constituent, component
组分分涂黏合剂	分施膠黏劑	separated application adhesive
组分指定图	成分指定圖	component assignment diagram
组合规则	組合規則	combining rule
组合项	組合項	combinatorial term
组态扩散	組態擴散	configurational diffusion
组织	組織	organization , tissue

祖国大陆名	台湾地区名	英　文　名
最大比生长速率	最大比生長速率	maximum specific growth rate
最大概率数	最可能數	most probable number
最大混合度	最大混合度	maximum mixedness
最大似然原理	最大相似原理	maximum likelihood principle
最低沸点共沸物	最低沸點共沸液	minimum boiling azeotrope
最低凝固点	最低凝固點	freezing minimum, minimum freezing point
最低容许收益率	最低容許收益率	minimum acceptable rate of return(MARR)
最陡上坡点	最陡上坡點	point of steepest ascent
最陡下坡点	最陡下坡點	point of steepest descent
最概然分布	最可能分布	most probable distribution
最高沸点共沸物	最高沸點共沸液	maximum boiling azeotrope
最高流速	尖峰流量	peak flow rate
最高允许温度	最高容許溫度	maximum allowable temperature
最高允许压力	最高容許壓[力]	maximum allowable pressure
最速上升法	最陡上升法	steepest ascent method
最速下降法	最陡下降法	steepest descent method
最小二乘法	最小平方法	least squares method
最小二乘分析	最小平方分析[法]	least squares analysis
最小回流	最小回流	minimum reflux
最小回流比	最小回流比	minimum reflux ratio
最小流化速度	最低流體化速度	minimum fluidization velocity, minimum fluidizing velocity
最小流化态	最低流體化	minimum fluidization
最小偏差	最小偏差	minimum deviation
最小塔板数	最少板數	minimum tray number
最小值原理	極小原理	minimum principle
最优操作条件	最適操作條件, 最適操作狀況	optimal operation condition, optimum operation condition
最优策略	最適策略	optimal policy, optimum policy
最优工况	最適條件, 最適狀況	optimal condition
最优化(=优化)		
最优回流比	最適回流比	optimum reflux ratio
最优回流[量]	最適回流	optimal reflux, optimum reflux
最优开关	最適開關, 最適切換	optimal switching, optimum switching
最优控制	最適控制	optimal control, optimum control
最优设计	最適設計	optimal design, optimum design
最优设计点	最適設定點	optimal set point, optimum set point
最优搜寻法	最適搜尋法	optimal searching method, optimum searching method

祖国大陆名	台湾地区名	英 文 名
最优调谐	最適調諧	optimal tuning, optimum tuning
最优途径	最適途徑	optimal path, optimum path
最优温度	最適溫度	optimal temperature, optimum temperature
最优系统	最適系統	optimal system, optimum system
最优性	最適性	optimality
最优值	最適值	optimal value, optimum value
最优转化率	最適轉化率	optimal conversion, optimum conversion
最优组成	最適組成	optimal composition
最终产品	最終產物	end product
最终澄清器	最終澄清器, 最終沉降槽	final clarifier
最终处理	最終處理	ultimate disposal
最终控制元件	最終控制原件	final control element
最终条件	最終條件	final condition
最终周期	最終週期	ultimate period
最终周期解	最終週期解	ultimate periodic solution
最终周期应答	最終週期應答	ultimate periodic response
最终状态	最終狀態	final state
醉椒	胡椒油	kava oil
左旋糖	左旋醣	levo-rotatory sugar
作用信号	引動信號	actuating signal
坐标系	坐標系	coordinate system

副　篇

A

英　文　名	祖国大陆名	台湾地区名
abatement	减量	降低
abherent	防黏剂	防黏著劑
ablation	烧蚀	融磨
abnormal reaction	非正常反应	異常反應
abrasion	磨损	磨耗, 磨損
abrasion index	磨耗指数	磨耗指數
abrasion resistance	耐磨性	磨耗阻力
abrasion test	磨耗试验	磨耗試驗
abrasive	磨料	磨(擦)料, 磨擦物
abrasive cleaner	擦洁剂	擦潔劑
abrasiveness	磨耗度	抗磨性
abrasive paper	砂纸	砂紙, 磨擦紙
absinthium	苦艾	苦艾
absolute alcohol	无水乙醇	無水酒精, 絕對酒精
absolute boiling point	绝对沸点	絕對沸點
absolute color value	绝对色值	絕對色值
absolute counting	绝对计数	絕對計數
absolute deviation	绝对偏差	絕對偏差
absolute dry fiber	绝对干纤维	絕對乾纖維
absolute entropy	绝对熵	絕對熵
absolute humidity	绝对湿度	絕對濕度
absolute juice	原汁	原汁
absolute moisture content	绝对湿含量	絕對水分含量
absolute pressure	绝对压力	絕對壓力
absolute pressure gauge	绝对压力计	絕對壓力計
absolute reaction rate	绝对反应速率	絕對反應速率
absolute specificity	绝对专一性	絕對特異性
absolute stability	绝对稳定性	絕對穩定性
absolute temperature	绝对温度	絕對溫度

英 文 名	祖国大陆名	台湾地区名
absolute thermodynamic temperature scale	绝对热力学温标	絕對熱力溫標
absolute turbidity	绝对浊度	絕對濁度
absolute value	绝对值	絕對值
absolute viscosity	绝对黏度	絕對黏度
absolute zero	绝对零度	絕對零度
absorbance	吸光度	吸收度,吸光度
absorbed dose	吸收剂量	吸收劑量
absorbency curve	吸收度曲线	吸收率曲線
absorbent	吸收剂	吸收劑
absorbent bandage	吸收绷带	吸收繃帶
absorbent bed	吸收层	吸收床
absorbent cellulose	吸收性纤维素	吸收性纖維素
absorbent filter	吸收过滤器	吸收過濾器
absorber	吸收器	吸收塔,吸收槽
absorbing apparatus	吸收装置	吸收裝置
absorption	吸收	吸收
absorption band	吸收谱带	吸收帶,吸光帶
absorption base	吸收基质	吸收基質
absorption cell	吸收池	吸收槽
absorption chromatography	吸收色谱	吸收層析
absorption coefficient	吸收系数	吸收係數
absorption column	吸收柱	吸收管,吸收塔
absorption correction	吸收校正	吸收修正
absorption countercurrent	逆流吸收	逆流吸收
absorption cross-section	吸收截面	吸收截面
absorption curve	吸收曲线	吸收曲線
absorption discontinuity	吸收不连续性	吸收不連續性
absorption factor	吸收因子	吸收因數,吸收因子
absorption field	吸收场	吸收場
absorption heat	吸收热	吸收熱
absorption index	吸收指数	吸收指數
absorption indicator	吸收指示剂	吸收指示劑
absorption isotherm	吸收等温线	吸收等溫線
absorption number	吸收数	吸收數
absorption oil	吸收油	吸收油
absorption ointment base	吸收性软膏基	吸收性軟膏基
absorption power	吸收能力	吸收能力
absorption process	吸收过程	吸收程序

英　文　名	祖国大陆名	台湾地区名
absorption rate	吸收速率	吸收速率
absorption ratio	吸收率	吸收比
absorption reaction	吸收反应	吸收反應
absorption refrigeration	吸收制冷	吸收製冷
absorption spectrum	吸收光谱	吸收譜
absorption system	吸收系统	吸收系統
absorption test	吸收试验	吸收試驗
absorption tower	吸收塔	吸收塔
absorptive-type filter	吸收式过滤器	吸收式濾器
absorptivity	吸收率	吸收率
abundance	丰度	豐度
abundance curve	丰度曲线	豐度曲線, 含量曲線
abutment	支座	拱台
acaricide	杀螨剂	殺壁虱劑
accelerated aging	加速老化	加速老化
accelerated cement	速凝水泥	快乾水泥, 速凝水泥
accelerated gum	速成胶质	速成膠
acceleration error coefficient	加速度误差系数	加速誤差係數
accelerator	①加速剂②加速器	①催速劑②加速器
accelerator of vulcanization	硫化加速剂	硫化催速劑
accelerometer	加速度计	加速計
accentric factor	偏心因子	偏離因數
acceptable quality level	合格质量标准	合格品質水準
acceptable return	可接受报酬	可接受(的)報酬
acceptance sampling	验收抽样	驗收取樣
acceptance sampling plan	验收抽样计划	驗收取樣計畫
acceptance test	验收试验	驗收試驗
acceptor	受体	接受體, 接收器
acclimation	驯化	環境適應
acclimation sludge	驯化污泥	環境污泥
accumulation rate	累积速率	累積速率
accumulative crystallization	聚集结晶	累積結晶
accuracy	准确度	準確度, 準確性
acetal	醛缩醇	縮醛類
acetal resin	缩醛树脂	縮醛樹脂
acetate film	醋酸酯薄膜	醋酸酯膠膜, 乙酸酯膠膜
acetate rayon	醋酸人造丝	醋酸嫘縈, 乙酸嫘縈
acetic acid bacteria	醋酸菌	醋酸菌

英　文　名	祖国大陆名	台湾地区名
acetonitrile	乙腈	乙腈
acetylation	乙酰化	乙醯化
acetyl cellulose	醋酸纤维素	醋酸纖維素
acetylene black	乙炔黑	乙炔黑
acetyl value	乙酰值	乙醯值
achromycin	四环素	鉑黴素
acid acceptor	酸受体	酸受體
acid catalysis	酸催化	酸催化[作用]
acid chrome dye	酸性铬媒染料	酸性鉻媒染料
acid color	酸性染料	酸性染料
acid component	酸性成分	酸性成分
acid cooler	酸冷却器	酸冷卻器
acid dye	酸性染料	酸性染料
acid feed box	加酸器	飼酸箱
acid fertilizer	酸性肥料	酸性肥料
acid firebrick	酸性耐火砖	酸性耐火磚
acid-forming bacteria	成酸菌	生酸菌
acid fume	酸烟	酸煙
acid gas	酸性气体	酸氣
acidic catalysis	酸性催化[作用]	酸性催化[作用]
acidic catalyst	酸性催化剂	酸性觸媒
acidification	酸化	酸化[作用]
acidifier	①酸化剂②酸化器	酸化劑
acidity	①酸度②酸性	①酸度②酸性
acidity index	酸度指数	酸度指數
acidity test	酸度检定	酸度試驗
acid measuring tank	量酸槽	量酸槽
acid mist	酸雾	酸霧
acid number	酸值, 酸价	①酸值②酸價
acid process	酸法	酸法
acid-proof brick	耐酸砖	耐酸磚
acid-proof cement	耐酸水泥	耐酸水泥
acid-proof enamel	耐酸搪瓷	耐酸搪瓷
acid-proof paper	耐酸纸	耐酸紙
acid-proof slab	耐酸板	耐酸板
acid pump	酸泵	酸泵
acid rain	酸雨	酸雨
acid refractory	酸性耐火材料	酸性耐火物
acid refractory brick	酸性耐火砖	酸性耐火磚

英　文　名	祖国大陆名	台湾地区名
acid resistance	耐酸性	抗酸性
acid resistant	耐酸物	抗酸劑
acid resisting alloy	耐酸合金	抗酸合金
acid sludge	酸性污泥	酸性污泥
acid tank	酸槽	酸液儲槽
acid tank car	酸槽车	酸槽車
acidulant	酸化剂	酸化劑
acidulating tank	酸化槽	加酸槽
acidulation	酸化	加酸
acidulator	酸化器	加酸器
acorn oil	橡实油	櫟實油
acousimeter	测声计	測音計
acoustic control	声控	聲控
acoustic flow	声速流动	聲速流動
acoustics	声学	聲學
acoustic velocity	音速	音速
acrylic ester	丙烯酸酯	丙烯酸酯
acrylic ester rubber	丙烯酸酯橡胶	丙烯酸酯橡膠
acrylic fiber	聚丙烯腈纤维	聚丙烯腈纖維
acrylic resin	丙烯酸树脂	壓克力樹脂, 丙烯酸酯 樹脂
acrylonitrile	丙烯腈	丙烯腈
acrylonitrile-butadiene-styrene resin	丙烯腈-丁二烯-苯乙烯 [树脂]	丙烯腈-丁二烯-苯乙烯 [樹脂]
actinometer	化学光度计	光量計, 露光計
actinomycetin	白放线菌素	放線菌素
actinomycin	放线菌素	放線菌
activated adsorption	活性吸附	活化吸附
activated alumina	活性矾土	活性礬土
activated aluminum oxide	活性氧化铝	活性氧化鋁
activated biological filtration	活性生物过滤	活性生物過濾
activated carbon	活性炭	活性碳
activated carbon column	活性炭柱	①活性碳管②活性碳塔
activated char	活性炭	活性焦
activated charcoal	活性炭	活性[木]炭
activated chemisorption	活化化学吸附	活化化學吸附
activated complex	活化络合物	活化錯合物
activated molecule	活化分子	活化分子
activated silica	活化硅石	活化矽石

英　文　名	祖国大陆名	台湾地区名
activated sludge	活性污泥	活性污泥
activated sludge plant	活性污泥处理场	活性污泥處理場
activated sludge process	活性污泥法	活性污泥法
activation	活化[作用]	活化[作用]
activation analysis	活化分析	活化分析
activation cross section	活化截面	活化截面
activation energy	活化能	活化能
activation volume	活化体积	活化體積
activator	活化剂	活化劑
active agent	活性剂	活性劑
active alkali	活性碱	活性鹼
active alumina	活性矾土	活性礬土
active carbon	活性炭	活性碳
active catalyst	活性催化剂	活性觸媒
active catalytic center	活性催化中心	觸媒活性中心
active center	活性中心	活性中心
active-center complex	活性中心络合物	活性中心錯合物
active clay	活性黏土	活性土
active complex	活性络合物	活性錯合物
active dry yeast	活性干酵母	活性乾酵母
active oxygen	活性氧	活性氧
active point	活性点	活性點
active site	活性部位	活性部位
active-site theory	活性部位理论	活性部位理論
active sludge	活性污泥	活性污泥
active solid	活性固体	活性固體
active sulfur	活性硫	活性硫
active surface area	活性表面积	活性表面積
activity	①活度②活性	活性
activity coefficient	活度系数	活性係數
activity correction factor	活性校正因子	活性校正因數
activity cracking	裂解活性	裂解活性
activity decay	活性衰变	活性衰變
activity distribution	活性分布	活性分布
activity factor	活性因子	活性因數
activity index	活性指数	活性指數
actual lift	实际上升	實際上升
actual plate	实际[塔]板	實際板
actual plate number	实际塔板数	實際板數

英 文 名	祖国大陆名	台湾地区名
actuating signal	作用信号	引動信號
actuator	执行机构	引動器
adaptive control	自适应控制	適應控制
adaptive decoupler	自适应去耦器	適應解偶器
adapt tuning	适应调谐	適應調諧
added value	增值	附加價值
addition polymerization	加成聚合	加成聚合(反應)
addition reaction	加成反应	加成反應
additive	添加剂	添加劑
additive decolorization	加成脱色	加成脫色
additive law	加成定律	加成律
additive property	加合性, 加成性	加成性
adduct	加合物	加成物
adductive crystallization	加成结晶	加成結晶
adhesion	黏附	附著, 黏著
adhesional wetting	黏附润湿	附著性潤濕
adhesion energy	黏附能	附著能
adhesion force	黏附力	附著力
adhesion strength	黏附强度	附著強度, 黏著強度
adhesion test	黏附试验	黏著試驗
adhesion work	黏附功	附著功
adhesive agent	胶黏剂	黏著劑, 黏合劑
adhesive attraction	黏合吸引	黏著吸引
adhesive formulation	黏合剂配方	黏著劑配方
adhesives	胶黏剂	黏合劑
adhesive strength	黏附强度	黏著強度
adhesive tape	胶带	膠帶
adhesive tension	黏附张力	黏著張力
adhesive test	黏附试验	黏著試驗
adiabatic apparatus	绝热仪器	絕熱裝置
adiabatic calorimeter	绝热式量热计	絕熱卡計
adiabatic chemical reaction	绝热化学反应	絕熱化學反應
adiabatic column	绝热塔	絕熱塔
adiabatic compression	绝热压缩	絕熱壓縮
adiabatic condition	绝热条件[状态]	絕熱狀況
adiabatic contraction	绝热收缩	絕熱收縮
adiabatic conversion	绝热转化	絕熱轉化
adiabatic cooling	绝热冷却	絕熱冷卻
adiabatic cooling curve	绝热冷却曲线	絕熱冷卻曲線

英　文　名	祖国大陆名	台湾地区名
adiabatic cooling line	绝热冷却线	絕熱冷卻線
adiabatic drying	绝热干燥	絕熱乾燥
adiabatic evaporation	绝热蒸发	絕熱蒸發
adiabatic expansion	绝热膨胀	絕熱膨脹
adiabatic factor	绝热因子	絕熱因數
adiabatic flow	绝热流	絕熱流動
adiabatic humidification	绝热增湿	絕熱增濕
adiabatic operating line	绝热操作线	絕熱操作線
adiabatic operation	绝热操作	絕熱操作
adiabatic plot	绝热图	絕熱圖
adiabatic process	绝热过程	絕熱程序
adiabatic reaction	绝热反应	絕熱反應
adiabatic reactor	绝热反应器	絕熱反應器
adiabatic rectification	绝热精馏	絕熱精餾
adiabatic saturation	绝热饱和	絕熱飽和
adiabatic saturation temperature	绝热饱和温度	絕熱飽和溫度
adiabatic system	绝热系统	絕熱系統
adiabatic temperature rise	绝热温升	絕熱溫升
adiabatic throttling device	绝热节流装置	絕熱節流裝置
adiabatic throttling process	绝热节流过程	絕熱節流程序
adiabatic twin calorimeter	双层绝热量热器	絕熱雙卡計
adiabatic wall	绝热壁	絕熱壁
adjacency matrix	相邻矩阵	相鄰矩陣
adjustable bearing	可调轴承	可調軸承
adjustable orifice	可调锐孔	可調孔口
adjustable parameter	可调参数	可調參數
adjustable weight	可调节权	可調節權
adjustment	调整	調整
administrative cost	行政成本	行政成本
admiralty test	干湿陈化试验	乾濕老化試驗
admissible roughness	容许粗糙度	容許粗糙度
adsorbability	吸附能力	吸附度
adsorbate	吸附质	吸附质
adsorbed complex	吸附络合物	被吸附(之)錯合物
adsorbent	吸附剂	吸附劑
adsorber	吸附器	吸附器,吸附塔
adsorption	吸附	吸附
adsorption capacity	吸附容量	吸附容量
adsorption center	吸附中心	吸附中心

英　文　名	祖国大陆名	台湾地区名
adsorption chart	吸附图表	吸附圖表
adsorption chromatography	吸附色谱法	吸附層析法
adsorption coefficient	吸附系数	吸附係數
adsorption column	吸附柱	吸附塔
adsorption control	吸附控制	吸附控制
adsorption-desorption	吸附-脱附	吸附-脱附
adsorption equilibrium	吸附平衡	吸附平衡
adsorption filtration	吸附过滤	吸附過濾
adsorption isobar	吸附等压线	吸附等壓線
adsorption isostere	吸附等容线	吸附等容線
adsorption isotherm	吸附等温线	吸附等溫線
adsorption method	吸附法	吸附法
adsorption potential	吸附势	吸附勢
adsorption rate	吸附速率	吸附速率
adsorption site	吸附部位	吸附部位
adsorption surface	吸附表面	吸附表面
adsorption tube	吸附管	吸附管
adsorption wave	吸附波	吸附波
adsorptive agent	吸附剂	吸附劑
adsorptive capacity	吸附量	吸附容量
adsorptivity	吸附能力	①吸附性②吸附度
advanced treatment	高级处理	高級處理
advanced wastewater treatment	高级废水处理	高級廢水處理[法]
adverse pressure gradient	逆压力梯度	反壓力梯度
aerated flow	掺气流	通氣流動, 曝氣流動
aerated lagoon	曝气塘	通氣污水塘, 曝氣污水塘
aerated stabilization pond	曝气稳定化池	通氣穩定化池, 曝氣穩定化池
aerated system	曝气系统	通氣系統, 曝氣系統
aerated water	汽水	碳酸水
aeration	①充气②曝气	①通氣②曝氣
aeration candle	通气管	通氣管
aeration device	充气装置	通氣裝置
aeration number	充气数	曝氣數, 通氣數
aeration rate	充气率	通氣速率
aeration tank	充气槽	通氣槽, 曝氣槽
aeration tube	充气管	通氣管
aeration wheel	充气轮	通氣輪

英　文　名	祖国大陆名	台湾地区名
aerator	充气器	通氣器
aerial condenser	空气冷凝器	空氣冷凝器
aerial contamination	空气污染	空氣污染
aerobic bacteria	好氧[细]菌, 好气[细]菌	好氧菌, 需氧菌
aerobic biological oxidation	好氧生物氧化	需氧生物氧化[法]
aerobic culture	好氧培养	嗜氧培養
aerobic decomposition	好氧分解	好氧分解
aerobic digestion	好氧消化	需氧消化
aerobic fermentation	好氧发酵	需氧發酵
aerobic oxidation	好氧氧化	需氧氧化
aerobic sludge digestion	好氧污泥消化	需氧污泥消化
aerobic treatment	好氧处理	需氧處理
aerodynamic diameter	空气动力直径	空氣動力直徑
aerodynamics	空气动力学	空氣動力學
aerophilic bacteria	好气细菌	好氧菌
aerosol	气溶胶	①氣溶膠②霧劑
aerosol bomb	喷雾弹	噴霧彈
aerosol propellant	喷雾推进剂	推噴劑
aerosol valve	喷雾阀	噴霧閥
affinage	精炼	精煉
affination	①[离心]精炼法②洗糖法	①精煉法②洗糖
affination number	洗糖数	洗糖數
affination value	洗糖值	洗糖值
affinity	亲和势	親和性, 親和力
affinity adsorption	亲和吸附	親和吸附
affinity chromatography	亲和色谱[法]	親和色譜法
affinity interaction	亲和作用	親和作用
affinity labeling	亲和标记	親和標記
affinity membrane	亲和膜	親和膜
affinity precipitation	亲和沉淀	親和沈澱
affinity ultrafiltration	亲和超滤	親和過濾
after burner	后燃装置	後燃裝置
after burning	后期燃烧	後燃
after contraction	后收缩	後收縮
after cooler	后置冷却器	後冷卻器
after fermentation	后发酵	後發酵
after filter	后过滤器	後濾器
after filtration	后过滤	後過濾

英　文　名	祖国大陆名	台湾地区名
after hardening	后硬化	後硬化
after product	后产物	後産物
after purification	后净化	後純化
after shrinkage	后收缩	後收縮
after treatment	后处理	後處理
after vulcanization	后硫化作用	後硫化
agar	琼脂	洋菜
agar diffusion	琼脂扩散	瓊脂擴散
agar diffusion technology	琼脂扩散技术	瓊脂擴散技術
agar electrophoresis	琼脂电泳	瓊脂電泳
agar filtration	琼脂过滤	瓊脂過濾
agar medium	琼脂培养基	洋菜培基
agarose	琼脂糖	瓊脂糖
agarose gel	琼脂糖胶	瓊脂糖膠
age distribution	年龄分布	年齡分布
age distribution function	年龄分布函数	年齡分布函數
age hardening	时效硬化	經時硬化
agglomerate	团块	團聚
agglomerating value	团聚值	團聚值
agglomeration	团聚	團聚
aggregate	聚集体	聚集體
aggregation	聚集	聚集[作用]
aggregation number	聚集数	聚集數
aggregative fluidization	聚式流态化	聚集流體化
aging	陈化	陳化, 老化
aging equipment	老化装置	老化裝置
aging machine	熟化器	老化器
aging oven	老化箱	老化箱
aging test	老化试验	老化試驗
aging test machine	老化试验机	老化試驗機
agitated crystallizer	搅拌结晶器	攪拌結晶器
agitated drier	搅拌干燥器	攪拌乾燥器
agitated reactor	搅拌反应器	攪拌反應器
agitated vessel	搅拌槽	攪拌槽
agitating	搅拌	攪拌
agitating tank	搅拌槽	攪拌槽
agitation	搅拌	攪拌
agitator	搅拌器	攪拌器
agitator drier	搅拌干燥器	攪拌乾燥器

英　文　名	祖国大陆名	台湾地区名
agricultural waste	农业废物	農業廢棄物
AI(＝artificial intelligence)	人工智能	人工智能
AIChE (＝American Institute of Chemical Engineers)	美国化学工程师学会	美國化學工程師學會
air agitation	充气搅拌	空氣攪拌
air agitator	充气搅拌器	空氣攪拌器
air bag	气囊	安全氣囊
air bell	气泡	氣泡
air binding	空气黏合	氣結
air blasting	空气鼓风	鼓風
air blower	鼓风机	鼓風機
air bubble system	气泡系统	氣泡指示系統
air cap	气帽	氣罩
air compressor	空气压缩机	空氣壓縮機
air conditioner	空调器	空氣調節器
air conditioning	空调	空氣調節
air conditioning unit	空调机组	空氣調節單元
air conveyor	气动运输机	空氣運送機
air-cooled heat exchanger	空气冷却器	氣冷式熱交換器
air cooled jacket	气冷夹套	氣冷套
air cooler	空气冷却器	空氣冷卻器
air cooling	空气冷却	空氣冷卻
air diffuser	空气扩散器	空氣擴散器
air discharge orifice	排气锐孔	空氣排放孔口
air displacement pump	排气泵	排氣泵
air displacement system	排气系统	排氣系統
air distillation	常压蒸馏	空氣蒸餾
air dryer	空气干燥器	空氣乾燥器
air drying	风干	風乾
air-drying loss	风干失重	風乾損失
air drying tower	空气干燥塔	空氣乾燥塔
air ejector	空气喷射器	空氣喷射器
air elutriation	空气扬析	空氣淘析
air-entraining agent	空气雾沫剂	空氣霧沫劑
air equilibrium distillation	常压平衡蒸馏	空氣平衡蒸餾
air evaporation	空气蒸发	空氣蒸發
air expander	空气膨胀器	空氣膨脹機
air false	漏气	漏氣
air film	空气膜	空氣膜

英　文　名	祖国大陆名	台湾地区名
air filter	空气过滤器	空氣過濾器
air filtration	空气过滤	空氣過濾
air flow	空气流动	空氣流動
air flow meter	空气流量计	空氣流量計
air-fuel ratio	空气-燃料比	空氣-燃料比
air furnace	空气炉	空氣爐
air gage	气压计	空氣計
air gap	气隙	空氣隙
air gas	风煤气	風煤氣
air heater	空气加热器	空氣加熱器
air hose	空气软管	空氣軟管
air-lift	气升	①氣提②氣升器
air-lift agitation	气升搅拌	氣升攪拌
air-lift agitator	气升搅拌器	氣升攪拌器
air-lift pump	气升泵	氣動泵
air lock	气闸	氣閘
air mixing depth	空气混合深度	空氣混合深度
air nozzle	空气喷头	空氣噴嘴
air-operated valve	气动阀	氣動閥
air permeability	透气性	透氣性
air pollutant	空气污染物	空氣污染物
air pollution	空气污染	空氣污染
air pollution abatement	降低空气污染	空氣污染減少
air pollution control	空气污染控制	空氣污染控制
air pollution engineering	空气污染工程	空氣污染工程
air preheater	空气预热器	空氣預熱器
air pressure	空气压力	空氣壓力
air pressure drop	空气压力降	空氣壓力降
air pressure dynamics	气压动态学	氣壓動態[學]
air pressure gage	气压计	氣壓計
air pump	气泵	空氣泵
air purge	空气吹扫	空氣沖洗
air quality	大气品位	空氣品質
air quality index	大气品位指数	空氣品質指數
air regulator	空气调节器	空氣調節器
air release valve	排气阀	釋氣閥
air relief valve	排气阀	釋氣閥
air ring	空气冷却环	空氣冷卻環
air saturator	空气饱和器	空氣飽和器

英 文 名	祖国大陆名	台湾地区名
air scrubber	空气洗涤器	空氣洗滌器
air separation	空气分离	空氣分離
air separation tank	空气分离槽	空氣分離槽
air separation tower	空气分离塔	空氣分離塔
air separator	空气离析器	空氣分離器
air shrinkage	风干收缩	風乾收縮
air slide	空气溜槽	空氣溜槽
air sparger	空气鼓泡器	空氣噴佈器
air sterilizer	空气灭菌器	空氣滅菌器
air stream	气流	氣流
air suction filter	空气吸滤器	空氣吸濾器
air suction rate	吸气速率	吸氣速率
air supply	空气供应	空氣供應
air-swept mill	气吹磨	風掃磨
air washing	空气洗涤	空氣洗滌
alarm	报警器	警報器
albite	钠长石	鈉長石
alcoholase	醇酶	酒精酶
alcoholysis	醇解	醇解
aldol	羟醛	醛醇
aleurites ardata oil	罂子桐油	子桐油
aleurites fordii oil	光桐油(中国三年桐的油)	光桐油(中國三年桐)
algae	藻类	藻
algin	藻胶	褐藻素
alginate fiber	藻酸纤维	褐藻素纖維
algin fiber	藻酸纤维	褐藻素纖維
algorithm	算法	算法
algorithmic synthesis technique	算法合成技术	算法合成技術
alignment chart	列线图	列線圖
aliphatic hydrocarbon	脂肪族烃	脂肪族烴
alizarine dyestuff	茜素染料	茜素染料
alizarine lake	茜素色淀	茜素色澱
alizarine oil	太古油	茜草油
alkali	碱	鹼
alkali cellulose	碱纤维素	鹼纖維素
alkali lignin	碱木质素	鹼木質素
alkalimeter	碱量计	鹼計
alkaline flooding	碱液泛	鹼液泛流[法]

英　文　名	祖国大陆名	台湾地区名
alkaline glaze	碱釉	鹼性釉
alkaline pulping	碱法制浆	鹼式制漿
alkalinity	碱度	鹼度
alkali process	碱法	鹼法
alkali resistance	抗碱性	抗鹼法
alkali silicate	碱性硅酸盐	鹼矽酸鹽
alkaloid	生物碱	生物鹼
alklalimetry	碱测定法	鹼測定法
alkyd resin	醇酸树脂	酸醇樹脂
alkylation	烷基化	烷化
alkylbenzene sulfonate	烷基苯磺酸盐	烷[基]苯磺酸鹽
allocation of resource	资源分配	資源配置
allowance	容差	容許量
alloy	合金	合金
alloy steel	合金钢	合金鋼
almond oil	杏仁油	杏仁油
aloxite	铝砂	鋁砂
alternate operating column	交替操作塔	交替操作塔
alternating copolymerization	交替共聚合	交替共聚合
alternating system	交替系统	交替系統
alternation	交替现象	交替[現象]
alternative copolymer	交替共聚物	交替共聚物
alternative hypothesis	备择假设	備擇假設
alternative investment	备择投资	交替投資
alternative operation	交替操作	交替操作
alum	明矾	明礬, 礬
alumina	矾土	礬土
alumina brick	高铝砖	高鋁磚, 礬士磚
aluminous cement	高铝水泥	高鋁水泥
aluminum foil	铝箔	鋁箔
aluminum mordant	铝媒染剂	鋁媒染劑
aluminum pipe	铝管	鋁管
aluminum soap	铝皂	鋁皂
alum tanning	[明]矾鞣[革]	明礬鞣制
alundum	刚铝石	剛[鋁]石
alundum filter	刚铝石过滤器	剛[鋁]石過濾器
Amagat law	阿马加定律	阿馬加定律
amalgam	汞齐	汞齊
amber	琥珀	琥珀

英　文　名	祖国大陆名	台湾地区名
ambient pressure	环境压力	周圍壓力
ambient temperature	环境温度	周圍溫度
American Institute of Chemical Engineers (AIChE)	美国化学工程师学会	美國化學工程師學會
American Petroleum Institute(API)	美国石油学会	美國石油協會
American Standards Association (ASA)	美国国家标准协会	美國國家標準協會
amination	氨基化	胺化
amino acid	氨基酸	胺基酸
amino resin	氨基树脂	胺基樹脂
ammonia extraction process	氨萃取过程	氨萃取法
ammonia liquor	氨水	氨水
ammonia reactor	氨反应器	氨反應器
ammonia saturator	氨饱和器	氨飽和器
ammonia-soda process	氨碱法	氨鹼法
ammonia still	蒸氨塔	氨蒸餾器
ammonia synthesis	氨合成	氨合成
ammoniated brine	氨盐水	氨化鹽水
ammoniated fertilizer	氨化肥料	氨化肥料
ammoniated superphosphate	含铵过磷酸钙	氨化過磷酸鈣
ammoniation	氨化[作用]	氨化
ammonia washer	氨洗涤器	洗氨器
ammonia water	氨水	氨水
ammonium alum	铵矾	銨礬
ammonium soap	铵皂	銨皂
ammonolysis	氨解[作用]	氨解
amorphism	非晶形[现象]	非晶體
amorphous carbon	无定形碳	非晶[形]碳
amorphous polymer	无定形聚合物	無定形聚合物, 非晶質聚合物
amorphous silica	非晶二氧化硅	非晶[形]矽石
amortization	分期偿还	分期償還
amosite	铁石棉[填充剂]	石綿礦
amperometric titration	电流滴定[法]	電流滴定[法]
amphotericin	两性霉素	抗黴酮
ampicillin	氨苄青霉素	胺苄青黴素
amplification	放大	放大
amplifier	放大器	放大器, 增輻器
amplitude	振幅	振幅
amplitude ration	振幅比	振幅比

英 文 名	祖国大陆名	台湾地区名
amylase	淀粉酶	澱粉酶
amylolysis	淀粉水解	澱粉水解
anaerobe	厌氧[细]菌	厭氣菌
anaerobic bacteria	厌氧[细]菌, 厌气[细]菌	厭氣[細]菌
anaerobic breakdown	厌氧分解	厭氣分解
anaerobic corrosion	微生物腐蚀	厭氣腐蝕
anaerobic culture	厌氧培养	厭養培養
anaerobic decomposition	厌氧分解	厭氣分解
anaerobic digester	厌氧消化槽	厭氣消化槽
anaerobic digestion	厌氧消化	厭氣消化
anaerobic fermentation	厌氧发酵	厭氣發酵
anaerobic oxidation	厌氧氧化	厭氣氧化
anaerobic sludge	厌氧污泥	厭氣污泥
anaerobic sludge digestion	厌氧污泥消化	厭氣污泥消化
anaerobic treatment	厌氧处理	厭氣處理
analog computation	模拟计算	類比計算[法]
analog computer	模拟计算机	類比計算機
analog simulation	模拟仿真	類比模擬
analogy	类比	類似
analogy system	类比系统	類比系統
analogy theory	类比理论	類似理論
analysis mode	分析型	分析型
analysis of reaction network	反应网络分析	反應網路分析
analytical balance	分析天平	分析天平
analytical distillation	分析蒸馏	分析蒸餾
analytical method	解析法	解析法
analytical solution of group contribution method	ASOG 法	ASOG 法
analyzer	分析器	分析儀, 分析器
anchorage-dependent cell	贴壁细胞	貼壁細胞
anchor agitator	锚式搅拌器	錨式攪拌器
andalusite	红柱石	紅柱石
anemometer	风速计	①流速計②風速計
anemoscope	风速仪	風向計
anergy	炕, 无效能	無效能
aneroid barometer	无液气压计	無液氣壓計
aneroid flowmeter	无液流量计	無液流量計
aneroid manometer	无液压力计	無液壓力計
anesthetic	麻醉剂	麻醉劑

英　文　名	祖国大陆名	台湾地区名
angle factor	角系数	角度因數
angle of arrival	到达角	到達角
angle of attack	冲角	攻角
angle of deflection	偏离角	偏向角
angle of difference	差角	差角
angle of fall	落角	落角
angle of inclination	倾角	傾角
angle of internal friction	内磨擦角	內磨擦角
angle of nip	夹角	挾角
angle of release	释放角, 开角	釋放角
angle of respose	休止角	休止角
angle of slide	滑动角	滑動角
angle of spatula	刮铲角	刮鏟角
angle of wall friction	壁摩擦角	壁摩擦角
angle valve	角阀	角閥
angular deformation	角[向]变形	角形變
angular distortion	角畸变	角畸變
angular impulse	角冲量	角衝量
angular momentum	角动量	角動量
angular scattering pattern	角散射模式	角散射模式
angular velocity	角速度	角速度
aniline black	苯胺黑	苯胺黑
aniline blue	苯胺蓝	苯胺藍
aniline dye	苯胺染料	苯胺染料
aniline green	苯胺绿	苯胺綠
aniline orange	苯胺橙	苯胺橙
aniline point	苯胺点	苯胺點
aniline red	苯胺红	苯胺紅
aniline violet	苯胺紫	苯胺紫
animal glue	动物胶	動物膠
animal hide	动物皮	動物皮
animal manure	动物厩肥	動物廄肥
animal oil	动物油	動物油
anion	阴离子	陰離子
anion exchanger	阴离子交换剂	①陰離子交換器②陰離子交換劑
anion exchange resin	阴离子交换树脂	陰離子交換樹脂
anionic exchange resin	阴离子交换树脂	陰離子交換樹脂
anionic polymerization	阴离子[催化]聚合	陰離子聚合[反應]

英 文 名	祖国大陆名	台湾地区名
anionic starch	阴离子淀粉	陰離子澱粉
anionic surface-active agent	阴离子型表面活性剂	陰離子界面活性劑
anionic surfactant	阴离子型表面活性剂	陰離子界面活性劑
anisotropic absorption	各向异性吸收	[各]向異性吸收
anisotropic membrane	各向异性膜	各向異性膜
anisotropy	各向异性	各向異性
ANN(=artificial neural network)	[人工]神经网络	[人工]神經網絡
annealing	退火	退火, 徐冷
annealing point	退火点	徐冷點
annuity table	年金表	年金表
annular flow	环状流	環狀流動
annular reactor	环状反应器	環狀反應器
annular space	环隙	環隙
annulus	环隙	環隙
anode	阳极	陽極
anode oxidation	阳极氧化	陽極氧化
anode potential	阳极电位	陽極電位
anodic corrosion	阳极腐蚀	陽極腐蝕
anodic reaction	阳极反应	陽極反應
anomalous viscosity	反常黏度	反常黏度
anthophyllite	直闪石	斜方角閃石
anthracene oil	蒽油	蒽油
anthracite	无烟煤	無煙煤
anthracite coal	无烟煤	無煙煤
anthraquinone dye	蒽醌染料	蒽醌染料
anthraquinone vat dye	蒽醌还原染料	蒽醌缸染料
antibiotic	抗生素, 抗菌素	抗生素
antiblock agent	防结块剂	防阻塞劑
antibody	抗体	抗體
anticaking agent	防结块剂	防結塊劑
anticipatory control	预先控制	預先控制
anticoagulant agent	抗凝剂	抗凝劑
anticoagulant effect	抗凝效应	抗凝效應
anticorrosion	防腐蚀	耐蝕
anticorrosive agent	防蚀剂	防銹劑, 防蝕劑
anticorrosive paint	防蚀漆	防銹劑, 防蝕漆
antifoam agent	防沫剂	防沫劑
antifreeze	防冻	抗凍劑
antifreeze mixture	抗冻混合物	抗凍混合物

英 文 名	祖国大陆名	台湾地区名
antifreezer	防冻剂	①防凍器②防凍劑
antifreezing agent	防冻剂	抗凍劑
antigelation extraction	反凝胶萃取	反凝膠萃取
antigen	抗原	抗原
antiknock	抗爆	抗震
antiknocking compound	抗爆化合物	抗震化合物
antiknock value	抗爆值	抗震值
antimicrobial action	抗微生物作用	抗微生物作用
antimicrobial activity	抗微生物活性	抗微生物活性
antimicrobial drug	抗微生物剂	抗微生物藥物
antimonial soap	锑皂	銻皂
antimony electrode	锑电极	銻電極
antimony white	锑白	銻白
antimony yellow	锑黄	銻黃
antioxidant	抗氧[化]剂	抗氧化劑
antiozonant	抗臭氧剂	抗臭氧[老化]劑
antirusting grease	防锈剂	防銹脂
antisoftener	防软剂	軟化防止劑
antistaining agent	防污染剂	防汙劑
antistatic agent	抗静电剂	抗靜電劑
antisurge control	防喘振控制	抗波動控制
antiwindup	抗卷绕	抗繞緊
Antoine equation	安托万方程	安托因方程
apatite	磷灰石	磷灰石
API (＝American Petroleum Institute)	美国石油学会	美國石油協會
API gravity	API 比重指数	美制[石油]比重
API hydrometer	API 比重计	美制[石油]比重計
apoenzyme	脱辅[基]酶	酶蛋白
apparatus	仪器	儀器, 裝置
apparent activation energy	表观活化能	視活化能
apparent activity	表观活度	視活性
apparent composition	表观组成	視組成
apparent density	视密度, 表观密度	視密度
apparent diffusivity	表观扩散系数	視擴散係數
apparent gravity	表观重力	視重力
apparent mass	表观质量	視質量
apparent order	表观阶断	視階
apparent porosity	表观孔隙率	視孔隙度
apparent purity	表观纯度	視純度

英　文　名	祖国大陆名	台湾地区名
apparent shear rate	表观剪切率	视剪率
apparent shear stress	表观剪切应力	视剪應力
apparent solubility	表观溶解度	视溶解度
apparent specific gravity	表观比重	视比重
apparent specific heat	表观比热	视比熱
apparent stress	表观应力	视應力
apparent viscosity	表观黏度	视黏度
apparent volume	表观体积	视體積
apparent weight	表观重量	视重量
apple seed oil	苹果子油	蘋果子油
applicability	适用性	應用性
application	应用	應用
applied chemistry	应用化学	應用化學
applied microbiology	应用微生物学	應用微生物學
applied thermodynamics	应用热力学	應用熱力學
appricot kernel oil	杏仁油	杏仁油
approximate method	近似法	近似法
approximate solution	近似解	近似解
approximation	近似	①近似②近似法
approximation formula	近似式	近似式
apron carrier	裙式输送机	裙式輸送機
apron conveyor	裙式运输机	裙式運送機
apron feeder	裙式给料机	裙飼機
aprotic solvent	非质子溶剂	無質子溶劑
aqua ammonia	氨水	氨水
aqua fortis	硝酸	硝酸
aqua regia	王水	王水
aqueous ammonia	氨水	氨水
aqueous extract	水萃取物	水相萃取物
aqueous solution	水溶液	水溶液
aqueous two-phase extraction	双水相萃取	雙水相萃取
aqueous two-phase system	双水相系统, 双水相体系	雙水相系統
aquolization process	加水裂化过程	加水裂化程序
araeometer	[液体]比重计	液體比重計
arc furnace	电弧炉	電弧爐
arch brick	拱砖	拱磚
arc resistance	耐电弧性	耐電弧性
arctic sperm oil	北极鲸蜡油	抹香鯨油, 北極鯨蠟油

英 文 名	祖国大陆名	台湾地区名
area-averaged equation	面积平均方程式	面積平均方程式
area flowmeter	面积流量计	面積流量計
area meter	面积计	面積計
area type flowmeter	面积流量计	面積流量計
arithmetic mean	算术平均	算術平均
arithmetic mean temperature difference	算术平均温差	算術平均溫差
arithmetic mean value	算术平均值	算術平均值
armature	电枢族	電樞
aromatics	芳[香]化合物	芳香烴
aromatization	芳构化	芳香[烴]化
Arrhenius equation	阿伦尼乌斯方程	阿瑞尼斯方程式
arrow diagram	箭头[流程]图	箭頭[流程]圖
arsenical insecticide	砷杀虫剂	砷殺蟲劑
arsenical rodenticide	砷杀鼠剂	砷殺鼠劑
artificial aging	人工老化	人工老化
artificial drier	人工干燥室	人工乾燥室
artificial fertilizer	人工肥料	人造肥料
artificial fiber	人造纤维	人造纖維
artificial fuel	人造燃料	人造燃料
artificial intelligence(AI)	人工智能	人工智能
artificial kidney	人工肾脏	人工腎臟
artificial leather	人造革	人造皮
artificial neural network(ANN)	[人工]神经网络	神經網絡
artificial organ	人工器官	人造器官
artificial silk	人造丝	人造絲
artificial sweetener	人工甜味剂	人工甜味料
art-printing paper	艺术印刷纸	美術印刷銅版紙
ASA(＝American standards Association)	美国国家标准协会	美國國家標準協會
asbestos	石棉	石綿
asbestos filter	石棉过滤器	石綿過濾器
ascending chromatography	上行色谱法	上升層析法
ascending-descending chromatography	升降色谱法	升降層析法
ascending kiln	上行窑	階級窯
ascending pipe	上行管	上升管
aseptic operation	无菌操作	無菌操作
aseptic seal	无菌密封	無菌密封
ash	灰分	灰,灰分
ash analysis	灰分分析	灰分分析
ash diffusion control	灰分扩散控制	灰[分]擴散控制

英 文 名	祖国大陆名	台湾地区名
ashless filter paper	无灰滤纸	無灰濾紙
ash-seed oil	槐子油	槐子油
ash zone	灰区	灰區
aspect number	形状数	形狀數
aspect ratio	①长宽比②高径比	縱橫比
aspergillus	菌	菌
asphalt	沥青	柏油
asphalt paint	沥青漆	柏油漆
asphaltum	[石油]沥青	瀝青
aspirating relay	吸气式替续器	吸引式替續器
aspirator	吸气器	[噴水]抽氣器
assembly line	装配线	裝配線
assembly of independent particles	独立粒子系集	獨立粒子系集
assembly of interacting particles	非独立粒子系集	非獨立粒子系集
assembly of localize particles	定域粒子系集	定域粒子系集
assembly of non-localized particles	非定域粒子系集	非定域粒子系集
assessment	评估	評估
asset accounting	资产会计	資產會計
asset type	资产型式	資產型式
assimilation	同化	同化
associated solution model	缔合溶液模型	締合溶液模型
asymmetrical flow	不对称流动	不對稱流動
asymmetric membrane	非对称膜	非對稱膜
asymptote	渐近线	漸近線
asymptotic plot	渐近线图	漸近線圖
asymptotic solution	渐近解	漸近解
asymptotic stability	渐近稳定性	漸近穩定性
atactic polymer	无规立构聚合物	無聚合物
athermal solution	无热溶液	恒焓溶液
atmolysis	微孔[透壁]分气法	細孔氣體分離法
atmospheric condenser	空气冷凝器	空氣冷凝器
atmospheric cooling tower	空气冷却塔	空氣冷卻塔, 風冷塔
atmospheric diffusion	大气扩散	大氣擴散
atmospheric drier	常压干燥器	空氣乾燥器
atmospheric flash tower	常压闪蒸塔	常壓驟餾塔
atmospheric fractionator	常压分馏器	常壓分餾器, 常壓分餾塔
atmospheric mashing	常压芽浆糖化	常壓製醪
atmospheric operation	常压操作	常壓操作

英　文　名	祖国大陆名	台湾地区名
atmospheric pressure	大气压	大氣壓力
atmospheric steam	常压蒸汽	常壓蒸汽
atometer	蒸发速度测定器	蒸發速率測定器
atomic heat capacity	原子热容	原子熱容量
atomic interaction	原子[相]互作用	原子相互作用
atomic mass	原子质量	原子質量
atomic number	原子序[数]	原子序
atomic spacing	原子间距	原子間距
atomic weight	原子量	原子量
atomic weight table	原子量表	原子量表
atomization	雾化	霧化
atomized oil	雾化油	霧化油
atomizer	雾化器	霧化器
atomizing air	雾化[用]空气	霧化空氣
atomizing burner	雾化燃烧器	霧化燃油器
atomizing nozzle	雾化喷嘴	霧化噴嘴
atomizing spray nozzle	雾化喷管	霧化噴嘴
atomizing steam	雾化蒸汽	霧化噴蒸汽
attenuation	衰减	衰減[作用]
attenuation effect	衰减效应	衰減效應
attenuation factor	衰减因子	衰減因數
attenuator	衰减器	衰減器
attractant	引诱剂	誘引劑
attractive force	吸引力	吸引力
attractive potential	吸引势	相吸勢
attrition	磨损	磨耗
attrition mill	碾磨机	銼磨機
attrition of catalyst	催化剂磨损	觸媒磨耗
attrition resistance	抗磨损	抗磨阻力
attrition resistant	抗磨剂	抗磨劑
auger machine	螺旋挤出机	螺旋擠製機
aureomycin	金霉素	金黴素
autoacceleration	自加速作用	自加速[作用]
auto-agglutination	自胶合	自膠合
autocatalysis	自催化	自催化反應
autoclave	加压灭菌器	熱壓[反應]器, 高壓釜, 高壓鍋
autoclave expansion test	加压膨胀试验	熱壓膨脹試驗
auto-coagulation	自动凝结	自凝聚

英 文 名	祖国大陆名	台湾地区名
autogenous mill	自磨机	自磨機
autoignition temperature	自燃温度	燃點, 自燃溫度
auto-induced reaction	自诱导反应	自誘發反應
autolysis	自溶	自消化, 自分解
autolytic lipase	自溶脂肪酶	自解脂酶
auto-manual switch	自动手控开关	自動手動開關
automatic control	自动控制	自動控制
automatic controller	自动控制器	自動控制器
automatic control system	自动控制系统	自動控制系統
automatic feed	自动进料	自動進料
automatic feeder	自动加料器	自動進料器
automatic filter	自动过滤器	自動過濾器
automatic process controller	自动过程控制器	自動程序控制器
automatic regulator	自动调节器	自動調節器
automatic thermostat	自动恒温器	自動恒溫器
autonomous system	自治系统	自發系統
auto-polymerization	自[动]聚合	自聚合
auto recorder	自动记录器	自動記錄器
auto-selector	自动选择器	自動選擇器
auto-selector control	自动选择器控制	自動選擇器控制
auto-stabilization	自动稳定	自穩定
autothermal reaction	自热反应	自熱反應
auto-vulcanization	自硫化	自硫化
autoxidation	自氧化	自氧化
auxiliary absorber	辅助吸收器	輔助吸收器
auxiliary electrode	辅[助]电极	輔助電極
auxochrome	助色团	助色團
availability	烟, 有效能	可用性
availability analysis	烟分析, 有效能分析	可用性分析
available chlorine	有效氯	有效氯
average boiling point	平均沸点	平均沸點
average deviation	平均偏差	平均偏差
average error	平均误差	平均誤差
average molecular weight	平均分子量	平均分子量
average residence time	平均停留时间	平均滯留時間
average specific filtration resistance	平均过滤比阻	平均過濾比阻
average temperature difference	平均温差	平均溫差
average velocity	平均速度	平均速度
averaging control	均匀调节	平均控制

英 文 名	祖国大陆名	台湾地区名
averaging level control	均匀液位控制	平均液位控制
aviation fuel	航空燃料	航空燃料
aviation gasoline	航空汽油	航空汽油
axial angle	轴角	軸角
axial dispersion	轴向分散	軸向分散
axial dispersion coefficient	轴向分散系数	軸向分散係數
axial flow	轴向流	軸向流動
axial-flow compressor	轴流式压缩机	軸向流動壓縮機
axial-flow fan	轴流[式]风机	軸向風扇
axial flow pump	轴流泵	軸向流泵
axial-flow reactor	轴流反应器	軸流反應器
axial mixing	轴向混合	軸向混合
axial symmetric flow	轴对称流	軸向對稱流動
axial symmetry	轴向对称	軸向對稱
axial velocity	轴向速度	軸向速度
axis of symmetry	对称轴	對稱軸
azeotrope	共沸物, 恒沸物	共沸物
azeotrope tower	共沸塔	共沸塔
azeotropic composition	共沸组成	共沸组成
azeotropic copolymer	恒[组]分共聚物	共沸共聚物
azeotropic copolymerization	恒[组]分共聚物	共沸共聚合
azeotropic distillation	共沸蒸馏, 恒沸蒸馏	共沸蒸餾
azeotropic drying	共沸干燥	共沸乾燥
azeotropic line	共沸线	共沸線
azeotropic mixture	共沸混合物	共沸混合液
azeotropic parameter	共沸参数	共沸參數
azeotropic phenomenon	共沸现象	共沸現象
azeotropic point	共沸点	共沸點
azeotropic process	共沸[蒸馏]过程	共沸程序
azeotropic property	共沸性质	共沸性質
azeotropic range	共沸范围	共沸範圍
azeotropic state	共沸状态	共沸狀態
azeotropic system	共沸系统	共沸系統
azeotropic temperature	共沸温度	共沸溫度
azeotropy	共沸[性]	共沸法
azo component	偶氮成分	偶氮成分
azo dye	偶氮染料	偶氮染料
azoic dye	偶氮染料	偶氮染料
azoic dyeing	偶氮染色	偶氮染色

英　文　名	祖国大陆名	台湾地区名
azotobacteria	固氮菌	固氮菌
azotometer	定氮仪	氮量計
azurite	蓝铜矿	藍銅礦

B

英文名	祖国大陆名	台湾地区名
bacillus	杆菌	桿菌
bacillus subtilis	枯草杆菌	枯草桿菌
bacitracin	杆菌肽	桿菌
back diffusion	反向扩散	回擴散
backfiller	回填机	回填機
backflow	倒流	反流
backflushing	反洗	反洗
backlash	齿隙	背隙[齒輪螺紋], 空檔
backmix reactor	返混反应器	返混反應器
back pressure	背压	反壓, 背壓
back pressure valve	止回阀	止回閥
back-propagation algorithm	反向传播算法	反向傳播算法
backtrapping	反截留	逆陷
backtrapping of bubble-cap tray	泡罩塔盘的反截留	泡罩盤之逆陷
backward feed	逆向进料	逆向進料
backward reaction	逆向反应	逆反應
backwashing	反洗	逆洗
backwash rate	反洗速率	逆洗速率
back water	回水	回水
bacteria	细菌	細菌
bacteria bed	细菌床	細菌床
bacteria content	细菌含量	細菌含量
bacteria count	细菌数	菌數
bacteria filter	滤菌器	濾菌器
bacterial fermentation	细菌[性]发酵	細菌[性]發酵
bacterial oxidation	细菌氧化	細菌氧化
bacteriologic filtration	细菌过滤	除菌過濾
baffle	档板, 折流板	擋板, 折流板, 緩衝板
baffle chamber	挡板室	擋板室
baffle column	挡板塔	擋板塔
baffle column mixer	挡板塔混合器	擋板塔混合器

英 文 名	祖 国 大 陆 名	台 湾 地 区 名
baffled evaporator	折流蒸发器	擋板蒸發器
baffle mark	挡板痕	擋板痕
baffle mixer	挡板混合器	擋板式混合器
baffle nozzle system	挡板喷嘴系统	擋板噴嘴系統
baffle pan	挡板塔盘	擋板盤
baffle plate	防冲板	擋板,折流板
baffle plate column mixer	挡板混合塔	擋板柱混合器
baffle plate tower	挡板塔	擋板塔
baffler	折流板	擋板器
baffle separation	折流分离	擋板分離
baffle tower	挡板塔	擋板塔
bagasse	蔗渣	蔗渣
bagasse board	蔗[渣]板	蔗[渣]板
bagasse digester	蔗渣蒸煮罐	蔗渣蒸煮器
bagassillo	蔗屑	蔗屑
bag filter	袋滤器	袋濾器
bag filter separator	袋式过滤分离器	袋形過濾分離器
bag moulding	袋模塑	袋式成型
bag type dust collector	袋式集尘器	袋式集塵器
bakelite	酚醛树脂	①酚醛樹脂②電木
baking furnace	烘炉	烘爐
baking oven	烘箱	焙箱
baking soda	碳酸氢钠	①焙用鹼②碳酸氫鈉
baking varnish	清烤漆	烤清漆
balance bellow	均衡伸缩囊	均衡伸縮囊
balanced growth	均衡生长	均衡生長
balanced value	均衡值	均衡值
balance equation	平衡方程式	均衡方程式
balance fertilizer	均衡肥料	均衡肥料
balance sheet	资产负债表	資產負債表,平衡表
balance water	平衡水	均衡水
balancing motor	平衡电动机	均衡電動機
balancing ring	平衡环	均衡環
ball bearing	滚珠轴承	球軸承
ball clay	球土	球狀黏土
balling	成球	成球
ball mill	球磨机	球磨機
ball mill reactor	球磨反应器	球磨反應器
ball mill refiner	球磨精制机	球磨精製機

英 文 名	祖 国 大 陆 名	台 湾 地 区 名
Banbury mixer	[班伯里]密[闭式混]炼机	密閉式混煉機, 班伯里混煉機
band conveyor	皮带运输机	帶運機
banded matrix	带状矩阵	帶式矩陣
bandwidth	带宽	頻帶寬度
bang-bang control	继电控制	砰砰控制
bank-note paper	钞票纸	鈔票紙
bank screen	筛组	篩組
bare pipe	裸管	裸管
bare surface	裸面	裸面
bare thermocouple	裸热电偶	裸熱電偶
bare tube	裸管	裸管
barite	重晶石	重晶石
barium alum	钡矾	鋇礬
barking drum	桶式去皮机	桶式去皮機
barograph	自记气压计	氣壓記錄器
barometer	气压计	氣壓計
barometer tube	气压管	氣壓管
barometric condenser	大气冷凝器	氣壓冷凝器
barometric discharge pipe	大气排放管	大氣排泄管
barometric effect	气压效应	氣壓效應
barometric equation	气压方程式	氣壓方程式
barometric height	气压高度	氣壓高度
barometric leg	大气腿	氣壓真空柱
barometric leg pump	气压柱泵	氣壓式泵
barometric pipe	大气腿	氣壓管
barometric pressure	大气压	大氣壓力
barometrograph	气压记录器	氣壓溫度記錄器
barrel mill	桶磨机	桶磨機
barrel sprayer	圆桶喷雾器	圓桶噴霧器
barrier function	闸函数	障壁函數
base	碱	鹼
base catalysis	碱催化	鹼催化[作用]
base-level override	基准位超越	基準位超越
base-level resonance	基准位共振	基準位共振
basic catalyst	碱性催化剂	鹼性觸媒
basic dye	碱性染料	鹼性染料
basic firebrick	碱性耐火砖	鹼性耐火磚
basicity	碱性	鹼度

英文名	祖国大陆名	台湾地区名
basic solid	碱性固体	鹼性固體
basket centrifuge	篮式离心机	籃式離心機
basket elevator	篮式升降机	籃式升降機
basket extractor	篮式提取器	籃式萃取器
basket filter	篮式过滤器	籃式(過)濾器
basket reactor	吊篮式反应器	籃式反應器
basket-type evaporator	悬筐蒸发器	籃式蒸發器
batch agitator	间歇搅拌器	批式攪拌器
batch centrifuge	间歇离心机	批式離心機
batch charger	间歇加料机	批式加料機
batch coke still	间歇焦化蒸馏器	批式焦化蒸餾器
batch control	间歇控制	批式控制
batch crystallization	间歇结晶	批式結晶
batch crystallizer	间歇结晶器	批式結晶器
batch culture	间歇培养	批式培養
batch culture	分批培养	批式培養
batch distillation	间歇蒸馏	批式蒸餾
batch drier	间歇干燥器	批式乾燥器
batch extraction	间歇萃取	批式萃取
batch extractor	间歇萃取器	批式浸取器
batch feeder	间歇进料器	批式進料器
batch filter	间歇过滤器	批式過濾機
batch filtration	间歇过滤	批式過濾
batch fractionation	间歇分馏	批式分餾
batch mixer	间歇混合器	批式混合器
batch number	批号	批號
batch operation	间歇操作	批式操作
batch process	间歇过程	批式程序
batch processing	间歇加工	批式加工[法]
batch production	间歇生产	批式生產
batch purification	间歇净化	批式純化
batch reactor	间歇反应器	批式反應器
batch rectification	间歇精馏	批式精餾
batch settling	间歇沉降	批式沉降
batch size	批量大小	批式產能
batch still	间歇蒸馏器	批式蒸餾器
batch still distillation	间歇釜式蒸馏	批式蒸餾
batch system	间歇系统	批式系統
batch vaporization	间歇汽化	批式汽化

英文名	祖国大陆名	台湾地区名
battery of CSTR's	连续搅拌反应釜串联组	串聯連續攪拌槽反應器組
bauxite	铝土矿	鋁礬土
bauxitic clay	铝质黏土	鋁氧黏土
bayberry oil	月桂子油	月桂子油
bay oil	[月]桂叶油	香葉油
bearing	轴承	軸承
bearing ball	轴承滚珠	球軸承
bearing box	轴承箱	軸承箱
bearing bracket	轴承托座	軸承架
bearing bushing	轴承衬套	軸承襯套
bearing collar	轴承圈	軸承圈
bearing housing	轴承箱	軸承箱
bearing mixer	轴承式混合器	軸承式混合器
beater	打浆机	打漿機, 攪合機
bed-collapsing technique	床层塌落技术	床層塌落技術
bed grid support	床栅支架	床柵支架
bed support	床层支架	床支架
beer	啤酒	啤酒
beet sugar	甜菜糖	甜菜糖
bee wax	蜂蜡	蜂蠟
bell and hopper	进料器	進料器
bell and spigot joint	承插接头	承插接頭
bell flowmeter	钟形流量计	鐘流量計
bell manometer	浮钟压力计	鐘式壓力計
bellows	风箱	①風箱②伸縮囊
bellows manometer	波纹管式压力计	伸縮囊壓力計
bellows type flowmeter	波纹管式流量计	伸縮囊式流量計
belt conveyor	皮带输送机	帶式運送機, 帶運機
belt dryer	带式干燥器	帶式乾燥器
belt extractor	转带浸取器	轉帶浸取器
belt filter	带式过滤机	帶濾機
bench scale reactor	台架反应器	實驗室規模反應器
bench scale test	台架试验, 模型试验	實驗室規模試驗
bend	弯管	彎管
bend equivalent pipe length	弯管当量管长	彎管相當管長
bending stress	弯[曲]应力	抗彎應力
bending test	弯曲试验	彎曲試驗
Benedict-Webb-Rubin equation	BWR 方程	BWR 方程

英文名	祖国大陆名	台湾地区名
Benson's solubility coefficient	本森[溶解度]系数	本森係數
benzofuran	苯并呋喃	苯并呋喃, 熏草呀
benzoin	安息香	安息香
benzoyl peroxide	过氧化苯甲酰	過氧化苯甲醯
bergamot oil	香柠檬油	香柑油
Berl saddle	弧鞍填料	弧鞍填料
Bernoulli equation	伯努利方程	白努利方程
beryl	绿柱石	鈹土
bias	偏压	偏, 偏向, 偏流, 偏壓
bias in proportional	比例控制偏离	比例控制的偏離
bicontinuous structure	双连续结构	雙連續結構
bidispersion	双分散	雙分散
bifurcation	分叉, 分支	分歧
bifurcation analysis	分叉分析	分歧分析
bifurcation point	歧点	歧點
bilateral slit	双向狭缝	雙向狹縫
bimetallic catalyst	双金属催化剂	雙金屬觸媒
bimetallic cluster	双金属簇	雙金屬團
bimetallic strip	双金属片	雙金屬片
bimetallic thermometer	双金属温度计	雙金屬溫度計
bimetallic thermoregulator	双金属调温计	雙金屬調溫計
bimodal	双峰的	雙模式
bimodal pore-size distribution	双峰孔径分布	雙模式孔徑分布
bimolecular reaction	双分子反应	雙分子反應
bin activator	料仓松动器	料倉鬆動器
binary azeotrope	二元共沸液	二元共沸液
binary catalyst	二元催化剂	二元觸媒
binary cycle	双循环	二元循環, 雙循環
binary distillation	二元蒸馏	二元蒸餾
binary mixture	二元混和物, 双组分混合物	二元混和物
binary system	二元系[统], 二组分系统	二元系, 雙成分系統
binder	黏合剂	黏結劑
binder power	黏合能力	結合能力
binding agent	黏合剂	黏結劑
binding energy	结合能	結合能
binding force	结合力	結合力
bin discharger	料仓卸料器	倉式卸料器
Bingham fluid	宾厄姆流体	賓漢流體

英文名	祖国大陆名	台湾地区名
binodal solubility curve	双结点溶度曲线	雙結溶解度曲線
bioassay	生物测定	生物測定
bioavailability	生物可利用率	生物可利用率
biocatalysis	生物催化	生物催化[作用]
biocatalyst	生物催化剂	生物催化劑
biocatalytic reaction	生物催化反应	生物催化反應
biocell	生物电池	生物電池
biochemical combustion	生物燃烧	生化燃燒
biochemical engineering	生化工程	生化工程
biochemical oxygen demand (BOD)	生化需氧量	生化需氧量, 生物需氧量
biochemical reaction	生化反应	生化反應
biochemical reaction engineering	生化反应工程	生化反應工程
biochemical reactor	生化反应器	生化反應器
biochemical separation	生化分离	生化分離
biochemical thermodynamics	生化热力学	生化熱力學
biochemistry	生物化学	生物化學
bioconversion	生物转化	生物轉化
biodegradability	生物降解性	生物降解性, 生物分解性
biodegradable detergent	生物降解洗涤剂	生物降解清潔劑
biodegradable plastic	生物降解塑料	生物可分解塑膠
biodegradation	生物降解	生物分解, 生物降解
biodeterioration	生物变质	生物腐蝕
bioenergetics	生物能学	生物能量學
bioengineering	生物工程	生物工程學
biofluid mechanics	生物流体力学	生物流體力學
biofunctional reagent	生物功能试剂	生物功能試劑
biogas	沼气	[微]生物生成氣
bioleaching	生物浸取	生物浸取
biological agent	生物制剂	生物製劑
biological dispersion	生物分散	生物分散
biological film	生物膜	生物薄膜
biological filter	生物滤器	生物濾器
biological flocculation	生物絮凝	生物絮凝
biological material	生物材料	生物材料
biological membrane	生物膜	生物[薄]膜
biological oxidation	生物氧化	生物氧化
biological polymer	生物聚合物	生物聚合物

英文名	祖国大陆名	台湾地区名
biological process	生物过程	生物程序
biological reactor	生物反应器	生物反應器
biological solid	生物固体	生物性固體
biological time lag	生物延时	生物時間落後
biological tower	生物塔	生物塔
biological treatment	生物处理	生物處理[法]
biological treatment process	生物处理过程	生物處理法
bioluminescence	生物发光	生物發光
biomacromolecule	生物大分子	生物大分子
biomass	生物质	生質
biomass conversion	生物质转化	生質轉化
biomass conversion process	生物质转化过程	生質轉化程序
biomass energy	生物质能	生質能
biomass fuel	生物质燃料	生質燃料
biomass gasification	生物质气化	生質氣化
biomass pyrolysis	生物质热解	生質高溫分解
biomaterial	生物材料	生物材料
biomedical device	生物医药器件	生醫裝置
biomedical engineering	生物医学工程	生醫工程
biomedical polymer	医用高分子	生醫聚合物
biomedical technology	生物医学技术	生醫技術
biomedicine	生物医学	生物醫學
bio-osmosis	生物渗透	生物滲透
biopolymer	生物聚合物	生物聚合物
bioprocess	生物过程	生物過程
bioreactor	生物反应器	生物反應器
biosensor	生物传感器	生物感測器
bioseparation	生物分离	生物分離
biospecifically binding	生物特异性连结	生物特異性連結
biosynthesis	生物合成	生物合成
biosynthetic reaction	生物合成反应	生物合成反應
biota	生物群	生物相
biotechnology	生物技术	生物技術
Biot number	毕奥数	畢奧數
biotoxicity	生物毒性	生物毒性
biotoxicity test	生物毒性试验	生物毒性試驗
biotransformation	生物转化	生物轉化
bipartite graph	偶图	偶圖
bipolar electrode	双极性电极	雙極電極

英文名	祖国大陆名	台湾地区名
bipolar moment	双极矩	雙極矩
birefraction	双折射	雙折射
birefringent fluid	双折射流体	雙折射流體
birotation	双旋光	雙旋光
birth rate	生成速率	生成速率
bistable device	双稳定装置	雙穩定裝置
bituminite	烟煤	煙煤
biuret	缩二脲	縮二脲
black ash	黑灰	黑灰
black body	黑体	黑體
black body coefficient	黑体系数	黑體係數
black box	黑箱	黑箱
black box method	黑箱法	黑箱法
black-box model	黑箱模型	黑箱模型
black liquor	黑液	黑液
black radiation	黑体辐射	黑體輻射
black water	浊水	濁水
blade	叶片	葉片
blade agitator	浆叶搅拌器	葉片攪拌器
bladed rotor	叶片转子	葉片轉子
blast	鼓风	①鼓風②爆炸
blast blower	鼓风机	鼓風機
blast burner	喷灯	噴燈
blast furnace	①高炉②鼓风炉	①高爐②鼓風爐
blast furnace coke	高炉焦	高爐焦
blast furnace gas	高炉煤气	高爐氣, 鼓風爐氣
blast furnace slag	高炉渣	高爐渣, 鼓風爐渣
blasticidin	杀稻瘟菌素	殺稻瘟菌素
blasting compound	爆炸化合物	爆炸化合物
bleaching	漂白	漂白
bleaching action	漂白作用	漂白作用
bleaching agent	漂白剂	漂白劑
bleaching liquor	漂白液	漂白液
bleaching powder	漂白粉	漂白粉
bleach powder	漂白粉	漂白粉
bleeder	泄放器	泄放器
blend	①掺合②掺合物	①掺合, 掺配②掺合物
blended gasoline	调和汽油	掺配汽油
blended paint	调和漆	調合漆

英文名	祖国大陆名	台湾地区名
blender	掺合机	掺合機
blending	掺合	掺合, 掺配
blending agent	掺合剂	掺合劑
blending mixer	掺合式混合器	掺合式混合器
blending process	掺合过程	掺配程序
blending system	掺合系统	掺配系統
blending tank	掺合槽	掺配槽
block copolymer	嵌段共聚物	段連共聚物, 嵌段共聚物
block diagram	框图	方塊圖, 塊解圖
block flow diagram	程序方框流程图	程序方塊圖
block polymerization	嵌段聚合	段連聚合[反應], 嵌段聚合[反應]
block tridiagonal matrix	方块三对角矩阵	方塊三對角矩陣
blotting paper	吸墨纸	吸墨紙
blow cock	排气栓	排氣栓
blowdown	排料	泄料, 排放
blowdown tank	泄料箱	排放槽
blower	鼓风机	鼓風機, 送風機
blow hole	[铸件等]气孔	氣孔
blowing agent	发泡剂	發泡劑
blow mold	吹模	吹模
blow molding	吹塑[成型]	吹氣成型法
blown glass	吹制玻璃	吹製玻璃
blown oil	吹制油	吹煉油
blow pipe	吹管	吹管
blow pit	泄料池	泄料池
blow tank	放空罐	泄料槽
blue gas	水煤气	水煤氣
blue vitriol	蓝矾	藍礬
BOD(＝biochemical oxygen demand)	生化需氧量	生化需氧量, 生物需氧量
Bodenstein number	博登施泰数	博登施泰數
BOD load	生化需氧量负载	生化需氧量負載
body-centered lattice	体心点格	體心晶格
boehmite	软水铝石	水鋁土
boiled oil	熟油	熟[煉]油
boiler	锅炉	鍋爐
boiler feed water	锅炉给水	鍋爐給水

英文名	祖国大陆名	台湾地区名
boiling	沸腾	沸腾
boiling-off	脱胶	脱膠, 脱色, 脱脂
boiling point	沸点	沸點
boiling point apparatus	沸点测定器	沸點測定器
boiling point-composition diagram	沸点-组成图	沸點-組成圖
boiling point curve	沸点曲线	沸點曲線
boiling point depression	沸点降低	沸點下降
boiling point diagram	沸点图	沸點圖
boiling point elevation	沸点升高	沸點上升
boiling point index	沸点指数	沸點指數
boiling point lowering	沸点降低	沸點下降
boiling point rise	沸点升高	沸點上升
boiling range	沸腾范围	沸點範圍
boiling starch	沸腾淀粉	糊化澱粉
bolometer	辐射热计	輻射熱測定器
bolt	螺栓	螺栓
bolt and nut	螺栓和螺帽	螺栓及螺帽
bolting cloth	筛布	篩布
bolting mill	筛选机	篩選機
bolting reel	转筒筛	轉筒篩
bolting silk	筛用丝	篩用絲
Boltzmann distribution	玻耳兹曼分布	波茲曼分布
bomb aging	弹内[加氧]陈化	彈內[加氧]老化
bomb calorimeter	弹式量热器	彈卡計
bomb reactor	弹式反应器	彈式反應器
bond	键	鍵
bond angle	键角	鍵角
bond energy	键能	鍵能
bonding	黏合	結合, 鍵結, 黏結
bonding force	键结力	鍵結力
bonding strength	黏合强度	鏈結強度, 結合強度
bone black	骨炭	骨黑
bone char	骨炭	骨碳
book value	账面价值	帳面價值
booster	升压器	增壓器
booster compressor	增压压缩机	增壓壓縮機
booster ejector	升压喷射器	助力噴射器
booster pump	升压泵	增壓泵
boost gage	升压压力表	升壓壓力錶

英文名	祖国大陆名	台湾地区名
boracite	方硼石	方硼石
Borad ring	[博拉德]双层网环	雙層網環
borate	硼酸盐(或酯)	硼酸鹽
borax	硼砂	硼砂,四硼酸鈉
borax glass	硼玻璃	硼玻璃
bornite	斑铜矿	斑銅礦
boron carbide	碳化硼	碳化硼
borosilicate glass	硼硅玻璃	硼矽酸玻璃
Bose-Einstein distribution	玻色-爱因斯坦分布	玻式愛因斯坦分布
bottleneck	瓶颈	瓶頸
bottom discharge	底部出料	底部卸料
bottom-level control	底部液位控制	塔底液位控制
bottom phase	底相,下相	底相
bottom plate	底板	底板
bottom product	底部产物	底部產物
bottom steam	底部蒸汽	塔底蒸汽
bottom valve	底阀	塔底閥
bottom yeast	沉底酵母	沈底酵母
boundary condition	边界条件	邊界條件
boundary layer	边界层	邊界層
boundary layer concentration	边界层浓度	邊界層濃度
boundary layer equation	边界层方程	邊界層方程式
bound moisture	结合水分	結合水分
Bourdon gauge	弹簧管压力计	彈簧管壓力計
bowl mill	碗形磨	碗形磨
branch and bound method	分支定界法	分支定界法
branched chain	支链	支鏈
break	破裂	破裂
breakthrough capacity	穿透容量	失效時容量
breakthrough curve	穿透曲线	失效曲線
breakthrough point	穿透点	失效點
breakthrough time	穿透时间	失效時間
breakup of emulsion	破乳	乳化液之分解
breathing tank	呼吸式油罐	通氣槽
brewery	酿造厂	釀造業
brewer yeast	啤酒酵母	釀造酵母
brewing	酿造	釀造
brick	砖	磚
bridge circuit	电桥	電橋

英文名	祖国大陆名	台湾地区名
brightness	亮度	亮度
brine	盐水	鹽水
brine wash	盐洗	鹽洗
briquetting	压块	壓塊
British pharmacopoeia	英国药典	英國藥典
British thermal unit (BTU)	英热单位	英熱單位
brittle material	脆性物料	脆性物料
brittleness	脆性	脆度, 脆性
brittle point	脆点	脆化點
bromination	溴化	溴化
bronze	青铜	青銅
broth dilution	稀胶法	稀膠法
Brownian diffusion	布朗扩散	布朗擴散
Broyden method	布罗伊登法	布羅伊登法
Brunauer-Emmett-Teller equation	BET 方程	BET 方程
BTU (＝British thermal unit)	英热单位	英熱單位
bubble	气泡	氣泡
bubble aeration	气泡曝气	氣泡通氣
bubble cap absorption column	泡罩吸收塔	泡罩吸收塔
bubble cap column	泡罩塔	泡罩塔
bubble cap contactor	泡罩接触器	泡罩接觸器
bubble cap plate	泡罩板	泡罩板
bubble cap plate distillation tower	泡罩板蒸馏塔	泡罩板蒸餾塔
bubble cap tower	泡罩塔	泡罩塔
bubble cap tray	泡罩板	泡罩板
bubble cloud	气泡云, 气泡晕	氣泡雲
bubble coalescence	气泡聚并	氣泡聚集
bubble column	鼓泡塔	鼓風塔
bubble counter	气泡计数器	氣泡計數器
bubble film	[气]泡膜	氣泡膜
bubble flow	气泡流	氣泡流動
bubble forming agent	气泡剂	氣泡劑
bubble fractionation	气泡分馏	氣泡分餾[法]
bubble gage	气泡指示计	氣泡指示計
bubble plate	泡罩板	泡罩板
bubble plate column	泡罩塔	泡板塔
bubble point	泡点	起泡點, 初沸點
bubble point pressure	泡点压力	起泡壓力
bubble point temperature	泡点温度	起泡溫度

英文名	祖国大陆名	台湾地区名
bubble reactor	气泡反应器	氣泡反應器
bubble tower	泡罩塔	泡罩塔
bubble tower bottom	泡罩塔底	泡罩塔底
bubble tower overhead	泡罩塔顶馏分	泡罩塔頂餾分
bubble tray	鼓泡塔盘	泡罩板
bubble type viscometer	气泡型黏度计	泡式黏度計
bubbling	鼓泡	氣泡
bubbling absorber	鼓泡式吸收器	氣泡吸收器
bubbling bed	沸腾床	氣泡床
bubbling cap	泡罩	泡罩
bubbling fluidization	鼓泡流态化	氣泡流態化
bubbling fluidized bed	鼓泡流化床	氣泡流體化床
bubbling hood	泡罩	泡罩
bucket conveyor	斗式运输机	斗式運送機, 斗運機
bucket elevator	斗式提升机	斗式升降機
bucket-elevator extractor	提斗浸取器	斗式浸取器
budding	芽殖	出芽
budding cell	芽殖细胞	出芽細胞
budding fungi	芽殖真菌	出芽真菌
buffer	缓冲剂	緩衝劑
buffer action	缓冲作用	緩衝作用
buffer capacity	缓冲能力	緩衝能力
buffer index	缓冲指数	緩衝指數
buffer layer	缓冲层	緩衝層
buffer region	缓冲区	緩衝區[域]
buffer solution	缓冲液	緩衝液
buffer tank	缓冲罐	緩衝罐
buffer zone	缓冲区	緩衝區
buhrstone mill dressing	石磨修整	石磨刻鑿
builder	填充剂	填充劑, 補助劑
bulk aerator	整体通气器	整體通氣器
bulk analysis	全分析	整體分析
bulk concentration	整体浓度	整體濃度
bulk density	堆密度	體密度
bulk diffusion	体扩散	體擴散
bulk flow	整体流	整體流動
bulk index	整体指数	整體指數
bulking filler	填充剂	增容填料
bulk intensity	整体强度	整體強度

英文名	祖国大陆名	台湾地区名
bulk modulus	本体模量	整體模數
bulk phase	本体相	整體相
bulk property	本体性质	整體性質
bulk storage	散装储存	散裝儲存
bulk temperature	本体温度	整體溫度
bulk viscosity	体[积]黏性	整體黏度
bulk weight	整重	整體重量
buoyancy	浮力	浮力
buoyant effect	浮力效应	浮力效應
buoyant force	浮力	浮力
burner	燃烧器	燃燒器
burning point	燃点	燃燒點, 火點
burning rate	燃烧率	燃燒速率
burning shrinkage	燃烧收缩	燃燒收縮
burnish	抛光	擦光
burn-out	燃尽	燃盡
burnt ocher	煅赭石	煆赭石
burnt sienna	煅富铁黄土	煆富鐵黃土
burst	爆裂	爆裂
bursting pressure	爆裂压力	爆裂壓力
bursting strength	爆裂强度	爆裂強度
bush	轴衬	軸襯, 襯套
business automation	企业自动化	商業自動化, 企業自動化
butterfly valve	蝶[形]阀	蝶型閥
butt joint	对接	對接
butyl cellulose	丁基纤维素	丁基纖維素
butyl rubber	丁基橡胶	丁基橡膠
BWR equation	BWR 方程	BWR 方程
bypass	旁路, 侧流	旁路, 分路
bypass control	旁通控制	旁路控制
bypass filter	旁通过滤器	旁通過濾器
bypassing	走旁路	旁流, 側流
bypassing effect	旁通效应	旁通效應
bypass pipe	旁通管	旁通管
bypass ratio	旁通比	旁通比
bypass valve	旁通阀	旁通閥
by-product	副产物	副產物

C

英　文　名	祖国大陆名	台湾地区名
CAD(＝computer-aided design)	计算机辅助设计	計算機輔助設計
cadmium cell	镉电池	鎘電池
cadmium soap	镉皂	鎘皂
cadmium yellow	镉黄	鎘黃
CAE(＝computer-aided engineering)	计算机辅助工程	計算機輔助工程
cage effect	笼效应	籠子效應
cage hydraulic press	笼式水压机	籠式水壓機
CAI(＝computer-aided instruction)	计算机辅助教学	計算機輔助教學
cake	滤饼	濾餅
cake conveyor	滤饼运输机	濾泥輸送機
caking	结块	成餅, 結塊
caking power	结块能力	結塊能力
caking property	结块性	結塊性
calandria	排管	排管
calandria pan	排管式蒸发罐	排管式蒸發罐
calandria type evaporator	中央循环管蒸发器, 排管蒸发器	中央循環管蒸發器
calcimeter	二氧化碳测定计	二氧化碳測定計
calcination	煅烧	煅燒
calcinator	煅烧炉	煅燒機
calcined clay	煅烧黏土	煅燒黏土
calcined coke	煅烧焦	煅燒焦
calcined plaster	煅石膏	煅石膏
calciner	煅烧炉	煅燒爐
calcining zone	煅烧区	煅燒區
calcite	方解石	方解石
calcium carbide	碳化钙	碳化鈣
calcium carbonate	碳酸钙	碳酸鈣
calcium lactate	乳酸钙	乳酸鈣
calcium soap	钙皂	鈣皂
calculation	计算	計算
calculator	计算器	計算器
calculus	微积分	微積分
calculus of variations	变分学	變分學, 變分法
calender	压延机	壓延機, 砑光機, 光機軋

英 文 名	祖国大陆名	台湾地区名
calender effect	压延效应	壓延效應
calibration	校准	校正
calibration chart	校准表	校正圖
calibration curve	校正曲线	校正曲線
calomel cell	甘汞电池	甘汞電池
calomel electrode	甘汞电极	甘汞電極
calorie	卡[路里]	卡洛里, 卡
calorimeter	量热计	卡計, 熱量計
calorimetry	量热法	熱量測定法
CAM (＝computer-aided manufacturing)	计算机辅助制造	計算機輔助製造
camphor oil	樟脑油	樟腦油
cane sugar	蔗糖	蔗糖
canned-motion pump	屏蔽泵	屏蔽泵
cannery waste water	罐头厂废水	罐頭廠廢水
canonical coordinates	正则坐标	正準座標
canonical ensemble	正则系综	正則系綜
canonical equation	正则方程	正準方程式
canonical form	正则形式	正準形式
canonical matrix	正则矩阵	正準矩陣
canonical partition function	正则配分函数	正準分配函數
canonical transformation	正则变换	正準變換
capacitance	电容	①電容②容率
capacitance element	电容元件	電率元件
capacitance liquid level gage	电容液面计	電容[式]液面計, 電容 [式]液位計
capacitance moisture meter	电容湿度计	電容[式]溼度計
capacitance pressure transducer	电容压力转换器	電容[式]壓力轉換器
capacitance probe	电容探针	電容探針
capacitor impulse type tachometer	电容脉动式转速计	電容器脈衝[式]轉速計
capacity	容量	容量
capacity estimate	容量估计	容量估計
capacity factor	容量因子	容量因數
capillarimeter	毛细测液器	毛細計
capillary	毛细管	毛細管
capillary absorption	毛细吸收	毛細吸收
capillary adsorption	毛细吸附	毛細吸附
capillary ascent	毛细上升	毛細上升
capillary attraction	毛细吸引	毛細吸引
capillary condensation	毛细冷凝	毛細冷凝

英　文　名	祖国大陆名	台湾地区名
capillary constant	毛细管膜常数	毛细常数
capillary depression	毛细下降	毛细下降
capillary effect	毛细效应	毛细效應
capillary elevation	毛细上升	毛细上升
capillary flask	毛细瓶	毛細瓶
capillary force	毛管力	毛細力
capillary module	毛细管膜组件	毛細管模組
capillary number	毛细管准数	毛細數
capillary penetration	毛细穿透	毛細穿透
capillary pressure	毛细压力	毛細壓力
capillary tension	毛细张力	毛細張力
capillary tube	毛细管	毛細管
capillary tube method	毛细管法	毛細管法
capillary-type bulb	毛细管型球	毛細管式球
capillary viscometer	毛细管黏度计	毛細[管]黏度計
capital	资本	資本
capital expenditure	基建费用	資本支出
capital gain	资本所得	資本所得
capital investment	资本投入	資本投資
capital investment cost factor	资本投资成本因数	資本投資成本因數
capitalized cost	核定资本值	資金成本
capital ratio	资本比率	資本比率
capital recovery factor	资本回收因数	資本回收因數
caprolactam	己内酰胺	己内醯胺
capsule	胶囊	膠囊
caramel	焦糖	焦糖
carbazole	咔唑	咔唑
carbide brick	碳化硅砖	碳化矽磚
carbide silicon	碳化硅	碳化矽
carbitol	卡必醇	卡必醇
carbohydrate	碳水化合物	碳水化合物
carbonaceous fuel	含碳燃料	含碳燃料
carbon adsorption	碳吸附	碳吸附
carbonation process	碳酸气饱充法	碳酸法
carbon balance	碳平衡	碳均衡
carbon black	炭黑	碳黑
carbon brick	炭砖	碳磚
carbon brush	炭刷	碳刷
carbon cycle	碳循环	碳循環

英　文　名	祖国大陆名	台湾地区名
carbon electrode	炭电极	碳電極
carbon fiber	碳纤维	碳纖維
carbon fiber reinforced plastic	碳纤维强化塑料	碳纖維強化塑膠
carbon filter	[活性]炭过滤器	碳濾器
carbon filtration	炭滤	碳濾[法]
carbonium dyes	阳碳离子染料	碳離子型染料
carbonium ion	阳碳离子	碳離子
carbonium mechanism	阳碳离子机理	碳離子機構
carbonless copy	无碳复写	無碳複本
carbon paper	复写纸	複寫紙
carbon refractory	碳质耐火材料	碳[質]耐火物
carbon steel	碳素钢	碳鋼
carbonylation	羰基化	羰化[反應]
carboxylase	羧基酶	羧基酶
carboxylate	羧酸盐	羧酸鹽
carboxylation	羧基化	羧化[作用]
carboxymethyl cellulose	羧甲基纤维素	羧甲纖維素
carburetted gas	增碳燃气	增碳燃氣
carburetted water gas	增碳水煤气	增碳水煤氣
carnallite	光卤石	光鹵石
carnauba wax	巴西棕榈蜡	棕櫚蠟, 卡拿巴蠟
carnotite	钒钾铀矿	釩酸鉀鈾礦
carrier	载体	①輸送機②載體
carrier distillation	载体分馏	載體蒸餾
carrier gas	载气	載體氣體
cascade	串级	①串級②階式
cascade combination	串级组合	串級組合
cascade compensation	串级补偿	串級補償
cascade concentration plate	串级浓缩板	串級濃縮板
cascade concentrator	串级蒸浓器	串級濃縮器
cascade control	串级控制	串級控制
cascade cooler	串级冷却器	串級冷卻器
cascade cycle	级联循环	級聯循環
cascade evaporator	串级蒸发器	串級蒸發器
cascade reactor	级联反应器	級聯反應器
cascade refrigeration	串级冷冻	串級冷凍
cascade ring	阶梯环	階梯環
cascade system	串级系统	串級系統
cascade tower	串级塔	串級塔

英　文　名	祖国大陆名	台湾地区名
cascade-type plate	串级分馏板	串級分餾板
case hardening	表面硬化	表面硬化
casein	酪蛋白	酪蛋白
casein adhesive	酪蛋白黏合剂	酪蛋白黏合劑
cash-flow diagram	现金流通图	現金流通圖
cashmilon	开斯米纶	開司米龍
cash ratio	现金比率	現金比率
cash recovery period	现金回收期	現金回收期
casing	机壳	護罩
casing head gas	油井气	井口天然氣
casing head gasoline	天然汽油	井口汽油
cassava starch	木薯淀粉	樹薯[澱粉]
cassia oil	桂皮油	桂皮油, 肉桂油
casting	①浇铸②铸件	①鑄造②鑄件
casting mold	铸模	鑄模
cast iron	铸铁	鑄鐵
castor oil	蓖麻油	蓖麻油
cast polymerization	铸塑聚合	鑄造聚合[反應]
cast resin	铸塑树脂	鑄模樹脂
cast steel	铸钢	鑄鋼
catabolism	分解代谢	細胞分解作用
catalysis	催化	催化[作用]
catalyst	催化剂	觸媒, 催化劑
catalyst activation	催化剂活化	觸媒活化
catalyst activity	催化剂活性	催化活性
catalyst attrition	催化剂磨损	觸媒磨耗
catalyst bed	催化剂床[层]	觸媒床
catalyst binder	催化剂黏结剂	觸媒黏結劑
catalyst burn-off still	催化剂烧尽釜	觸媒清燒蒸餾器
catalyst carrier	催化剂载体	觸媒載體, 催化劑載體
catalyst circulation	催化剂循环	觸媒循環
catalyst coking	催化剂结焦	觸媒結焦
catalyst contamination index	催化剂污染指数	觸媒污染指數
catalyst deactivation	催化剂失活	觸媒去活化
catalyst decay	催化剂衰变	觸媒衰變
catalyst dehydrogenation	脱氢催化剂	脱氫觸媒
catalyst diluent	催化剂稀释剂	觸媒稀釋劑
catalyst effectiveness	催化剂有效性	觸媒有效性
catalyst extender	催化增效剂	觸媒增效劑

英 文 名	祖国大陆名	台湾地区名
catalyst formulation	催化剂配方	觸媒[之]配方
catalyst fouling	催化剂污染	觸媒積垢
catalyst geometry	催化剂[几何]形状	觸媒[幾何]形狀
catalyst impregnation	催化剂浸渍	觸媒[之]浸漬
catalyst inhibitor	催化[剂]抑制剂	觸媒抑制劑
catalyst iron-alumina	铁铝氧催化剂	鐵鋁氧觸媒
catalyst iron-ammonia	铁氨催化剂	鐵氨觸媒
catalyst life	催化剂寿命	觸媒壽命
catalyst particle	催化剂粒子	觸媒粒子
catalyst pellet	催化剂球粒	觸媒粒
catalyst poisoning	催化剂中毒	觸媒中毒
catalyst pore structure	催化剂细孔结构	觸媒細孔結構
catalyst porosity	催化剂孔隙度	觸媒孔隙度
catalyst reactivation	催化剂再生	觸媒再活化
catalyst receiver	催化剂接收器	觸媒接收器
catalyst regeneration	催化剂再生	觸媒再生
catalyst stabilizer	催化剂稳定剂	觸媒安定劑
catalyst support	①催化剂载体②触媒架	①觸媒載體②觸媒架
catalytic action	催化作用	觸媒作用, 催化作用
catalytic activity	催化活性	催化活性
catalytic agent	催化剂	催化劑
catalytic area	催化面积	催化面積
catalytic carrier	催化剂载体	觸媒載體
catalytic chemisorption	催化化学吸附	催化化學吸附
catalytic conversion	催化剂转化	觸媒轉化
catalytic converter	催化剂转化器	觸媒轉化器
catalytic cracking	催化裂化	[觸]媒裂[解]法
catalytic cracking process	催化裂化过程	觸媒裂解法, 煤裂法
catalytic cycle	催化循环	催化循環
catalytic decomposition	催化分解	催化分解
catalytic gauze	催化剂网	觸媒網
catalytic-gauze reactor	催化剂网反应器	觸媒網反應器
catalytic kinetics	催化动力学	催化動力學
catalytic metal	催化金属	催化性金屬
catalytic oxidation	催化氧化	催化氧化[反應]
catalytic polymerization	催化聚合	催化聚合[反應]
catalytic process	催化过程	催化程序
catalytic promoter	催化促进剂	觸媒促進劑, 助催化劑
catalytic reaction	催化反应	催化反應, 觸媒反應

英　文　名	祖国大陆名	台湾地区名
catalytic reactor	催化反应器	觸媒反應器
catalytic reforming	催化重整	觸媒重組
catalytic reforming process	催化重整过程	觸媒重組法, 媒組法
catalytic site	催化部位	催化部位
catalytic specificity	催化专一性	催化特定性
catalytic surface	催化面	催化表面
catalytic system	催化系统	觸媒系統
catastrophe theory	突变论	災難理論
catastrophic model	突变模型	災難模型
cathode	阴极	陰極
cathodic reaction	阴极反应	陰極反應
cation	阳离子	陽離子
cation exchange	阳离子交换	陽離子交換
cation exchanger	①阳离子交换剂②阳离子交换器	①陽離子交換劑②陽離子交換器
cation exchange resin	阳离子交换树脂	陽離子交換樹脂
cationic polymerization	阳离子聚合	陽離子聚合[反應]
cationic starch	阳离子淀粉	陽離子澱粉
cationic surface active agent	阳离子型表面活性剂	陽離子界面活性劑
cationic surfactant	阳离子型表面活性剂	陽離子界面活性劑
causality	因果性	因果性
causticization	苛化	苛性化
caustic soda	氢氧化钠, 烧碱	氫氧化鈉, 燒鹼
cavitation	汽蚀	空洞現象
cavitation corrosion	空蚀	空洞腐蝕
cavity radiation	空穴辐射	空洞輻射
cell culture	细胞培养	細胞培養
cell cycle	细胞周期	細胞周期
cell debris	细胞碎片	細胞碎片
cell density	细胞密度	細胞密度
cell disruption	细胞破碎	細胞破碎
cell extract	细胞抽取物	細胞抽取物
cell fusion	细胞融合	細胞融合
cell harvesting	细胞收集	細胞收集
cell homogenate	细胞匀浆	細胞匀漿
cell house	电槽室	電槽室
cell loading	细胞负载	細胞負載
cell lysis	细胞溶解	細胞分解
cell model	胞腔模型	細胞模型

英　文　名	祖国大陆名	台湾地区名
cellophane	赛璐玢	賽珞凡
cell separation	细胞分离	細胞分離
cell suspension culture	细胞悬浮培养	細胞懸浮培養
cellular convection	蜂窝状对流	蜂巢形對流
cellular material	微孔材料	海綿狀材料
cellulase	纤维素酶	纖維素酶
celluloid	赛璐珞	賽璐珞
cellulose	纤维素	纖維素
cellulose acetate	醋酸纤维素	醋酸纖維素
cellulose acetate butyrate	乙酸-丁酸纤维素	乙酸-丁酸纖維素
cellulose adhesive	纤维素黏合剂	纖維素黏合劑
cellulose colloid	纤维素胶体	纖維素膠體
cellulose derivative	纤维素衍生物	纖維素衍生物
cellulose lacquer	纤维素喷漆	纖維素噴漆
cellulose nitrate	硝酸纤维素	硝化纖維素, 硝酸纖維 素
cellulose nitrate plastic	硝酸纤维素塑料	硝化纖維素塑膠, 硝酸 纖維素塑膠
cellulose propionate	丙酸纤维素	丙酸纖維素
cellulose xanthate	黄原酸纤维素	黃酸纖維素
cement	水泥	水泥
cementation	渗碳	①膠接②滲碳③硬化
cementation index	[水泥]硬化率	[水泥]硬化指數
cementation process	渗碳法	①膠接程序②滲碳程序
cement pump	水泥输送泵	水泥泵
center of asymptote	渐近线中心	漸進線中心
center of buoyancy	浮[力中]心	浮力中心
center of mass	质[量中]心	質量中心
centipoise	厘泊	厘泊 (黏度單位)
centistokes	厘泡	厘拖 (重力黏度單位)
centri cleaner	锥形除渣器	淨漿機
centrifugal	离心机	離心機
centrifugal blower	离心	離心鼓風機, 離心送風 機
centrifugal clarification	离心澄清	離心澄清
centrifugal collector	离心收集器	離心收集器, 離心集塵 器
centrifugal compressor	离心压缩机	離心壓縮機
centrifugal cyclone	离心旋风器	離心旋風器

英 文 名	祖国大陆名	台湾地区名
centrifugal dehydrator	离心脱水器	離心脱水機
centrifugal dryer	离心干燥器	離心乾燥機
centrifugal extractor	离心萃取器	離心萃取器
centrifugal filter	离心过滤机	離心過濾機
centrifugal force	离心力	離心力
centrifugalization	离心分离	離心化
centrifugal liquid separator	离心式液体分离器	離心式液體分離器
centrifugal mill	离心碾磨机	離心磨[機]
centrifugal molding	离心成型	離心成型
centrifugal propeller	离心式螺旋桨	離心式螺旋槳
centrifugal pump	离心泵	離心泵
centrifugal purification	离心纯化	離心純化
centrifugal screen	离心筛	離心篩
centrifugal scrubber	离心涤气机	離心洗氣器
centrifugal separator	离心分离机	離心分離器
centrifugal settling	离心沉降	離心沉降
centrifugal solid separator	离心式固体分离器	離心式固體分離器
centrifugal spray	离心喷雾	離心噴霧
centrifugal spray tower	离心喷淋塔	離心噴霧塔
centrifugation	离心[分离]	離心[分離]
centrifuge	离心机	離心機
centrifuge basket	离心机盘	離心機籃
centrifuge blade	离心机叶片	離心機葉片
centrifuge head	离心机转头	離心機高差
centrifuge tube	离心[机]管	離心管
centripetal acceleration	向心加速度	向心加速度
centripetal force	向心力	向心力
centroid	质[量中]心	質心
cephalin	脑磷脂	腦磷脂
ceramic	陶瓷制品	陶瓷,陶器
ceramic glaze	陶瓷釉	陶瓷釉
ceramic refractory	陶瓷耐火材料	陶瓷耐火物
ceramics	陶瓷	陶業,陶藝,陶瓷學
ceramic tile	瓷砖	磁磚
ceramic tube	陶瓷管	陶瓷管
ceramic ware	陶瓷器皿	陶器
cermet	陶瓷金属	金屬瓷料
cermet resistor	陶瓷金属电阻器	金屬陶瓷電阻器
cetane index	十六烷指数	十六烷指數

英　文　名	祖国大陆名	台湾地区名
cetane number	十六烷值	十六烷值
chain conveyor	链式输送机	鏈式運送機, 鏈運機
chain depolymerization	链解聚[作用]	鏈解聚合[反應]
chain initiation	链引发	連鎖反應引發
chain polymerization	链聚合	連鎖聚合[反應]
chain propagation	链增长	連鎖反應增長
chain reaction	链反应	連鎖反應
chain termination	链终止	連鎖反應終止
chain termination reaction	链终止反应	連鎖終止反應
chain transfer	链转移	連鎖反應轉移
chalcolite	辉铜矿	輝銅礦
chalcopyrite	黄铜矿	黃銅礦
chalk	白垩	白堊
chamber filter press	箱式压滤机	廂式壓濾機
chamber process	铅室法	鉛室法
chamotte	①熟料②黏土砖	燒粉
change-can mixer	搅浆机	攪漿機
changeover cost	更迭成本	更迭成本
changeover time	更迭时间	更迭時間
channel	①槽②通道	①槽②通道
channel black	槽[法炭]黑	槽黑
channel flow	渠道流	渠流
channeling	沟流	溝流
channeling effect	沟流效应	溝流效應
channel open	明渠	明渠
channel type flowmeter	槽式流量计	槽式流量計
Chao-Seader's method	赵[广绪]-西得方法	趙-西德方法
char	炭	炭
characteristic constant	特性常数	特性常數
characteristic curve	特性曲线	特性曲線
characteristic equation	特性方程	特性方程式
characteristic factor	特性因数	特性因數
characteristic function	特性函数	特性函數
characteristic length	特征长度	特性長度
characteristic reaction time	特性反应时间	特性反應時間
characteristic root	特性根	特性根
characteristic time	特征时间	特性時間
characteristic value	特征值	特性值
characterization	特性化	特性化, 示性

英 文 名	祖国大陆名	台湾地区名
characterization factor	特性因素	特性化因數
charcoal	木炭	木炭
charge conveyor	加料运输机	加料運送機
charged membrane	荷电膜	荷電膜
charge elevator	加料升降机	加料升降機
charge hopper	加料斗	加料斗
charger	加料机	加料機
chasing method	追赶法	追趕法
check list	检查报表	檢查報表
check valve	止逆阀, 单向阀	止回閥, 單向閥
chelating agent	螯合剂	螯合劑, 鉗合劑
chemical	化学品	化學品
chemical absorbent	化学吸收剂	化學吸收劑
chemical absorption	化学吸收	化學吸收
Chemical Abstracts	化学文摘	化學摘要
chemical adsorption	化学吸附	化學吸附
chemical bond	化学键	化學鍵
chemical bonding	化学成键	化學鍵結
chemical change	化学变化	化學變化
chemical clarifier	化学澄清器	化學澄清器
chemical coagulation	化学凝聚	化學凝聚
chemical conditioning	化学调理	化學調理[法]
chemical diffusivity	化学扩散系数	化學擴散係數
chemical dynamics	化学动态学	化學動態學
chemical energy	化学能	化學能
chemical engineer	化学工程师	化學工程師
chemical engineering	化学工程	化學工程
chemical engineering analysis	化工分析	化工分析
chemical engineering construction cost index	化学工程建造成本指数	化學工程營建成本指數
chemical engineering kinetics	化工动力学	化工動力學
chemical engineering process analysis	化工过程分析	化工程序分析
chemical engineering science	化学工程学	化工科學
chemical engineering thermodynamics	化工热力学	化工熱力學
chemical equation	化学方程式	化學方程式
chemical equilibrium	化学平衡	化學平衡
chemical feedstock	化学原料	化學原料
chemical flooding	化学液泛	化學氾流[法]
chemical kinetics	化学动力学	化學動力學

英　文　名	祖国大陆名	台湾地区名
chemical operation	化学操作	化學操作
chemical oscillation	化学振荡	化學振盪
chemical oscillator	化学振荡器	化學振盪器
chemical oxygen demand (COD)	化学需氧量	化學需氧量
chemical parameter	化学参数	化學參數
chemical physics	化学物理[学]	化學物理[學]
chemical plant	化[学]工厂	化學工廠
chemical polarization	化学极化	化學極化
chemical potential	化学势, 化学位	化學勢, 化學位能
chemical precipitation	化学沉淀	化學沈澱[法]
chemical process	化学过程	化學程序
chemical process industry	化学[过程]工业	化學程序工業
chemical promoter	化学促进剂	化學促進劑
chemical property	化学性质	化學性質
chemical pulping process	化学制浆法	化學製漿法
chemical reaction	化学反应	化學反應
chemical reaction engineering	化学反应工程	化學反應工程
chemical reaction equilibrium	化学反应平衡	化學反應平衡
chemical reaction stability	化学反应稳定性	化學反應穩定性
chemical reactor	化学反应器	化學反應器
chemical reactor stability	化学反应器稳定性	化學反應器穩定性
chemical seal	化学密封	化學密封
chemical separation	化学分离	化學分離
chemical smoke	化学烟雾	化學煙幕
chemical species	化学物种	化學物種
chemical stability	化学稳定性	化學穩定性, 化學安定性
chemical transport	化学输送	化學輸送
chemical treatment	化学处理	化學處理
chemical treatment process	化学处理过程	化學處理法
chemical vapor	化学蒸气	化學蒸氣
chemical vapor deposition	化学气相沉积	化學蒸氣沉積[法]
chemiluminescence	化学发光	化學發光
chemisorption	化学吸附	化學吸附
chemistry of hydrogen energy	氢能化学	氫能化學
chemohydrodynamics	化学流体动力学	化學流體動力學
chemostat	恒化器	恆化器
china	瓷[器]	瓷
china clay	瓷土	瓷土

英　文　名	祖国大陆名	台湾地区名
chinaware	瓷器	瓷器
chip drier	木片干燥器	木片乾燥器
chiral separation	手性分离	對掌性分離
chloral	三氯乙醛	三氯乙醛
chloramine	氯胺	氯胺
chloranil	四氯苯醌	四氯[苯]醌
chloride contact chamber	氯化物接触室	氯化物接觸室
chlorinated hydrocarbon	氯化烃	氯[化]烴
chlorinated paraffin	氯化石蜡	氯[化]蠟
chlorinated phenol	氯化酚	氯化酚
chlorinated quinone	氯化苯醌	氯化[苯]醌
chlorinated rubber	氯化橡胶	氯化橡膠
chlorination	氯化	氯化[反應]
chloromycetin	氯霉素	氯黴素
chlorophyll	叶绿素	葉綠素
chloroprene rubber	氯丁橡胶	氯平橡膠
chlorosulfonated polyethylene	氯磺化聚乙烯	氯磺化聚乙烯
chlorosulfonation	氯磺酰化	氯磺酸化[反應]
chlorosulfonic polyethylene	氯磺化聚乙烯	氯磺化聚乙烯
choke	噎塞	阻塞
choke crushing	阻塞压碎	阻塞壓碎
choked nozzle	闸喉喷嘴	閘喉噴嘴
choke flow	扼流	阻流
choking	噎塞	噎塞
cholesteric phase	螺状相	膽固醇型液晶相
cholesteric state	胆甾状态	膽固醇型液晶態
cholesterol	胆固醇	膽固醇
chord line	弦线	弦線
chromatoelectrophoresis	色谱电泳	電泳層析
chromatofocusing	色谱聚焦	聚焦層析
chromatogram	色谱图	層析圖
chromatograph	色谱仪	層析儀
chromatographic column	色谱柱	層析柱
chromatographic column pulse reactor	色谱柱式脉冲反应堆	層析柱式脈動反應器
chromatographic method	色谱法, 层析法	層析法
chromatographic separation	色谱分离	層析分離
chromatography	色谱法, 层析法	層析法
chrome alum	钾铬矾	鉻礬
chrome brick	铬砖	鉻磚

英　文　名	祖国大陆名	台湾地区名
chrome dye	铬染料	鉻媒染料
chrome green	铬绿	鉻綠
chrome leather	铬革	鉻革, 鞣皮
chrome-magnesite brick	铬镁砖	鉻美磚
chrome mordant dyeing	铬媒染色	鉻媒染色
chrome orange	铬橙	鉻橙
chrome pigment	铬颜料	鉻顏料
chrome plating	镀铬	鍍鉻
chrome red	铬红	鉻紅
chrome refractory	铬质耐火材料	鉻[質]耐火物
chrome steel	铬钢	鉻鋼
chrome tannage	铬鞣	鉻鞣
chrome tanning	铬鞣制	鉻鞣製
chrome yellow	铬黄	鉻黃
chromia-alumina catalyst	氧化铬-氧化铝催化剂	鉻鋁氧觸媒
chromogen	发色体	色素原
chromometer	比色计	比色計
chromophore	发色团	發色團, 色基
chrysoberyl	金绿宝石	金綠寶石
chute	滑槽	滑槽
cinematic model	影式模型	影式模型
cinnamon oil	肉桂油	桂皮油, 肉桂油
circular chart	圆形图	圓狀圖
circularity	圆形度	圓形度
circular scale indicator	圆形标度指示计	圓形標[度指]示計
circulating criterion	循环判据	循環準則
circulating drop	循环滴	循環流動滴
circulating fluidized bed	循环流化床	循環流化床
circulating pump mixer	循环泵混合器	循環泵混合器
circulating reactor	环流反应器	循環反應器
circulating reflux	循环回流	循環回流
circulation	环流	循環, 流通
circulation flow	循环流动	循環流動
circulation layer	循环层	循環層
circulation method	循环法	循環法
circulation mixer	循环式混合器	循環式混合器
circulation pipe	循环管	循環管
circulation tube	循环管	循環管
circumferential velocity	圆周速度	圓周速度

英　文　名	祖国大陆名	台湾地区名
cis-butadiene rubber	顺[式]丁[二烯]橡胶	丁二烯橡膠
cis-form	顺式	順式
cis-isoprene rubber	顺式异戊二烯橡胶	順異戊二烯橡膠
citronellal	香茅醛	香茅醛
citronella oil	香茅油	香茅油
citronellol	香茅醇	香茅醇
Clapeyron-Clausius equation	克拉珀龙-克劳修斯方程	克来匹隆-克勞修斯方程
clarification	澄清	澄清[作用]
clarification capacity	澄清能力	澄清能力
clarification filtration	澄清过滤	澄清過濾
clarification process	澄清过程	澄清程序
clarification zone	澄清区	澄清區
clarifier	澄清器	澄清器
clarifier filter	澄清过滤器	澄清器
clarifying capacity	澄清能力	澄清能力
classical fluidization	经典流态化	古典流態化
classification	分级	類析[法], 分類
classifier	分级器	類析器
clathrate	笼合物	晶籠化合物
clathration	笼合[作用]	晶籠[作用]
Clausius inequality	克劳修斯不等式	克勞修斯不等式
clay	黏土	黏土
cleaning process	清洗过程	清洗程序
clearance	间隙	間隙
clearance meter	间隙测量计	間隙計
clearance volume	间隙体积	間隙體積
climbing-film evaporator	升膜蒸发器	升膜蒸發器
clinging cavity	黏着空洞	附著空洞
clone	克隆	克隆
close cut distillate	精密分馏馏出物	精分餾出物
close cut fraction	精密分馏馏分	精分餾分
close cut separation	精密分馏分离	精分分離
closed boundary	闭式边界	閉式邊界
closed circuit	闭路	閉路
closed circuit crushing	闭路压碎	閉路壓碎, 循環壓碎
closed circuit grinding	闭路研磨	閉路研磨, 循環研磨
closed conduct	封闭导管	封閉導管
closed impeller	封闭式叶轮	隱閉葉輪

英　文　名	祖国大陆名	台湾地区名
closed loop	闭环	閉環
closed loop control	闭环控制	閉環控制
closed loop frequency response	闭环频率应答	閉環頻率應答
closed loop stability	闭环稳定性	閉環穩定性
closed loop system	闭环系统	閉環系統
closed loop transfer function	闭环传递函数	閉環循環函數
closed reactor	密闭反应器	密閉反應器
closed system	封闭系统	封閉系統
closed vessel	闭式容器	密閉容器
cloth filter	布滤机	布濾機
cluster of particles	颗粒簇	顆粒團簇
coagulant	凝聚剂	凝聚劑
coagulant aid	助凝剂	助凝劑
coagulating agent	凝聚剂	凝聚劑
coagulation tank	凝聚槽	凝聚槽
coal accelerator	卸煤加速器	卸煤加速器
coal bed	煤床	煤床
coal conversion	煤转化	煤轉化
coal-derived gas	煤衍生气	煤衍生氣
coal-derived liquid	煤衍生液[体]	煤衍生液[體]
coal devolatilization	煤的脱挥发作用	煤之去揮發作用
coal distillation	煤蒸馏	煤蒸餾
coalescence	聚并, 凝并	聚結
coalescence model	聚并模式	聚結模式
coalescer	聚并器	聚結器
coal field	煤田	煤田
coal friability	煤脆性	煤脆性
coal gas	煤气	煤礦氣
coal gasification	煤气化	煤氣化[法]
coal gasification process	煤气化过程	煤氣化程序
coal gasifier	煤气化炉	煤氣化器
coal liquefaction	煤液化	煤液化[法]
coal liquefaction process	煤液化过程	煤液化程序
coal porosity	煤孔隙度	煤孔隙度
coal processing technology	煤加工技术	煤處理技術
coal pyrolysis	煤热解	煤之高溫分解
coal tar	煤焦油	煤溚, 煤焦油
coal-water slurry	水煤浆	煤水漿
coaptation	参数推算	參數推算

英　文　名	祖国大陆名	台湾地区名
coarse crusher	粗级压碎机	粗級壓碎機
coarse particle	粗颗粒	粗顆粒
coarse screen	粗筛[网]	粗篩
coarse suspension	粗悬浮液	粗懸浮液
coastal pollution	海岸污染	海岸污染
coating technology	涂料技术	塗裝技術
co-catalyst	助催化剂	助催化劑
cock	旋塞	活栓,旋塞
coconut oil	椰子油	椰子油
cocurrent flow	并流,同向流	同向流動
cocurrent operation	并流操作	同向流操作
cocurrent process	并流过程	同流程序
COD(＝chemical oxygen demand)	化学需氧量	化學需氧量
code	编码	代號,規範,簡碼
cod-liver oil	鱼肝油	鱈魚肝油
coefficient contraction	收缩系数	收縮係數
coefficient of performance (COP)	性能系数	性能係數
coefficient of variation	变动系数	變異係數
coenzyme	辅酶	輔酵素
coexistence equation	共存方程	共存方程式
coextraction	共萃取	共萃取
cofactor	辅因子	輔因子
coffee mill	咖啡碾磨机	咖啡磨粉機
cohesion	内聚力	內聚力
cohesion work	内聚功	內聚功
cohesive density	内聚能密度	內聚能密度
coil	蛇管	旋管,蛇管,盤管
coil condenser	蛇管冷凝器	旋管冷凝器
coiled pipe	蛇形管	旋管
coiled radiator	蛇管散热器	旋管散熱器
coil evaporator	蛇管式蒸发器	旋管蒸發器
coil expansion pipe	盘形膨胀管	盤形膨脹管
co-ion	共离子	共離子
coke	焦炭	焦炭
coke drum	结焦槽	結焦槽
coke knocker	除焦机	除焦機
coke number	焦值	焦值
coke oven	[炼]焦炉	煉焦爐
coke oven gas	焦炉煤气	焦爐氣

英 文 名	祖国大陆名	台湾地区名
coke oven tar	焦炉油	焦爐溚
coke porosity	焦炭孔隙度	焦炭孔隙度
coker	焦化器	煉焦器
coke still	焦化蒸馏器	焦化蒸餾器
coking	结焦,结炭	①煉焦,煉焦煤②結焦
coking coal	[炼]焦煤	煉焦煤
coking process	炼焦过程	煉焦法
coking reaction	焦化反应	焦化反應
cold junction	冷模合	低溫接點
cold reflux	冷回流	冷回流
cold rubber	冷制橡胶	冷製橡膠
cold-soda process	冷碱法	冷鹼法
cold vulcanization	冷硫化	冷硫化法
colistin	黏菌素	黏菌素
collagen	胶原	膠原
collecting ring	收集环	收集環
collection efficiency	捕集效率	捕集效率
collector	收集剂	收集器
collector electrode	集尘极	集[板]電擊
colligative property of solution	溶液的依数性	溶液的依數性
colloid	胶体	膠體
colloidal clay	胶态白土	膨土,漿土
colloidal dispersion	胶态分散	膠態分散
colloidal mill	胶体磨	膠體磨機
colloidal particle	胶粒	膠體粒子
colloidal phenomenon	胶体现象	膠體現象
colloidal platinum	胶体铂	膠體鉑
colloidal precipitation	胶体沉淀	膠體沈澱
colloidal solution	胶体溶液	膠體溶液
colloidal state	胶体状态	膠體狀態
colloidal suspension	胶体悬浮液	膠體懸浮液
colloid chemistry	胶体化学	膠體化學
colloidization	胶化	膠體化
colloid solution	胶体溶胶	膠體溶液
colloid stabilization	胶体稳定化	膠體穩定化
color comparator	比色器	比色器
color densitometer	比色密度计	比色密度計
colorimeter	比色计	比色計
column	①柱②塔	柱

英 文 名	祖国大陆名	台湾地区名
column capacity	柱容量	塔柱容量
column characterization factor	柱特性因素	塔柱示性因數
column chromatography	柱色谱法	柱式層析法
column internals	塔内件	塔内件
column packing	塔填充物	塔填充物
column plate	塔板	塔板
column scrubber	洗涤塔	洗滌塔
column still	柱式蒸馏器	柱式蒸餾器
column tray	塔板	塔板
combinatorial term	组合项	組合項
combining rule	组合规则	組合規則
combustible liquid code	可燃液体规范	可燃液體規範
combustion	燃烧	燃燒
combustion chamber	燃烧室	燃燒室
combustion characteristics	燃烧特性	燃燒特性
combustion curve	燃烧曲线	燃燒曲線
combustion efficiency	燃烧效率	燃燒效率
combustion engine	内燃机	内燃機
combustion front	燃烧面	燃燒峰面
combustion initiator	引燃剂	引燃劑
combustion mechanism	燃烧机理	燃燒機構
combustion process	燃烧过程	燃燒程序
combustion rate	燃烧速率	燃燒速率
combustion reaction	燃烧反应	燃燒反應
combustion reactor	燃烧反应器	燃燒反應器
combustion zone	燃烧区	燃燒區
commensalism	共生现象	①共棲②共存
commercial development	商业发展	商業發展
comminuting machine	粉碎机	粉碎機
comminution	粉碎	粉碎
comminutor	粉碎机	粉碎機
commodity chemical	大宗化学品	大宗化學品
commodity polymer	大宗聚合物	大宗聚合物
common stock	普通原料	普通原料,共用原料
common utility	公共设施	共用設施
compact heat exchanger	紧凑型换热器	緊緻熱交換器
compaction	压紧	壓緊
compactness	紧密度	緊密度
comparator	比较器	比較器

英　文　名	祖国大陆名	台湾地区名
compartment drier	箱式干燥器	廂式乾燥計
compartmented tank reactor	多室槽式反应器	分倉式反應器
compatibility	相容性	互適性, 相容性
compensated thermometer system	补偿温度计系统	補償溫度計系統
compensation	补偿	①補償②調整
compensation effect	补偿效应	補償效應
compensator	补偿器	補償器
competitive adsorption	竞争吸附	競爭吸附
competitive inhibition	竞争性抑制	競爭性抑制
competitiveness	竞争性	競爭性
competitive reaction	竞争反应	競爭反應
complementary feedback	互补反馈	互補回饋
complete mixing	全混	全混
complete-mixing digester	完全混合消化	完全混合消化
complete mixing flow	全混流	全混流
complex balancing	络合物平衡	錯合物均衡
complex conjugate root	复数共轭根	複數共軛根
complex method	复合形法	複合形法
complexometric titration	络合滴定	錯離子滴定法
complex pipe system	复合管道系统	複合管路系統
complex plane	复平面	複數平面
complex potential	复位势	複勢
complex process	复合过程	複合程序
complex reaction	复杂反应	複雜反應
complex reaction network	复杂反应网络	複雜反應網路
complex transition	过渡络合物	過渡錯合物
complex variable	复变量	複變數
complex viscosity	复数黏度	複數黏度成分
component	组分	成分
component assignment diagram	组分指定图	成分指定图
composite	复合材料	複合物
composite-averaged equation	复合平均方程式	複合平均方程式
composite liquids	复合液体	複合液體
composite material	复合材料	複合材料
composite membrane	复合膜	複合膜
composite sample	复合样品	複合樣品
composite transfer coefficient	复合传递系数	複合傳送係數
composite wall	复合壁	複合壁
composition	组成	組成

英　文　名	祖国大陆名	台湾地区名
composition control	组成控制	組成控制
composition specification	组成规格	組成規格
composition trajectory	组成轨道	組成軌跡
compositive wall	综合壁	複合壁
compost	堆肥	堆肥
composting	堆肥法	堆肥法
compound	化合物	化合物
compound ball mill	复合球磨机	複合球磨機
compound distillating apparatus	复式蒸馏器	複式蒸餾器
compound interest	复利	複利率法
compound interest factor	复利因子	複利率因數
compound rectifying column	复式精馏塔	複式精餾塔
compressed air	压缩空气	壓縮空氣
compressed air sprayer	压缩空气喷雾器	氣壓噴霧器
compressibility	可压缩性	壓縮度
compressibility coefficient	压缩系数	壓縮係數
compressibility criterion	压缩率判据	壓縮度準則
compressibility factor	压缩因子	壓縮因數
compressible filter cake	可压缩滤饼	可壓縮濾餅
compressible flow	可压缩流动	可壓縮流動
compressible fluid	可压缩流体	可壓縮流體
compression	压缩	壓縮
compression-permeability technique	压缩渗透技巧	壓縮滲透技術
compression process	压缩过程	壓縮程序
compression ratio	压缩比	壓縮比
compression wave	压缩波	壓縮波
compression work	压缩功	壓縮功
compression zone	压缩区	壓縮區
compressive force	压缩力	壓縮力
compressive pressure	压缩压力	壓縮壓力
compressive strength	压缩强度	抗壓強度
compressor	压缩机	壓縮機
compressor specification	压缩机规格	壓縮機規格
computation	计算	計算
computer-aided design (CAD)	计算机辅助设计	計算機輔助設計
computer-aided engineering (CAE)	计算机辅助工程	計算機輔助工程
computer-aided instruction (CAI)	计算机辅助教学	計算機輔助教學
computer-aided manufacturing (CAM)	计算机辅助制造	計算機輔助製造
computer control	计算机控制	計算機控制

英　文　名	祖国大陆名	台湾地区名
computer program	计算机程序	計算機程式
computer simulation	计算机模拟	計算機模擬
concentrate	浓缩液	濃縮物
concentration	浓度	濃度
concentration boundary layer	浓度边界层	濃度邊界層
concentration bulk	整体浓度	整體濃度
concentration cell	浓差电池	濃度差電池
concentration difference	浓度差	濃度差
concentration diffusion	浓度扩散	濃度擴散
concentration gradient	浓度梯度	濃度梯度
concentration polarization	浓差极化	濃度極化
concentration profile	浓度[分布]剖面[图]	濃度剖面圖
concentration-rate curve	浓度速率曲线	濃度速率曲線
concentrator	浓缩器	濃縮器
concentric cylinder viscometer	同心柱黏度计	同軸圓柱黏度計
concentric dial	同心刻度盘	同心儀
concentric float	同心浮球	同心浮子
concentric orifice plate	同心孔口板	同心孔口板
concentric reducer	同心变径管	同心漸縮管
concentric ring	同心环	同心環
concentric-tube column	同心管柱	同心管柱
concrete	混凝土	混凝土
concrete pipe	水泥管	水泥管
concurrent reaction	并发反应	併發反應
condensate	冷凝液	冷凝液
condensation coefficient	缩合系数	縮合係數
condensation point	冷凝点	冷凝點
condensation polymerization	缩聚[反应]	縮聚合反應
condensation reaction	缩合反应	縮合反應
condensation tube	冷凝管	冷凝管
condensed phase	凝结相	凝結相
condenser	①冷凝器②电容器	①冷凝器②電容器
condenser paper	电容器纸	積電紙
condensing chamber	冷凝室	冷凝室
condensing temperature	冷凝温度	冷凝溫度
condensing tube	冷凝管	冷凝管
conditional stability	条件稳定	有條件的穩定性
conditioning agent	调节剂	調理劑
conductance	电导	電導度

英　文　名	祖国大陆名	台湾地区名
conductibility	传导性	傳導性
conduction	传导	傳導
conduction heat transfer	热传导	熱傳導
conductive heat transfer	热传导	熱傳導
conductive polymer	导电聚合物	導電聚合物
conductivity	电导率	傳導度
conductivity cell	电导池	導體電池
conductometer	电导计	熱導計, 電導計
conductometric titration	电导滴定	電導滴定法
conductor catalyst	导体催化剂	導體觸媒
conduit	导管	導管
cone-and-plate viscometer	锥板式黏度计	板錐黏度計
cone atomizer	锥形雾化器	錐形霧化器
cone crusher	圆锥破碎机	錐碎機
cone feeder	锥形给料器	錐形進料器
cone flow	锥形流动	錐形流動
cone roof tank	锥顶油罐	錐頂油槽
confidence level	置信水平	可信水平
confidence limit	置信限	可信度限制
confidence region	置信域	可信區域
configurational diffusion	组态扩散	組態擴散
configurational partition function	位形配分函数, 构型配分函数	組態分配函數
configurational property	位形性质, 构型性质	組態性質
conformal mapping	共形映射	保形映射, 保角映射
conformation	构象	構形
congealing point	冻凝点	凍凝點
congealing temperature	冻凝温度	凍凝溫度
congelation	冻凝	凍凝
congelation point	冻凝点	凍凝點
conjugate depth	共轭深度	共軛深度
conjugate pair	共轭对	共軛對
conjugate phase	共轭相	共軛相
conjugate solution	共轭溶液	共軛溶液
consecutive reaction	连串反应	連串反應
conservation equation	守恒方程式	守恆方程式
conservation law	守恒律	守恆律
conservation law for angular momentum	角动量守恒	角動量守恆
consistency	一致性	一致性

英　文　名	祖国大陆名	台湾地区名
consistency test	一致性试验	一致性試驗
consolute temperature	临界共溶温度	臨界共溶溫度
constant	常数	常數
constant[flow]rate filtration	恒速过滤	恆流率過濾
constant boiling	共沸	共沸, 恆沸
constant boiling point	共沸点	恆沸點
constant composition solution	恒组成溶液	恆组成溶液
constant head tank	定位槽	定高差槽
constant molar overflow	恒摩尔溢流	定莫耳溢流
constant phase angle curve	等相角曲线	等相角曲線
constant pressure	恒压	恆壓, 等壓, 定壓
constant pressure filtration	恒压过滤	恆壓過濾
constant rate drying period	恒速干燥[阶]段	恆速乾燥期
constant shear viscometer	恒剪力黏度计	恆剪力黏度計
constant speed pump	定速泵	恆速泵
constant temperature bath	恒温浴	轉速計槽
constant temperature oven	恒温烘箱	恆溫烘箱
constituent	组分	组成, 成分
constitutive equation	本构方程	本質常數
constraint	约束[条件]	限制
contact angle	接触角	接觸角
contact catalysis	接触催化	接觸催化作用
contact chamber	接触室	接觸室
contact condenser	接触冷凝器	接觸冷凝器
contact filtration	接触过滤	接觸過濾
contacting pattern	接触方式	接觸方式
contacting phase	接触相	接觸相
contact nucleation	接触成核	接觸成核
contactor	接触器	接觸器
contactor efficiency	接触器效率	接觸器效率
contact resistance	接触电阻	接觸阻力
contact stabilization process	接触稳定过程	接觸穩定程序
container	容器	容器
contaminant	污染物	污染物
contaminated soil	污染土壤	受污土壤
contamination	污染	污染
content	含量	含量
contingency factor	意外费用因数	經常開銷費用因數
continuity equation	连续方程	連續方程式

英　文　名	祖国大陆名	台湾地区名
continuity law	连续性定律	連續定律
continuous belt drum filter	连续皮带鼓式过滤器	連續輸送鼓式過濾器
continuous carbon filter	连续式碳滤器	連續式碳濾器
continuous carbon filtration	连续碳滤法	連續碳濾法
continuous cash flow	连续现金流动	連續現金流動
continuous centrifuge	连续离心机	連續離心機
continuous crystallization	连续结晶	連續結晶
continuous crystallizer	连续结晶器	連續結晶器
continuous culture	连续培养	連續式培養菌
continuous-cycling method	连续环圈法	連續環圈法
continuous decantation	连续倾析	連續傾析
continuous digester	连续蒸煮器	連續蒸煮器
continuous distillation	连续蒸馏	連續蒸餾
continuous drier	连续干燥器	連續乾燥器
continuous extraction	连续萃取	連續萃取
continuous extractor	连续萃取器	連續萃取器
continuous filter	连续过滤器	連續過濾器
continuous filtration	连续过滤	連續過濾
continuous flow reactor	连续[流动]反应釜	連續流動式反應器
continuous flow stirred tank reactor (CSTR)	连续流动式搅拌槽反应器	連續流動式攪拌槽反應器
continuous flow system	连续流动系统	連續流動系統
continuous fractionation	连续分馏	連續分餾
continuous grinder	连续研磨机	連續研磨機
continuous interest	连续利息	連續利率
continuous kiln	连续窑	連續窯
continuous operation	连续操作	連續操作
continuous phase	连续相	連續相
continuous polymerization	连续聚合	連續聚合
continuous process	连续过程	連續程序
continuous processing	连续加工	連續加工法
continuous reactor	连续式反应器	連續式反應器
continuous stirred reactor	连续搅拌反应器	連續攪拌反應器
continuous stripping still	连续汽提器	連續汽提器
continuous tunnel drier	连续隧道干燥器	連續隧式乾燥器
continuous tunnel kiln	连续隧道窑	連續隧式窯
continuous type process	连续过程	連續程序
continuum	连续介质	連體
contour plot	等值线图表	輪廓圖

英 文 名	祖国大陆名	台湾地区名
contraction coefficient	收缩系数	收縮係數
contraction equivalent pipe length	当量收缩管长	縮管相當管長
contraction loss	收缩损失	收縮損失
contraction stability	收缩稳定性	收縮穩定性
control	控制	控制
control action	控制作用	控制作用
control algorithm	控制算法	控制算則
control chart	控制图	管制圖
control configuration	控制构型	控制組態
control function	控制函数	控制函數
control interval	控制区间	控制區間
controllability	可控性	可控性
control law	控制律	控制律
controlled field	控制现场	控制區域
controlled release	控[制]释[放]	控制釋放
controlled variable	控制变量	受控變數
controller adaptation	控制器匹配	控制器調適
controller calibration	控制器校正	控制器校正
controller gain	控制器增益	控制器增益
controller mechanism	控制器机理	控制器機構
controller parameter	控制器参数	控制器參數
controller ringing	控制器振铃	控制器振鈴
controller setting	控制器设定	控制器設定
controller tuning	调节器参数整定	控制器調協
control limit	控制界限	管制界限
control loop	控制回路	控制環路
control panel	控制盘仪表	控制儀表板
control point	控制点	控制點
control-reactor	控制反应器	反應器控制
control scheme	控制线路	控制線路
control single-loop	控制单环路	單環路控制
control strategy	控制策略	控制策略
control surface	控制表面	限制表面
control system	控制系统	控制系統
control valve	控制阀	控制閥
control variable	控制变量	控制變數
control volume	控制体积	限制體積
convected coordinates	迁移坐标	對流座標
convection	对流	對流

英　文　名	祖国大陆名	台湾地区名
convection heat transfer	对流传热	熱對流
convection mass transfer	对流传质	對流質傳
convection section	对流段	對流段
convective acceleration	对流加速度	對流加速度
convective derivative	对流导数	對流導數
convective flux	对流通量	對流通量
convective heat transfer	对流传热	熱對流
convective heat transfer coefficient	对流传热系数	熱對流係數
convergence acceleration	收敛加速	加速收斂
convergence criterion	收敛判据	收斂判斷
convergent tube	渐缩管	漸縮管
converging-diverging nozzle	缩扩喷嘴	細腰噴嘴
conversion	转化	轉化
conversion coating	转换涂层	變換塗層
conversion constant	换算常数	換算常數
conversion constant force-mass	力质换算常数	力質量換算常數
conversion curve	换算曲线	轉換曲線
conversion equation	换算方程式	換算方程式
conversion factor	换算因子	換算因數
conversion level	转化深度	轉化程度
conversion parameter	转化参量	轉化參數
conversion principle	转化原理	轉化原理
conversion rate	转化率	轉化率
conversion table	换算表	換算表
conversion temperature	转化温度	轉化溫度
conversion-temperature chart	转化率-温度图	轉化率-溫度圖
conversion time	转化时间	轉化時間
converter	转化器	轉化器
conveyer	运输机	運送機
conveying	输送	運送
conveyor belt	运输带	運送機帶
convolution	卷积	摺積
convolution integral	卷积积分	摺積積分
cook	蒸煮	蒸煮
cooking acid	蒸煮酸	蒸煮酸
cooking liquor	蒸煮液	蒸煮液
cooking process	蒸煮过程	蒸煮程序
cooking tank	蒸煮槽	蒸煮槽
coolant	冷却剂	冷卻劑

英　文　名	祖国大陆名	台湾地区名
cooler	冷却器	冷卻器
cooling agent	冷却剂	冷卻劑
cooling area	冷却面积	冷卻面積
cooling coil	冷却蛇管	冷卻旋管
cooling curve	冷却曲线	冷卻曲線
cooling fin	散热片	散熱片
cooling medium	冷却介质	冷卻界質
cooling rate	冷却速率	冷卻速率
cooling surface	冷却面	冷卻表面
cooling system	冷却系统	冷卻系統
cooling time	冷却时间	冷卻時間
cooling tower	冷却塔	冷卻塔
cooling water	冷却水	冷卻水
cooling zone	冷却段	冷卻區
cool liming	冷浸灰法	冷汁加灰法
cooperative adsorption	合作吸附	合作吸附
coordinates covalence	配位共价	配位共價
coordinate system	坐标系	坐標系
coordination catalyst	络合催化剂	配位觸媒
coordination compound	配位化合物	配位化合物
coordination polymerization	配位聚合	配位聚合反應
coordination reaction	配位反应	配位反應
COP(＝coefficient of performance)	冷冻系数	性能係數
copolymer	共聚物	共聚物
copolymerization	共聚合	共聚合反應
coproduct	联产品	連產物
coreactant	共反应剂	共反應物
corner flow	拐角流	角形流動
corner frequency	角频率	角頻率
corner tap	转角接头	轉角接頭
corn oil	玉米油	玉米油
corn starch	玉米淀粉	玉米澱粉
corn sugar	玉米糖	玉米糖
corn syrup	玉米糖浆	玉米糖漿
corrective action	校正作用	校正作用
correlating data	相关数据	相關數據
correlation	关联, 关联式	相關式
correlation function	相关函数	相關函數
corresponding state	对应态	對應狀態

英 文 名	祖国大陆名	台湾地区名
corrodent	腐蚀剂	腐蝕質
corrosion	腐蚀	腐蝕
corrosion allowance	腐蚀裕量	腐蝕容許量
corrosion fatigue	腐蚀疲劳	腐蝕疲勞
corrosion inhibitor	缓蚀剂	腐蝕抑制劑
corrosion prevention	防腐蚀	防蝕
corrosion rate	腐蚀速率	腐蝕速率
corrosion test	腐蚀试验	腐蝕試驗
corrosive action	腐蚀作用	腐蝕作用
corrugated sheet	波纹板	浪板
corrugated tube	波纹管	波形管
corrugating paper	瓦楞纸	瓦楞紙
cortisone	可的松	皮質酮
cost	成本	成本
cost accounting	成本会计	成本會計
cost benefit analysis	成本效益分析	成本效益分析
cost curve	成本曲线	成本曲線
cost-effective analysis	成本效益分析	成本效益分析
cost-effective management	成本效益管理	成本效益管理
cost effectiveness	成本效益	成本效益
cost engineer	成本工程师	成本工程師
cost estimate	成本估计	成本估計值
cost estimation	成本估计	成本估計值
cost factor	成本因子	成本因數
cost index	成本指数	成本指數
cotton paper	棉纸	棉紙
cotton-seed oil	棉子油	棉子油
coulometric titration	电量滴定	電量滴定法
Coulter counter	库尔特粒度仪	庫爾特粒度儀
coumarone	苯并呋喃	苯并呋喃
coumarone-indene resin	苯并呋喃-茚树脂	苯并呋喃樹脂
counter	计数器	計數器
countercurrent absorption	逆流吸收	逆流吸收
countercurrent column process	逆流塔过程	逆流塔程序
countercurrent decantation	逆流倾析	逆流傾析
countercurrent extraction	逆流萃取	逆流萃取
countercurrent flow	逆流	逆流
countercurrent flow reactor	逆流反应器	逆流反應器
countercurrent jet condenser	逆流喷射式冷凝器	逆流噴凝器

英　文　名	祖国大陆名	台湾地区名
countercurrent network	逆流网络	逆流網路
countercurrent operation	逆流操作	逆流操作
countercurrent process	逆流过程	逆流程序
countercurrent washing	逆流洗涤	逆流洗滌
counter diffusion	逆扩散	逆向擴散
counter emf	反电动势	反電動勢
counterflow	逆流	逆流,逆向流動
counter range	计数器范围	計數器範圍
counter tube	计数管	計數管
couple	偶合	力偶,偶
coupled equations	偶合方程式	偶合方程式
coupled system	偶合系统	偶合系統
coupling	联管节	接頭偶合
coupling joint	[管]接头	接頭
cracked gas	裂化气	裂解氣
cracked gasoline	裂化汽油	裂解汽油
cracking	裂化	裂解,裂煉
cracking activity	裂化活性	裂解活性
cracking catalyst	裂化催化剂	裂解觸媒
cracking chamber	裂化室	裂解室
cracking distillation	裂化蒸馏	裂解蒸餾
cracking fractionator	裂化分馏塔	裂解分餾塔
cracking furnace	裂化炉	裂解爐
cracking reaction	裂化反应	裂解反應
cracking severity	裂化度	裂解度
cradle feeder	摇饲机	搖飼機
creep	蠕变	①潛變②蠕變
creeping flow	蠕流	蠕流
creeping motion	蠕变运动	蠕流運動
creep test	蠕变试验	潛變試驗
creosote	杂酚油	雜酚油
cresol	甲酚	甲酚
crevice corrosion	缝隙腐蚀	裂紋腐蝕
cricondenbar	临界凝析压力	臨界凝固壓
cricondentherm	临界凝析温度	臨界凝固溫度
cristobalite	方石英	白矽石
criterion	判据	準則
criterion for chemical reaction equilibrium	化学反应平衡判据	化學反應平衡準則

英　文　名	祖国大陆名	台湾地区名
critical back pressure ratio	临界反压比	臨界反壓比
critical compressibility factor	临界压缩因子	臨界壓縮因數
critical compressibility ratio	临界压缩比	臨界壓縮比
critical concentration	临界浓度	臨界濃度
critical condition	临界状态	臨界條件
critical constant	临界常数	臨界常數
critical damping	临界阻尼	臨界阻尼
critical density	临界密度	臨界密度
critical depth	临界深度	臨界深度
critical diameter	临界直径	臨界直徑
critical flow	临界流	臨界流動
critical flow capacity	临界流动量	臨界流動量
critical frequency	临界频率	臨界頻率
critical gain	临界增益	臨界增益
critical humidity	临界湿度	臨界濕度
critical isotherm	临界等温线	臨界等溫線
critical mass	临界质量	臨界質量
critical micelle concentration	临界胶束浓度	臨界微胞濃度
critical mixing temperature	临界混合温度	臨界混合溫度
critical moisture content	临界湿含量	臨界含水量
critical molar volume	临界摩尔体积	臨界莫耳體積
critical nucleus size	临界晶核尺寸	臨界晶核大小
critical packing size	临界填料大小	臨界填充大小
critical path	临界途径	臨界途徑
critical phenomenon	临界现象	臨界途徑法
critical point	临界点	臨界點
critical potential	临界势	臨界勢
critical pressure	临界压力	臨界壓力
critical pressure ratio	临界压力比	臨界壓力比
critical pressure relief device	临界泄压器件	臨界洩壓性質
critical property	临界性质	臨界性質
critical revolution	临界速率	臨界轉速
critical size	临界尺寸	臨界尺寸
critical slope	临界斜率	臨界斜率
critical solubility	临界溶度	臨界溶解度
critical solution point	临界溶解点	臨界溶解點
critical solution temperature	临界共溶温度	臨界溶解溫度
critical sonic property	临界音速性质	臨界音速性質
critical specific surface	临界比表面	臨界比表面

英　文　名	祖国大陆名	台湾地区名
critical specific volume	临界比容	臨界比容
critical speed	临界转速	臨界速率
critical temperature	临界温度	臨界溫度
critical temperature difference	临界温差	臨界溫差
critical throat velocity	临界喉道流速	臨界喉道流速
critical time	临界时间	臨界時間
critical value	临界值	臨界值
critical velocity	临界速度	臨界速度
critical volume	临界体积	臨界體積
crosscurrent	错流	橫流
crosscurrent extraction	错流萃取	橫流萃取
crosscurrent operation	错流操作	橫流操作
crossflow	错流	交叉流動, 交流
crossflow adsorber	错流吸附器	交叉流動吸附器
crossflow operation	错流操作	交流操作
cross linkage	交联	交联
cross-linked polymer	交联聚合物	交联聚合物
cross-linking	交联	交联
cross-linking agent	交联剂	交联劑
cross-linking density	交联密度	交联密度
crossover	交迭	交越
crossover frequency	交迭频率	交越頻率
crossover gain	交迭增益	交越增益
cross section	截面	橫切面, 截面
cross sectional area	截面积	截面積
crude	原油	原油
crude distillation	原油蒸馏	原油蒸餾
crude oil	原油	原油
crude resources	原油资源	原油資源
crush	破碎	壓碎
crusher	破碎机	壓碎機
cryogenic industry	低温工业	低溫工業
cryogenic process	深度冷冻	低溫程序
cryolite	冰晶石	冰晶石
crystal	晶体	晶體
crystal face	晶面	晶面
crystal form	晶形	晶形
crystal growth	晶体生长	晶體成長
crystal habit	晶体习性	晶體習性

英　文　名	祖国大陆名	台湾地区名
crystal lattice	晶格	晶格
crystalline polymer	结晶聚合物	晶質聚合物
crystalline silica	晶硅石	晶矽石
crystalline structure	晶形结构	晶形結構
crystalline transition	晶形转变	晶形轉變
crystallite	微晶	微晶
crystallization	结晶	結晶
crystallizer	结晶器	結晶器
crystallizing point	结晶温度	結晶點
crystallography	结晶学	結晶學
crystal nucleation	晶体成核	晶體成核
crystal seed	晶种	晶種
crystal shape	晶形	晶形
crystal size	晶体大小	晶體大小
crystal size distribution	晶体大小分布	晶體粒度大小
crystal system	晶系	晶系
crystal water	结晶水	結晶水
CSTR (＝continuous flow stirred tank reactor)	连续流动式搅拌反应器	連續流動式攪拌槽反應器
CSTR design	连续搅拌反应釜设计	連續流動攪拌槽反應器設計
CSTR's in series	重串连续搅拌反应釜组	串聯連續攪拌槽反應器組
cubic[al]expansion	立方膨胀	體膨脹
cubical expansion coefficient	立方膨胀系数	體脹係數
cubic average boiling point	立方平均沸点	立方平均沸點
cullet	碎玻璃	玻璃屑
cultivation	培养	培養
cultivator	培养器	培養器
culture	培养	培養菌, 培養種菌
cumene	异丙苯	異丙苯
cumene hydroperoxide	过氧化氢异丙苯	氫過氧化異丙苯
cumulative distribution	累积分布	累積分布
cumulative probability	累积概率	累積機率
cumulative screen analysis	累积筛析	累積篩析
cumulative weight distribution	累积重量分布	累積重量分布
cumulative weight percent oversize	累积重量百分率过粗部分	累積重量百分率過粗部分

英　文　名	祖国大陆名	台湾地区名
cumulative weight percent undersize	累积重量百分率过细部分	累積重量百分率過細部分
cumulative weight undersizer fraction	累积过细重量分率	累積過細重量分率
cup-mixing composition	杯混合平均组成	杯混合平均組成
cuprammonium rayon	铜铵人造纤维	銅氨縲縈
cuprite	赤铜矿	赤銅礦
cure meter	固化计	熟化計
curing test	熟化试验	硬化試驗
curing time	熟化时间	熟化時間
current asset	流动资产	流動資產
current meter	流速计	流速計
current ratio	流动比	流動比例
curvature	曲率	曲率
curve	曲线	曲線
curvilinear coordinates	曲线坐标	曲線座標
cushion plate	缓冲板	緩衝板
cut diameter	切面直径	切面直徑
cut-off frequency	截止频率	截止頻率
cutting oil	切削油	切削油
CVD (=chemical vapor deposition)	化学气相沉积	化學蒸氣沉積法
cycle length	周期时间	周期
cycle time	周期时间	周期
cyclic operation	周期操作	周期操作
cyclic process	循环过程	循環程序
cyclic reaction	周期反应	周期反應
cyclic reactor	周期式反应器	周期式反應器
cyclic stress	周期性应力	周期性應力
cycling	循环	循環
cyclization	环化	環化
cyclization reaction	环化反应	環化反應
cyclize rubber	环化橡胶	環化橡膠
cycloheximide	放线菌酮	環己醯亞胺素
cycloidal blower	摆旋鼓风机	擺旋鼓風機
cyclone	旋风器	旋風器
cyclone efficiency	旋风器效率	旋風器效率
cyclone evaporator	旋风蒸发器	旋風蒸發器
cyclone exhauster	旋风排气机	旋風排氣機
cyclone scrubber	旋风洗涤器	旋風洗氣器
cyclone separation	旋风分离	旋風分離

英　文　名	祖国大陆名	台湾地区名
cyclone separator	旋风分离器	旋風分離器
cycloserine	环丝氨酸	環絲胺酸
cylinder drier	圆筒干燥器	圓筒乾燥器
cylindrical coordinates	柱面坐标	圓柱座標
cytolysis	细胞溶解	細胞溶解

D

英　文　名	祖国大陆名	台湾地区名
damped oscillation	阻尼振荡	阻尼振盪
damped oscillator	阻尼振荡器	阻尼振盪器
damper	阻尼器	①阻尼器,制振器②氣閘③擋板
damping	阻尼	阻尼
damping coefficient	阻尼系数	阻尼係數
damping constant	阻尼常数	阻尼常數
damping fluid	阻尼流体	阻尼流體
damping force	阻尼力	阻尼力,制振力
damping ratio	阻尼比	阻尼比
dashpot	黏性制振器	黏性制振器
data	数据	數據
data analysis	数据分析	數據分析
data experimental	实验数据	實驗數據
data fitting	数据拟合	數據擬合
data inaccuracy	数据不准性	數據不準性
data log	数据日志	數據日誌
data processing	数据处理	數據處理
data reconciliation	数据调谐,数据校正	數據調和
data screening	数据筛选	數據篩選
datum	①基准②数据	①基準②數據
datum temperature	基准温度	基準溫度
daughter cell	子细胞	子細胞
daughter element	子元素	子元素
dB (=decibel)	分贝	分貝
D control (=derivative control)	微分控制	微分控制
DDT (=dichloro-diphenyl-trichloroethane)	滴滴涕	滴滴梯
deactivation	失活	去活化

英　文　名	祖国大陆名	台湾地区名
deactivation energy	减活化能	去活化能
deactivator	失活剂	去活化劑
dead band	死带	靜滯帶, 死帶
dead beat response	不摆应答	不攞應答
dead space	静滞区	靜滯區
dead state	死态	靜滯態
dead time	停滞时间	遲延時間, 延遲時間
deadwater region	死水区	死水區[域]
dead weight gage	静重仪	靜重儀
dead weight piston gage	静重活塞压力计	靜重活塞壓力計
dead zone	死区	靜滯區
deaeration	脱气	除氣
deaerator	脱气塔	除氣器
deaggregation	解聚集	去聚集[作用]
dealkylation	脱烷基化	脱烷, 去烷[反應]
deaminase	脱氨基酶	去胺酶
deamination	脱氨基	去胺[反應]
dearomatization	脱芳构化	去芳香化[反應]
deashed coal	去灰煤	去灰煤
deashing	去灰分	去灰分
deashing process	去灰分过程	去灰分程序
death phase	死亡期	死亡期
death rate	死亡速率	死亡速率
debug	检错	除錯
debutanizer	脱丁烷塔	去丁烷塔
decantation	倾析	傾析[法]
decanter	倾析器	傾析器
decarboxylation	脱羧	脱羧[反應]
decarburization	脱碳	去碳
decay	衰变	衰變, 衰退
decay-affected selectivity	衰变影响选择性	衰退影響之選擇性
decay constant	衰变常数	衰變常數
decaying catalyst	衰变催化剂	衰退觸媒
decay ratio	衰变比	衰退比, 衰變比
decentralized control	分散控制	分區控制
dechlorinator	脱氯塔	去氯器
decibel (dB)	分贝	分貝
decision making	决策	決策
decision making under risk	风险型决策	風險型決策

英 文 名	祖国大陆名	台湾地区名
decision making under uncertainty	不确定型决策	不確定型決策
decision tree	决策树	决策樹
decision variable	决策变量	决策變數
decline phase	衰亡期	衰減期
declining balance method	递减均衡法	遞減均衡法
decoking	除焦	除焦
decoking time	除焦时间	除焦時間
decolorization	脱色	脱色
decolorizer	脱色剂	脱色劑
decomposer	①分解器②分解剂	①分解器②分解劑
decomposition	分解	分解
decomposition-coordination method	分解协调法	分解協調法
decomposition efficiency	分解效率	分解效率
decomposition potential	分解电势	分解勢
decomposition pressure	分解压力	分解壓[力]
decomposition voltage	分解电压	分解電壓
decompressor	减压器	減壓器
decontaminant	①去污剂②净化剂	①去污劑②消毒劑
decontamination factor	去污指数, 净化指数	去污指數
decoupled loops	解偶环路	解偶環路, 去偶合環路
decoupled system	解偶系统	解偶系統, 去偶合系統
decoupler	解偶器	解偶器, 去偶合器
deep-well injection	深水井注入	深井注入[法]
deethanizer	脱乙烷塔	去乙烷塔
default value	缺陷值	隱含值
defecated juice	澄清汁	澄清汁
defecation	澄清	澄清
defecator	澄清器	澄清器
defect crystal	缺陷晶体	缺陷晶體
defect solid	缺陷固体	缺陷固體
deferred annuity	延期年金	延期年金
deflagration	爆燃	爆燃
deflagrator	爆燃器	爆燃器
deflection	偏离	偏轉, 偏向
deflection potentiometer	偏离电位计	偏轉電位計
deflection separator	折流分离器	折流分離器
deflocculant	抗絮凝剂	去絮凝劑
deflocculation	解絮凝, 反絮凝	去絮凝[作劑]
deflocculation agent	抗絮凝剂	去絮凝劑

英　文　名	祖国大陆名	台湾地区名
defluidization	反流态化	去流體化
defoaming agent	消泡剂	消泡劑
deformation	变形	變形
deformation coordinates	变形坐标	變形座標
deformation eutectic	变形共熔物	變形共熔物
deformation point	变形点	變形點
deformation range	变形范围	變形範圍
deformation rate	变形速率	變形速率
deformation work	变形功	變形功
degasification	脱气	除氣
degasifier	脱气器	除氣器
degeneracy	退化	退化
degeneration	退化	退化
degeneration form	退化形式	退化形式
degradable plastic	可降解塑料	可降解塑膠
degradation	降解	①降解②裂解③劣化
degrease	脱脂	脱脂
degree of consolidation	压实度	壓密度
degree of cross-linking	交联度	交聯度
degree of crystallization	结晶度	結晶度
degree of dispersion	分散度	分散度
degree of expansion	膨胀度	膨脹度
degree of freedom	自由度	自由度
degree of hardness	硬度	硬度
degree of interaction	相互作用度	相互作用度
degree of intermixing	互混度	互混度
degree of ionization	电离度	游離度
degree of mixing	混合程度	混合度
degree of penetration	渗入度	針入度
degree of polymerization	聚合度	聚合度
degree of refining	精炼度	精煉度
degree of reproducibility	再现度	再現度
degree of reversibility	可逆度	可逆度
degree of saturation	饱和度	飽和度
degree of segregation	离析度	隔離度
degree of stability	稳定度	穩定度, 安定度
degree of super saturation	过饱和度	過飽和度
degree of swelling	溶胀度	膨潤度
degree of turbulence	湍流度	紊流度

英 文 名	祖国大陆名	台湾地区名
dehalogenation	脱卤	脱鹵[反應]
dehumidification	减湿	除濕
dehumidifier	除湿器	除濕器
dehydrate	脱水物	脱水物
dehydrated alcohol	脱水酒精	脱水酒精
dehydration	脱水	脱水
dehydrator	脱水器	脱水器
dehydrocyclization	脱氢环化作用	脱氫環化[反應]
dehydrogenase	脱氢酶	去氫酶
dehydrogenation	脱氢	脱氫[反應]
dehydrogenation catalyst	脱氢催化剂	脱氫觸媒
deionization	去离子作用	去離子[作用]
deionized water	去离子水	去離子水
delaminated clay	层离黏土	剝層黏土
delamination	分层	剝層[作用]
delay	延迟	延遲, 遲延
delay action	延迟作用	遲延作用, 延遲作用
delay time	延迟时间	遲延時間, 延遲時間
delignification	去木素作用	去木質素
deliquescence	潮解	潮解
delusterant	去光剂	去光劑
delusting	去光	去光
demethanization	脱甲烷	去甲烷[反應]
demethanizer	甲烷馏除器	去甲烷塔
demineralization	去矿化	去礦[物]質
demineralized water	软化水	去礦[物]質水
demineralizer	软化器	去礦[物]質劑
demineral water	软化水	去礦[物]質水
demister	除雾器	除霧器
demonstration unit	示范装置	示範裝置
demulsification	反乳化	去乳化[作用]
denaturant	变性剂	變性劑
denaturation	变性	變性
denatured alcohol	变性酒精	變性酒精
denitration	脱硝	脱硝
denitration tower	脱硝塔	脱硝塔
denitrator	脱硝塔	脱硝塔
denitrification	反硝化作用	脱氮化[反應]
denitrifying bacteria	反硝化细菌	脱氮[化]菌

英　文　名	祖国大陆名	台湾地区名
denitrogenation	脱氮	脱氮[反應], 脱氮化[作用], 去氮
dense medium separation	密媒分离	濃媒分離
dense-phase	密相, 浓相	稠相
dense-phase flow	密相流动	稠相流動
dense-phase fluidized bed	密相流化床	稠相流體化床
dense soda ash	重苏打灰	重鹼灰
densimeter	密度计	密度計
densitometer	[感光]密度计	感光密度計
density	密度	密度
density gradient centrifugation	密度梯度离心	密度梯度離心
density index	密度指数	密度指數
deodorant	除臭剂	除臭劑
deodorization	除臭	除臭
deodorizer	除臭剂	除臭劑
deoxidation	脱氧	去氧化[反應]
deoxygenation	脱氧	去氧[反應]
departure function	偏离函数	偏離函數
dependent variable	因变量	因變數
depletion	耗尽	耗盡
depletion allowance	耗尽容许度	耗盡容許度
depolymerization	解聚	解聚合[作用]
depreciation	折旧	折舊, 贬值
depreciation cost	折旧费	折舊成本
depreciation reserve	折旧备用金	折舊準備金
depropagation reaction	负增长反应	去增長反應
derivative	①衍生物②导数	①衍生物②導數
derivative control (＝D control)	微分控制	微分控制
derivative dilatometry	导数膨胀计测定法	導數膨脹測定法
derivative thermogravimetric analysis (＝DTGA)	导数热重分析法	導數熱重分析法
derivative thermogravimetry	导数热重分析法	微分热重量法
derivative time	微分时间	微分時間
derivative time constant	微分时间常数	微分時間常數
derived unit	导出单位	導出單位
desalination	脱盐	淡化
desaturation	去饱和作用	去飽和作用
describing function	描述函数	描述函數
desiccant	干燥剂	乾燥劑

英　文　名	祖国大陆名	台湾地区名
desiccation apparatus	干燥器皿	乾燥器
desiccator	干燥器	乾燥器
design	设计	設計
design chart	设计图	設計圖
design criterion	设计准则	設計準則
design data	设计数据	設計數據
design engineer	设计工程师	設計工程師
design equation	设计方程式	設計方程式
design mode	设计模型	設計模型
design package	成套设计文件	成套設計文件
design parameter	设计参数	設計參數
design project	设计项目	設計專案
design standard	设计标准	設計標準
design tool	设计工具	設計工具
design variable	设计变量	設計變數
desired product	期望产物	期望產物
desired value	期望值	期望值
desolvation	去溶剂化	去溶劑[作用]
desolventizer	脱溶剂器	脫溶劑器
desorption	解吸	脫附
desorption control	脱附控制	脫附控制
desorption factor	解吸因子	脫附因子
desorption isotherm	脱附等温线	脫附等溫線
destabilization	去稳定作用	去安定[作用]
destructive distillation	分解蒸馏	分解蒸餾
desublimation	凝华作用	反昇華
desulfurization	脱硫	脫硫[反應]
desuperheat	去过热	去過熱
desuperheater	去过热器	去過熱器
detailed balancing	细致平衡	明細均衡
detailed engineering design	施工图设计	細部工程設計
detector	检测器	偵檢器
detention period	停滞期间	停留期間
detention time	停滞时间	停留時間
detergency	去垢力	清潔力
detergent	洗涤剂	清潔劑
determinant	行列式	行列式
determination	测定	測定
deterministic model	确定性模型	決定性模型

英　文　名	祖国大陆名	台湾地区名
detonation	起爆	起爆
detonation time	爆炸时间	爆炸時間
detoxification	解毒	去毒
detrition	磨耗	磨耗
developer	①显影剂②显色剂	①顯影劑②顯色劑
development project	开发项目	開發專案
deviation	偏差	偏差, 偏離
deviation form	偏差形式	偏差形式
deviatoric stress	偏应力	軸差應力, 偏向應力
device	装置	裝置
devitrification	脱玻化	①反玻化②結晶化③失透明[作用]
devulcanization	脱硫	去硫化
dewatering	脱水	脱水
dewaxing	脱蜡	脱蠟
dew point	露点	露點
dew point meter	露点计	露點計
dew point pressure	露点压力	露點壓力
dew point recorder	露点记录器	露點記錄器
dew point temperature	露点温度	露點溫度
dextran	葡聚糖	葡聚糖
dextrin	糊精	糊精
dextrose	葡萄糖	葡萄糖
diafiltration	渗滤	滲濾
dial indicator	刻度盘指示器	針盤指示計
dialysis	渗析, 透析	透析[作用]
dialysis cell	渗析槽	透析槽
dialysis culture	渗析培养, 透析培养	透析培養
dialyzate	渗析液, 透析液	透析液
dialyzator	渗析器, 透析器	透析儀
dialyzer	渗析器, 透析器	透析儀
diameter	直径	直徑
diaphragm	隔膜片	隔膜片
diaphragm box	隔膜盒	隔膜盒
diaphragm box level gauge	隔膜盒液位计	隔膜盒液位計
diaphragm cell	隔膜电解槽	隔膜[電解]槽
diaphragm chamber	隔膜室	隔膜室
diaphragm compressor	隔膜压缩机	隔膜壓縮機
diaphragm control valve	隔膜控制阀	隔膜控制閥

英 文 名	祖国大陆名	台湾地区名
diaphragm D/P cell	隔膜微压差	隔膜微壓差
diaphragm operated valve	隔膜操作阀	隔膜操作閥
diaphragm pressure gage	膜式压力计	隔膜壓力計
diaphragm pump	隔膜泵	隔膜泵
diaphragm screen	膜筛	隔膜篩
diaphragm-spring actuator	膜簧式驱动器	膜簧式引動器
diaphragm type pressure transducer	隔膜式压力转换器	隔膜壓力轉換器
diaphragm type strain gage	隔膜式应变计	隔膜應變計
diaphragm valve	隔膜阀	隔膜閥
diaspore	硬水铝石	水鋁石
diastase	淀粉糖化酶	澱粉酶
diathermal wall	透热壁	無熱阻壁
diatomaceous earth	硅藻土	矽藻土
diatomaceous-earth filter	硅藻土过滤器	矽藻土過濾器
diatomite	硅藻土	矽藻土
diazo compound	重氮化合物	重氮化合物
diazo dye	重氮染料	重氮染料
diazotization	重氮化	重氮化[反應]
dichloro-diphenyl-trichloroethane (DDT)	滴滴涕	滴滴梯
dichotomous search	对分搜索	二分法搜尋
die	模	模
die casting	压铸	壓鑄, 模鑄
die casting molding	压铸成型	壓鑄成型
dielectric	介电体	介電
dielectric absorption	介电吸收	介電吸收
dielectric constant	介电常数	介電常數
dielectric heating	介电加热	介電加熱
diesel engine	柴油机	柴油機
diesel fuel	柴油	柴油燃料
diesel knock	柴油爆震	柴油震爆
die swell	离模膨胀	模頭膨脹
difference equation	差分方程	差分方程式
differential absorption	差示吸收	微分吸收
differential adsorption	差示吸附	微分吸附
differential analyzer	差示分析仪	微分分析橨
differential centrifugation	差速离心[分离]	差速離心分離
differential condensation	微分冷凝	微分冷凝
differential contact equipment	微差接触设备	微差接觸設備
differential control	微分控制	微差控制

英 文 名	祖国大陆名	台湾地区名
differential dilatometry	示差热膨胀测量术	微差膨脹測定法
differential distillation	微分蒸馏	微分蒸餾
differential energy balance	微分能量平衡	微分能量均衡
differential equation	微分方程	微分方程式
differential gap	微差间隙	微差間隙
differential head	差异压头	微差高差
differential heat of dilution	微分稀释热	微分稀釋熱
differential heat of solution	微分溶解热	微分溶解熱
differential manometer	差示压力计	微差壓力計
differential mass balance	微分质量平衡	微分質量均衡
differential method	微分法	微分法
differential pressure	压[力]差	微差壓力, 差壓
differential pressure cell	差压池	微壓差計
differential pressure limit	差压界限	微差壓力極限
differential pressure transducer	差压传感器	微差壓力轉換器
differential reactor	微分反应器	微分反應器
differential scanning calorimetry	差示扫描量热法	差示掃描量熱法
differential screen analysis	差示筛析	微分篩析
differential settling	微分沉降	微分沉降
differential surface	微分表面	微分表面
differential technique	微分法	微分法
differential thermal analysis	差热分析法	微差熱分析
differential thermogravimetric analysis	差示热重分析法	差示熱重分析法
differential thermomechanical analysis	差示热机械分析法	差示熱機械分析法
differential thermometer	微差温度计	微差溫度計
differential thermometry	微差测温法	微差測溫法
differential U-tube	差示 U 形管	微差 U[形]管
differential valve	差动阀	差壓閥
differential weight distribution	微分重量分布	微差重量分布
differential yield	微分产率	微分產率
differentiator	微分器	微分器
diffraction	衍射, 绕射	繞射
diffuse layer	扩散层	擴散層
diffuser	扩散器	擴散器
diffuser casing	扩散器护罩	擴散器護罩
diffusion	扩散	擴散
diffusion barrier	扩散膜	擴散障壁
diffusion catalysis	扩散催化	擴散催化[作用]
diffusion coefficient	扩散系数	擴散係數

英　文　名	祖国大陆名	台湾地区名
diffusion constant	扩散常数	擴散常數
diffusion control	扩散控制	擴散控制
diffusion-controlled reaction	扩散控制反应	擴散控制反應
diffusion current	扩散电流	擴散電流
diffusion equation	扩散方程	擴散方程式
diffusion film	扩散薄膜	擴散薄膜
diffusion flux	扩散通量	擴散通量
diffusion mass transfer	扩散传质	擴散質傳
diffusion potential	扩散电位	擴散勢
diffusion pressure	扩散压力	擴散壓
diffusion process	扩散过程	擴散程序
diffusion pump	扩散泵	擴散泵
diffusion rate	扩散速率	擴散速率
diffusion resistance	扩散阻力	擴散阻力
diffusion ring	扩散环	擴散環
diffusion-thermo effect	扩散热效应	擴散熱效應
diffusion time	扩散时间	擴散時間
diffusion vacuum pump	扩散真空泵	擴散真空泵
diffusivity	扩散系数	擴散係數
digester	①蒸煮器②消化槽	①蒸煮器②消化槽
digestion	消化	消化
digital computer	数字计算机	數位計算機
digital control	数字控制	數位控制
digital recorder	数字记录器	數位記錄器
diglyceride	甘油二酯	二甘油酯
digraph	有向图	有向圖
dilatancy	胀塑性	膨脹性
dilatant fluid	胀塑性流体	膨脹性流體
dilatation	膨胀	膨脹
dilatational stress	膨胀应力	膨脹應力
dilation	膨胀	膨脹
dilatometer	膨胀计	膨脹計
dilatometry	膨胀测定法	膨脹測定法
diluent	稀释剂	稀釋劑
dilute phase	稀相	稀相
dilute-phase fluidized bed	稀相流化床	稀相流體化床
dilution	稀释	稀釋
dilution effect	稀释效应	稀釋效應
dimensional analysis	量纲分析, 因次分析	因次分析

英　文　名	祖国大陆名	台湾地区名
dimensional consistency	量纲一致性	因次一致性
dimensional homogeneity	量纲均匀性	因次均匀性
dimensional invariance	量纲不变性	因次不變性
dimensional stability	尺寸稳定性	尺寸穩定性
dimensionless group	无量纲数群	無因次群
dimensionless number	无量纲数	無因次數
dimensionless parameter	无量纲参数	無因次參數
dimerization	二聚	二聚合作用
diode	二极管	二極體
diode function generator	二极函数发生器	雙極函數產生器
dip bleaching	浸漂	浸漂
dip dyeing	浸染	浸染
dipeptidase	二肽酶	二肽酶
dip glazing	浸渍釉	浸釉
dipleg	料腿	料腿
dipole	偶极	偶極
dipole moment	偶极矩	偶極矩
dipping	浸渍, 浸泡	浸漬
dipping tank	浸渍槽	浸漬槽
dip pipe level gage	浸管液位计	浸管液位計
dip tinning	浸渍镀锡	浸漬鍍錫
dip tube	倾斜管	傾斜管
direct-acting controller	直接作用控制器	直接作用控制器
direct coal liquefaction	直接煤液化	直接煤液化[法]
direct control	直接控制	直接控制
direct cost	直接成本	直接成本
direct digital control	直接数字控制	直接數位控制
direct dye	直接染料	直接染料
direct-fired furnace	直接火焰炉	直接火加熱爐
direct gate	直接浇口	直接澆口
direct hydrogenation	直接氢化	直接氫化[反應]
direct liquefaction	直接液化	直接液化
direct method	直接法	直接法
direct production cost	直接生产成本	直接生產成本
direct pyrolysis	直接热解	直接高溫分解
direct reduction	直接还原	直接還原[法]
direct search method	直接搜索法	直接搜尋法
direct substitution	直接代入法	直接代入法
disaccharide	双糖	雙醣

英　文　名	祖国大陆名	台湾地区名
disc attrition mill	圆盘磨	盤式研磨
disc centrifuge	盘式离心机	盤式離心機
disc column	圆盘塔	盤式塔
disc crusher	盘式压碎机	盤碎機
disc evaporator	盘式蒸发器	盤式蒸發器
disc filter	盘滤机	盤濾機
disc flow condenser	圆盘流动式冷凝器	圆盤流動式冷凝器
discharge coefficient	①流量系数②孔流系数	排放係數
discharge head	排出压头	排放高差
discharge header	排放集管	排放集管[箱]
discharge loss	卸出损失	洩流損失, 排放損失
discharge pressure	排放压力	排放壓力
discharge time	卸料时间	排放時間
discharge valve	排出阀	排放閥
disc meter	盘式流量计	盤式流量計
discontinuous process	间歇操作法	不連續程序
discontinuous processing	间歇加工	不連續加工[法]
discount breakeven point	折现盈亏平衡点	折現損益均衡點
discounted cash flow rate of return	折现收益率	折現收益率
discount factor	折现因子	折現因數
discrete settling	个别沉降	個別沉降
disinfectant	消毒剂	消毒劑
disinfection	消毒	消毒
disintegration mechanism	衰变机理	蛻變機構
disintegrator	破碎机	散解機
disk	圆盘	圓盤
disk column absorber	盘柱吸收器	盤柱吸收器
disk dryer	圆盘干燥器	盤式乾燥器
disk feeder	圆盘给料机	盤飼機
disk filter	盘式过滤机	盤濾機
disk gate	圆盘浇口	圓盤澆口
dislocation	转位	移位
dispersant	分散剂	分散劑
dispersed-air flotation	曝气浮选	分散空氣浮選[法]
dispersed dye	分散[性]染料	分散染料
dispersed flow	分散流, 弥散流	分散流
dispersed medium	分散介质	分散介質
dispersed phase	分散相	分散相
dispersed slug flow	分散节涌流	分散塞流

英　文　名	祖国大陆名	台湾地区名
dispersed system	分散系统	分散系统
disperse phase	分散相	分散相
dispersibility	分散性	分散性
dispersion	分散	分散
dispersion coefficient	弥散系数	分散係數
dispersion curve	色散曲线	分散曲線
dispersion force	色散力	分散力
dispersion mill	分散磨	分散磨
dispersion model	弥散模型	分散模型
dispersion reagent	分散剂	分散劑
dispersive power	分散本领	分散能力
displacement	置换, 排代	取代, 置換
displacement float level gage	移标式液位计	移標式液位計
displacement flowmeter	位移流量计	位移流量計
displacement meter	位移计	位移[流量]計
displacement pressure gage	位移压力计	位移壓力計
displacement reaction	置换反应	取代反應, 置換反應
disposal	处置	①排列②配置③處理, 處置
dissimilation	异化	異化
dissipated energy	耗散能	散逸能
dissipation	耗散	散逸
dissipation function	耗散函数	散逸函數
dissociation temperature	解离温度	解離溫度
dissociative adsorption	解离吸附	解離吸附
dissolution	溶解	溶解
dissolved air	溶解空气	溶解空氣
dissolved-air flotation	溶气浮选	溶解空氣浮選[法]
dissolved oxygen	溶解氧	溶氧
dissolved oxygen analyzer	溶氧分析仪	溶氧分析儀
dissolved oxygen probe	溶氧探头	溶氧探針
dissolved solid	溶解固体	溶解固體
dissolver	溶解器	溶解器
distance factor	距离因数	距離因數
distance velocity lag	距速滞后	距離速度落後
distillate	馏出液	餾出物
distillating apparatus	蒸馏器	蒸餾器
distillation	蒸馏	蒸餾
distillation column	蒸馏塔	蒸餾塔

英　文　名	祖国大陆名	台湾地区名
distillation range	馏程	蒸餾範圍
distillation still	蒸馏釜	蒸餾鍋
distillation test	蒸馏试验	蒸餾試驗
distillation trap	蒸馏阱	蒸餾閘
distillation tray	蒸馏塔板	蒸餾板
distillation value	蒸馏值	蒸餾值
distillation with chemical reaction	反应蒸馏	反應蒸餾
distillation zone	蒸馏区	蒸餾區
distilled water	蒸馏水	蒸餾水
distiller	蒸馏器	蒸餾器
distilling apparatus	蒸馏器	蒸餾器
distilling condenser	蒸馏冷凝器	蒸餾冷凝器
distilling flask	蒸馏瓶	蒸餾瓶
distilling range	馏程	蒸餾範圍
distributed component	分布成分	分布成分
distributed control system	集散控制系统	分布控制系統
distributed lag	分布滞后	分布落後
distributed-parameter model	分布参数模型	分布參數模型
distributed-parameter system	分布参数系统	分布參數系統
distributed property	分布性质	分布性質
distributed system	分布系统	分布系統
distribution coefficient	分配系数	分布係數
distribution cost	分销成本	分銷成本
distribution curve	分布曲线	分布曲線
distribution function	分布函数	分布函數
distribution law	分配定律	分配定律
distribution plate	分布板	分配板
distribution ratio	分配比	分配比
distributive law	分配律	分配律
distributor	①分布器②配电器	①分布器②分配器
distributor arm	分布器臂	分配器臂
disturbance	扰动	擾動
diversion valve	转向阀	轉向閥
dividing surface	分界表面,界面相	界面相
Dixon ring	θ网环	網環
dolomite	白云石	白雲石
dolomite cement	白云石水泥	白雲石水泥
dolomite earthenware	白云石陶器	白雲石陶器
dolomitic lime	白云石石灰	白雲石石灰

英　文　名	祖国大陆名	台湾地区名
dolomitic limestone	白云石灰石	白雲石灰石
dominant eigenvalue	主要本征值	主要特性值
dominant pole	主导极点	主要極點
dominant root	主根	主根
Donor	给体	給予體, 施體
Dopant	掺杂剂	摻雜劑
Dope	掺入	摻雜
Doping	掺杂	摻雜
Dosage	剂量	劑量
Dose	剂量	劑量
Dosing	用剂	用劑
double acting pump	双动泵	雙動泵
double arm kneading mixer	双臂捏合机	雙臂捏合機
double column	双重塔	雙重塔
double cone blender	双锥掺合机	雙錐摻合機
double cone classifier	双锥分级机	雙錐類析器
double declining balance method	双倍定率递减折旧法	雙倍定率遞減折舊法
double decomposition	复分解	複分解
double distilled water	再蒸馏水	再蒸餾水
double effect evaporation	双效蒸发	雙效蒸發
double effect evaporator	双效蒸发器	雙效蒸發器
double entry book keeping	复式簿记	複式簿記
double filter	双式过滤器	雙式濾器
double-glazed insulating pane	双层隔热玻璃板	雙層隔熱玻璃板
double helical ribbon mixer	双螺带混合机	雙臂帶混合機
double motion agitator	双动搅拌器	雙動攪拌器
double paddle mixer	双桨混合器	雙漿混合器
double pipe cooler	套管冷却器	套管冷卻器
double-pipe heat exchanger	套管热交换器	套管熱交換器
double pipe reactor	套管反应器	套管反應器
double ported valve	双口阀	雙口閥
double promoted catalyst	双重促进催化剂	雙重促進化觸媒
double roll crusher	双辊轧碎机	雙輥軋碎機
double seat valve	双座阀	雙座閥
double sizing	双重上胶	雙重上膠
double slide wire bridge	双滑线电桥	雙滑線電橋
double solvent extraction	双溶剂萃取	雙溶劑萃取
double suction impeller	双吸叶轮机	雙吸取葉輪機
double superphosphate	富过磷酸钙	重過磷酸鈣

英 文 名	祖国大陆名	台湾地区名
doubling calendar	重合研光机	重合壓延機
doubling time	倍增时间	倍增時間
dough mixer	面团混合机	調麵機
downcomer	①降液管②落水管③溢流管	①降流管②下向流管
downcomer backup	降液管液柱高度	降液管液柱高度
downdraft kiln	倒焰窑	倒焰窯
downflow	降液	下向流
downflow reactor	下行式反应器	下向流反應器
down-pipe	降液管	下向流管
downstream	下游	下游
downstream pressure	下游压力	下游壓力
downstream processing	下游处理	下游處理
draft fan	通风扇	通風扇
draft gage	①通风计②差式风压计	①通風計②差式壓力計
draft tube	导流筒	通風管,流通管
draft tube baffle	导流筒挡板	通風管擋板
draft-tube-baffled crystallizer	导流筒挡板结晶器,DTB 结晶器	導流擋板結晶器
draft tube mixer	导流筒混合器	流通管混合器
drag coefficient	阻力系数	阻力係數
drag effect	曳引效应	阻力效應
drag force	曳力	①阻力②拉力
drag reduction	减阻	阻力降低
drag scraper	拖铲	拖刮機
drag torque type viscosimeter	曳力矩式黏度计	拖扭矩式黏度計
draining table	排水台	排水檯
drain line	排水管道	排洩管線
drain pipe	排泄管	排洩管
drain pump	疏水泵	排洩泵
drain trap	排水阱	排洩閘
drain valve	放泄阀	排洩閥
draw bar	拉杆	拉引棒
draw gang	切工组	切工組
drawing apparatus	绘图仪	製圖儀,繪圖儀
drawing paper	绘图纸	繪圖紙
drawoff cock	排出栓	排出栓
drier	①干燥器②催干剂	①乾燥器,乾燥機②乾燥劑

英　文　名	祖国大陆名	台湾地区名
drier bin	干燥箱	乾燥箱
drier cabinet	干燥橱	乾燥櫥
drift	漂移	漂移
drip	滴器口	滴口
drip condenser	水淋冷凝器	滴水冷凝器
drip dry	滴干法	滴乾
driver	驱动器	推進器
driver horsepower	驱动功率	驅動功率
driving force	推动力	驅動力
driving function	驱动函数	驅動函數
drop	①滴②差降	①滴②差降
drop breakup	液滴瓦解	液滴瓦解
droplet	微滴	小滴
drop phase	滴相	滴相
dropwise condensation	滴状冷凝	滴式冷凝
drum	筒	桶, 鼓
drum barker	桶式去皮机	桶式去皮機
drum boiler	桶式锅炉	桶式鍋爐
drum drier	鼓式干燥器	桶式乾燥器
drum filter	鼓型过滤机	桶式過濾器
drum separator	鼓式分离器	桶式分離器
drum sifting	转筒筛	圓筒篩選
drum type counter	鼓式计数器	鼓式計數器
drum type recorder	鼓式记录器	鼓式記錄器
drum type scale indicator	鼓式标度指示计	鼓式標[度指]示計
drum washing	转筒洗涤	轉筒洗滌
dry and wet thermometer	干湿球温度计	乾濕球溫度計
dry bulb temperature	干球温度	乾球溫度
dry bulb thermometer	干球温度计	乾球溫度計
dry cell	干电池	乾電池
dry combustion	干式燃烧	乾式燃燒
dry distillation	干馏	乾餾
dryer	①干燥器②干燥剂	①乾燥器, 乾燥機②乾燥劑
dry heat sterilization	干热灭菌	乾燥滅菌
drying	干燥	乾燥
drying agent	干燥剂	乾燥劑
drying cabinet	干燥橱	乾燥櫥
drying drum	干燥转鼓	乾燥桶

英 文 名	祖国大陆名	台湾地区名
drying filter	干滤器	乾式過濾器
drying hack	干燥架	乾燥架
drying house	干燥房	乾燥房
drying kiln	干燥窑	烘窯
drying oil	干性油	乾性油
drying oven	烘箱	乾燥烘箱
drying rate	干燥速率	乾燥速率
drying-rate curve	干燥速率曲线	乾燥速率曲線
drying room	干燥室	乾燥室
drying shed	干燥棚	乾燥棚
drying shrinkage	干燥收缩	乾燥收縮
drying stove	干燥炉	乾燥爐
drying tower	干燥塔	乾燥塔
drying tray	干燥盘	乾燥盤
drying tunnel	烘道	乾燥隧道
drying zone	干燥区	乾燥區
dry oxidation	干式氧化	乾式氧化法
dry process	干燥过程	乾法
dry spinning	干纺	乾紡[絲]
dry steam	干蒸汽	乾蒸汽
dry surface	干燥面	乾燥面
dual-flow tray	穿流塔板	雙流塔板
dual-function catalyst	双功能催化剂	雙功能觸媒
duality	对偶[性]	對偶
dual-medium filter	双层过滤器	雙介質過濾器
dual-mode control	双式控制	雙式控制
dual-mode system	双模式系统	雙模式系統
dual recirculation	双重循环	雙重循環
dual-site mechanism	双部位机理	雙部位機構
duct	导管	導管
ductile material	延性物料	延展性材料
dummy activity	虚拟活性	虛擬活性
dummy plate	隔板	隔板
dummy variable	哑变数	啞變數
dumped packing	散装填料	堆積填充
dumper	卸料器	傾卸車
duplex control	双用控制	雙用控制
duplex reciprocating pump	双缸往复泵	雙缸往復泵
duriron	杜里龙耐酸硅铁	高矽鐵

英　文　名	祖国大陆名	台湾地区名
durometer	硬度计	硬度計
dust	粉尘	塵, 灰塵, 粉塵
dust collection	集尘	集塵
dust collector	集尘器	集塵器
dust cyclone	旋风除尘器	粉塵旋風器
dust cyclone separator	粉尘旋风分离器	粉塵旋風分離器
dust equipment	粉尘装置	粉塵裝置
dustiness	含尘量	含塵量
dusting agent	隔离剂	防黏著劑
dust-laden gas	含尘气体	含塵氣體
dust particle	粉尘粒子	粉塵粒子
dust particle size	粉尘粒子大小	粉塵粒子大小
dust separator	粉尘分离器	粉塵分離器
duty factor	负载因数	負載因數
dye	染料	染料
dyeability	可染性	可染性
dye affinity chromatography	染料亲和色谱[法]	染料親和色層法
dyeing	染色	染色
dyeing machine	染色机	染色機
dye uptake method	染料摄入法	染料攝入法
dynamical equation	动力方程	動態方程式
dynamic analysis	动态分析	動態分析
dynamic behavior	动态行为	動態行為
dynamic compensation	动态补偿	動態補償
dynamic decoupler	动态去偶器	動態解偶器
dynamic element	动态元素	動態元素
dynamic equilibrium	动态平衡	動態平衡
dynamic error	动态误差	動態誤差
dynamic error coefficient	动态误差系数	動態誤差係數
dynamic lag	动态滞后	動態落後
dynamic model	动态模型	動態模式
dynamic modulus	动态模量	動態模數
dynamic optimization	动态优化	動態最適化
dynamic pressure	动压强	動壓[力]
dynamic process	动态过程	動態程序
dynamic programming	动态规划	動態規劃
dynamic reflectance spectroscopy	动态反射光谱[法]	動態反射光譜學, 動態反射光譜法
dynamic response	动态响应	動態應答

英 文 名	祖国大陆名	台湾地区名
dynamic rigidity	动态刚性	動態剛性
dynamics	动态学	①動態②動態學③動力學
dynamic similarity	动态相似	動態相似性
dynamic simulation	动态模拟	動態模擬
dynamic system	动态系统	動態系統
dynamic test	动态试验	動態試驗
dynamic themogravimetry	动态热重量分析法	動態熱重量法
dynamic thermomechanometry	动态热机械法	動態熱機械法
dynamic viscosity	动力黏度	動態黏度
dynamite	黄色炸药	代納邁炸藥

E

英 文 名	祖国大陆名	台湾地区名
earning rate	收益率	收益率
earthenware	陶器	陶器,土器
earthenware clay	陶土	陶土
earth pigment	土质颜料	土質顏料
earth wax	地蜡	地蠟
ebonite	硬质胶	硬橡膠
ebonite wax	乌木蜡	烏木蠟
ebony wax	乌木蜡	烏木蠟
ebullated bed reactor	沸腾床反应器	沸騰床反應器
ebullating bed	沸腾床	沸騰床
ebullating bed reactor	沸腾床反应器	沸騰床反應器
ebullition	沸腾	沸騰
eccentricity	偏心度	偏心率
eccentric orifice plate	偏心孔口板	偏心孔口板
eccentric reducer	偏心渐缩管	偏心漸縮管
eccentric ring	偏心环	偏心環
eccentric scale	偏心秤	偏心儀
economic evaluation	经济评价	經濟評估
economic feasibility	经济可行性	經濟可行性
economic impact	经济冲击	經濟衝擊
economic life	经济寿命	經濟壽命
economic pressure drop	经济压力降	經濟壓力降
economic reflux ratio	经济回流比	經濟回流比

英　文　名	祖国大陆名	台湾地区名
economizer	节能器	省熱器
ecosystem	生态系统	生態系統
E curve	E 曲线	E 曲線
eddy	[旋]涡	渦流
eddy absorption	涡流吸收	渦流吸收
eddy current tachometer	涡流转速计	渦流轉速計
eddy diffusion	涡流扩散	渦流擴散
eddy diffusivity	涡流扩散系数	渦流擴散係數
eddy flow	涡流	渦流
eddy kinematic viscosity	涡流运动黏度	渦流動黏度
eddy length	涡流长度	渦流長度
eddy loss	涡流损耗	渦流損失
eddy motion	涡流运动	渦流運動
eddy residence time	涡流停留时间	渦流滯留時間
eddy resistance	涡流阻力	渦流阻力
eddy shedding	涡流卸掉	渦流分離
eddy stress	涡流应力	渦流應力
eddy transport	涡流传递	渦流輸送
eddy viscosity	涡流黏度	渦流黏度
edge effect	边缘效应	邊緣效應
edge runner	轮碾机	輪輾機
edible fat	食用脂肪	食用脂肪
edible tallow	食用牛脂	食用牛脂
E distribution	E 分布	E 分布, 滯留時間分布
educt	浸提物	析出物
eductor	喷射器	引入器
effective aeration	有效曝气	有效通氣
effective concentration	有效浓度	有效濃度
effective density	有效密度, 修正密度	有效密度
effective diffusivity	有效扩散系数	有效擴散係數
effective grain size	有效粒度	有效顆粒大小
effective interest	有效利息	有效利息
effectiveness	有效性	有效度, 有效性
effectiveness coefficient	有效系数	有效係數
effectiveness factor	有效因子	有效因子
effective packing	紧束效应	緊束效應, 質量效應
effective particle diameter	有效颗粒直径	顆粒直徑
effective pore radius	有效孔半径	有效細孔半徑
effective power	有效功率	有效功率

英 文 名	祖国大陆名	台湾地区名
effective surface	有效表面	有效表面
effective thermal conductivity	有效导热系数	有效導熱係數
effective value	有效值	有效值
effective volatility	有效挥发度	有效揮發度
efficiency	效率	效率
efflorescence	风化	風化
effluence	流出	流出
effluent	流出物	流出物
effluent pipe	流出管	①放流管②出料管
effluent standard	排放标准	放流[水]標準
efflux	射流	①射流, 流出②流出量
efflux coefficient	射流系数	射流係數
efflux time	射流时间	射流時間
efflux velocity	射流速度	射流速度
efflux viscometer	流出式黏度计	射流黏度計
eigenfunction	特征函数	特徵函數
eigenvalue	特征值	特徵值
ejector	喷射器	噴射器
ejector arrangement	喷射器排列	噴射器排列
ejector capacity	喷射器容量	噴射器容量
ejector sump	喷射器油箱	射出器機油箱
elastic capsule	弹性胶囊	彈性膠囊
elastic energy	弹性能	彈性能
elastic fiber	弹性纤维	彈性纖維
elasticity	弹性	彈性
elasticity number	弹性数	彈性數
elasticity test	弹性试验	彈性試驗
elastic liquid	弹性液体	彈性液體
elastic modulus	弹性模量	彈性模數
elastic sulphur	弹性硫	彈性硫
elastomer	弹性体	彈性體
elastomer blend	弹性体共混物	彈性體摻合物
elastoplastic	碳塑性体	彈性體
elastopolymer	弹性聚合物	彈性聚合物
elbow	弯头	①彎頭②肘管
elbow nozzle	弯头喷嘴	彎頭噴嘴
elbow pipe	肘形管	肘形管
electric actuator	电动动器	電動引動器
electric analogy	电类比	電類比

英 文 名	祖国大陆名	台湾地区名
electric chemical potential	电化学势	電化學勢
electric conductivity	①电导率②电导性	①導電係數②電傳導性
electric conductivity coefficient	电导系数	導電係數
electric control	电动控制	電動控制
electric controller	电控制器	電動式控制器
electric energy	电能	電能
electric field	电场	電場
electric-impedance heated reactor	电热反应器	電熱反應器
electric installation	电气设施	電力裝置
electric osmosis	电渗现象	電滲透
electric pipe precipitator	电气管道除尘器	電管集塵器
electric plate precipitator	电极板除尘器	電板集塵器
electric porcelain	电瓷	絕電瓷
electric resistance pyrometer	电阻高温计	電阻高溫計
electric signal	电动信号	電動信號
electric transmission	电动传送	電動傳送
electric transmitter	电动传送器	電動傳送器
electric type flowmeter	电子流量计	電子流量計
electroaffinity	电亲和性	電親和性
electroanalysis	电解分析	電分析
electrobalance	电动天平	電動天平
electrocapillarity	电毛细管现象	電毛細管現象
electrocasting	①电涂装②电涂层	電鑄
electrocast mullite	电铸莫来石	電鑄富鋁紅柱石
electrochemical equivalent	电化当量	電化學當量
electrochemical industry	电化学工业	電化學工業
electrochemical reaction	电化学反应	電化學反應
electrochemical reaction engineering	电化学反应工程	電化學反應工程
electrochemical reactor	电化学反应器	電化學反應器
electrochemistry	电化学	電化學
electrochlorination	电氯化反应	電氯化[反應]
electrochromatography	电色谱	電色層分析
electrochromic display	电致变色显示	電色顯示
electrocleaning	电清洗	電解清洗
electrocoating	电涂	電塗, 電塗層
electroconductive polymer	导电高分子	導電高分子
electrocratic dispersion	电稳分散作用	電穩分散[作用], 電穩分散液
electrodeposit	电沉积物	電積層[法]

英　文　名	祖国大陆名	台湾地区名
electrodeposition	电沉积	電沉積[法]
electrodialysis	电渗析	電透析[作用]
electrodialysis process	电渗析过程	電透析法
electrodispersion	电分散	電分散
electroflotation	电浮选法	電浮選[法]
electrofluorination	电解氟化	電氟化[反應]
electroforming	电铸	電鑄
electrofusion	电融合	電融合
electrohydraulic actuator	电动液压引动器	電動液壓[式]引動器
electrohydrodimerization	电解加氢二聚合	電解加氫二聚合[反應]
electrokinetic phenomenon	动电现象	電動現象
electrokinetic potential	动电势	電動位
electrokinetics	动电力学	電動力學
electroless coating	无电涂	無電塗,無電塗裝
electroless plating	无电镀	無電鍍
electroluminescence	电致发光	電發光
electrolysis	电解	電解
electrolysis bath	电解槽	電解槽
electrolyte	电解质	電解質
electrolytic cell	电解池	電解槽
electrolytic conductance meter	导电计	導電計
electrolytic industry	电解工业	電解工業
electrolytic oxidation	电解氧化	電解氧化
electrolytic process	电解过程	電解法
electrolytic separation	电解分离	電解分離
electrolyzer	电解槽	電解槽
electromagnetic flowmeter	电磁流量计	電磁流量計
electromagnetic radiation	电磁辐射	電磁輻射
electromagnetic separation	电磁分离[法]	電磁分離[法]
electromagnetic separator	电磁离析器	電磁分離器
electromagnetic spectrum	电磁谱	電磁[波]譜
electrometallurgy	电冶金	電冶金學
electromigration	电迁移	電移
electromotive force (EMF)	电动势	電動勢
electromotive series	电动序	電動勢序
electronic amplifier	电子放大器	電子放大器
electronic computer	电子计算机	電子計算機
electronic control	电子控制	電子控制
electronic flowmeter	电子流量计	電子流量計

英　文　名	祖国大陆名	台湾地区名
electronic material	电子材料	電子材料
electronic partition function	电子配分函数	電子分配函數
electronic signal	电子讯号	電子訊號
electronic simulator	电子模拟器	電子模擬器
electroosmosis	电渗	電滲透
electrophoresis	电泳	電泳
electroplating	电镀	電鍍
electroplating engineering	电镀工程	電鍍工程
electropneumatic actuator	电动-气动引动器	電動氣動[式]引動器
electrostatic attraction	静电吸引	靜電吸引
electrostatic effect	静电效应	靜電效應
electrostatic precipitator	静电沉降器	靜電集塵器
electrostatic separation	静电分离	靜電分離
electrostatic separator	静电分离器	靜電分離器
electrothermal process	电热过程	電熱法
electroviscous effect	电黏效应	電黏性效應
element	[微]元	①元素②元件③要素
elementary reaction	元反应	基本反應
element primary	主要元素	主要元素
elimination reaction	消除反应	脫離反應
elongational flow	拉伸流动	伸長流動
elongational viscosity	拉伸黏度	伸長黏度
elongation rate	拉伸速率	伸長速率
eluant	洗提液	洗提液, 洗滌液, 洗析液
eluate	洗出液	洗出液
elution	洗脱	洗提, 洗析
elution chromatography	洗脱色谱法	洗析層析[法]
elutriation	扬析	淘析
elutriation constant	扬析常数	淘析常數
elutriator	淘析器	淘析器
emanation thermal analysis	射气热分析	放射性熱分析[法]
embossing	压花	壓紋
embossing calender	压花压光机	壓紋機
embossing machine	压花机	壓紋機
embryo	胚胎	胚胎
emerald	纯绿宝石	純綠寶石, 祖母綠, 翡翠
emerald green	翡翠绿	翡翠綠
emergency operation	应急操作	緊急操作
emery	金刚砂	剛砂

英　文　名	祖国大陆名	台湾地区名
emery cloth	砂布	砂布
emery wheel	砂轮	砂輪
EMF(＝electromotive force)	电动势	電動勢
emission inventory	排放物清单	排放物清單
emission spectrometer	发射光谱仪	發射分光計
emission spectrum	发射光谱	發射光譜
emission standard	排放标准	排放標準
emissive power	发射能力	發射能力
emissivity	发射率	發射率
emitter	发射体	發射體
empirical equation	经验方程式	實驗方程式,經驗方程式
empirical formula	经验式	經驗式
empirical method	经验法	經驗法
empirical model	经验模型	經驗模型
empirical rule	经验法则	經驗法則
emulsifiability	可乳化性	可乳化性
emulsification	乳化作用	乳化
emulsifier	①乳化器②乳化剂	①乳化器②乳化劑
emulsifying agent	乳化剂	乳化劑
emulsifying colloid	乳化胶体	乳化膠體
emulsifying ointment	乳化软膏	乳化軟膏
emulsion	乳液,乳浊液	乳液,乳膠,乳態
emulsion liquid membrane	乳化液膜	乳液薄膜
emulsion paint	乳化漆	乳化漆
emulsion phase	乳相	乳化相
emulsion polymerization	乳液聚合	乳化聚合
emulsion stability	乳化稳定	乳態穩定性
emulsoid	乳胶	乳化膠體,類乳化液
enamel drier	瓷漆干燥室	瓷漆乾燥室
enamel glaze	珐琅釉	琺瑯釉
enamel paint	瓷漆	瓷漆
enantiometric selectivity	对映体选择性	鏡像對應異構選擇性
enargite	硫砷铜矿	硫砷銅礦
encapsulation	胶囊封装	膠囊封裝
enclosed impeller	闭工叶轮	封閉葉輪
end effect	末端效应	端點效應
end gas	尾气	尾氣
endoenzyme	内酶	内酶

英　文　名	祖国大陆名	台湾地区名
endothermicity	吸热性	吸熱度
endothermic process	吸热过程	吸熱程序
endothermic reaction	吸热反应	吸熱反應
end point	终点	終點
end-point control	终点控制	終點控制
end product	最终产品	最終產物
energetic constitutive equation	能量本质方程式	能量本質方程式
energetic equation of state	能量状态方程	能量狀態方程式
energetics	能量学	能量學
energized molecule	①受激分子②高能化分子	①受激分子②高能化分子
energy	能量	能量
energy balance	能量衡算, 能量平衡	能量均衡
energy balance equation	能量平衡方程式	能量均衡方程式
energy barrier	能量障壁	能量障壁
energy conservation	能量守恒	能量守恆
energy consumption	能量损耗	能量消耗
energy content	能量含量	能量含量
energy conversion	能量转换	能量轉換
energy conversion engineering	能量转换工程	能量轉換工程
energy converter	能量转化器	①能量轉化器②能量轉換物
energy density	能量密度	能量密度
energy dissipation	能量耗散	能量散逸
energy distribution	能量分布	能量分布
energy efficiency	能量效率	能量效率
energy engineering	能源工程	能源工程
energy equation	能量方程	能量方程式
energy equivalent	能量当量	能量當量
energy exchanger	能量交换器	能量交換器
energy flow	能量流动	能量流
energy flux	能量通量	能量通量
energy generation curve	能量生成曲线	能量生成曲線
energy grade line	能量级线	能量級線
energy integration	能量集成	能量集成
energy level	能级	能階
energy loss	能量损失	能量損失
energy loss curve	能量损失曲线	能量損失曲線
energy producing reaction	生能反应	生能反應

英　文　名	祖国大陆名	台湾地区名
energy science	能源科学	能量科學
energy separating agent	能量分离剂	能量分離劑
energy sink	能汇	能量匯座
energy source	能源	能源
energy spectrum	能谱	能譜
energy state	能态	能量狀態
energy storage	能量储存	能量儲存
energy surface	能量面	表面能
energy technology	能源技术	能量技術, 能源技術
energy transfer	能量传递	能量傳送
energy transformation	能量转换	能量轉換
engineering	工程	①工程②工程學
engineering analysis	工程分析	工程分析
engineering design	工程设计	工程設計
engineering design cost	工程设计成本	工程設計成本
engineering experiment	工程实验	工程實驗
engineering material	工程材料	工程材料
engineering plastic	工程塑料	工程塑膠
enhanced oil recovery	提高采收率	增強石油回收[法]
enhancement factor	增强因子	增進因數
enhancer	增强子	①增味劑②增進劑
enlargement loss	扩大损失	擴大損失
enriching section	提浓段	增濃段
enrichment	富集	增濃
enrichment culture	富集培养	增濃培養
enrichment factor	富集因子	增濃因數
enstatite	顽辉石	頑火輝石
entanglement	缠结	纏繞
enthalpy	焓	焓
enthalpy-composition chart	焓-组成图	焓-組成圖
enthalpy-composition diagram	焓-组成图	焓-組成圖
enthalpy-concentration diagram	焓浓图	焓-濃度圖
enthalpy-entropy diagram	焓-熵图, H-i 图	焓-熵圖
enthalpy-humidity chart	焓-湿图	焓-濕度圖
enthalpy of activation	活化焓	活化焓
enthalpy of formation	生成焓	生成焓
enthalpy of reaction	反应焓	反應焓
entrained bed reactor	夹带床反应器	載送床反應器
entrainer	夹带剂	①進料器②霧沫器

英　文　名	祖国大陆名	台湾地区名
entrainer pump	夹带剂泵	進料泵
entrainment	[雾沫]夹带	霧沫
entrainment evaporator	雾沫蒸发器	霧沫蒸發器
entrainment mixer	雾沫混合器	霧沫混合器
entrainment trap	夹带物分离器	霧沫阱
entrainment velocity	夹带速度	霧沫速度
entrance effect	进口效应	進口效應
entrance region	进口区	進口區
entrapment	包埋	包埋
entropy	熵	熵
entropy balance	熵衡算	熵均衡
entropy change	熵变化	熵變化
entropy equation	熵方程式	熵方程式
entropy flow	熵流, 熵通量	熵流, 熵通量
entropy generation	熵产生	熵生成
entropy of activation	活化熵	活化熵
entropy production	熵产生	熵生成
entry length	进口长度	進口長度
enumeration algorithm	枚举法	列舉法
environment	环境	環境
environmental chemistry	环境化学	環境化學
environmental contamination	环境污染	環境污染
environmental cost	环境成本	環境成本
environmental engineering	环境工程	環境工程
environmental impact	环境冲击	環境衝擊
environmentally sponsored decay	环境引发衰变	環境引發衰退
environmental pollution	环境污染	環境污染
environmental protection	环境保护	環境保護
environmental quality	环境质量	環境品質
environmental science	环境科学	環境科學
environmental state	环境态	環境態
environmental technology	环境技术	環境技術
enzymatic analysis	酶法分析	酶分析法
enzymatic electrocatalysis	酶促电催化	酶電催化
enzymatic hydrolysis	酶法水解	酶法水解
enzymatic method	酶法	酶試驗法
enzymatic reaction	酶促反应	酶反應
enzymatic reaction kinetics	酶反应动力学	酶動力學
enzyme	酶	酶, 酵素

英　文　名	祖国大陆名	台湾地区名
enzyme activity	酶活力	酶活性
enzyme catalysis	酶催化	酶催化
enzyme catalyzed reaction	酶催化反应	酶催化反應
enzyme deactivation	酶失活	酶去活化
enzyme electrode	酶电极	酶電極
enzyme fermentation	酶发酵	酶發酵
enzyme immunoassay	酶免疫分析法	酶免疫分析法
enzyme-linked immunosorbent assay	酶联免疫吸附测定	酶聯免疫吸附測定
enzyme membrane	酶膜	酶膜
enzyme product complex	酶产物复合物	酶產物複合物
enzyme selectivity	酶选择性	酶選擇性
enzyme specificity	酶专一性	酶專一性
enzyme stabilization	酶稳定	酶穩定化
enzyme-substrate complex	酶-底物复合物	酶-底物複合物
enzyme-substrate reaction	酶底物反应	酶受質複合物反應
eosin dye	曙红染料	曙紅染料
eosin lake	曙红色淀	曙紅色澱
epichlorohydrin	环氧氯丙烷	環氧氯丙烷
epitaxial growth	外延生长	磊晶生長, 晶體同軸生長
epitaxy	外延	磊晶
epoxidation	环氧化	環氧化[反應]
epoxy resin	环氧树脂	環氧樹脂
equality constraint	等式约束	等式約束
equalization	相等	均化
equalization tank	平衡槽	均化槽
equalizing valve	平衡阀	均衡閥
equal-settling particle	等沉降粒子	等沉降粒子
equation	方程	方程式
equation of continuity	连续性方程	連續性方程式
equation of motion	运动方程	運動方程式
equation of state	状态方程	狀態方程
equilateral triangular diagram	等边三角图	等邊三角圖
equilibration	平衡化	平衡化
equilibration rate	平衡速率	平衡速率
equilibrium	平衡	平衡
equilibrium approach	平衡法	平衡法
equilibrium approximation	平衡近似	平衡近似[法]
equilibrium composition	平衡组成	平衡組成

英　文　名	祖国大陆名	台湾地区名
equilibrium concentration	平衡浓度	平衡濃度
equilibrium condensation	平衡冷凝	平衡冷凝
equilibrium constant	平衡常数	平衡常數
equilibrium conversion	平衡转化[率]	平衡轉化率
equilibrium criterion	平衡判据	平衡準則
equilibrium curve	平衡曲线	平衡曲線
equilibrium diagram	平衡图	平衡圖
equilibrium distillation	平衡蒸馏	平衡蒸餾
equilibrium distribution	平衡分布	平衡分布
equilibrium flash vaporization	平衡闪蒸	平衡驟汽化
equilibrium flash vaporization curve	平衡闪蒸曲线	平衡驟汽化曲線
equilibrium freezing curve	平衡冷凝曲线	平衡冷凍曲線
equilibrium line	平衡线	平衡線
equilibrium mixture	平衡混合物	平衡混合物
equilibrium moisture	平衡水分	平衡水分
equilibrium moisture content	平衡含水率	平衡含水量
equilibrium plot	平衡图	平衡圖
equilibrium point	平衡点	平衡點
equilibrium separation process	平衡分离过程	平衡分離程序
equilibrium stage	平衡级	平衡階
equilibrium still	平衡釜	平衡釜
equilibrium system	平衡系统	平衡系統
equilibrium vaporization	平衡汽化	平衡汽化
equilibrium yield	平衡收率	平衡產率
equilmolar counter diffusion	等摩尔逆向扩散	等莫爾逆向擴散
equipment cost	设备成本	裝置成本
equipment flow measuring	流量测量设备	流量測量裝置
equipment flowsheet	设备流程图	裝置流程圖
equipment reliability	设备可靠性	裝置可靠性
equipment safety	设备安全	裝置安全性
equipotential line	等势线	等位線
equivalent	①当量②等效	當量, 等值
equivalent conductance	①当量电导②等效电导	當量導電度
equivalent diameter	①当量直径②等效径	相當直徑
equivalent free-falling diameter	等效自由沉降直径	等效自由沉降直徑
equivalent isothermal temperature	等温温度相当值	等溫溫度相當值
equivalent length	当量长度	相當長度
equivalent non-circular duct diameter	非圆管当量直径	非圓管相當直徑
equivalent particle diameter	当量粒径	相當粒徑

英　文　名	祖国大陆名	台湾地区名
equivalent pipe length	等值管长度	相當管長
equivalent point	等当点	當量點
eradicator	消除器	去污劑, 除草劑
erepsin	肠蛋白酶	腸(肽)酶
erlenmeyer flask	锥形瓶	錐形瓶
erosion	磨蚀	浸蝕
error	误差	誤差
error constant	误差常数	誤差常數
error criterion	误差判据	誤差準則
error integral	误差积分	誤差積分
error propagation	误差传播	誤差傳遞
error ratio	误差比	誤差比
error signal	误差信号	誤差信號
error-squared control	误差平方控制	誤差平方控制
eruption	喷发	噴發
erythromycin	红霉素	紅黴素
erythrophyll	叶红素	葉紅素
Escherichia coli	大肠杆菌	大腸桿菌
essence recovery	香精回收	香精回收
essential oil	精油	香精油
ester	酯	酯
ester gum	甘油松香酯	松香甘油酯, 松香硬酯
esterification	酯化作用	酯化[反應]
ester number	酯值	酯值
ester value	酯值	酯值
estimate	估计量	估計值
estimated variance	估计方差	估計變異數
estimation	估值, 估算	估計
estimator	估计器	估計器
etchant	蚀刻剂	蝕刻液
etching	①蚀刻②浸蚀	①蝕刻②浸蝕
etching plasma	等离子蚀刻[法]	電漿蝕刻法
etching plasma-enhanced	等离子加强蚀刻	電漿加強蝕刻
ether	醚	醚
etherification	醚化	醚化[反應]
ether number	醚数	醚值
ether value	醚值	醚值
ethyl cellulose	乙基纤维素	乙基纖維素
ethylene-propylene rubber	乙丙橡胶	乙烯-丙烯三共聚物

英　文　名	祖国大陆名	台湾地区名
eugenol	丁子香酚	丁香酚
eutectic alloy	低共熔合金	共熔合金
eutectic equilibrium	低共熔平衡	共熔平衡
eutectic mixture	低共熔物	共熔混合物
eutectic point	[低]共熔点	共熔點
eutectics	低共熔物	共熔物
eutectic state	低共熔状态	共熔狀態
eutectic system	低共熔系统	共熔系統
eutectic temperature	低共熔温度	共熔溫度
evaluation	评估	評估
evaluation replacement	替换评估	替換評估
evaporating column	浓缩塔	蒸發塔
evaporating pipe	蒸发管	蒸發管
evaporation	蒸发	蒸發
evaporation coefficient	蒸发系数	蒸發係數
evaporation cooling	蒸发冷却	蒸發冷卻
evaporation efficiency	蒸发效率	蒸發效率
evaporation loss	蒸发损失	蒸發損失
evaporation rate	蒸发率	蒸發效率
evaporative condenser	蒸发冷凝器	蒸發冷凝器
evaporative cooling	蒸发冷却	蒸發冷卻
evaporative crystallization	蒸发结晶	蒸發結晶
evaporative crystallizer	蒸发结晶器	蒸發結晶器
evaporative refrigeration	蒸发冷冻	蒸發冷凍
evaporator	蒸发器	蒸發器
evaporator room	蒸发室	蒸發室
evaporator scale	蒸发器污垢	蒸發器積垢
evaporator with direct heating	直接加热型蒸发器	直接加熱型蒸發器
evaporator with horizontal tubes	水平列管蒸发器	平管式蒸發器
evaporator with submerged combustion	浸没燃烧蒸发器	浸沒燃燒蒸發器
evolutionary method	调优法	調優法
evolutionary operation	调优操作	調優操作
evolved gas analysis	逸出气体分析法	釋出氣體分析法
evolved gas detection	逸出气体检测	釋出氣體偵測法
exact solution	精确解	精確解
excess air	过量空气	過量空氣
excess chemical potential	超额化学势	過剩化學勢
excess concentration technique	超额浓度法	過量濃度法
excess enthalpy	超额焓	過剩焓

英 文 名	祖国大陆名	台湾地区名
excess entropy	超额熵	過剩熵
excess function	超额函数	過剩函數
excess Gibbs free energy	超额吉布斯自由能	過剩吉布斯自由能
excess property	超额性质	過剩性質
excess volume	超额体积	過剩體積
excess water	过量水	過量水
exchange adsorption	交换吸附	交換吸附
exchange energy	交换能	交換能
exchanger	①交换器②交换剂	①交換器②交換劑
excited state	激发态	激態
exergy	㶲, 有效能	有效能量
exergy analysis	㶲分析, 有效能分析	有效能分析
exergy balance	㶲衡算, 有效能衡算	有效能均衡
exergy loss	㶲损失	有效能損失
exhalation valve	排气阀	排氣閥
exhauster	排气机	排氣機
exhaust pipe	排气管	排氣管
exhaust steam	排汽	排放蒸汽
exit gas	出口气体	出口氣體
exit tube	排气管	排氣管
exosmosis	外渗	外滲透
exothermicity	放热性	放熱度
exothermic process	放热过程	放熱程序
exothermic reaction	放热反应	放熱反應
expanded bed	膨胀床	膨脹床
expanded bed contactor	膨胀床接触器	膨脹床接觸器
expander	膨胀器	膨脹器, 膨脹機
expansion	膨胀	擴大, 膨脹
expansion coefficient	膨胀系数	膨脹係數
expansion factor	膨胀因子	膨脹因子
expansion joint	膨胀节	脹縮接頭
expansion loop	膨胀弯管	伸縮圈
expansion loss	膨胀损失	膨脹損失
expansion work	膨胀功	膨脹功
expected value criterion	期望值判据	期望值判斷
experimental data	实验数据	實驗數據
experimental design	实验设计	實驗設計
experimental error	实验误差	實驗誤差
experimental reactor	实验用反应器	實驗用反應器

英 文 名	祖国大陆名	台湾地区名
expert system	专家系统	專家系統
explicit function	显函数	顯函數
explicit method	显式法	顯式法
explosion	爆炸	爆炸
explosion hazard	爆炸危险	爆炸傷害
explosion index	爆炸指数	爆炸指數
explosion limit	爆炸极限	爆炸界限
explosion potential	爆炸势	爆炸潛力, 爆炸勢
explosion pressure	爆炸压[力]	爆炸壓力
explosion proof	防爆	防爆
explosive	炸药	炸藥
explosive polymerization	爆聚[合]	爆聚合
exponential distribution	指数分布	指數分布
exponential function	指数函数	指數函數
exponential lag	指数落后	指數落後
exponential law	指数律	指數律
exposed center	曝露中心	曝露中心
exposure limit	曝露界限	曝露界限
exposure test	曝露试验	曝露試驗
exposure time	曝光时间	曝露時間
expression	压榨, 挤出	壓榨
expression constant	压榨常数	壓榨常數
extended aeration	延时曝气	延時曝氣
extended surface tube	扩展式表面管	延伸表面管
extensibility	可扩展性	延伸性
extension	伸长	延伸
extensional flow	延伸流动	延伸流動
extensional viscosity	拉伸黏度	延伸黏度
extension neck	延伸颈	延伸頸
extensive property	广度性质	外延性質
extent of decomposition	分解程度	分解程度
extent of reaction	反应进度	反應程度
external area	外表面积	外表面積
external capacity	外在容积	外在容積
external control loop	外控制环路	外控制環路
external coordinates	外部坐标	外部座標
external diameter	外径	外徑
external diffusion	外扩散	外擴散
external disturbance	外扰[动]	外來擾動

英 文 名	祖国大陆名	台湾地区名
external energy	外能	外能
external flow	外流	外部流動
external force	外力	外力
external heat exchanger	外换热器	外熱交換器
external heating	外热法	外熱法
external pressure	外压力	外壓力
external recycle	外循环	外循環
external recycle reactor	外循环反应器	外循環反應器
external reflux	外回流	外回流
external sizing	外部施胶	外部上膠
external surface	外表面	外表面
external surface concentration	外表面浓度	外表面濃度
extinction	熄灭	熄滅
extinction coefficient	消光系数	消光係數
extinction temperature	熄灭温度	熄滅溫度
extracellular enzyme	胞外酶	胞外酶
extract	萃取液	萃取液
extraction	萃取	萃取
extraction battery	萃取器组	萃取器組
extraction column	萃取塔	萃取塔
extraction factor	萃取因子	萃取因數
extraction process	萃取过程	萃取程序
extraction rate	萃取率	萃取速率
extraction tower	萃取塔	萃取塔
extractive crystallization	萃取结晶	萃取結晶
extractive crystallizer	萃取结晶器	萃取結晶器
extractive distillation	萃取蒸馏	萃取蒸餾
extractive ultrafiltration	萃取超滤	萃取性超過濾
extractor	萃取器	萃取器
extrapolation	外推	外插法
extruder	挤出机	擠壓機
extrusion	挤出	擠壓
exudation	渗出	渗出, 泌出

F

英 文 名	祖国大陆名	台湾地区名
fabrication	制造	製造

英　文　名	祖国大陆名	台湾地区名
fabric filter	织物过滤器	織物過濾器
factor	因子	因子
factorial design	析因设计	因子設計
factorial experiment	析因实验	因子實驗
facultative aerobic bacteria	兼性需氧细菌	兼性需氧細菌
facultative bacteria	兼性细菌	兼性菌
facultative lagoon	兼性污水塘	兼性污水塘
facultative respiration	兼性呼吸	兼性呼吸
fading memory material	记忆衰退材料	記憶衰退材料
failure diagnosis	故障诊断	故障診斷
failure mode and effect analysis	故障形式和影响分析	故障形式和影響
falling ball method	落球法	落球法
falling ball viscometer	落体黏度计	落球黏度計
falling body viscometer	落球黏度计	落球黏度計
falling film	降膜	[下]降膜
falling-film evaporator	降膜蒸发器	降膜蒸發器
falling film type absorber	降膜吸收器	降膜式吸收器
falling needle viscometer	落针黏度计	落針黏度計
falling rate drying period	降速干燥[阶]段	減速乾燥期
falling rate period	降速期	減速期
falling sphere viscometer	落球黏度计	落球黏度計
fan	排风机	風扇
fan blower	扇形排风机	通風扇
faraday	法拉第	法拉第
fast fluidization	快速流态化	快速流體化
fast fluidized bed	快速流化床	快速流體化床
fast reaction	快速反应	快速反應
fatigue strength	疲劳强度	疲勞強度
fatigue test	疲劳试验	疲勞試驗
fats and oils	油脂	油脂
fatty acid	脂肪酸	脂肪酸
fatty alcohol	脂肪醇	脂肪醇
fault diagnosis	故障诊断	故障診斷
F curve	F 曲线	F 曲線
feasibility	可行性	可行性
feasibility analysis	可行性分析	可行性分析
feasibility study	可行性研究	可行性研究
feasibility survey	可行性调查	可行性調查
feasible path method	可行路径法	可行路徑法

英　文　名	祖国大陆名	台湾地区名
feasible region	可行域	可行區
fed batch culture	流加培养	流加培養
feed	进料	①進料②餇入③食用
feedback	反馈	回饋
feedback bellow	反馈伸缩囊	回饋伸縮囊
feedback compensation	反馈补偿	回饋補償
feedback complementary	互补反馈	互補回饋
feedback control	反馈控制	回饋控制
feedback controller	反馈控制器	回饋控制器
feedback control system	反馈控制系统	回饋控制系統
feedback element	反馈元件	回饋元件
feedback loop	反馈回路	回饋環路
feedback mechanism	反馈机理	回饋機構
feedback negative	负反馈	負回饋
feedback positive	正反馈	正回饋
feedback signal	反馈信号	回饋信號
feedback system	反馈系统	回饋系統
feed belt	进料带	進料帶
feed composition	进料组成	進料組成
feed conveyor	进料输送机	進料輸送器
feed disk	盘式进料器	進料盤
feed drum	筒式进料机	進料鼓
feeder	进料器	①進料器,飼機②取食者
feedforward	前馈	前饋
feedforward control	前馈控制	前饋控制
feedforward system	前馈系统	前饋系統
feed hole	进料孔	進料孔
feed pipe	进料管	進料管
feed quality	进料质量	進料品質
feed rate	加料速度	進料速率
feed ratio	进料比	進料比
feed regulator	进料调节器	進料調節器
feed roller	进料滚柱	進料輥
feed stock	原料	原料
feed stream	进料流	進料流
feed system	进料系统	進料系統
feed tank	加料槽	進料槽
feed temperature	进料温度	進料溫度

英　文　名	祖国大陆名	台湾地区名
feed tray	进料板	進料板
feed tray location	进料板位置	進料板位置
feed valve	进料阀	進料閥
feed water	给水	給水
feed water treatment	给水处理	給水處理
feed well	给水井	進料井
feldspar	长石	長石
feldspathic glaze	长石釉	鈉鈣長石
Fenske packing	螺线圈填料, 芬斯克填料	芬斯基填料
Fenske's equation	芬斯克方程	芬斯基方程式
fermentation	发酵	發酵[法]
fermentation apparatus	发酵器械	發酵裝置
fermentation cellar	发酵窖	發酵窖
fermentation chamber	发酵室	發酵室
fermentation efficiency	发酵效率	發酵速率
fermentation mechanism	发酵机理	發酵機構
fermentation medium	发酵培养基	發酵基, 發酵介質
fermentation pathway	发酵途径	發酵途徑
fermentation pattern	发酵模式	發酵方式
fermentation process	发酵过程	發酵程序
fermentation tank	发酵槽	發酵槽
fermentation technology	发酵技术	發酵技術
fermentation tube	发酵管	發酵管
fermentation tube test	发酵管试验	發酵管試驗
fermentation vat	发酵槽	發酵缸, 發酵槽
fermenter	发酵罐	發酵槽
fermentor	发酵罐	缸式發酵槽
Fermi-Dirac distribution	费米-狄拉克分布	費米-笛拉克分布
ferro-alloy	铁合金	鐵合金
ferromanganese	铁锰合金	錳鐵齊
ferrophophorus	磷铁	磷鐵齊
ferro-silico-manganese	硅锰铁合金	矽錳鐵齊
ferrosilicon alloy	硅铁合金	矽鐵齊
fertilization	施肥	施肥
fertilizer	肥料	肥料
F factor	F 因子	F 因子
fiber reinforced plastics	纤维增强塑料	纖維強化塑膠
Fibonacci search method	斐波那契搜索法	費布納西搜尋法

英　文　名	祖国大陆名	台湾地区名
fibre	纤维	纖維
Fick's law	菲克定律	費克定律
filament	单丝	單絲, 絲狀纖維
filamentous bacteria	丝状细菌	絲狀細菌
filler	填料	填料
filling	填充	填料, 裝填
film boiling	膜状沸腾	薄膜沸騰
film coefficient	膜系数	薄膜係數
film condensation	膜状冷凝	薄膜冷凝
film diffusion	膜扩散	薄膜擴散
film diffusion control	膜扩散控制	薄膜擴散控制
film drier	薄膜干燥机	薄膜乾燥器
film heat transfer coefficient	传热膜系数	薄膜熱傳係數
film phase	膜相	膜相
film resistance	膜阻力	薄膜阻力
film scrubber	膜式洗涤器	膜式洗滌器
film strength	膜强度	薄膜強度
film temperature	膜温度	薄膜溫度
film test	膜试验	薄膜試驗
film theory	膜理论	薄膜理論
film type condensation	膜式冷凝	薄膜冷凝
film type evaporator	膜式蒸发器	薄膜蒸發器
filmwise condensation	膜状冷凝	薄膜冷凝
film yeast	膜酵母	膜酵母
filter aid	助滤剂	助濾器
filter area	过滤面积	過濾面積
filter bag	滤袋	濾袋
filter box	滤箱	濾箱
filter cake	滤饼	濾餅
filter capacity	过滤器容量	過濾器容量
filter clogging	过滤器堵塞	過濾器阻塞
filter cloth	滤布	濾布
filter compressor	过滤器压缩机	過濾器壓縮機
filter depth	滤池深度	濾池深度
filter drum	滤桶	濾桶
filter efficiency	过滤效率	過濾器效率
filter flowsheet	过滤流程	過濾器流程
filter frame	滤框	濾框
filter liquor	滤液	濾液

英　文　名	祖国大陆名	台湾地区名
filter paper	滤纸	濾紙
filter performance	过滤性能	過濾器性能
filter plate	滤板	濾板
filter press	压滤机	壓濾機
filter press cake	压滤饼	壓濾餅
filter press cloth	压滤布	壓濾布
filter sand	滤砂	濾砂
filtration	过滤	過濾
filtration area	过滤面积	過濾面積
filtration cycle	过滤周期	過濾周期
filtration medium	过滤介质	過濾介質
filtration membrane	滤膜	濾膜
filtration rate	过滤速率	過濾速率
filtration sterilization	过滤灭菌	過濾除菌
fin	翅片	鰭片
final boiling point	终沸点	終沸點
final clarifier	最终澄清器	最終澄清器, 最終沉降槽
final condition	最终条件	最終條件
final control element	最终控制元件	最終控制原件
final state	最终状态	最終狀態
final-value theorem	终值原理	終值定理
financing cost	财务会计成本	財務會計成本
fine crusher	细碎机	細級壓碎機
fine dispersion	精细分散	精細分散
fine particle	细颗粒	細顆粒
fine screen	精筛	精篩
finishing	精整	最後加工, 整理
finite boundary element	有限边界元素	有限邊界原素
finite difference	有限差分	有限差分
finite disturbance	有限扰动	有限擾動
finite element method	有限元法	有限原素法
finite-time stability	有限时间稳定性	有限時間穩定性
finned pipe	翅片管	鰭片管
finned surface	翅片表面	鰭片表面
finned tube	翅片管	鰭片管
finned-tube heat exchanger	翅管式热交换器	鰭片管熱交換器
fire and explosion index	火灾及爆炸指数	火災及爆炸指數
firebrick	耐火砖	耐火磚

英　文　名	祖国大陆名	台湾地区名
fireclay	耐火黏土	耐火黏土
fireclay brick	耐火砖	耐火黏土磚
fired furnace	明火加热炉	加熱爐
fired heater	明火加热器	加熱器
fired-tubular reactor	火管式反应器	火管式反應器
fire extinguisher	灭火器	滅火器
fire hazard	火灾	火災危害
fire protection	防火	火災防護, 防火
fire retardant	阻燃剂	阻燃劑, 防火劑
fire-retarding paint	防火漆	防火漆
first law of thermodynamics	热力学第一定律	熱力學第一定律
first order control system	一级控制系统	一階控制系統
first order kinetics	一级动力学	一階動力學
first order lag	一级落后	一階落後
first order phase transition	一级相变	一階相變
first order process	一级过程	一階程序
first order reaction	一级反应	一階反應
first order response	一级应答	一階應答
first order system	一级系统	一階系統
first order transition temperature	一级转变温度	一階轉移溫度
fission	裂变	分裂
fissionability	可裂变性	可分裂性
fissionable material	可裂变物质	可分裂材料
fission bomb	原子弹	分裂彈
fission energy	裂变能	分裂能
fission product	裂变产物	分裂產物
fission radioactive	放射性裂变	放射性分裂
fissure	裂缝	裂縫
fitting equivalent length	配件当量长度	配件相當長度
fitting equivalent pipe length	配件当量管长	配件相當管長
fitting flowsheet	配件流程图	配件流程圖
fitting reducing	渐缩管件	漸縮管件
fitting resistance	配件阻力	配件阻力
fixation	固定	固定
fixation bath	固定浴	固定浴
fixed agent	固定剂	固定劑
fixed bed	固定床	固定床
fixed bed process	固定床过程	固定床法
fixed bed reactor	固定床反应器	固定床反應器

英　文　名	祖国大陆名	台湾地区名
fixed capital	固定资本	固定資本
fixed capital investment	固定资本投资额	固定資本投資額
fixed catalyst bed	固定催化剂床	固定觸媒床
fixed charge	固定费用	固定費用
fixed cost	固定成本	固定成本
fixed-growth system	固定成长系统	固定成長系統
fixed nitrogen	固定氮	固定氮
fixed oil	不挥发油	不揮發油
fixed tube-sheet heat exchanger	固定管板换热器	固定管板熱交換器
flagella	鞭毛	鞭毛
flame	火焰	火燄
flame reactor	火焰反应器	火燄反應器
flame resisting agent	阻燃剂	耐燃劑
flame spectrometer	火焰分光仪	火燄分光劑
flame stability	火焰稳定性	火燄固定性
flame temperature	火焰温度	火燄溫度
flammability	可燃性	可燃性,可燃度
flange	法兰	凸緣法蘭
flange connection	法兰连接	凸緣法蘭連接
flange coupling	法兰接头	凸緣法蘭接頭
flange joint	法兰联管节	凸緣法蘭接頭
flange tap	法兰旋塞	凸緣分接頭
flapper	挡板	擋板
flapper nozzle mechanism	挡板喷嘴机构	擋板噴嘴機構
flash calculation	闪蒸计算	驟沸計算
flash chamber	闪蒸室	驟沸室
flash chilling	骤冷	驟冷
flash chlorination	骤氯化	驟氯化[反應]
flash concentrator	闪蒸浓缩器	急驟濃縮器
flash distillation	闪速蒸馏	驟餾
flash distillation column	闪蒸塔	驟餾塔
flash drier	急骤干燥器	驟乾器
flash drum	闪蒸槽	驟沸桶
flash drying	闪速干燥	驟沸乾燥
flash evaporation	闪蒸	驟蒸發
flash film evaporator	薄膜闪蒸器	[急]驟[薄]膜蒸發器
flashing liquid	闪蒸液	驟沸液
flashing mixer	骤混器	驟混器
flashing mixing	骤混	驟混

英 文 名	祖国大陆名	台湾地区名
flash pasteurizer	瞬时杀菌器	瞬間殺菌器
flash point	闪点	閃[火]點
flash point apparatus	闪点测定器	閃[火]點測定器
flash roaster	闪速焙烧炉	急驟焙燒爐
flash tower	闪蒸塔	驟餾塔
flash vapor	闪蒸气	驟蒸氣
flash vaporization	闪蒸	驟汽化
flash vaporization curve	闪蒸曲线	驟汽化曲線
flask	烧瓶	燒瓶
flat paddle	平板桨	平板漿
flat plate	平板	平板
flat-plate approximation	平板近似法	平板近似法
flat-plate flow	平板流动	平板流動
flavoring materials	调味剂	調味料
flax	亚麻	亞麻
flax fiber	亚麻纤维	亞麻纖維
flax-seed oil	亚麻子油	亞麻仁油
flexitray	挠性盘	活動盤
flexural stress	弯曲应力	塑流應力
flickering flame	闪烁火焰	閃爍火燄
flight conveyor	刮板输送机	梯板運送機, 梯運機
flint	燧石	燧石
flint clay	硬质黏土	燧石黏土
flint fireclay	硬质耐火黏土	燧石耐火黏土
flint glass	火石玻璃	燧石耐火玻璃
flint pebble	燧卵石	燧卵石
float	浮标	浮子, 浮桶, 浮標
float glass	浮法玻璃	飄浮玻璃
float indicator	浮标指示计	浮標指示計
floating action	浮动作用	浮動作用
floating control	浮动控制	浮動控制
floating-cover digester	浮盖消化槽	浮[動]蓋消化槽
floating head heat exchanger	浮头换热器	浮[動]頭熱交換器
floating pressure control	浮动压力控制	浮動壓力控制
floating pressure operation	浮动压力操作	浮動壓力操作
floating rate	浮动速率	浮動速率
floating roof tank	浮顶槽	①浮頂油槽②浮頂槽
floating soap	浮水皂	浮皂
floating speed	浮动速率	浮動速率

英　文　名	祖国大陆名	台湾地区名
floating type bell gage	浮式钟型压力计	浮式鐘型壓力計
floating valve tray	浮阀板	浮閥板
floating zone method	悬浮区法	浮動帶域[純化]法
floatsam	浮料	浮料
float-type manometer	浮式压力计	浮標式壓力計
float valve	浮球阀	浮閥
floc	絮凝物	絮凝體
flocculant	絮凝剂	絮凝劑
flocculate	絮凝物	絮凝物
flocculating mechanism	絮凝机理	絮凝機構
flocculation	絮凝	絮凝作用
flocculation reaction	絮凝反应	絮凝反應
flocculation tank	絮凝槽	絮凝槽
flocculator	絮凝器	絮凝器
flocculent	絮凝剂	絮凝劑
flocculent particle	絮凝粒子	絮凝粒子
flocculent settling	絮凝沉降	絮凝沉降
flocculent suspension	絮凝悬浮液	絮凝懸浮液
floc growth	絮凝物成长	絮凝體成長
floc particle	絮凝粒子	絮凝粒子
floc size	絮凝物大小	絮凝體大小
floc test	絮凝测验	絮凝測驗
flooded suction line	溢流吸入管线	氾流吸入管線
flooding	液泛	氾流
flooding point	泛点	氾流點
flooding velocity	泛点速度	氾流速度
Flory Huggins theory	弗洛里-哈金斯理论	佛洛里-哈金斯理論
flotation	浮选	浮選[法]
flotation process	浮选过程	浮選法
flotation tank	浮选槽	浮選槽
flow	流[动]	流動
flow behavior index	流动行为指数	流動行為指數
flow calorimeter	流动量热计	流動卡計
flow control	流量控制	流量控制
flow diagram	①流程图②程序框图	流程圖
flow diagram symbol	流程图符号	流程圖符號
flow gage	流量计	流量計
flow indicator	流量指示计	流量指示計
flow integration	流量积分	流量積分

英 文 名	祖国大陆名	台湾地区名
flow integrator	流量积分器	流量積分器
flow lag	流动落后	流動落後
flow line	流动线	流動線
flow-line interception	流线截取	流線截取
flow measuring equipment	流量测量装置	流量量測装置, 流血测量装置
flowmeter	流量计	流量計
flow method	流动法	流動法
flow mixer	流动混合器	流動混合器
flow net	流动网	流動網
flow nozzle	流量喷嘴	流動噴嘴
flow number	[搅拌]流量数	流量數
flow pattern	流型	流動型式
flow process	流水作业	流動程序
flow rate	流率	①流[動]率②流量
flow reactor	流动反应器	流動反應器
flow resistance	流动阻力	流動阻力
flowsheet	流程图	流程圖
flowsheet design	流程图设计	流程圖設計
flowsheeting	流程模拟	流程模擬
flowsheet symbol	流程图符号	流程圖符號
flowsheet synthesis	流程综合	流程綜合
flow system	流动系统	流動系統
flow visualization	流动显示	流動觀測
flow work	流动功	流動功
flue	烟道	煙道
flue gas	烟道气	煙道氣
fluid	流体	流體
fluid bed	流化床	流體床
fluid-bed conversion	流化床转化	流體床轉化
fluid-bed reactor	流化床反应器	流體床反應器
fluid dynamics	流体动力学	流體動力學
fluid energy mill	流能磨	流體能量研磨機
fluid flow	流体流动	流體流動
fluid flow resistance	流体流动阻力	流體流動阻力
fluid-fluid reaction	流体-流体反应	流體-流體反應
fluid friction	流体摩擦	流體摩擦
fluid head	流体高差	流體高差
fluidics	流控技术	流控學

英　文　名	祖国大陆名	台湾地区名
fluidizable particle size	可流态化颗粒大小	可流體化粒大小
fluidization	流态化	流體化
fluidization number	流化数	流體化數
fluidized bed	流化床	流體化床反應器
fluidized bed adsorber	流化床吸附器	流體化床吸附器
fluidized bed boiler	流化床锅炉	流體化床鍋爐
fluidized bed combustion	流化床燃烧	流體化床燃燒
fluidized bed combustor	流化床燃烧器	流體化床燃燒器
fluidized bed dryer	流化床干燥器	流體化床乾燥器
fluidized bed gasifier	流化床气化器	流體化床氣化器
fluidized bed reactor	流化床反应器	流體化床
fluidized catalyst bed	流化催化剂床	流體化觸媒床
fluidized distillation	流化蒸馏	流體化蒸餾
fluidized mixing	流化混合	流體化混合
fluidized reactor	流化反应器	流體化反應器
fluidized roaster	流化焙烧炉	流體化焙燒爐
fluidizing velocity	流化速度	流體化速度
fluid kinematics	流体运动学	流體運動學
fluid mechanics	流体力学	流體力學
fluid phase	流体相	流體相
fluid-solid reaction	流体-固体反应	流體-固體反應
fluid velocity	流体速度	流體速度
fluorapatite	氟磷灰石	氟磷灰石
fluorescein	荧光黄	螢光黃
fluorescence	荧光	螢光
fluorescence spectroscope	荧光分光镜	螢光譜儀
fluorescent crystal	荧光晶体	螢光晶體
fluorescent dye	荧光染料	螢光染料
fluorescent spectrum	荧光光谱	螢光光譜
fluorescent whitening agent	荧光增白剂	螢光增白劑
fluorocarbon	氟碳化物	氟碳化物
fluorocarbon resin	氟碳树脂	氟碳樹脂
fluorspar	荧石	氟石, 螢石
flush filter plate	平槽压滤板	平槽壓濾板
flushing	涌料	沖流
flush plate	平槽滤板	平槽濾板
flux	通量	①通量②助溶劑
flux density	通量密度	通量密度
flux density vector	通量密度矢量	通量向量密度

英　文　名	祖国大陆名	台湾地区名
foam	泡沫	泡體, 泡沫
foam agent	发泡剂	發泡劑
foam breaker	消泡器	除泡機
foam concrete	泡沫混凝土	泡混凝土
foam fractionation	泡沫分馏	泡沫分餾
foaming	发泡	發泡
foam plastic	泡沫塑料	泡沫塑膠
foam rubber	泡沫橡胶	泡沫橡膠
foam separation	泡沫分离	泡沫分離
foam stabilizer	稳泡剂	泡沫穩定劑
focus	焦点	焦點
fog nozzle	雾喷嘴	霧噴嘴
food processing	食品加工	食品加工
food-to-microorganism ratio	食物-微生物比	食物-微生物比
foot valve	底阀	底閥, 腳閥
force balance	力平衡	力均衡
force-current analogy	力-电流类比	力-電流類比
forced circulation evaporation	强制循环蒸发	強制循環蒸發
forced circulation evaporator	强制循环蒸发器	強制循環蒸發器
forced convection	强制对流	強制對流
forced convection evaporation	强制对流蒸发	強制對流蒸發
forced convection heat transfer	强制对流传热	強制對流熱傳送
forced diffusion	强制扩散	強制擴散
forced draft	强制通风	強制通風, 鼓風
forced draft aerator	鼓风通气机	鼓風通氣器
forced draft cooling tower	强制通风凉水塔	強制通風冷卻塔
forced draft fan	鼓风机	鼓風扇, 強制通風扇
forced vortex	强制涡旋	強制漩渦
force-mass conversion constant	力-质量换算常数	力-質量換算常數
force oscillation	强制振荡	強制振盪
force potential	力势	力勢
force-voltage analogy	力-电压类比	力-伏特類比
forcing function	强制函数	強制函數
formalin	福尔马林	福馬林, 甲醛液
formal report	正式报告	正式報告
formamide	甲酰胺	甲醯胺
formamide process	甲酰胺法	甲醯胺法
form drag	形状阻力	形狀阻力
formula	①[化学]式②公式	①化學式②[公]式

英　文　名	祖国大陆名	台湾地区名
formulation	数式化	①配方②公式化
forsterite	镁橄榄石	矽酸鎂石
forsterite brick	镁橄榄石砖	矽酸鎂石磚
forsterite porcelain	镁橄榄石瓷	矽酸鎂石瓷
forward feed	顺向进料	順向進料
forward reaction	正反应	正反應
fossil	化石	化石
fossil fuel	化石燃料	化石燃料
fouling	污垢, 结垢	積垢
fouling coefficient	污垢系数	積垢係數
fouling factor	污垢系数	積垢因數
fourdrinier machine	长网[成型]机	長綱[製紙]機
Fourier number	傅里叶数	傅利葉數
four-way valve	四通阀	四通閥
fractal	碎形	分形
fraction	①分数②分率	①分數②分率
fractional absorption	分级吸收	分吸收
fractional adsorption	分级吸附	分吸附
fractional analysis	分段分析	分段分析
fractional column	分馏塔	分餾塔
fractional condensation	分凝作用	分級冷凝
fractional condenser	分凝器	分凝器
fractional condensing tube	分凝管	分凝管
fractional conversion	转化率	轉化分率
fractional crystallization	分步结晶	分段結晶
fractional dialysis	分级渗析	分級透析
fractional distillation	分馏	分餾
fractional efficiency	分级效率	分級效率
fractional expansion	分步膨胀	分步膨脹
fractional extraction	分馏萃取	分[步]萃取
fractional life	部分寿命	部分壽命
fractional-life method	部分衰减法	部分衰減法
fractional-order reaction	分数级反应	分數階次反應
fractional precipitation	分级沉淀	分級沈澱
fractionating column	分馏柱	分餾塔
fractionating tower	分馏塔	分餾塔
fractionation	分馏	分餾
fractionation assembly	分馏装置	分餾裝置
fractionator	分馏器	分餾器

英　文　名	祖国大陆名	台湾地区名
fraction of coverage	覆盖率	覆蓋率
fracture mechanics	断裂力学	破壞力學
fracture strength	断裂强度	破壞強度
fracture stress	断裂应力	破壞應力
fragility	易碎性	易碎性
fragmentation	碎裂	碎片化[作用]
frame	框架	框,架
frame and plate filter	板框[式]过滤机	框板過濾機
frame filter press	框式压滤机	框式壓濾機
frame of reference	参考坐标	參考座標
franklinite	锌铁尖晶石	鋅鐵礦
free acid	游离酸	自由酸
free alkali	游离碱	自由鹼
freeboard	自由空间, 分离空间	自由空間, 分離空間
free-centered lattice	面心晶格	面心晶格
free convection	自由对流	自然對流
free crushing	自由压碎	自然壓碎
free energy	自由能	自由能
free energy-composition diagram	自由能组成图	自由能組成圖
free enzyme	游离酶	游離酶
free falling velocity	自由沉降速度	自由沉降速度
free flow	自由流动	自由流動
free flow head	自由流动压头	自由流動高差
free moisture	游离水分, 自由水分	自由水分
free moisture content	自由水含量	自由水含量
free path	自由程	自由徑
free radical	自由基	自由基
free radical mechanism	自由基机理	自由基機構
free radical polymerization	自由基聚合	自由基聚合[反應]
free radical reaction	自由基反应	自由基反應
free settling	自由沉降	自由沉降
free site concentration	自由部位浓度	自由部位濃度
free stream	自由流	自由流
free surface	自由表面	自由表面
free vortex	自由涡旋	自由漩渦
free water	自由水	自由水
freeze coagulation	冷冻凝器	冷凍凝聚
freeze drier	冷冻干燥机	冷凍乾燥機
freeze drying	冷冻干燥	冷凍乾燥

英　文　名	祖国大陆名	台湾地区名
freezer	冷冻器	冷凍器
freezing minimum	最低凝固点	最低凝固點
freezing point	凝固点	凝固點, 冰點
freezing point depression	凝固点降低	凝固點下降, 冰點下降
freezing point lowering	凝固点降低	凝固點下降
freezing process	冷冻过程	冷凍法, 冷凍程序
freon	氟氯烷	氟氯烷
frequency	频率	頻率
frequency band	频带	頻率帶
frequency domain	频域	頻率領域
frequency factor	频率因子	①頻率因數②指數前因數
frequency response	频率响应	頻率應答
frequency response analysis	频率响应分析	頻率應答分析
frequency shift	频移	頻率移動, 頻率移位
frequency test	频率测试	頻率試驗法
frequency tracer	频率示踪器	頻率示蹤器
friability	脆性	脆性
friction	摩擦	摩擦
frictional resistance	摩擦阻力	摩擦阻力
friction coefficient	摩擦系数	摩擦係數
friction corrosion	摩擦腐蚀	摩擦腐蝕
friction coupling	摩擦接头	摩擦接頭
friction drag	摩擦阻力	摩擦阻力
friction drop	摩擦压力降	摩擦壓力降
friction factor	摩擦因子	摩擦因數
friction head	摩擦压头	摩擦高差
frictionless flow	无摩擦流动	無摩擦流動
friction loss	摩擦损失	摩擦損失
friction loss factor	摩擦损失因子	摩擦損失因數
friction separator	摩擦分离器	摩擦分離器
friction test	摩擦试验	摩擦試驗
frit	玻璃料	玻璃料
fritted glass filter	多孔玻璃过滤器	燒結玻璃過濾器
frontal analysis	锋面分析	鋒面分析
frosted glass	磨砂玻璃	磨砂玻璃
froth	泡沫	泡沫
froth depth	泡沫深度	泡沫深度
froth flotation	泡沫浮选	泡沫浮選

英 文 名	祖国大陆名	台湾地区名
fructose	果糖	果糖
F-test	F 检验	F 試驗
fuel	燃料	燃料
fuel-air ratio	燃料空气比	燃料空氣比
fuel alcohol	燃料酒精	燃料酒精
fuel bed	燃料床	燃料床
fuel cell	燃料电池	燃料電池
fuel cost	燃料成本	燃料成本
fuel industry	燃料工业	燃料工業
fuel oil	燃料油	燃料油
fuel ratio	燃料比	燃料比
fugacity	逸度	逸壓
fugacity coefficient	逸度系数	逸壓係數
fugitive accelerator	短效加速剂	短效催速劑
fuller's earth	漂白土	漂白土
fully developed flow	充分发展流	完全發展流動
fumagillin	烟曲霉素	煙黴素
fumigant	熏蒸剂	燻蒸劑
functional equation	泛函方程式	泛函方程式
functional group	官能团	官能基, 功能基
functional polymer	功能高分子	功能性聚合物
functional relationship	函数关系	函數關係[式]
function generator	函数发生器	函數波產生器
fundamental principle	基本原理	基本原理
fundamental property relation	基本性质关系式	基本性質關係式
fundamentals	基本原理	基本原理
fundamental unit	基本单位	基本單位
fungi	真菌	真菌
fungicide	杀菌剂	殺菌劑
furan plastic	呋喃塑料	呋喃塑造
furan resin	呋喃树脂	呋喃樹脂
furnace	[窑]炉	爐
furnace black	炉黑	爐黑
fusel oil	杂醇油	雜醇油
fusion	①熔融②聚变	熔化, 熔合
fusion curve	熔化曲线	熔化曲線
fusion point	熔点	熔點

G

英 文 名	祖国大陆名	台湾地区名
gage number	规号	規號
gage pressure	表压	錶壓
gain	增益	增益
gain compensation	增益补偿	增益補償
gain crossover	增益交互	交越增益
gain crossover frequency	增益分隔频率	增益交越頻率
gain crossover point	增益交点	增益交越點
gain margin	增益裕量	增益邊限
gain matrix	增益矩阵	增益矩陣
gain-phase plot	增益相图	增益-相位圖
galvanic action	电流作用	電流作用
galvanic cell	原电池	電池
galvanic corrosion	电偶腐蚀	電流腐蝕
galvanic protection	镀锌保护	鍍鋅保護
galvanization	镀锌	鍍鋅
galvanometer	检流计	電流計
gangue	煤矸石	礦渣, 礦石
gas	气[体]	①氣②瓦斯
gas analyzer	气体分析仪	氣體分析儀
gas burner	煤气喷灯	煤氣爐, 煤氣燈
gas calorimeter	气体量热计	煤氣卡計
gas chromatogram	气相色谱图	氣相層析圖
gas chromatograph	气相色谱仪	氣相層析儀
gas chromatography	气相色谱法	氣相層析法
gas constant	气体常数	氣體常數
gas distributor	气体分布器	氣體分布器
gaseous emission standard	气体发射标准	氣體排放標準
gaseous floatation	气体浮选	氣體浮選
gaseous fuel	气体燃料	氣體燃料
gaseous solution	气体溶液	氣體溶液
gas field	气田	天然氣田
gas film	气膜	氣膜
gas film contactor	气膜接触器	氣膜接觸器
gas film control	气膜控制	氣膜控制
gas-forming reaction	成气反应	成氣反應

英　文　名	祖国大陆名	台湾地区名
gas holdup	持气率	持氣率
gasification	气化	氣化
gasification process	气化过程	氣化程序
gasification reaction	气化反应	氣化反應
gasification technology	气化技术	氣化技術
gasifier	气化炉	氣化爐, 氣化器
gasifying reaction	气化反应	氣化反應
gasket groove	垫圈槽	墊圈溝
gasket material	垫片材料	墊片材料
gas law	气体定律	氣體定律
gas-liquid absorption	气液吸收	氣液吸收
gas-liquid contacting	气液接触	氣液接觸
gas-liquid contactor	气液接触器	氣液接觸器
gas-liquid equilibrium	气液平衡	氣液平衡
gas-liquid equilibrium process	气液平衡过程	氣液平衡程序
gas-liquid reaction	气液反应	氣液反應
gas-liquid reactor	气液反应器	氣液反應器
gas-liquid separation	气液分离	氣液分離
gas-liquid-solid reaction	气液固反应	氣液固反應
gas-liquid-solid reactor	气液固反应器	氣液固反應器
gas main	气体总管	氣體總管
gas measuring tube	量气管	量氣管
gas meter	煤气表	氣量計
gasohol	酒精汽油	酒精汽油
gas oil	瓦斯油	製氣油, 柴油
gas-oil ratio	气油比	氣-油比
gasoline	汽油	汽油
gasoline engine	汽油机	汽油引擎
gasoline trap	汽油阱	汽油阱
gas permeability	透气性	透氣性
gas permeation	气体渗透	氣體滲透
gas phase	气相	氣相
gas phase control	气相控制	氣相控制
gas phase loading factor	气相动能因子	氣相動能因子
gas phase reaction	气相反应	氣相反應
gas phase solid	气相固体	氣態固體
gas producer	煤气发生炉	發生爐
gas receiver	气体收集器	收氣器
gas reforming	气体重整	氣重組

英　文　名	祖国大陆名	台湾地区名
gas regulator	气体调节器	氣體調節器
gas scrubber	气体洗涤器	洗氣器
gas shift	水煤气变换	水煤氣轉化
gas shift process	水煤气变换过程	水煤氣轉化程序
gas shift reaction	水煤气变换反应	水煤氣轉化反應
gas-solid equilibrium	气固平衡	氣固平衡
gas-solid reaction	气固反应	氣固反應
gas-solid reactor	气固反应器	氣固反應器
gas stripping	气提	氣提
gas thermometer	气体温度计	氣體溫度計
gas trap	气阱	氣阱
gas turbine	燃气轮机	氣渦輪機
gas turbine engine	燃气轮发动机	氣渦輪引擎
gas venting system	排气系统	排氣系統
gate	闸门	①閘②澆口
gate valve	闸阀	閘閥
gauge	表	表
Gaussian elimination	高斯消元法	高斯消去法
gauze catalyst	网状催化剂	觸媒綱
gaylussite	斜钠钙石	單斜鈉石灰
gear	齿轮	齒輪
gear mixer	齿轮式混合器	齒輪式混合器
gear pump	齿轮泵	齒輪泵
gear train	齿轮组	齒輪系
gel	凝胶	凝膠
gelatin	明胶	明膠
gelatination	胶凝[作用]	膠凝
gelatin dynamite	胶质炸药	膠質代納邁炸藥
gelatining agent	胶凝剂	膠化劑
gelation	胶凝化[作用]	膠化
gel chromatography	凝胶色谱[法]	凝膠色譜分析[法]
gel entrapment	胶体包埋	膠體包埋
gel filtration	凝胶过滤	凝膠過濾
gel filtration chromatography	凝胶过滤色谱[法]	凝膠過濾層析法
gel immunoelectrophoresis	凝胶免疫电泳	凝膠免疫電泳
gene	基因	基因
generalization	普适化	通式化, 一般化
generalized coordinates	广义坐标	廣義座標
generalized correlation	通式	通式, 一般化關係式

英　文　名	祖国大陆名	台湾地区名
generalized equation	普适方程	通式
generalized fluidization	广义流态化	廣義流態化
generalized Newtonian fluid	广义牛顿流体	廣義牛頓流體
generalized Z chart	压缩因子通用图	Z值通用圖, 壓縮比通用圖
general purpose detergent	通用清洁剂	通用清潔劑
general reduced gradient method	通用简约梯度法	通用簡約梯度法
general service check list	通用服务检查表	一般服務檢查表
generation time	传代时间	傳代時間
genetically engineered cell	基因工程细胞	基因工程細胞
genetic engineering	基因工程, 遗传工程	基因工程
geometric factor	几何因数	幾何因數
geometric isomer	几何异构物	幾何異構物
geometric mean	几何平均	幾何平均
geometric mean area	几何平均面积	幾何平均面積
geometric mean temperature difference	几何平均温差	幾何平均溫差
geometric programming	几何规划	幾何規劃
geometric similarity	几何相似	幾何相似性
geometric surface area	几何表面积	幾何表面積
germicide	杀菌剂	殺菌劑
Gibbs Duhem equation	吉布斯-杜安方程	吉布斯-杜安方程
Gibbs free energy	吉布斯自由能, 自由焓	吉布斯自由能, 自由焓
gibbsite	三水铝石	水礬土, 水鋁氣
gibbsite fireclay	三水铝石耐火土	水礬土耐火土
glacial acetic acid	冰醋酸	冰醋酸
glass	玻璃	玻璃
glass blowing	玻璃吹制	玻璃吹製
glass ceramics	玻璃陶瓷	玻璃陶瓷
glass drawing	玻璃拉制	玻璃拉製
glass electrode	玻璃电极	玻璃電極
glass enamel	玻璃搪瓷	玻璃搪瓷
glass fiber	玻璃纤维	玻璃纖維
glass fiber reinforced plastics (GFRP)	玻璃纤维强化塑料	玻璃纖維強化塑膠
glass filter	玻璃过滤器	玻璃過濾器
glass gage	玻璃液位计	玻璃液位計, 玻璃量規
glass-lined pipe	衬玻璃管	玻璃襯管
glass lining	玻璃衬里	玻璃襯裡
glass pipe	玻璃管	玻璃管
glass pipe and fitting	玻璃管及配件	玻璃管子及配件

英　文　名	祖国大陆名	台湾地区名
glass reinforced plastics	玻璃强化塑料	玻璃強化塑膠
glass stem thermometer	玻璃温度计	玻璃溫度計
glass textile	玻璃纤维	玻璃纖維
glass transition temperature	玻璃转变温度	玻璃轉移溫度
glass tube	玻璃管	玻璃管
glass wool	玻璃棉	玻璃絨, 玻璃綿
glassy state	玻璃态	玻璃狀態
glassy zone	玻璃区	玻璃區
glauberite	钙芒硝	鈣芒硝
glaze	釉	釉
glazing	上釉	浸釉[法]
global optimum	总体最优[值]	總體最適[值]
global rate	总体速率	總速率
global reaction	总体反应	總體反應
global reaction rate	总体反应速率	總體反應速率
global stability	总体稳定性	總體穩定性
globe valve	截止阀	阻塞閥
glost kiln	釉烧窑	釉窯
glucan	葡聚糖	葡聚糖
glucose	葡萄糖	葡萄糖
glue	胶	膠
glutamicus micrococcus	麸酸球菌	麩酸球菌
gluten	面筋	麩質, 麵筋
glutin	明胶蛋白	明膠蛋白
glyceride	甘油酯	甘油酯
glycerine	甘油	甘油
glycogen	肝糖	肝糖
glycol	乙二醇	乙二醇
glycolysis	糖酵解	醣解
glycolytic pathway	糖酵解途径	糖酵解途徑
glyptal resin	甘酞树脂	酞酸樹脂
golden section searching method	黄金分割法	黃金分割搜尋法
Gouy Stodola theorem	古伊-斯托多拉定理	高伊-斯托多拉定理
governing equation	支配方程	統御方程式
gradient	梯度	梯度
gradientless contactor	无梯度接触器	無梯度接觸槽
gradientless reactor	无梯度反应器	無梯度反應器
gradient method	梯度法	梯度法
Graetz number	格雷茨数	格雷茲數

英　文　名	祖国大陆名	台湾地区名
graft copolymer	接枝共聚物	接枝聚合物
grafting	接枝[法]	接枝[法]
graft polymer	接枝聚合物	接枝共聚物
graft polymerization	接枝聚合	接枝聚合[反應]
grain	晶粒	粒
grain effect	晶粒效应	壓紋效應
grain growth	晶粒长大	顆粒成長
grain shape	[晶]粒形[状]	粒形
grain size	晶粒大小	顆粒大小
grainsize analysis	粒度分析	粒徑分析
grainsize analyzer	粒度分析仪	粒徑分析儀
gramicidin	短杆菌肽	短桿菌素, 滅革蘭菌素
grand canonical ensemble	巨正则系综	巨正則係綜
grand canonical partition function	巨正则配分函数	巨正則配分函數
granite	花岗岩	花崗岩
granular bed filter	颗粒层过滤器	顆粒層過濾器
granular material	颗粒材料	粒狀物質
granular solid filter	粒状固体过滤器	粒狀固體過濾機
granulation	造粒	造粒
granule	颗粒体	顆粒
granulometer	粒度分析仪	粒徑分析儀
granulometry	粒度分析	粒徑分析
graphical analysis	图解分析	圖解分析法
graphical calculation	图解计算	圖解計算
graphical integration	图解积分	圖解積分法
graphical method	图解法	圖解法
graphical solution	图解	圖解
graphic panel	图解仪表板	圖解儀表板
graphite	石墨	石墨纖維
graphite crucible	石墨坩埚	石墨坩堝
graphite electrode	石墨电极	石墨電極
graphite fiber reinforced plastics	石墨纤维强化塑料	石墨纖維強化塑膠
graphite grease	石墨润滑脂	石墨滑脂
graphite refractory	石墨耐火材料	石墨耐火物
graphitic fiber	石墨纤维	石墨纖維
graphitization	石墨化	石墨化
graph theory	图论	圖論
Grashof number	格拉斯霍夫数	葛瑞斯何夫數
gravimetric analysis	重量分析[法]	重量分析[法]

英　文　名	祖国大陆名	台湾地区名
gravitation	引力	重力
gravitational constant	引力常量	重力常數
gravitational conversion factor	重力换算因数	重力換算因數
gravitational field	引力场	重力場
gravitational filter	重力过滤器	重力過濾器
gravitational force	重力	重力
gravitational separation	重力分离	重力分離
gravitation settling	重力沉降	重力沉降
gravity	重力	重力
gravity cell	重力电池	重力電池
gravity circulation	重力循环	重力循環
gravity filter	重力式过滤器	重力過濾器
gravity filtration	重力过滤	重力過濾
gravity flow	重力流动	重力流動
gravity head	重力压头	重力高差
gravity meter	重差计(测定比重用)	比重计
gravity segregation	重力分凝	重力分凝, 重力離析
gravity separation	重力分离	重力分離
gravity separator	重力分离器	重力分離器
gravity settler	重力沉降器	重力沉降器
gravity settler separator	重力沉降分离器	重力沉降分離器
gravity settling	重力沉降	重力沉降
gravity thickener	重力增稠器	重力增稠器
gravity thickening	重力增稠	重力增稠
gray body	灰体	灰體
gray surface	灰面	灰面
gray system	灰色系统	灰色系統
grease	润滑脂	[潤]滑脂
greenhouse effect	温室效应	溫室效應
grey body	灰体	灰體
GRG method	通用简约梯度法	通用簡約梯度法
grid agitator	框式搅拌器	框式攪拌器
grind	研磨	研磨
grindability	可磨性	可磨性
grindability index	可磨性指数	可磨性指數
grinder	磨床	磨床, 研磨機
grinding	研磨	研磨
grinding additive	研磨辅料	研磨添加劑
grinding aids	研磨辅料	助磨劑

英　文　名	祖国大陆名	台湾地区名
grinding ball	磨球	磨球
grinding machine	磨床	磨床
grinding medium	研磨介质	研磨介質
grizzly	格筛	柵篩
gross earnings cost	毛利成本	毛利成本
gross error	过失误差	總誤差
gross error identification	过失误差检出	總誤差鑑定
gross heating value	总热值	總熱值
groundwater	地下水	地下水
groundwater pollution	地下水污染	地下水污染
groundwater treatment	地下水处理	地下水處理
groundwood pulp	磨木浆	磨木漿
group activity coefficient	基团活度系数	基團活性係數
group fraction in solution	溶液中基团分率	溶液中基團分率
group specificity	基团专一性	基團異性
growth curve	生长曲线	生長曲線
growth factor	生长因子	生長因子
growth kinetics	生长动力学, 增殖动力学	生長動力學
growth parameter	生长参数	生長參數
growth rate	生长率	生長速率
growth yield	生长收率	生長產量
growth yield coefficient	生长收率系数, 增殖收率系数	生長產量係數
guanase	鸟嘌呤酶	鳥嘌呤酶
guanine	鸟嘌呤	鳥嘌呤
guano	海鸟粪	[海]鳥糞
guano phosphate	鸟粪磷酸盐	[海]鳥糞磷礦
guaranteed quality	保证质量	保證品質
guard bed	护床	護床
guard catalyst	防护催化剂	防護觸媒
gum	树胶	膠
gun cotton	硝棉	強[硝化]棉
gypsite	土石膏	土[狀]石膏
gypsum	石膏	石膏
gyrating screen	回转振动筛	偏旋篩
gyratory break	回转轧碎机	迴轉破碎機
gyratory crusher	回转破碎机	迴轉壓碎機
gyroscopic flowmeter	回转流量计	迴轉流量計

H

英　文　名	祖国大陆名	台湾地区名
hair hygrometer	毛法湿度计	毛法濕度計
hair-pin tube	发夹管	髮夾管
hairpin tube heat exchanger	U 型管换热器	髮夾管換熱器
hair spring	发状弹簧	髮狀彈簧
half body	半体	半體
half life	半衰期	半衰期
half life method	半衰期法	半衰期法
half life of enzyme	酶半衰期	酶半衰期
halite	石盐	岩鹽
halogenation	卤化	鹵化[反應]
hammer crusher	锤[式破]碎机	鎚碎機
hammer mill	锤[式]磨[碎]机	鎚磨機
handling facility	装卸设备	裝卸設備
hard detergent	硬性洗涤剂	硬性清潔劑
hardened oil	硬化油	硬化油
hardener	硬化剂	硬化劑
hardening	硬化	硬化
hardness	硬度	硬度
hardness scale	硬度标度	硬度尺標
hardness test	硬度试验	硬度試驗
hard rubber	硬橡胶	硬橡膠
hard sphere	硬球	硬球
hard sphere theory	①硬球理论②碰撞理论	①硬球理論②碰撞理論
hardware	硬件	硬體
hard water	硬水	硬水
hard wax	硬蜡	硬蠟
harmonic analysis	谐波分析	諧波分析
harmonic cycling	谐和循环	諧和循環
harmonic mean	谐和平均	諧和平均
harmonic oscillator	谐振子	諧波振盪器
harmonic response	谐波应答	諧波應答
hastelloy	哈斯特镍基合金	赫史特合金
Hatta number	八田数	八田數
hay filter	枯草过滤器	乾草過濾器
hazard evaluation	危险性评估	危害性評估

英　文　名	祖国大陆名	台湾地区名
hazard index	危险指数	危險指數
hazardous waste	危险废物	危害性廢棄物
head	扬程	高差, 揚程
head condenser	顶冷凝器	頂冷凝器
head effect	高差效应	高差效應
header	集管	集管[箱]
head loss	压头损失	高差損失, 落差損失
heap and dump leaching	堆[积]浸[取]	堆積浸取
hearth	炉缸	爐
heat	热	熱
heat accumulator	蓄热器	蓄熱器
heat balance	热[量]平衡	熱量均衡
heat capacity	热容	熱容量
heat capacity at constant pressure	等压热容	等壓熱容量
heat capacity at constant volume	等容热容	等容熱容量
heat capacity flow rate	热容流率	熱容流率
heat carrier	载热体	載熱體
heat content	热含量	熱含量
heat convection	热对流	熱對流
heat dissipation	热耗散	熱散逸
heat duty	热负载	熱負載
heat effect	热效应	熱效應
heat efficiency	热效率	熱效率
heat emission	热发射	熱發射
heat energy	热能	熱能
heat engine	热机	熱機
heater	加热器	加熱器, 加熱爐
heat exchange	换热	熱交換
heat exchanger	换热器, 热交换器	熱交換器
heat exchanger network	换热器网路	熱交換器網路
heat flow	热流[量]	熱流動
heat flow	热流	熱流動
heat flux	热通量	熱通量
heat flux	热流通量	熱通量
heat generation	热生成	熱生成
heat generation curve	热生成曲线	熱生成曲線
heat generation line	热生成线	熱生成線
heating	加热	加熱
heating area	加热面积	加熱面積

英 文 名	祖国大陆名	台湾地区名
heating coil	加热盘管	加熱旋管
heating cost	加热成本	加熱成本
heating medium	载热体	熱媒, 加熱介質
heating rate	加速速率	加熱速率
heating time	加热时间	加熱時間
heating unit	加热单元	加熱裝置, 加熱單元
heating value	热值	熱值
heat insulating material	隔热材料	保溫材料
heat insulation	隔热	隔熱
heat integration	热集成	熱整合
heat interchanger	热交换器	熱交換器
heat-liberating reaction	放热反应	放熱反應
heat loss	热损失	熱損失
heat of absorption	吸收热	吸收熱
heat of adhesion	附着热	附著熱
heat of adsorption	吸附热	吸附熱
heat of coagulation	凝结热	凝聚熱
heat of combination	化合热	化合熱
heat of combustion	燃烧热	燃燒熱
heat of compression	压缩热	壓縮熱
heat of condensation	冷凝热	冷凝熱
heat of conversion	转化热	轉化熱
heat of crystallization	结晶热	結晶熱
heat of decomposition	分解热	分解熱
heat of dilution	稀释热	稀釋熱
heat of evaporation	蒸发热	蒸發熱
heat of formation	生成热	生成熱
heat of fusion	熔化热	熔化熱
heat of hydration	水合热	水合熱
heat of liquefaction	液化热	液化熱
heat of mixing	混合热	混合熱
heat of neutralization	中和热	中和熱
heat of reaction	反应热	反應熱
heat of solidification	固化热	固化熱
heat of solution	溶解热	溶解熱
heat of sublimation	升华热	昇華熱
heat of transition	转变热	轉移熱, 轉變熱
heat of vaporization	汽化热	汽化熱
heat-pipe	热管	熱管

英　文　名	祖国大陆名	台湾地区名
heat-pipe exchanger	热管换热器	熱管交換器
heat pollution	热污染	熱污染
heat pump	热泵	熱泵
heat rate	热功转化率	熱功轉化率
heat recovery	热量回收	熱回收
heat regeneration	加热再生	熱再生
heat regenerator	蓄热器	蓄熱器
heat removal curve	热量去除曲线	熱量去除曲線
heat removal line	热量去除曲线	熱量去除曲線
heat reservoir	储热器	貯熱器
heat resistance	耐热性	耐熱性
heat sealing adhesive	热封黏合剂	熱封黏合劑
heat sink	热阱	熱匯座, 熱壑
heat source	热源	熱源
heat transfer	传热, 热量传递	熱傳送
heat transfer area	传热面积	熱傳面積
heat transfer coefficient	传热系数	熱傳係數
heat transfer equipment	传热设备	熱傳裝置
heat transfer medium	载热体	熱傳送介質
heat transfer rate	传热速率	熱傳速率
heat transfer resistance	传热阻力	熱傳阻力
heat transport	热传递	熱輸送
heat treatment	热处理	熱處理
heat value	热值	熱值
heavy chemical	重化学品	重化學品
heavy duty detergent	高效型洗涤剂	強力清潔劑
heavy ends	重尾馏分	重餾分
heavy fuel oil	重燃料油	重燃料油
heavy hydrocarbon	重烃	重烴
heavy hydrogen	重氢	重氫
heavy oil	重质油	重油
heavy straight-run	重直馏油	重直餾油
heavy water	重水	重水
height equivalent of a theoretical plate	等[理论]板高度, 理论板当量高度	理論板相當高度
height of a heat transfer unit	传热单元高度	熱傳單元高度
height of a mass transfer unit	传质单元高度	質傳單元高度
helical agitator	螺旋搅拌器	螺旋攪拌器
helical coil	螺旋管	螺旋管

英　文　名	祖国大陆名	台湾地区名
helical fin	螺旋翅片	螺旋鰭片
helical pressure spring	螺旋压力弹簧	螺旋壓力彈簧
helical ribbon agitator	螺带搅拌器	螺帶攪拌器
helical slidewire	螺旋滑线	螺旋滑線
heliotrope oil	葵花油	葵花油
Helmholtz free energy	亥姆霍兹自由能,自由能	亥姆霍兹自由能,自由能
hematite	赤铁矿	赤鐵礦
hemicellulose	半纤维素	半纖維素
hemp seed oil	麻油	木麻油
Henry's law	亨利定律	亨利定律
herbicide	除草剂	除草劑
Hess's law	赫斯定律	赫斯定律,蓋斯定律
heteroazeotrope	非均相共沸混合物	不均勻相共沸液
heterofermentative bacteria	异型发酵菌	雜發酵菌
heterogeneity	①不均匀性②多相性	①不勻性②多相性③不勻相
heterogeneous azeotrope	非均相共沸混合物	不勻相共沸液
heterogeneous catalysis	非均相催化	不勻相催化(作用)
heterogeneous equilibrium	多相平衡	不勻相平衡
heterogeneously fluidized bed	非均一流化床	不勻相流體化床
heterogeneous nucleation	异相成核	不勻相成核
heterogeneous polymerization	非均相聚合	不勻相聚合(反應)
heterogeneous population	杂菌群	雜菌群
heterogeneous reaction	非均相反应	不勻相反應
heterogeneous reactor	非均相反应器	不勻相反應器
heterogeneous state	非均相状态	不勻相狀態
heterogeneous system	非均相系统	不勻系統
heteropolyacid catalyst	杂多酸	異聚酸觸媒
heuristic method	直观推断法,启发式方法	啟發式法
heuristic rule	直观推断法则,启发式规则	啟發式法則
hexagonal closest packed lattice	六方最密堆积晶格	六方最密晶格
hexose phosphate shunt pathway	磷酸己糖旁路途径	磷酸己糖旁路途徑
hidden layer	隐蔽层	隱含層
hierarchical control	递阶控制	階梯控制
high alumina brick	高铝砖	高鋁磚
high alumina cement	高铝水泥	高鋁水泥

英 文 名	祖国大陆名	台湾地区名
high alumina clay	高铝黏土	高鋁黏土
high alumina glass	高铝玻璃	高鋁玻璃
high alumina refractory	高铝耐火材料	高鋁耐火物
high antiknocking fuel	高抗爆性燃料	高抗震爆燃料
high-Btu gas	高热量气体	高熱量氣體
high density polyethylene	高密度聚乙烯	高密度聚乙烯
high heating value	高热值	高熱值
high impact polystyrene	高抗冲聚苯乙烯	耐衝擊聚苯乙烯
high iron Portland cement	高铁[卜特兰]水泥	高鐵[卜特蘭]水泥
high lift device	增升装置	舉升裝備
high-pass filter	高通滤波器	高通濾波器
high performance liquid chromatograph	高效液相色谱仪	高性能液相層析儀
high polymer	高聚物	聚合物, 高分子化合物
high-rate digester	高速率消化槽	高速率消化槽
high selector switch	高选择器开关	高選擇器開關
high silica iron	高硅铁	高矽鐵
high speed carbon steel	高速碳钢	高速碳鋼
high speed evaporator	高速蒸发器	高速蒸發器
high strength portland cement	高强硅盐水泥	高強度[卜特蘭]水泥
high styrene rubber	高苯乙烯橡胶	高苯乙烯橡膠
high tensile strength	高抗拉强度	高抗張力
high test molasses	高级糖蜜	高級糖蜜
high transmission glass	高透光玻璃	高透光玻璃
high vacuum distillation	高真空蒸馏	高真空蒸餾
high vacuum operation	高真空操作	高真空操作
hindered sedimentation	受阻沉降	受阻沉降
hindered settling	受阻沉降	受阻沉降
histogram	直方图	直立圖, 矩形圖
hoist	起重机	起重機
holding period	保持期间	貯留期間
holding pond	滞留池	貯留池
holding tank	滞留槽	貯留槽
holding time	保留时间	貯留時間
hold-up	滞留量	貯留量
hold-up time	滞留时间	貯留時間
hollander	打浆机	打漿機
hollow agitator	自吸搅拌器, 空心叶轮搅拌器	空心葉輪攪拌器
hollow-fiber module	中空纤维组件	中空纖維組件

英 文 名	祖国大陆名	台湾地区名
holocellulose	全纤维素	全纖維素
homoazeotrope	均相共沸混合物	匀相共沸液
homofermentative bacteria	同型发酵菌	純種發酵菌
homogenate	匀浆	均質
homogeneity	均匀性	均匀性
homogeneous azeotrope	均匀共沸混合物	匀相共沸液
homogeneous body	均匀体	均匀體
homogeneous catalysis	均相催化	匀相催化[作用]
homogeneous combustion	均相燃烧	匀相燃燒
homogeneous equilibrium	均相平衡	匀相平衡
homogeneously fluidized bed	均匀流化床	匀相流體化床
homogeneous nucleation	均相成核	匀相成核
homogeneous phase	均相	匀相
homogeneous reaction	均相反应	匀相反應
homogeneous reactor	均相反应器	匀相反應器
homogeneous system	均相系统	均匀系統
homogeneous turbulence	均匀湍流	均匀紊流
homogenization	匀化	均質化
homogenizer	匀化器	均質機
homologous compound	同系化合物	同系物
homologue	同系物	同系物
homopolymer	均聚物	同元聚合物
homopolymerization	均聚反应	同元聚合[反應]
hood	防护罩	通風櫥, 排氣罩
hoop stress	环向应力	環應力
horizontal centrifugal screen	卧式离心筛	卧式離心篩
horizontal digester	卧式蒸煮器	卧式蒸煮器
horizontal furnace	卧式炉	卧式爐
horizontal mixer	卧式混合器	卧式混合器
horizontal spray chamber	横式喷雾室	卧式噴霧室
horizontal tube bank	平管排	平管排
horizontal tube evaporator	水平管式蒸发器	平管蒸發器
horsepower (hp)	马力	馬力
hose	胶管	軟管
hot blast heater	热风炉	熱風器
hot dipping	热浸	熱浸漬
hot floor drier	热坑干燥室	熱坑乾燥室
hot gas recycle process	热气再循环过程	熱氣循環程序
hot junction	热接点	熱接點

英　文　名	祖国大陆名	台湾地区名
hot lime process	热石灰法	熱石灰法
hot liming	热汁加灰法	熱汁加灰法
hot molding	加热成型	加熱成型
hot press	热压机	熱壓機
hot pressing	热压	熱壓
hot process	高温法	高溫法
hot reflux	热回流	熱回流
hot rubber	高温橡胶	熱製橡膠
hot runner	热流道	熱澆道
hot spot	热点	熱點
hot vapor	热汽	熱汽
hot vulcanization	热硫化	熱硫化
hp (＝horse power)	马力	馬力, 功率
human serum albumin	人血清清蛋白	人血清清蛋白
humectant	润湿剂	保濕劑
humid chart	湿度表	溼度表
humid heat	湿热	溼熱
humidification	增湿	增溼
humidification process	增湿过程	增溼程序
humidifier	增湿器	增溼器
humidifying	增湿	增溼
humidifying chart	增湿图	增溼圖
humidity	湿度	溼度
humidity chart	湿度表	溼度表
humidity ratio	湿度比	溼度比
humidizer	增湿剂	增溼器, 增溼劑
humid volume	湿体积	溼體積
humus	腐殖质	堆肥
hurdle	栅格	栅格
hybrid computer	数字模拟计算机	拼合計算機, 複合計算機
hybridization	杂交	雜交
hybrid tumor	杂交瘤	雜交瘤
hydrant water	消防水	消防水
hydrase	水化酶	水合酶
hydrate	水合物	水合物
hydrated lime	熟石灰	水合石灰, 熟石灰
hydrated salt	水合盐	水合鹽
hydrate of methane	甲烷水合物	甲烷水合物

英　文　名	祖国大陆名	台湾地区名
hydration	水合	水合
hydrator	水合器	水合器
hydraulic actuator	液压制动器	液壓引動器
hydraulically smooth pipe	水力平滑管	水力平滑管
hydraulically smooth wall	水力平滑壁	水力平滑壁
hydraulic cement	水硬水泥	水凝水泥
hydraulic centrifuge	液压离心机	液壓離心機
hydraulic classification	水力分类	水力類析
hydraulic control	液压控制	液壓控制
hydraulic controller	液压控制器	液壓控制器
hydraulic conveying	液压运送	液壓運送
hydraulic cylinder	液压筒	液壓筒
hydraulic diameter	水力直径	水力直徑
hydraulic efficiency	水力效率	水力效率
hydraulic fluid	液压流体	液壓流體
hydraulic fracturing	液压破裂法	液壓破裂[法]
hydraulic grade line	水力坡度线	水力級線
hydraulic head	水力压头	水力高差
hydraulic jig	水簸机	水簸器
hydraulic jump	水跃	水躍
hydraulic lime	水硬石灰	水凝石灰
hydraulic load	水力负载量	水力負載量
hydraulic mean diameter	水力平均直径	水力平均直徑
hydraulic overload	水力超载量	水力超載量
hydraulic press	水压机	水壓機
hydraulic pressure	液压	水壓
hydraulic pump	水力泵	水力泵
hydraulic radius	水力半径	水力半徑
hydraulic radius for fluid flow	流体流动水力半径	流體流動水力半徑
hydraulic radius for heat transfer	传热水力半径	熱傳水力半徑
hydraulic resistance	水力阻力	水力阻力
hydraulic resonance	水力共振	水力共振
hydraulics	水力学	水力學
hydraulic sensor	水力传感器	水力感測器
hydraulic separation	水力分离	水力分離
hydraulic separator	水力分离器	水力分離器
hydraulic tachometer	水力转速计	水力[式]轉速計
hydraulic turbine	水轮机	水力渦輪機
hydraulic valve	水力阀	水力閥

英 文 名	祖国大陆名	台湾地区名
hydraulic water	水压用水	水壓用水
hydrazine	肼	肼, 聯胺
hydrocarbon	碳氢化合物	烴, 碳氫化合物
hydrocarbon black	烃黑	烴黑
hydrocarbon fermentation	烃发酵	烴發酵
hydrochloride rubber	氢氯化橡胶	氫氯化橡膠, 鹽酸化橡膠
hydrocolloid	水胶体	親水膠體
hydrocracker	加氢裂化器	加氫裂解爐
hydrocracking	加氢裂化	加氫裂解[反應]
hydrocracking unit	加氢裂化单元	加氫裂解單元
hydrocyclone	旋液分离器	流體旋風器
hydrodealkylation	加氢脱烷基化	加氫脫烷[反應]
hydrodemetallation catalyst	加氢脱金属催化剂	加氫脫金屬觸媒
hydrodenitrogenation	加氢脱氮	加氫脫氮[反應]
hydrodesulfurization	加氢脱硫	加氫脫硫[反應]
hydrodynamically smooth surface	流体力学平滑表面	流體動力平滑表面
hydrodynamic boundary layer	流体动力边界层	流體動力邊界層
hydrodynamic effect	流体动力效应	流體動力效應
hydrodynamic interaction	流体动力相互作用	流體動力相互作用
hydrodynamics	流体动力学	流體動力學
hydrodynamic stability	流体动力稳定性	流體動力穩定性
hydroformate	加氢重整生成物	加氫重組物
hydroforming	加氢重整	加氫重組[反應]
hydroforming catalyst	加氢重整催化剂	加氫重組觸媒
hydroforming process	加氢重整过程	加氫重組程序
hydroformylation	加氢甲酰基化	加氫甲醯化[反應]
hydrogasification	加氢气化	加氫氣化[反應]
hydrogasifier	加氢气化器	加氫氣化器
hydrogenase	氢化酶	氫酶
hydrogenated fat	氢化脂	氫化脂
hydrogenated oil	氢化油	氫化油
hydrogenation	加氢	氫化(反應)
hydrogenation catalyst	氢化催化剂	氫化觸媒
hydrogenation reaction	氢化反应	氫化反應
hydrogen bomb	氢弹	氫彈
hydrogen bubble technique	氢气泡法	氫氣泡法
hydrogen discharge tube	氢放电管	氫放電管
hydrogen electrode	氢电极	氫電極

英　文　名	祖国大陆名	台湾地区名
hydrogen explosion	氢气爆炸	氫氣爆炸
hydrogen fuel	氢燃料	氫燃料
hydrogen fuel cell	氢燃料电池	氫燃料電池
hydrogen ion comparator	氢离子比长仪	氫離子比色計
hydrogenolysis	氢解	氫解(反應)
hydrokinematics	流体运动学	流體運動學
hydrolase	水解酶	水解酶
hydroliquefaction	加氢液化	加氫液化[反應]
hydrolysate	水解产物	水解產物
hydrolysis	水解	水解(作用)
hydrolytic adsorption	水解吸附	水解吸附
hydromechanics	流体力学	流體力學
hydrometallurgy	湿法冶金[学]	溼法冶金[學]
hydrometer	液体密度计	比重計
hydrometry	液体密度测定法	液體比重測定法
hydrophile	亲水物	親水物
hydrophile-lyophile balance	亲水亲油平衡	親水性-親油性均衡
hydrophilic colloid	亲水胶体	親水膠體
hydrophilic fiber	亲水纤维	親水纖維
hydrophilic particle	亲水性颗粒	親水性顆粒
hydrophobe	疏水物	①疏水性②疏水物
hydrophobic chromatography	疏水色谱[法]	疏水層析法
hydrophobic colloid	疏水胶体	疏水膠體
hydrophobic fiber	疏水纤维	疏水纖維
hydrophobic particle	疏水性颗粒	疏水性顆粒
hydropyrolysis	加氢热解	加氫高溫分解
hydroreforming	加氢重整	加氫重組
hydroreforming catalyst	加氢重整催化剂	加氫重組觸媒
hydrosol	水溶胶	水溶膠
hydrostatic condition	流体静压条件	液體靜壓狀態
hydrostatic force	流体静力	液體靜力
hydrostatic force meter	流体静力计	液體靜力計
hydrostatic head	液柱静压头	液體靜力高差
hydrostatic pressure	液体静压[强]	液體靜壓[力]
hydrostatics	液体静力学	液體靜力學
hydrosulfite	亚硫酸氢盐	二亞硫酸鹽
hydrosulfite bleaching	亚硫酸氢盐漂白	二亞硫酸鹽漂白
hydrotreater	加氢处理装置	加氫處理器
hydrotreating	加氢处理	加氫處理

英　文　名	祖国大陆名	台湾地区名
hydrotreatment	加氢处理	加氫處理
hydrotrope	水溶助长剂	增水溶剂
hydrotropic agent	水溶助长剂	增溶剂
hydrotropy	水溶助长性	增水溶性
hydroxide ion	氢氧根离子	氫氧離子
hydroxyl group	羟基	羥基
hydroxyl value	羟值	羥值
hygrometer	湿度计	溼度計
hygrometry	测湿法	測溼法
hygromycin	潮霉素	溼球菌黴素
hygroscopicity	吸湿性	吸溼能力, 吸溼性
hygroscopic water	湿存水	溼水分
hyperbolic type[kinetic]equation	双曲线型[动力学]方程	雙曲線型[動力學]方程
hyperelastic material	超弹性材料	超彈性物質
hyperfiltration	超过滤	超微過濾
hyperforming	超重整	超重組
hypersorber	超吸附器	超吸附器
hypersorption	超吸附[法]	超吸附[法]
hypertonic solution	高渗溶液	高滲壓溶液
hypochlorite bleaching	次氯酸盐漂白	次氯酸鹽漂白
hypothesis	假设	假說
hypotonic solution	低渗溶液	低滲壓溶液
hyrolytic enzyme	水解酶	水解酶
hysteresis	滞后	①遲滯②磁滯

I

英　文　名	祖国大陆名	台湾地区名
ice bath	冰浴	①冰浴②冰槽
ice point	冰点	冰點
I control (＝integral control)	积分控制	積分控制
ideal contact stage	理想接触级	理想接觸階
ideal cycle	理想周期	理想周期
ideal flow	理想流动	理想流動
ideal fluid	理想流体	理想流體
ideal gas	理想气体	理想氣體
ideal gas constant	理想气体常数	理想氣體常數
ideal gas enthalpy	理想气体焓	理想氣體焓

英　文　名	祖国大陆名	台湾地区名
ideal gas law	理想气体定律	理想氣體定律
ideal gas state	理想气体状态	理想氣體狀態
ideal gas temperature scale	理想气体温标	理想氣體溫標
ideal mixture	理想混合物	理想混合物
ideal plate	理想板	理想板
ideal reactor	理想反应器	理想反應器
ideal solution	理想溶液	理想溶液
ideal stage	理想级	理想階
ideal tray	理想板	理想板
ideal tubular reactor	理想管式反应器	理想管式反應器
ideal work	理想功	理想功
igneous quartz	火成石英	火成石英
igneous rock	火成岩	火成岩
ignition	起燃	點火, 燃燒
ignition point	燃点	燃點
ignition time	点火时间	點火時間
illite	伊利石	白雲母石
ilmenite	钛铁矿	鈦鐵礦
immersed surface	浸没表面	浸沒表面
immersion error	浸没误差	浸入誤差
immiscibility	不互溶性	不互溶性
immiscible flow	不互溶流动	不互溶流動
immiscible liquid	不互溶液体	不互溶液體
immiscible phase	不互溶相	不互溶相
immiscible solvent	不互溶溶剂	不互溶溶劑
immiscible system	不互溶系统	不互溶系統
immobilization	固相化	固定化
immobilization technology	固定化技术	固定化技術
immobilized cell reactor	固定化细胞反应器	固定化細胞反應器
immobilized enzyme	固定化酶	固定酶
immobilized liquid membrane	支撑液膜	支撐液膜
immune eletrophoresis	免疫电泳	免疫電泳
immunoadsorption	免疫吸附	免疫吸附
impact	冲击	衝擊
impact crusher	冲击式破碎机	衝擊式壓碎機
impact head	冲击压头	衝擊高差
impact number	冲击数	衝擊數
impactor	冲击器	衝擊機
impact pressure	冲击压力	衝擊壓[力]

英 文 名	祖国大陆名	台湾地区名
impact strength	冲击强度	衝擊強度
impact test	冲击试验	耐衝擊試驗
impedance	阻抗	阻抗
impeded settling	阻滞沉降	阻滯沉降
impeller	叶轮	槳,葉輪
impeller agitator	叶轮搅拌器	葉輪攪拌器
impeller diameter	叶轮直径	葉輪直徑
impeller mixer	叶轮混合器	葉輪混合器
impeller shaft	叶轮轴	葉輪軸
impeller type flowmeter	叶轮式流量计	葉輪流量計
impermeability	不渗透性	不透性
impermeability membrane	不透性膜	不透性薄膜
impermeable partition	不透性隔板	不透性隔板
impingement	碰撞	碰撞
impingement centrifugal separator	冲击式离心分离器	衝射離心分離器
impingement corrosion	冲击腐蚀	衝射腐蝕
impingement scrubber	冲击式涤气器	衝射洗氣器
implicit enumeration method	隐枚举法	隱枚舉法
implicit function	隐函数	隱函數
implicit method	隐式法	隱式法
impregnation	浸渍	浸漬
impulse	①脉冲②冲量	①脈衝②衝量
impulse disturbance	脉冲搅动	脈衝擾動
impulse function	脉冲函数	脈衝函數
impulse response	脉冲响应	脈衝應答
impulse tachometer	脉冲转速计	脈衝轉速計
impulse type	脉冲型	轉速計
impurity	杂质	不純物,雜質
inactivation	失活	失活
inactivator	惰化剂	惰化劑
incandescent light	白热光	白熱光,白熱燈
incidence matrix	关联矩阵	關聯矩陣
incineration	灼烧	焚化,灰化
incinerator	焚烧炉	焚化爐
incipient fluidization	起始流化态	起始流化態
incipient fluidizing velocity	起始流化速度	起始流化速度
inclined grate cooler	斜栅冷却器	斜栅冷卻器
inclined manometer	斜管压力计	斜管壓力計
inclined plate	斜板	斜板

英　文　名	祖国大陆名	台湾地区名
inclined tube evaporator	斜管蒸发器	斜管蒸發器
income tax	所得税	所得稅
incompressibility	不可压缩性	不可壓縮性
incompressible filter cake	不可压缩滤饼	不可壓縮濾餅
incompressible flow	不可压缩流动	不可壓縮流動
incompressible fluid	不可压缩流体	不可壓縮流體
incondensable gas	不冷凝气体	不冷凝氣體
incorporating machine	捏合机	捏合機
incremental investment	增资	增資
incubation period	培养期	培養期
incubator	恒温箱	保溫箱
indanthrene blue	阴丹士林蓝	陰丹士林藍
indanthrene dye	阴丹士林染料	陰丹士林染料
independent deactivation	独立失活	獨立失活
independent reaction	独立反应	獨立反應
independent variable	自变数	自變數
indicator	①指示剂②指示计	①指示劑②指示計
indigo blue	靛蓝	靛藍
indigo brown	靛棕	靛棕
indigo carmine	食用靛蓝	靛胭脂
indigo dyeing	靛染	靛染
indigo dye vat	靛染缸	靛染缸
indigo white	靛白	靛白
indirect condenser	间接式冷凝器	間接[式]冷凝器
indirect cost	间接成本	間接成本
indirect diaphragm valve	间接隔膜阀	間接隔膜閥
indirect evaporation	间接蒸发	間接蒸發
indirect heat exchange	间接热交换	間接熱交換
indirect heating	间接加热	間接加熱
indirect hydrogenation	间接氢化	間接氫化[反應]
indirect liquefaction	间接液化	間接液化
indirect liquid level measurement	间接液位测量	間接液位量測
indirect method	间接法	間接法
induced dipole	诱导偶极	感應偶極
induced draft	引风	抽氣通風
induced draft cooling tower	引风式凉水塔	抽氣通風冷卻塔
induced drag	诱发阻力	感應阻力
induced fit theory	[酶]诱导契合学说	[酶]誘導契合學說
induced reaction	诱发反应	誘導反應

英　文　名	祖国大陆名	台湾地区名
induction furnace	感应炉	感應爐
induction period	诱导期	誘導期
industrial alcohol	工业酒精	工業酒精
industrial chemistry	工业化学	工業化學
industrial gas	工业煤气	工業用氣體
industrial instrument	工业仪器	工業儀器
industrial process	工业过程	工業程序
industrial reactor	工业反应器	工業反應器
industrial stoichiometry	工业化学计量学	工業化學計量學
industrial-use detergent	工业用洗涤剂	工業用清潔劑
industrial waste	工业废料	工業廢棄物
industrial wastewater	工业废水	工業廢水
industrial wastewater treatment	工业废水处理	工業廢水處理
industrial water	工业用水	工業用水
industrial water treatment	工业水处理	工業用水處理
industry	工业	工業
inert	惰性物	惰性物
inert gas	惰性气体	惰性氣體
inertia	惯性	慣性
inertial centrifugal separator	惯性离心分离器	慣性離心分離器
inertial device	惯性装置	慣性裝置
inertial force	惯性力	慣力
inertial settling	惯性沉降	慣性沉降
inferential control	推理控制	推算控制
infiltration	渗滤	滲濾
infinite dilution	无限稀释	無限稀釋
infinite reflux	无限回流	無限回流
inflammability	易燃性	可燃性
inflating agent	发泡剂	發泡劑
inflection point	拐点	反曲點, 拐點
influent	流入物	流入物
informal report	非正式报告	非正式報告
information	信息	資訊
informational polymer	信息聚合物	資訊聚合物
information board	信息板	資訊板
information flow diagram	信息流图	資訊流圖
information-handling	信息处理	資訊處理
information storage	信息储存	資訊儲存
infrared analysis	红外分析	紅外線分析

英　文　名	祖国大陆名	台湾地区名
infrared analyzer	红外线分析仪	紅外線分析儀
infrared drier	红外线干燥器	紅外線乾燥器
infrared drying	红外干燥	紅外乾燥
infrared spectrophotometer	红外分光光度计	紅外線分光光度計, 紅外線光譜儀
infrared spectrophotometry	红外光谱	紅外線光譜測定法
infrared spectroscopy	红外光谱法	紅外線光譜學, 紅外線光譜法
ingot iron	生铁	生鐵
inherent viscosity	比浓对数黏度	固有黏度
inhibition	抑制	抑制[作用]
inhibitor	抑制剂	抑制劑
inhomogeneity	非均匀性	不匀性
initial boiling point	初沸点	初沸點
initial condition	初始条件	初始條件
initial rate	初速率	初速率
initial relaxed system	初始松弛系统	初始鬆弛系統
initial state	初[始状]态	初始狀態
initial value theorem	初值定理	初值定理
initial velocity	初速[度]	初速度
initiation of chain reaction	链反应引发	連鎖反應[的]起始
initiator	引发剂	啟發劑, 引發劑
injection molding	注射成型	射出成型
injector	注射器	注射器, 射出器
inlet	入口	進口
in-line filter	管线过滤器	管線過濾器
in-line mixer	管路混合器	管路混合器
in-line tube arrangement	直列管排	直列管排
inner tube	内胎	內胎
inner valve	内阀	內閥
inoculation	接种	接種
inoculum	接种物	接種液
input	输入	輸入
input disturbance	输入搅动	輸入擾動
input function	输入函数	輸入函數
input node	输入节点	輸入節點
input-output	投入产出	輸入投出
input signal	输入信号	輸入信號
input variable	输入变量	輸入變數

英　文　名	祖国大陆名	台湾地区名
insecticide	杀虫剂	殺蟲劑
inside diameter	内径	内徑
in-situ combustion	就地燃烧	就地燃燒
in-situ processing	就地加工	就地加工
in-situ recovery	就地回收	就地回收
in-situ regeneration	就地再生	就地再生
insolubilization	不溶解化	不溶解化
instability	不稳定性	不穩定性, 不安定性
instability constant	不稳定常数	不穩定常數
installation	安装	安裝
installation cost	安装成本	安裝成本
instantaneous fractional yield	瞬时产量分率	瞬間產量分率
instantaneous reaction	瞬时反应	瞬間反應
instantaneous reaction rate	瞬时反应速率	瞬間反應速率
instantaneous system	瞬时系统	瞬間系統
instrument	仪器	儀器
instrumental analysis	仪器分析	儀器分析
instrumentation	仪表配置	儀器配置, 儀器規劃
instrumentation diagram	仪表配置图	儀器配置圖
instrument flowsheet	仪表流程图	儀器流程圖
instrument servomechanism	仪器伺服机构	儀器伺服機構
insulated column	隔热[蒸馏]塔	隔熱柱, 保溫柱
insulated pipe	绝热管	隔熱管, 保溫管
insulated runner	绝热浇道	隔熱澆道
insulating brick	绝热砖	隔熱磚
insulating efficiency	绝热效率	絕緣效率, 隔熱效率
insulating firebrick	绝热耐火砖	隔熱耐火磚
insulating lining	隔热衬里	絕緣襯裏
insulating material	绝缘材料	絕緣材料, 隔熱材料, 保溫材料
insulating paper	绝缘纸	絕緣紙
insulating refractory	绝热耐火材料	隔熱耐火物
insulation	隔热	①保溫, 隔熱②絕緣
insulator	绝缘体	絕緣體
insulator catalyst	绝缘体催化剂	絕緣體觸媒
insulin	胰岛素	胰島素
insurance	保险	保險
intake pressure	进气压力	進氣壓[力]
intake screen	进料滤网	進料篩網

英　文　名	祖国大陆名	台湾地区名
intake stroke	进气行程	吸氣衝程
intalox saddle	矩鞍形填料	矩鞍形填料
integer programming	整数规划	整數規劃
integral	积分	積分[式]
integral action	积分作用	積分作用
integral control action	积分控制作用	積分控制作用
integral control (I control)	积分控制	積分控制
integral distribution curve	积分分布函数	累積分布曲線
integral energy equation	积分能量方程	積分能量方程式
integral feedback	积分反馈	積分回饋
integral heat of solution	积分溶解热	積分溶解熱
integral method	积分法	積分法
integral momentum equation	积分动量方程	積分能量方程式
integral of absolute error	绝对误差积分	絕對誤差積分
integral of square error	误差平方积分	誤差平方積分
integral reactor	积分反应器	積分反應器
integral test	积分检验[法]	積分試驗(法)
integral time	积分时间	積分時間
integral time constant	积分时间常数	積分時間常數
integral windup	积分绕紧	積分繞緊
integral yield	累积产率	累積產率
integrated error	累积误差	累積誤差
integrated process	集成过程	整合程序
integration	集成, 整合	整合
integrator	积分仪	積分器
intensity	强度	強度
intensity of turbulence	湍流强度	紊流強度
intensive property	强度性质, 内含性质	內含性質
intensive state variable	强度状态变量	內含狀態變數
interacting loops	相互作用环路	相互作用環路
interacting system	相互作用系统	相互作用系統
interaction	相互作用	相互作用
interaction coefficient	相互作用系数	交互作用係數
interaction energy	相互作用能	相互作用能
interaction factor	相互作用因数	相互作用因數
interaction force	相互作用力	相互作用力
interaction index	相互作用指数	相互作用指數
interaction of control systems	控制系统相互作用	控制系統相互作用
interaction potential	相互作用势	相互作用勢

英　文　名	祖国大陆名	台湾地区名
intercoagulation	相互凝聚	相互凝聚
intercondenser	中间冷凝器	中間冷凝器
intercooler	中间冷却器	中間冷卻器
interdiffusion	相互扩散	相互擴散
interest	利息	利息
interest recovery period	利息回收期	利息回收期
interface	界面	界面
interface composition	界面组成	界面組成
interface control	界面控制	界面控制
interfacial agent	界面活性剂	界面活性劑
interfacial area	界面面积	界面面積
interfacial concentration	界面浓度	界面濃度
interfacial energy	界面能	界面能
interfacial engineering	界面工程	界面工程
interfacial film	界面薄膜	界面薄膜
interfacial phenomenon	界面现象	界面現象
interfacial potential	界面势	界面勢
interfacial property	界面性质	界面性質
interfacial resistance	界面阻力	界面阻力
interfacial shape	界面形状	界面形狀
interfacial temperature	界面温度	界面溫度
interfacial tensimeter	界面张力计	界面張力計
interfacial tension	界面张力	界面張力
interfacial turbulence	界面湍流	界面紊流
interference pattern	干涉图样	干涉圖型
interferon	干扰素	干擾素
intergranular corrosion	晶间腐蚀	粒間腐蝕
interlaminar stress	层间应力	層間應力
intermediate	中间体	中間產物
intermediate clarifier	中间澄清器	中間澄清器,中間沉降槽
intermediate crusher	中间压碎机	中級壓碎機
intermediate product	中间产物	中間產物
intermediate storage	中间产物储存	中間產物儲存
intermetallic compound	金属间化合物	金屬間化合物
intermittent drier	间歇干燥室	間歇乾燥室
intermixing	互混	互混
intermolecular force	分子间力	分子間力
internal adsorption	内吸附	內吸附

英 文 名	祖国大陆名	台湾地区名
internal capacity	内在容量	内在容量
internal combustion	内燃	内燃
internal combustion engine	内燃机	内燃機
internal coordinates	内坐标	内部座標
internal diffusion	内扩散	内擴散
internal disturbance	内扰	内在擾動
internal energy	内能	内能
internal flow	内流	内部流動
internal friction	内耗	内摩擦
internal heating	内部加热	内部加熱
internal mixer	密炼机	密閉混合器
internal pressure	内压力	内壓(力)
internal rate of return	内部收益率	内部收益率
internal recirculation reactor	内循环反应器	内循環反應器
internal recycle	内循环	内循環
internal reflux	内回流	内回流
internal reflux control	内回流控制	内回流控制
internals	内[部]构件	内[部]構件
internal surface	内表面	内表面
international practical temperature scale	国际实用温标	國際實用溫標
interparticle diffusion	粒间扩散	粒間擴散
interphase diffusion	相间扩散	相間擴散
interphase exchange coefficient	相间交换系数	相間交換係數
interphase-intraphase diffusion	相间相内扩散	相間相内擴散
interphase-intraphase effectiveness	相间相内有效性	相間相内有效度
interphase mass transfer	相际传质	相間質傳
interphase reaction	相间反应	相間反應
interphase temperature	相间温度	相間溫度
interpolation	内插	插值法, 内插法
interstage cooling	级间冷却	階間冷卻
interstage feed injection	级间注射进料	階間(注射)進料
interstage heating	级间加热	階間加熱
interstage recirculation	级间循环	階間循環
interstitial velocity	空隙速度	間隙速度
interval halving	半区间法	半區間法
intracellular enzyme	胞内酶	胞内酶
intramolecular force	分子内力	分子内力
intraparticle diffusion	粒内扩散	粒内擴散
intraparticle diffusivity	粒内扩散系数	粒内擴散係數

英 文 名	祖国大陆名	台湾地区名
intraphase effectiveness	相内有效性	相內有效度
intraphase temperature	相内温度	相內溫度
intraphase yield	相内产率	相內產率
intrinsic activation energy	固有活化能	固有活化能
intrinsic angular momentum	固有角动量	固有角能量
intrinsic energy	固有能	固有能
intrinsic heat	固有热	蘊熱, 固有熱
intrinsic kinetics	本征动力学	固有動力學
intrinsic pressure	蘊压	蘊壓[力]
intrinsic rate	固有速率	固有速率
intrinsic safety	固有安全	固有安全
intrinsic viscosity	特性黏度	極限黏度
invariant	不变式	不變式
invention	发明	發明
inventory	[库]存量, 藏量	存貨, 盤存
inventory control	库存控制	存貨控制
inverse derivative control	反微分控制	反微分控制
inversed solubility curve	逆溶度曲级	反溶解度曲線
inverse emulsion	反相乳液	反乳液
inverse lever arm rule	反杠杆法则	反損桿法則
inverse response	反应答	反應答
inverse transformation	逆变换	逆變換
inversion point	转化点	轉化點, 反轉點
inversion temperature	转化温度	轉化溫度
invertase	转化酶	轉化酶
inverted manometer	反向压力计	反向壓力計
invert sugar	转化糖	轉化糖
investment capital	投资资本	投資資本
investment recovery	投资回收	投資回收
inviscid flow	非黏流动	無黏性流動
in vitro	体外	體外
iodine value	碘值	碘價, 碘值
ion-beam deposition	离子束沉积法	離子束沉積法
ion-beam etching	离子束蚀刻法	離子束蝕刻法
ion exchange	离子交换	離子交換
ion exchange agent	离子交换剂	離子交換劑
ion exchange capacity	离子交换容量	離子交換容量
ion exchange chromatography	离子交换色谱[法]	離子交換層析法
ion exchange equilibrium	离子交换平衡	離子交換平衡

英　文　名	祖国大陆名	台湾地区名
ion exchange membrane	离子交换膜	離子交換薄膜
ion exchange process	离子交换过程	離子交換法
ion exchanger	离子交换剂	離子交換器, 離子交換剂
ion exchange resin	离子交换树脂	離子交換樹脂
ion exchange separation	离子交换分离	離子交換分離[法]
ion exclusion	离子排斥	離子排斥
ion flotation	离子浮选	離子浮選[法]
ionic diffusion	离子扩散	離子擴散
ionic migration	离子迁移	離子遷移
ionic mobility	离子迁移率	離子移動率
ionic polymerization	离子聚合	離子聚合[反應]
ionic potential	离子电位	離子電位
ionic reaction	离子反应	離子反應
ionic strength	离子强度	離子強度
ionic surfactant	离子型表面活性剂	離子界面活性劑
ion implantation	离子注入	離子植入[法]
ionization	电离	游離
ionization constant	电离常数	游離常數
ionization energy	电离能	游離能
ionization potential	电离电位	游離電位
ionomer	离子交联聚合物	多離子聚合物
iron-alumina catalyst	铁铝氧催化剂	鐵鋁氧觸媒
iron-ammonia catalyst	铁氨催化剂	鐵氨觸媒
iron bacteria	铁细菌	鐵細菌
iron catalyst	铁催化剂	鐵觸媒
iron mordant	铁媒染剂	鐵媒染劑
iron pyrite	黄铁矿	黄鐵礦
iron tannage	铁鞣	鐵鞣
irreversibility	不可逆性	不可逆性
irreversible catalysis	不可逆催化	不可逆催化[作用]
irreversible cell	不可逆电池	不可逆電池
irreversible chemical reaction	不可逆化学反应	不可逆化學反應
irreversible process	不可逆过程	不可逆程序
irreversible reaction	不可逆反应	不可逆反應
irreversible thermodynamics	不可逆[过程]热力学	不可逆熱力學
irreversible work	不可逆功	不可逆功
irrigation rate	润湿率	潤濕率
irritant	刺激物	刺激物

英　文　名	祖国大陆名	台湾地区名
irrotational flow	无旋流	不旋轉流動
irrotational motion	无旋运动	不旋轉運動
isenthalpic compression	等焓压缩	等焓壓縮
isenthalpic expansion	等焓膨胀	等焓膨脹
isenthalpic process	等焓过程	等焓程序
isentropic compression	等熵压缩	等熵壓縮
isentropic expansion	等熵膨胀	等熵膨脹
isentropic flow	等熵流动	等熵流動
isentropic process	等熵过程	等熵程序
isobar	等压线	等壓線
isobaric process	等压过程	等壓程序
isobaric superheating	等压过热	等壓過熱
isochore	等容线	等容線
isochoric process	等容过程	等容程序
isocline	等斜线	等斜率線
isoconcentration curve	等浓度曲线	等濃度曲線
isocyanate adhesive	异氰酸酯黏合剂	異氰酸酯黏合劑
isocyanate polymer	异氰酸酯聚合物	異氰酸酯聚合物
isoelectric focusing	等电聚焦	等電聚焦
isoelectric point	等电点	等電點
isoelectric precipitation	等电沉淀	等電沈澱
isoelectric separation	等电点分离	等電點分離
isoforming	异构重整	異構重組
isoionic point	等离点	等離點
isokinetic point	等动力学点	等動力點
isokinetic temperature	等动力学温度	等動力溫度
isolated system	隔离系统,孤立系统	隔離系統
isolating oil	绝缘油	絕緣油
isomer	异构体	異構物
isomerization	异构化	異構化
isometric flowsheet	等容流程图	等距流程圖,等角流程圖
isometric process	等容过程	等容程序
isometrics	等容线	等容線
isopiestic method	等压法	等蒸汽壓法
isoprene rubber	异戊二烯橡胶	異戊二烯橡膠,異平橡膠
isopropylbenzene	异丙苯	異丙苯
isostere	同电子排列体	等容線

英　文　名	祖国大陆名	台湾地区名
isotachophoresis	等速电泳	等速電泳
isotactic polymer	全同立构聚合物, 等规聚合物	同规立構聚合物
isotactic polypropylene	全同立构聚丙烯	同规立構聚丙烯
isotherm	等温线	等溫線
isothermal adsorption	等温吸附	等溫吸附
isothermal compressibility	等温压缩性	等溫壓縮性
isothermal compression	等温压缩	等溫壓縮
isothermal diffusion	等温扩散	等溫擴散
isothermal distillation	等温蒸馏	等溫蒸餾
isothermal effectiveness	等温有效性	等溫有效度
isothermal environment	等温环境	等溫環境
isothermal expansion	等温膨胀	等溫膨脹
isothermal flow	等温流动	等溫流動
isothermal line	等温线	等溫線
isothermal multiplicity	等温多重性	等溫多重性
isothermal operation	等温操作	等溫操作
isothermal process	等温过程	等溫程序
isothermal reaction	等温反应	等溫反應
isothermal reactor	等温反应器	等溫反應器
isothermal thermogravimetry	等温热重量分析法	等溫熱重量法
isotherm migration	等温线迁移	等溫線遷移
isotonic concentration	等渗压浓度	等滲壓濃度
isotonic solution	等渗溶液	等滲壓溶液
isotope	同位素	同位素
isotope separation	同位素分离	同位素分離
isotopic abundance	同位素丰度	同位素含量, 同位素豐度
isotopic method	同位素法	同位素法
isotopic tracer	同位素指示剂	同位素示蹤劑
isotropic flow	各向同性流动	各向同性流動
isotropic solid	各向同性固体	各向同性固體
isotropic turbulence	各向同性湍流	各向同性紊流
isotropy	各向同性	各向同性
iterative method	迭代法	疊代法

J

英　文　名	祖国大陆名	台湾地区名
jacket	夹套	套, 夾套
jacketed autoclave	夹套加压釜	夾套高壓鍋
jacketed crystallizer	夹套结晶器	夾套結晶器
jacketed evaporator	夹套蒸发器	夾套蒸發器
jacketed kettle	夹套锅	[夾]套鍋
jacketed seamless kettle	无缝夹套锅	無縫[夾]套鍋
jacketed still	夹套蒸馏器	夾套蒸餾器
jacketed wall	夹套壁	夾套壁
jacket heating	夹套加热	夾套加熱
Jacobian matrix	雅可比矩阵	雅可比矩陣
Japan	日本[天然]漆	高清漆, 黑漆
Japan black	黑漆	黑漆
jar fermentor	缸式发酵槽	缸式發酵槽
jaw crusher	颚[式破]碎机	顎碎機
jet compression	喷射压缩	噴射壓縮
jet condenser	喷射冷凝器	噴凝器
jet drier	喷射干燥器	噴射乾燥器
jet engine	喷气式发动机	噴射引擎
jet exit pressure	喷射出口压	噴射出口壓[力]
jet flooding	喷射液泛	噴射氾流
jet fuel	喷气燃料	噴射機燃油
jet mill	气体粉碎机	氣流粉碎機
jet penetration length	射流穿透长度	射流穿透長度
jet pump	喷射泵	噴射泵
jet reactor	射流反应器	射流反應器
jetsam	沉料	沉料
jetter	喷洗器	噴洗器
jet tray	舌形板	舌形板
jet type washer	喷射式洗涤器	噴射式洗滌器
j-factor	j 因子	j 因數
j_D-factor	j_D 因子, 传质 j 因子	j_D-因數
j_H factor	j_H 因子, 传热 j 因子	j_H-因數
jigged fluidized bed	跳汰流化床	跳汰流化床
jigger	辘轳	拉坯盤車
jigging	簸动	簸選

英　文　名	祖国大陆名	台湾地区名
jigging screen	簸动筛	簸選篩, 選礦篩
joint	[管]接头	接頭, 接合
joint efficiency	接合效率	接頭效率
joule	焦[耳]	焦耳
Joule Thomson coefficient	焦耳-汤姆孙系数	焦耳-湯姆遜係數
Joule Thomson effect	焦耳-汤姆孙效应	焦耳-湯姆遜效應
junction	接合	接點, 接面, 接頭
junction point	接点	接點
junction potential	接点电位	接點電位
jute	黄麻	黄麻

K

英　文　名	祖国大陆名	台湾地区名
kainite	钾盐镁矾	鉀鎂礬, 鉀瀉鹽
kali salt	钾盐	鉀鹽
kaolin	高岭土	高嶺土
kaolinite	高岭石	高嶺石
kapok oil	木棉	木棉子油
karbate	无孔碳	卡珀(炭精製品)
kava oil	醉椒	胡椒油
kaya oil	榧子油	榧子油
kephalin	脑磷脂	腦磷脂
keratin	角蛋白	角蛋白
kerosene	煤油	煤油
kerosene cracking	煤油裂解	煤油裂解
kerosene degreasing	煤油脱脂	煤油脱脂
kerosene emulsion	煤油乳剂	煤油乳劑
kerosene fuel	煤油燃料	煤油燃料
kerosine	煤油	煤油
kettle	锅	鍋
key component	关键组分	關鍵成分, 主成分
key word	关键字	關鍵字
kieselguhr	硅藻土	矽藻土
kieselguhr dynamite	硅藻土炸药	矽藻土炸藥
kieserite	水镁矾	硫酸鎂石
killed carbon steel	镇静钢	全靜碳鋼
killed steel	镇静钢	全靜鋼

英　文　名	祖国大陆名	台湾地区名
kiln	窑	窯
kinase	激酶	激酶
kinematic property	运动性质	運動性質
kinematics	运动学	運動學
kinematic similarity	运动相似	運動相似性
kinematic viscometer	运动黏度计	動黏度計
kinematic viscosity	运动黏度	動黏度
kinetic analysis	动力分析	動力分析
kinetic control	动力学控制	動力學控制
kinetic energy	动能	動能
kinetic expression	动力学式	動力式
kinetic head	动压头	動壓頭
kinetic model	动力模式	動力模式
kinetic parameter	动力参数	動力參數
kinetic property	动力性质	動力性質
kinetics	动力学	動力學
kinetic theory	动力学理论	動力理論
Kirchhoff's law	基尔霍夫定律	基爾霍夫定律
kneader	捏合机	捏揉機
kneading	捏合	捏揉
knife barker	刮刀工去皮机	刀式去皮機
knock characteristic test	爆震试验	震爆試驗
knockout drum	分离鼓	液氣分離器
knot screen	结筛	結篩
Knudsen diffusion	克努森扩散	努生擴散
Knudsen number	克努森数	努生數
Kolmogorov's scale	科尔莫戈罗夫尺度	科爾莫戈羅夫尺度
konimeter	测尘器	測塵器
kraft paper	牛皮纸	牛皮紙
kraft process	硫酸盐法	牛皮紙漿法, 硫酸鹽法
kraft pulp	硫酸盐浆	牛皮紙漿
Kremser's diagram	克伦舍尔图	克倫舍爾圖
Kühni extractor	屈尼萃取塔	屈尼萃取塔
kurtosis	峰度	峰態, 峭度
K value	K 值	汽液平衡比例, K 值

L

英　文　名	祖国大陆名	台湾地区名
labile region	不稳定区	易變區(域)
labile zone	不稳定区	易變區
laboratory manual	实验室手册	實驗室手冊
laboratory reactor	实验室反应器	實驗室反應器
labor cost	人工成本	人工成本
lac	紫胶	蟲漆
lacquer	漆	噴漆
lactalbumin	乳清蛋白	乳白蛋白
lactam	内酰胺	內醯胺
lactase	乳糖酶	乳糖酶
lactic acid	乳酸	乳酸
lactobiose	乳二糖	乳糖
lactone	内酯	內酯
lactose	乳糖	乳糖
ladle	钢水包	澆斗
lag compensation	滞后补偿	落後補償
lag element	落后元件	落後元件
lagged pipe	隔热管	保溫管
lagged surface	隔热面	保溫面
lagging	隔热层	保溫層
lagoon	氧化塘	污水塘
lag phase	滞后期	遲滯期
lake	色淀	色澱, 沈澱色料
lamina	薄层	薄片, 薄層, 一板
laminar boundary layer	层流边界层	層流邊界層
laminar flow	层流, 滞流	層流
laminar flow reactor	层流反应器	層流反應器
laminar jet absorber	层流式喷流吸收器	層式噴流吸收器
laminar region	层流区	層流區[域]
laminar sublayer	层流底层	層流次層
laminate	层压材料	積層板
laminated safety glass	层压安全玻璃	積層安全玻璃
laminated wall	层压板壁	積層板壁
lamination	层压	積層
lamination press	层压压制机	積層壓製機

英　文　名	祖国大陆名	台湾地区名
lamp black	灯黑	燈煙, 燈黑
land disposal	废弃物陆上处理	[廢棄物]陸上處理[法]
landfarming	陆地废水处理	陸地[廢水]處理
landfill	废渣埋填	[廢棄物]掩埋
land pollution	土地污染	土地污染
lanolin	羊毛脂	羊毛脂
lantern gland	灯笼填函盖	燈籠填函蓋
lap joint	搭接接头	搭接
lard oil	猪油	豬油
latent energy	潜能	潛能
latent heat	潜热	潛熱
latent heat of vaporization	汽化潜热	汽化潛熱
lateral stress	侧向应力	側向應力
latex	胶乳	乳膠
latex paint	胶乳漆	乳膠漆
lattice	点阵, 晶格	格子, 晶格
laundry detergent	洗衣洗涤剂	洗衣清潔劑
laundry soap	洗涤皂	洗衣皂
laundry soda	洗衣碱	洗衣鹼
laurel oil	月桂油	月桂油
law	定律	定律, 律
law of corresponding state	对应态律	對應狀態定律
law of definite proportions	定比定律	定比定律
law of mass action	质量作用定律	整體作用定律
layout	布置	布置
layout check list	布置检查表	布置檢查表
layout diagram	布置图	布置圖
leachate	浸取液	瀝取液
leaching	浸取	瀝取
leaching process	浸取法	瀝取法
leaching tank	浸取槽	瀝取槽
lead	铅	鉛
lead alkali glass	铅碱玻璃	鉛鹼玻璃
lead-base grease	铅基润滑脂	鉛基(潤)滑脂
lead chamber	铅室	鉛室
lead compensation	超前补偿	超前補償
leaded gasoline	含铅汽油	加鉛汽油
leaded zinc	含铅氧化锌	含鉛鋅白
lead glass	铅玻璃	鉛玻璃

英 文 名	祖国大陆名	台湾地区名
leading edge	前沿	前緣
lead pipe	铅管	鉛管
lead soap	铅皂	鉛皂
lead storage battery	铅蓄电池	鉛蓄電池
lead white	铅白	鉛白
lead wire	导线	導線
lead wire error	导线误差	導線誤差
leaf filter	叶滤机	葉濾器
leakage	泄漏	洩漏
leakage test	泄漏试验	漏洩試驗
leak detector	检漏器	偵漏器
lean gas	贫气	貧氣
lean oil	贫油	貧油
lean solution	贫溶液	貧溶液
learning control system	学习控制系统	學習控制系統
least squares analysis	最小二乘分析	最小平方分析[法]
least squares method	最小二乘法	最小平方法
leather	皮革	皮革
leathery state	皮革态	皮革態
lecithin	卵磷脂	蛋黃態
legal liability	法定责任	法定責任
lehr	退火炉	退火窯
lemon-grass oil	柠檬草油	檸檬草油
leuco base	隐色基	無色基
leuco compound	隐色体	無色化合物
leucomycin	柱晶白霉素	白黴素
level control	液位控制	液位控制
level controller	液位控制器	液位控制器
level gauge	液位计	液位計
level transmitter	液位传送器	液位傳送器
lever arm principle	杠杆原理	槓桿原理
lever arm rule	杠杆法则	槓桿法則
lever safety valve	杠杆安全阀	槓桿安全閥
levo-rotatory sugar	左旋糖	左旋醣
levulose	果糖	果糖
lift	①升降机②上升	①升降機,升液器②上升
lift coefficient	上升系数	上升係數
light ash	轻灰	輕純鹼

英　文　名	祖国大陆名	台湾地区名
light duty detergent	轻垢洗涤剂	輕污清潔劑
light ends	轻馏分	輕餾分
light fuel oil	轻燃料油	輕燃料油
light hydrocarbon	轻烃	輕烴
light key component	轻主要组分	輕主成分
light naphtha	轻石脑油	輕石油腦
light oil	轻质油	輕油
light straight run	轻直馏油	輕直餾油
lignification	木质化	木質化
lignin	木素	木質素
ligninase	木质酶	木質酶
lignin sulphonic acid	木质磺酸	木質磺酸
lignite	褐煤	褐煤
lignitic bituminite	褐烟煤	褐煙煤
lignocellulose	木质纤维素	木質纖維素
lignocellulosic anthracite	木质纤维素无烟煤	木質纖維素無煙煤
lime	石灰	石灰
lime feldspar	钙长石	鈣長石
lime glass	钙玻璃	鈣玻璃
lime kiln	石灰窑	石灰窯
lime milk	石灰乳	石灰乳
lime soap	石灰皂	鈣皂
lime-soda process	石灰苏打法	石灰蘇打法
limestone	石灰石	[石]灰石
liming	浸灰	加灰法, 浸灰法
liming process	浸灰法	浸灰法
limit control	限度控制	限度控制
limit current	极限电流	極限電流
limit cycle	极限环	①極限環圈②極限循環
limiter	限制器	限制器
limiting batch size	极限批式尺寸	極限批式產能
limiting component	限量成分	限量成分
limiting cycle time	极限周期	極限周期
limiting factor	①极限因子②限制因素	①極限因數②限制因素
limiting reflux condition	极限回流条件	極限回流條件
limiting viscosity	极限黏度	極限黏度
limonite	褐铁矿	褐鐵礦
linear absorption coefficient	线性吸收系数	線性吸收係數
linear analysis	线性分析	線性分析

英 文 名	祖国大陆名	台湾地区名
linear azeotrope	线性共沸液	線性共沸液
linear control	线性控制	線性控制
linear control valve	线性控制阀	線性控制閥
linearity	线性	線性
linearization	线性化	線性化
linearized system	线性化系统	線性化系統
linear lead	线性超前	線性超前
linear low density polyethylene	线型低密度聚乙烯	線型低密度聚乙烯
linear model	线性模型	線性模式
linear momentum principle	线性动量原理	線性動量原理
linear notch	线性凹槽	線性凹槽
linear programming (LP)	线性规划	線性規劃
linear regression analysis	线性回归分析	線性迴歸分析
linear system	线性系统	線性系統
linear valve	线性阀	線性閥
linear velocity	线速度	線速度
linear viscoelasticity	线性黏弹性	線性黏彈性
lined pipe	衬里管	襯管
line loss	线路损失	線路損失
linen	亚麻布	亞麻布
line sink	线汇座	線匯座, 線壑
line source	线源	線源
lining	衬里	襯裏
linoleum	油毡	油甃, 油布
linseed oil	亚麻子油	亞麻仁油
linter	精制棉短绒	棉絨
lipase	脂肪酶	脂酶
liquefaction	液化	液化
liquefaction point	液化点	液化點
liquefaction process	液化过程	液化程序
liquefaction reactor	液化反应器	液化反應器
liquefied fuel	液化燃料	液化燃料
liquefied gas	液化气体	液化氣体
liquefied natural gas	液化天然气	液化天然氣
liquefied petroleum gas	液化石油气	液化石油氣
liquefier	液化器	液化器
liquid	液体	液體, 液相
liquid air	液态空气	液態空氣
liquid ammonia	液氨	液氨

英 文 名	祖国大陆名	台湾地区名
liquid chromatogram	液相色谱图	液相層析圖
liquid chromatograph	液相色谱仪	液相層析儀
liquid chromatography	液相色谱法	液相層析[法]
liquid column gage	液柱压力计	液柱壓力計
liquid crystal	液晶	液晶
liquid crystal display	液晶显示	液晶顯示
liquid crystal polymer	液晶聚合物	液晶聚合物
liquid crystal transition	液晶转变	液晶轉變
liquid cyclone	水力旋流器	液體旋風器
liquid cyclone separator	水力旋流分离器	液體旋風分離器
liquid extract	流浸膏剂	液體萃取物
liquid film	液膜	液膜
liquid film coefficient	液膜系数	液膜係數
liquid film contactor	液膜接触器	液膜接觸器
liquid film resistance	液膜阻力	液膜阻力
liquid flotation	液体浮选	液體浮選[法]
liquid fluidized bed	液体流化床	液體流體化床
liquid fuel	液体燃料	液體燃料
liquid-gas reaction	液气反应	液氣反應
liquid head	液体压头	液體高差
liquid hourly space velocity	液态空速	每小時之液體空間速度
liquid interface control	液体界面控制	液體界面控制
liquid level	液位	液位, 液面
liquid level control	液面控制	液位控制
liquid level gage	液面计	液位計
liquid level indicator	液位指示器	液位指示計
liquid-liquid equilibrium	液体平衡	液液平衡
liquid-liquid extraction	液液萃取	液液萃取
liquid-liquid reaction	液液反应	液液反應
liquid-liquid separator	液液分离器	液液分離器
liquid membrane	液膜	液[態薄]膜
liquid nitrogen	液氮	液態氮
liquid oxygen	液氧	液態氧
liquid paraffin	液体石蜡	液態石蠟
liquid phase	液相	液相
liquid phase cracking	液相裂解	液相裂解
liquid phase reaction	液相反应	液相反應
liquid phase reactor	液相反应器	液相反應器
liquid propellant	液体火箭燃料	液體推進劑

英　文　名	祖国大陆名	台湾地区名
liquid reaction	液态反应	液態反應
liquid relief valve	安全泄液阀	液體釋放閥
liquid ring compressor	液环压缩机	液環壓縮機
liquid seal	液封	液體密封
liquid-solid cyclone	液固旋风器	液固旋風器
liquid-solid reaction	液固反应	液固反應
liquid solution	液体溶液	液體溶液
liquid thermometer	液体温度计	液體溫度計
liquid trap	捕液器	液阱
liquid-vapor equilibrium	液汽平衡	液汽平衡
literature	文献	文獻
literature survey	文献调查	文獻調查
lithium-base grease	锂基润滑脂	鋰皂基潤滑脂
lithography	平版印刷术	石板印刷法, 石印法, 印型法
lithopone	锌钡白	鋅鋇白
litmus paper	石蕊试纸	石蕊試紙
lixiviation	浸滤	瀝濾
load	负载	負載, 負荷
load change	负荷变动	負荷變化
load disturbance	负荷搅动	負荷擾動
loading effect	负荷效应	負荷效應
loading point	载点	負荷點
local acceleration	局部加速度	局部加速度
local composition	局部组成	局部組成
local equilibrium	局部平衡	局部平衡
local optimum	局部最优[值]	局部最適[值]
local stability	局部稳定性	局部穩定性
lock hopper	闭锁式料斗	閉鎖式料斗
locus	轨迹	軌跡
locus of constant phase	等相位轨迹	等相位軌跡
locus of critical point	临界点轨迹	臨界點軌跡
loft drying	风干	風乾
logarithmic mean	对数平均	對數平均
logarithmic mean area	对数平均面积	對數平均面積
logarithmic mean mole fraction	对数平均摩耳分率	對數平均莫耳分率
logarithmic mean radius	对数平均半径	對數平均半徑
logarithmic mean temperature difference	对数平均温差	對數平均溫差
logarithmic phase	对数生长期	對數生長期

英　文　名	祖国大陆名	台湾地区名
logarithmic plot	对数图	對數圖
log normal distribution function	对数正态分布	對數常態分布函數
longitudinal dispersion coefficient	纵向分散系数	縱向分散係數
longitudinal fin	纵向翅片	縱向鰭片
longitudinal stress	纵向应力	縱向應力
long sweep elbow	长弯头	長彎頭
long tube evaporator	长管蒸发器	長管蒸發器
loop	环路, 回路	環路, 圈
loop configuration	环路构型	環路組態
looped pipe system	环状管路系统	環狀管路系統
loop reactor	环流反应器	環流反應器
loop response	环路应答	環路應答
loop transfer function	环路转移函数	環路轉移函數
loss	损耗	損失
loss tangent	损耗角正切	損失模數比, 損失正切
lost work	损失功	損失功
louver type baffle	百叶窗挡板	百葉窗擋板
low-BTU gas	低热值煤气	低熱量氣體
low density polyethylene	低密度聚乙烯	低密度聚乙烯
lower bound	下限	下限
lower limit	下限	下限
low pass filter	低通滤波器	低頻濾波器
low selector switch	低选择器开关	低選擇器開關
low temperature carbonization	低温碳化	低溫碳化[法]
LP(＝linear programming)	线性规划	線性規劃
lubricant	润滑剂	潤滑劑
lubricating grease	润滑脂	潤滑脂
lubricating oil	润滑油	潤滑油
lubrication	润滑	潤滑
lubrication mechanics	润滑力学	潤滑力學
lubricity	润滑性	潤滑性
luminescence	①发光②冷光	①發光②冷光
luminescent dye	发光染料	發光染料
luminous paint	发光涂料	發光漆
lumped parameter model	集总参数模型	塊集參數模式
lumped parameter system	集总参数系统	塊集參數系統
lumped system	集总系统	塊集系統
lumping	参数集总	塊集
lumping kinetics	集总动力学	塊集動力學

英　文　名	祖国大陆名	台湾地区名
lustering agent	上光剂	發光劑
luwa evaporator	薄膜蒸发器	薄膜蒸發器
lye	碱液	鹼水
lyolysis	液解	液解[作用]
lyophilic colloid	亲液胶体	親液膠體
lyophilization	冷冻干燥	冷凍乾燥
lyophobic colloid	疏液胶体	疏液膠體
lyotropic liquid crystal	感胶液晶	向液性液晶
lysine	赖氨酸	離胺酸
lysozyme	溶菌酶	溶菌酶

M

英　文　名	祖国大陆名	台湾地区名
maceration	浸渍	滲浸
machine molding	机械成型	機械成型[法]
Mach number	马赫数	馬赫數
macro emulsion	粗滴乳状液	巨乳液
macrofluid	宏观流体	巨觀流體
macrokinetics	宏观动力学	巨觀動力學
macromixing	宏观混合	巨觀混合
macromolecule	大分子	巨分子
macropore	大孔	巨觀細孔
macroscale behavior	宏观规模行为	巨標[度]行為
macroscopic balance	宏观平衡	巨觀均衡
macroscopic reversibility	宏观可逆性	巨觀可逆性
madder	茜草	茜草, 茜草染料
madder root	茜草根	茜草根
magazine grinder	箱式研磨机	箱式研磨機
magnesia	氧化镁	氧化鎂, 鎂氧, 鎂礬土
magnesia brick	镁砖	鎂磚
magnesite	菱镁矿	菱鎂礦
magnesite brick	镁砖	鎂磚
magnesite-chrome brick	镁铬砖	鎂鉻磚
magnesite refractory	镁质耐火材料	鎂氧耐火物
magnetically stabilized fluidized bed	磁稳流化床	磁穩流化床
magnetic data storage	磁式数据储存	磁式資料儲存
magnetic disk	磁盘	磁碟

英　文　名	祖国大陆名	台湾地区名
magnetic field	磁场	磁場
magnetic filtration	磁力过滤	磁力過濾
magnetic flowmeter	电磁流量计	磁力流量計
magnetic impurity	磁性不纯物	磁性不純物
magnetic moment	磁矩	磁矩
magnetic separation	磁分离	磁力分離
magnetic separator	磁力分离器	磁力分離器
magnetic stirrer	[电]磁搅拌器	磁力攪拌器
magnetic susceptibility	磁化率	磁化率
magnetic tape	磁带	磁帶
magnetite	磁铁矿	磁鐵礦
magnetochemistry	磁化学	磁化學
magneto fluidization	磁力流态化	磁力流態化
magnetohydrodynamics	磁流体动力学	磁性流體力學
magnitude ratio	大小比值	大小比值, 量比值
magnitude scaling	大小量度	大小量度
maintenance	维修	維護, 保養
maintenance cost	维修费	維護成本
maintenance downtime	停修时间	維護停工時間
maintenance management	维修管理	維護管理
major decoupling	主要解偶	主要解偶
malachite	孔雀石	孔雀石
malachite green	孔雀绿	孔雀綠
malathion	马拉硫磷	馬拉松[農藥]
maldistribution	不良分布	不良分布
malic acid	苹果酸	蘋果酸
malonic acid	丙二酸	丙二酸
malt	麦芽	麥芽
maltase	麦芽糖酶	麥芽酶
maltose	麦芽糖	麥芽糖
manganite	水锰矿	水錳礦
manhole	人孔	人孔
man hour	工时	人時, 工時
manifold	歧管	歧管
man-made fiber	人造纤维	人造纖維
manometer	液柱压力计	[流體]壓力計
manometric fluid	测压液	測壓液
manometric head	测压压头	壓力高差
manual operation	人工操作	人工操作

英　文　名	祖国大陆名	台湾地区名
manual reset	手动重整	手動重整
manufacture	制造	製造
manufacturing cost	制造成本	製造成本
manufacturing technology	制造技术	製造技術
mapping	映射	映射
marcasite	白铁矿	白鐵礦
margarine	人造黄油	人造奶油
marginal profit	边际利润	邊際利潤
margin gain	边界增益	邊限增益
margin of stability	稳定界限	穩定邊限
margin phase	边界相	邊限相位
Margules equation	马居尔方程	馬居爾方程
market cost	市场成本	市場成本
market research	市场研究	市場研究
MARR(＝minimum acceptable rate of return)	最低容许收益率	最低容許收益率
marsh gas	沼气	沼氣
Martin Hou equation [of state]	马丁-侯[虞钧]方程	馬丁-侯[虞鈞]方程
mash	麦芽汁	醪
masking power	掩蔽能力	遮蓋能力
mass	质量	質量
mass absorption coefficient	质量吸收系数	質量吸收係數
mass action kinetics	质量作用动力学	質量作用動力學
mass average velocity	质量平均速度	質量平均速度
mass balance	质量衡算	質量均衡
mass concentration	质量浓度	質量濃度
mass conservation	质量守恒	質量守恆
mass diffusivity	质量扩散系数	質量擴散係數
mass flow	质量流	質量流動
mass flowmeter	质量流量计	質量流量計
mass flow rate	质量流率, 质量流量	質量流量
mass flux	质量通量	質量通量
mass fraction	质量分率	質量分率
mass generation curve	质量生成曲线	質量生成曲線
massive bed reactor	宽床反应器	寬床反應器
mass polymerization	本体聚合[反应]	本體聚合[反應]
mass removal curve	质量去除曲线	質量去除曲線
mass separating agent	物质分离剂	質量分離劑
mass spectrogram	质谱图	質譜圖

英　文　名	祖国大陆名	台湾地区名
mass spectrometer	质谱仪	質譜儀
mass spectrometry	质谱法	質譜法
mass spectrum	质谱	質譜
mass transfer	传质, 质量传递	質量傳送
mass transfer coefficient	传质系数	質傳係數
mass transfer control	传质控制	質傳控制
mass transfer equipment	传质装置	質傳裝置
mass transfer operation	传质操作	質傳操作
mass transfer rate	传质速率	質傳速率
mass transfer resistance	传质阻力	質傳阻力
mass transfer zone (MTZ)	传质区	質傳區
mass transport	质量迁移	質量輸送
mass velocity	质量流速	質量速度
master controller	主令控制器	主控制器
matching of streams	流股匹配	流股匹配
material	物料	材料
material and energy balance	物料与能量平衡	質能均衡
material balance	物料衡算, 物料平衡	物質均衡
material balance flowsheet	物料衡算流程图	物質均衡流程圖
material cost	材料成本	材料成本
material cost index	材料成本指数	材料成本指數
material derivative	物质导数	物質導數
material design	材料设计	材料設計
material function	物质函数	物質函數
material manufacturing	材料制造	材料製造
material of construction	建筑材料	營建材料
material processing	材料加工	材料加工
material seal	料封	料封
materials engineering	材料工程	材料工程
materials science	材料科学	材料科學
material synthesis	材料合成	材料合成
mathematical model	数学模型	數學模型
mathematical modeling	数学模拟	數學模式化
mathematical programming	数学程序	數學規劃
matrix	①矩阵②基体	①矩陣②基體
matrix arithmetics	矩阵算术	矩陣算術
matrix inversion	矩阵求逆	矩陣反轉
matter	物质	物質
matt glaze	无光釉	無光釉

英　文　名	祖国大陆名	台湾地区名
matt paint	无光油漆	無光油漆
maximax criterion	大中取大判据	大中取大判據
maximin utility criterion	小中取大效用判据	小中取大效用判據
maximum allowable pressure	最高允许压力	最高容許壓[力]
maximum allowable temperature	最高允许温度	最高容許溫度
maximum boiling azeotrope	最高沸点共沸物	最高沸點共沸液
maximum likelihood principle	最大似然原理	最大相似原理
maximum mixedness	最大混合度	最大混合度
maximum principle	极大值原理	極大原理
maximum specific growth rate	最大比生长速率	最大比生長速率
Maxwell relation	麦克斯韦关系	馬克斯威爾關係
McMahon packing	网鞍填料	網鞍填料
MD method	分子动态法	分子動態法
MD tray	多降液管塔板	多降液管塔板
mean	均值	①平均②平均值
mean deviation	平均偏差	平均偏差
mean error	平均误差	平均誤差
mean free path	平均自由程	平均自由徑
mean radius	平均半径	平均半徑
mean residence time	平均停留时间	平均滯留時間
mean square error	均方误差	均方誤差
mean temperature difference	平均温差	平均溫差
measurability	可测量性	可量測性
measurable property	可测量性质	可測性質
measurement	测量	量測,量度,測量
measurement lag	测量滞后	量測落後
measuring device	测量装置	量測裝置
measuring element	测量元件	量測元件
measuring flask	量瓶	量瓶
measuring lag	测量滞后	量測落後
measuring means	测量方法	量測方法
mechanical agitation	机械搅拌	機械攪拌
mechanical automation	机械自动化	機械自動化
mechanical centrifugal separator	机械离心分离器	機械離心分離器
mechanical controller	机械式控制器	機械式控制器
mechanical design	机械设计	機械設計
mechanical disintegration	机械破碎	機械散解
mechanical draft	机械通风	機械通風
mechanical draft cooling tower	机械通风凉水塔	機械通風冷卻塔

英　文　名	祖国大陆名	台湾地区名
mechanical energy	机械能	機械能
mechanical energy balance	机械能量平衡	機械能均衡
mechanical energy equation	机械能方程式	機械能方程式
mechanical equivalent of heat	热功当量	熱功當量
mechanical filter	机械过滤器	機械過濾器
mechanical flotation cell	机械浮选池	機械浮選池
mechanical flow diagram	工程流程图	機械流程圖
mechanical flowsheet	工程流程图	機械流程圖
mechanical loss	机械损失	機械損失
mechanical-physical separation process	机械-物理分离过程	機械-物理分離程序
mechanical plating	机械镀	機械壓紋
mechanical pulp	机械纸浆	機械紙漿
mechanical pulping	机械制浆	機械製漿
mechanical pulping process	机械制浆法	機械製漿法
mechanical seal	机械密封	機械密封
mechanical separation	机械分离	機械分離
mechanical separator	机械分离器	機械分離器
mechanical work	机械功	機械功
mechanics	力学	力學
mechanism deficiency	机理缺陷	機構缺陷
mechanism of catalysis	催化作用机理	觸媒作用機構, 催化機構
mechanism structure	机理结构	機構結構
mechanistic equation	机械学方程式	機構方程式
mechano-chemical polishing	机械化学磨光	機械化學磨光
mechano-chemical process	化学动力过程	機械化學[製漿]法
median	中位数	中點, 中值
median selector	中值选择器	中點選擇器
medium	介质	①介質②培養基
medium crushing	中级压碎	中級壓碎
medium filter	介质过滤器	介質過濾器
medium grinding	中级研磨	中級研磨
medium screening	中级筛选	中級篩選
medium sizing	中级筛选	中級篩選
medium sweep elbow	中弯头	中彎頭
melamine adhesive	三聚氰胺黏合剂	三聚氰胺黏合劑, 美耐皿黏合劑
melamine formaldehyde resin	三聚氰胺-甲醛树脂	三聚氰胺甲醛樹脂
melamine resin	蜜胺树脂	三聚氰胺樹脂

英 文 名	祖国大陆名	台湾地区名
melanogen dye	黑素原染料	硫化染料
melanoid	类黑素	類黑素
melanoidin	蛋白黑素	類黑精
Mellapak packing	板波纹填料	板波紋填充
melt	①熔化②熔化物	①熔態②熔體
melt flow index	熔体流动指数	熔化指數
melt index	熔体指数	熔化指數
melting point	熔点	熔點
melting point depression	熔点下降	熔點下降
melting range	熔矩	熔點範圍
melting temperature	熔化温度	熔化溫度
melt spinning	熔纺	熔紡[絲]
melt viscosity	熔融黏度	熔體黏度
membrane	膜	薄膜
membrane bioreactor	膜生物反应器	薄膜生物反應器
membrane distillation	膜蒸馏	薄膜蒸餾
membrane extraction	膜萃取	薄膜萃取
membrane filter	膜滤器	薄膜過濾器
membrane module	膜组件	薄膜組件
membrane permeability	膜通透性	薄膜透過性
membrane permeation	膜渗透	薄膜穿透
membrane pump	膜泵	薄膜泵
membrane reactor	膜反应器	薄膜反應器
membrane separation	薄膜分离	薄膜分離
membrane separation technology	薄膜分离技术	薄膜分離技術
membrane strength	膜强度	薄膜強度
membrane support	膜支撑物	薄膜支撐物
membrane technology	膜技术	薄膜技術
membrane vesicle	膜囊	薄膜囊
memory cell	①记忆细胞②存储单元	記憶格
memory fluid	记忆流体	記憶流體
memory function	记忆函数	記憶函數
menthol	薄荷脑	薄荷腦
mercaptan	硫醇	硫醇
mercerization	丝光处理	絲光化
mercerized cotton	丝光棉	絲光棉
mercerized yarn	丝光纱	絲光紗
mercerizing	丝光处理	絲光處理
mercury cathode cell	汞阴极电解槽	汞陰極電解槽

英 文 名	祖国大陆名	台湾地区名
mercury cell	水银电解槽	汞電池
mercury decomposer	分汞器	分汞器
mercury diffusion pump	汞扩散泵	汞擴散泵
mercury float manometer	汞浮子压力计	汞浮子壓力計
mercury pollution	汞污染	汞污染
mercury thermometer	水银温度计	水銀溫度計
mesh	①网孔②筛目	①篩孔②網目
mesh analysis	筛析	篩析
mesh efficiency	筛目效率	篩孔效率
MESH equations	MESH 方程组	MESH 方程組
mesh filter	筛网过滤器	篩網過濾器
mesh screening	网筛	網篩
mesh separator	筛网分离器	篩孔分離器
mesomorphic state	介晶态	液晶狀態
mesophase	中间相	液晶相
mesophile	中温菌	嗜溫菌
mesophilic bacteria	中温菌	嗜溫菌
mesopore	细孔	細孔
mesoscale behavior	介标行为	介標[度]行為
metabolic control	代谢控制	代謝控制
metabolic rate	代谢速率	代謝速率
metabolism	代谢[作用]	代謝作用,新陳代謝
metabolite	代谢物	代謝物
metal	金属	金屬
metal complex acid dye	金属络合酸性染料	金屬錯鹽酸性染料
metal finishing	金属表面处理	金屬表面處理
metal ion inactivator	金属离子惰化剂	金屬離子惰化劑
metallic catalyst	金属催化剂	金屬觸媒
metallic glass	金属玻璃	金屬玻璃
metallic mordant	金属媒染剂	金屬媒染劑
metallic soap	金属皂	金屬皂
metallized dye	金属配位染料	金屬化染料
metallized plastic	金属塑料	金屬化塑膠
metallizing	喷镀金属	金屬化
metallography	金相学	金相學
metallorganic chemical vapor deposition	金属有机化学蒸气沉积	金屬有機化學蒸氣沈積法
metallurgical coke	冶金焦	冶金焦
metallurgical process	冶金过程	冶金程序

英　文　名	祖国大陆名	台湾地区名
metallurgy	冶金学	冶金學
metal plating	金属电镀	金屬電鍍
metal spraying	金属喷镀	金屬噴霧
metastability	亚稳定性	介穩定性
metastable limit	亚稳界限	介穩定界限
metastable region	亚稳区	介穩定區[域]
metastable state	亚稳态, 介稳态	介穩定狀態
metathesis	复分解	複分解
metathetical reaction	复分解反应	複分解反應
metering	计量	計量
metering pump	计量泵	計量泵
methanation	甲烷化	甲烷化
methanation reaction	甲烷化反应	甲烷化反應
methanator	甲烷转化器	甲烷化器
methane bacteria	甲烷菌	甲烷菌
methane-forming bacteria	甲烷菌	甲烷菌
methane lean gas	甲烷贫气	甲烷貧氣
methane rich gas	甲烷富气	甲烷富氣
methanol	甲醇	甲醇, 木精
method	方法	方法
method of excess	过量法	過量法
method of finite difference	有限差分法	有限差分法
method of initial rate	初速率法	初速率法
method of isolation	隔离法	隔離法, 單離法
method of moments	力矩法	矩法
method of moments matching	配矩法	配矩法
methyl alcohol	甲醇	甲醇, 木精
methylated spirit	变性酒精	加甲醇酒精
methylating agent	甲基化剂	甲基化劑
methylation	甲基化作用	甲基化
methyl cellulose	甲基纤维素	甲基纖維素
methylene blue	亚甲蓝	亞甲藍
methyl methacrylate-acrylonitrile-but adiene-styrene copolymer	甲基丙烯酸甲酯-丙烯腈-丁二烯-苯乙烯共聚物	甲基丙烯酸甲酯-丙烯腈-丁二烯-苯乙烯共聚物
methyl orange	甲基橙	甲基橙
methyl parathion	甲基对硫磷	甲[基]巴拉松
methyl red	甲基红	甲基紅
methyl rubber	甲基橡胶	甲基橡膠

英 文 名	祖国大陆名	台湾地区名
methyl violet	甲基紫	甲基紫
methyl yellow	甲基黄	甲基黄
metric system	公制	公制, 米制
mica	云母	雲母
micellar catalysis	胶束催化	微胞催化[作用]
micellar flooding	胶束泛滥	微胞氾流[法]
micelle	胶束	微胞
micellization	胶束化	微胞化
Michaelis Menton constant	米氏常数	米氏常數
Michaelis Menton equation	米氏方程	米氏方程
Michaelis Menton kinetics	米氏动力学	米氏動力學
microanalysis	微量分析	微量分析
microbe	微生物	微生物
microbial contamination	杂菌感染	微生物感染
microbial corrosion	微生物腐蚀	微生物[引起的]腐蚀
microbial dynamics	微生物动力学	微生物動態[學]
microbial fermentation	微生物发酵	微生物發酵
microbial film	微生物膜	微生物薄膜
microbial fouling	微生物结垢	微生物積垢
microbial kinetics	微生物动力学	微生物動力學
microbial process	微生物法	微生物程序, 微生物法
microbial reaction	微生物反应	微生物反應
microbial reactor	微生物反应器	微生物反應器
microbiological process	微生物过程	微生物程序, 微生物法
microbiology	微生物学	微生物學
microcanonical ensemble	微正则系综	微正則系綜
microcanonical partition function	微正则配分函数	微正則配分函數
microcapsule	微胶囊	微膠囊
microcarrier	微载体	微載體
microcatalytic reactor	微型催化反应器	微觸媒反應器
microchemistry	微量化学	微量化學
microcircuit	微电路	微電路
microcomputer	微型计算机	微算機, 微電腦
microcrystal	微晶	微晶
microcrystalline	微晶	微晶
microcrystalline structure	微晶结构	微晶結構
microcrystalline wax	微晶蜡	微晶蠟
microemulsion	微乳	微乳液
microencapsulation	微囊化	微膠囊化

英　文　名	祖国大陆名	台湾地区名
microfiltration	微滤	微孔過濾
microfluid	微观流体	微觀流體
micromeritics	微粒学	微粒學
micrometer	测微计	测微計, 分厘卡
micromixing	微观混合	微觀混合
micron	微米	微米
micronizer	微粉磨	微磨機
microorganism	微生物	微生物
microphotometer	显微光度计	微光度計
micropore	微孔	微孔
microporous filter	微孔过滤器	微孔過濾器
microporous membrane	微孔滤膜	微孔膜
microporous separator	微孔分离器	微孔分離器
microreaction engineering	微型反应工程	微反應工程
microreactor	微型反应器	微反應器
microreversibility	微观可逆性	微觀可逆性
microscale phenomenon	微量现象	微標[度]現象
microscopic balance	微观平衡	微觀均衡
microscopic reversibility	微观可逆性	微觀可逆性
microscopy	显微术	顯微術, 顯微法
microsphere	微球	微球
microstate	微观态	微觀狀態
microstrainer	微滤器	濾微器
microstraining	微滤	微[過]濾
microstructure	微结构	微結構
microstructured material	微结构材料	微結構化材料
microwave	微波	微波
microwave drying	微波干燥	微波乾燥
microwave spectroscopy	微波波谱学	微波光譜學
microwave spectrum	微波谱	微波譜
middle oil	中油	中油
midlife activity	半生活性	半生活性
migration	迁移	泳動
migration potential	迁移电位	遷移電位
mild steel	软钢	軟鋼
milky glass	乳白玻璃	乳白玻璃
mill	碾磨机	磨碾機
mill capacity	研磨能力	研磨能力
mill efficiency	研磨效率	研磨效率

英 文 名	祖国大陆名	台湾地区名
milligram	毫克	毫克
millimeter	毫米	毫米
milling capacity	研磨能力	壓榨能力
millivoltmeter	毫伏特计	毫伏特計
mill roller	研磨滚柱	壓榨輥
mill room	研磨室	粉碎室
mineral acid	无机酸	礦酸
mineral dye	无机染料	礦物性染料
mineral fertilizer	矿物肥料	礦物肥料
mineralogy	矿物学	礦物學
mineral oil	矿物油	礦油
mineral resources	矿物资源	礦物資源
mineral wax	矿蜡	礦蠟
mingler	拌和机	拌和機
miniature instrument	微型仪器	小型儀器
minicomputer	小型计算机	小型電腦, 小型計算機
minimax regret criterion	大中取小遗憾判据	大中取小遺憾判據
minimum acceptable rate of return (MARR)	最低容许收益率	最低容許收益率
minimum boiling azeotrope	最低沸点共沸物	最低沸點共沸液
minimum deviation	最小偏差	最小偏差
minimum fluidization	最小流化态	最低流體化
minimum fluidization velocity	最小流化速度	最低流體化速度
minimum fluidizing velocity	最小流化速度	最低流體化速度
minimum freezing point	最低凝固点	最低凝固點
minimum principle	最小值原理	極小原理
minimum reflux	最小回流	最小回流
minimum reflux ratio	最小回流比	最小回流比
minimum tray number	最小塔板数	最少板數
mining explosive	矿用炸药	礦用炸藥
minisupercomputer	小型超级计算机	小型超級計算機
minium	铅丹	鉛丹, 鉛紅
MINLP(=mixed integev nonlinear programming)	混合整数非线性规划	混合整數非線性規劃
minor loop	副回路	次環路
mirror colorimeter	镜式比色计	鏡式比色計
mirror filter reflectometer	滤镜反光计	濾鏡反光計
mirror polishing	镜面磨光	鏡面磨光
mirror spectrophotometer	镜式光谱仪	鏡式光譜計

英　文　名	祖国大陆名	台湾地区名
miscalibration	误校正	誤校正
miscibility	互溶性	互溶性
miscible flooding	互溶泛滥	互溶氾流[法]
miscible system	互溶系统	互溶系統
mist	雾	霧
mist collector	集雾器	集霧器
mist flow	雾状流	霧沫流動
mist particle	雾粒	霧粒
mist separator	油雾分离器	分霧器
miter elbow	斜弯头	斜彎頭
miticide	杀螨剂	殺蟎劑
mixed acid	混合酸	混合酸
mixed catalyst	混合催化剂	混合觸媒
mixed crystal	混晶	混合晶體
mixed feeding	混合进料	混合進料
mixed fertilizer	混合肥料	複合肥料
mixed flow	混流	混合流動
mixed flow pump	混流泵	混流泵
mixed flow reactor	混合流动反应器	混合流動反應器
mixed integer nonlinear programming (MINLP)	混合整数非线性规划	混合整數非線性規劃
mixed juice	混合汁	混合汁
mixed liquor	混合液	混合液
mixed-liquor suspended solid	混合液悬浮固体	混合液懸浮固體
mixed-liquor volatile suspended solid	混合液挥发性悬浮固体	混合液揮發性懸浮固體
mixedness	混合度	混合度
mixed node	混合节点	混合節點
mixed order	混合序列	混合階
mixed production	混合生产	混合產物
mixed reactor	混合反应器	混合[流動]反應器
mixed suspension	混合悬浮	混合懸浮
mixer	混合器	混合器, 混合機
mixer power	混合器功率	混合器功率
mixer-settler	混合澄清器, 混合沉降器	混合器-沈降器
mixing	混合	混合
mixing agitator	混合搅拌机	混合攪拌機
mixing baffle	混合用挡板	混合用擋板
mixing capacity	混合能力	混合能力

英　文　名	祖国大陆名	台湾地区名
mixing chamber	混合室	混合室
mixing column	混合柱	混合柱
mixing cup temperature	混合杯温	混合杯溫
mixing depth	混合深度	混合深度
mixing device	混合装置	混合裝置
mixing draft tube	气流混合管	混合流通管
mixing drum	混合罐	混合桶
mixing entrainment	混合雾沫	混合霧沫
mixing index	混合指数	混合指數
mixing intensity	混合强度	混合強度
mixing length	混合长度	混合長度
mixing paddle	混合桨叶	混合槳葉
mixing pattern	混合型式	混合型式
mixing rate	混合速率	混合速率
mixing rule	混合规则	混合法則
mixing tank	混合槽	混合槽
mixing temperature	混合温度	混合溫度
mixing time	混合时间	混合時間
mixture	混合物	混合物
mobile phase	流动相	移動相
mobile reactor	流动反应器	移動反應器
mobility	迁移率	①移動率②機動性
modacrylic fiber	改性聚丙烯腈纤维	副丙烯腈纖維
model	模型	①模型②模式
model building	模型建立	模式建立
model identification	模型辨识	模型辨識
modeling	建模	模式化
model parameter	模型参数	模型參數
model scale	模型比例	模型比例
model test	模型试验	模型試驗
modified alkyd resin	改性醇酸树脂	改質醇酸樹脂
modified starch	改性淀粉	改質澱粉
modulation	调制	調變
module	模块	模組,組件
modulus	模量	模數
modulus ratio	模量比	模數比
moist material	湿物料	濕物料
moisture	含湿量	①水分②溼氣
moisture content	湿含量	含水量

英　文　名	祖国大陆名	台湾地区名
moisture determination apparatus	湿度测定器	水分测定器
moisture factor	湿度因数	水分因數
moisture meter	湿度计	溼度計
moisture separator	去湿器	溼氣分離器
molal average boiling point	摩尔平均沸点	重量莫耳平均沸點
molal concentration	重量摩尔浓度	重量莫耳濃度
molal heat capacity	重量摩尔热容	重量莫耳熱容
molality	重量摩尔浓度	重量莫耳濃度
molal solution	摩尔溶液	重量莫耳溶液
molar average boiling point	摩尔平均沸点	莫耳平均沸點
molar average diffusivity	摩尔平均扩散系数	莫耳平均擴散係數
molar average velocity	摩尔平均速度	莫耳平均速度
molar concentration	摩尔浓度	莫耳濃度
molar flux	摩尔通量	莫耳通量
molar heat capacity	摩尔热容	莫耳熱容量
molar humidity	摩尔湿度	莫耳溼度
molar solution	摩尔溶液	莫耳溶液
molar volume	摩尔体积	莫耳體積
molasses	糖蜜	糖蜜
molasses alcohol	糖蜜酒精	糖蜜酒精
molasses graining	糖蜜晶粒	糖蜜起晶
molasses medium	糖蜜塔养基	糖蜜培養基
molasses sugar	废糖蜜中的糖	含蜜糖
mold amylase	霉菌淀粉酶	黴菌澱粉酶
molding	模塑	①模製②成型
molding composition	模压成分	模製配料
molding compression	模压	模[製]壓[縮]
molding press	压模机	模壓機
mold-release agent	脱模剂	脱模劑
mole balance	摩尔衡算	莫耳均衡
molecular beam deposition	分子束沉积法	分子束沈積法
molecular beam epitaxy	分子束外延	分子束磊晶
molecular biology	分子生物学	分子生物學
molecular configuration	分子构型	分子組態
molecular diffusion	分子扩散	分子擴散
molecular diffusivity	分子扩散系数	分子擴散係數
molecular dispersion	分子分散	分子分散
molecular distillation	分子蒸馏	分子蒸餾
molecular dynamic method	分子动态法	分子動態法

英　文　名	祖国大陆名	台湾地区名
molecular interaction	分子相互作用	分子相互作用
molecularity	反应分子数	①分子度②分子性
molecular parameter	分子参数	分子參數
molecular partition function	分子配分函数	分子配分函數
molecular rearrangement	分子重排	分子重排
molecular self-assembly	分子自集	分子自集
molecular sieve	分子筛	分子篩
molecular-sieve catalyst	分子筛催化剂	分子篩觸媒
molecular-sieve zeolite	分子筛沸石	分子篩沸石
molecular simulation	分子模拟	分子模擬
molecular species	分子物种	分子物種
molecular structure	分子结构	分子結構
molecular symmetry	分子对称性	分子對稱性
molecular thermodynamics	分子热力学	分子熱力學
molecular transformation	分子转换	分子轉換
molecular transport	分子输送	分子輸送
molecular velocity	分子速度	分子速度
molecular weight	分子量	分子量
molecular weight distribution (MWD)	分子量分布	分子量分布
molecule	分子	分子
Mollier diagram	莫利尔图	莫利爾圖
molten-bath gasifier	熔浴气化器	熔浴氣化器
molybdenum catalyst	钼催化剂	鉬觸媒
moment	矩	矩
moment of force	力矩	力矩
moment of inertia	转动惯量	慣性力矩
momentum	动量	動量
momentum balance	动量平衡	動量均衡
momentum conservation	动量守恒	動量守恆
momentum diffusivity	动量扩散系数	動量擴散係數
momentum equation	动量方程	動量方程式
momentum flux	动量通量	動量通量
momentum principle	动量原理	動量原理
momentum separator	动量分离器	動量分離器
momentum transfer	动量传递	動量傳送
momentum transfer coefficient	动量传递系数	動量傳遞係數
momentum transport	动量输送	動量輸送
monazite	独居石	獨居石
monitor	监视器	監測器

英　文　名	祖国大陆名	台湾地区名
monitoring	监视	監測
monitoring station	监测站	監測站
monoazo dye	单偶氮染料	單偶氮染料
monochromatic emissive power	单色发射能力	單色發射能力
monoclonal antibody	单克隆抗体	單克隆抗體
monococcus	单球菌	單球菌
Monod growth kinetics	莫诺生长动力学	莫諾生長動力學
monodisperse	单分散	單分散
monodispersion	单分散性	單分散
monofilament	单丝	單絲纖維
monoglyceride	甘油单酯	單甘油酯
monolithic catalyst	整装催化剂	整裝催化劑
monomer	单体	單體
monomolecular reaction	单分子反应	單分子反應
monopad filter	单垫过滤机	單墊濾機
monosaccharide	单糖	單醣
Monte Carlo simulation	蒙特卡罗模拟	蒙地卡羅模擬
monticellite	钙镁橄榄石	鈣橄欖石
montmorillonite	蒙脱石	微晶高嶺石
mordant	媒染剂	媒染劑
mordant assistant	媒染助剂	媒染助劑
mordant azo dye	媒染偶氮染料	媒染偶氮染料
mordant color	媒染色料	媒染色料
mordant dye	媒染染料	媒染染料
mordanting process	媒染法	媒染法
mortar	砂浆	灰泥
most probable distribution	最概然分布	最可能分布
most probable number	最大概率数	最可能數
mother cell	母细胞	母細胞
mother liquor	母液	母液
motion	运动	運動
motionless mixer	无动混合器	無動混合器
motor	电动机	馬達,電動機
motor actuator	电动机驱动器	馬達引動器
motor constant	电动机常数	馬達常數
mould	霉菌	霉菌
mounting	装配	安裝
moving bed	移动床	移動床
moving bed adsorber	移动床吸附器	移動床吸附器

英　文　名	祖国大陆名	台湾地区名
moving bed gasifier	移动床气化器	移動床氣化器
moving bed process	移动床法	移動床法
moving bed reactor	移动床反应器	移動床反應器
moving boundary	移动界面	移動界面
moving catalyst bed	移动催化剂床	移動觸媒床
moving coil galvanometer	圈转电流计	圈轉電流計
moving phase	移动相	移動相
MTZ(＝mass transfer zone)	传质区	傳質區
mud cake	滤饼	濾餅
mud press	滤泥机	濾泥機
mud settler	泥浆沉降器	泥漿沈降器
mullite	莫来石	富鋁紅柱石
mullite porcelain	莫来石瓷	富鋁紅柱石瓷
mullite refractory	莫来石耐火材料	富鋁紅柱石耐火物
multiaxial stress	多轴应力	多軸應力
multibed reactor	多床反应器	多床反應器
multiblade fan	多叶片风机	多葉風扇
multiblade mixer	多叶片混合器	多葉混合器
multicasing turbine	多室涡轮机	多室渦輪機
multichamber centrifuge	多室离心机	多室離心機
multichannel analyzer	多道分析仪	多通道分析儀
multicomponent	多组分	多成分,多元
multicomponent absorption	多组分吸收	多成分吸收
multicomponent azeotrope	多组分共沸物	多成分共沸液
multicomponent diffusion	多组分扩散	多成分擴散
multicomponent mixture	多元混合物,多组分混合物	多成分混合物
multicomponent separation sequence	多元分离序列	多成分分離順序
multicomponent separation sequencing	多元分离序列	多成分分離定序
multicomponent system	多元系[统],多组分系统	多成分系統
multicone separator	多锥管分离器	多錐管分離器
multicyclone	多管旋风分离器	多旋風分離器
multidowncomer tray	多降液管塔板	多降液管塔板
multieffect evaporation	多效蒸发	多效蒸發
multifunctional catalyst	多功能催化剂	多功能觸媒
multifunctional enzyme	多功能酶	多功能酶
multilayer adsorption theory	多层吸附理论	多層吸附理論
multilevel method of optimization	多层次优化法	多層次優化法
multiloop control system	多回路控制系统	多環路控制系統

英　文　名	祖国大陆名	台湾地区名
multiloop system	多回路系统	多環路系統
multi-objective programming	多目标规划	多目標規劃
multipass condenser	多程冷凝器	多程冷凝器
multipass exchanger	多程换热器	多程熱交換器
multipass heater	多程加热器	多程加熱器
multiphase flow	多相流	多相流
multiphase reactor	多相反应器	多相反應器
multiphase system	多相系统	多相系統
multiple batch extraction	多级批式萃取	多階批式萃取
multiple bed reactor	多床反应器	多床反應器
multiple blade mixer	多叶混合器	多葉混合器
multiple distributed component	多分布组分	多分布成分
multiple evaporation	多效蒸发	多效蒸發
multiple evaporator	多效蒸发器	多效蒸發器
multiple extraction	多次萃取	多次萃取
multiple feed stream	多重进料流	多重進料流
multiple hearth furnace	多室反应器	複床爐
Multiple-input-multiple-output-system	多输入多输出系统	多輸入多輸出系統
multiple-input system	多重输入系统	多[重]輸入系統
multiple loop system	多回路系统	多環路系統
multiple output control	多重输出控制	多重輸出控制
multiple output system	多重输出系统	多[重]輸出系統
multiple pipe system	多管系统	多管系統
multiple product column	多重产物塔	多重產物塔
multiple reactions	多重反应	多重反應
multiple reactor system	多重反应器系统	多反應器系統
multiple reboiler	多重再沸器	多重再沸器
multiple split point	多分隔点	多分隔點
multiple stability	多重稳态	多重穩態
multiple stage centrifuge	多级离心机	多階離心機
multiple stage compression	多级压缩	多階壓縮
multiple stage compressor	多级压缩机	多階壓縮機
multiple steady state	多重稳态	多重穩態
multiplet	多重态	多重態
multiplication factor	多重性因子	倍增因數
multiplicity	多重态	多重態, 多重性
multiplicity of steady states	多稳态性	多穩態性
multiplier	乘法器	相乘器
multiplying manometer	倍示压力计	倍示壓力計

英 文 名	祖国大陆名	台湾地区名
multipoint recorder	多点记录器	多點記錄器
multiport system	多进出口系统	多進出口系統
multipurpose batch plant	万能批量工厂	目標批式工廠
multi-region model	多区模型	多區模型
multiribbon blender	多带掺合机	複帶掺合機
multistage centrifugal pump	多级离心泵	多階離心泵
multistage compression	多级压缩	多階壓縮
multistage compressor	多级压缩机	多階壓縮機
multistage diffusion	多级扩散	多階擴散
multistage distillation	多级蒸馏	多階蒸餾
multistage ejector	多级喷射泵	多階射出器
multistage fluidized bed	多级流化床	多階流化床
multistage grinding	多级研磨	多階研磨
multistage process	多段过程	多階程序
multistage system	多段体系	多階系統
multitube reactor	多管反应器	多管反應器
multivariable control system	多变量控制系统	多變數控制系統
multivariable optimization	多变量最优化	多變數最適化
multivariable system	多变量系统	多變數系統
municipal waste	城市垃圾	都市廢棄物
municipal wastewater	城市污水	都市污水
muriate of potash	氯化钾	氯化鉀
muriatic acid	盐酸	鹽酸
Murphree efficiency	默弗里效率	默弗里效率
muscovado sugar	混糖	黃砂糖
muscovite	白云母	白雲母
mushroom distilling column	蕈状蒸馏塔	菇形蒸餾塔
mustard oil	芥子油	芥子油
mutagen	诱变剂	突變劑
mutation	突变	突變
mutation frequency	突变频率	突變頻率
mutual solubility	互溶度	互溶度
MWD(＝molecular weight distribution)	分子量分布	分子量分布

N

英 文 名	祖国大陆名	台湾地区名
nanodroplet	纳米液滴	奈米液滴

英 文 名	祖国大陆名	台湾地区名
nanofiltration	纳米过滤	奈米過濾
nanomaterials	纳米材料	奈米材料
nanometer	纳米	奈米
nanoparticle	纳米粒子	奈米粒子
nano-plymer materials	纳米聚合材料	奈米聚合[物]材料
nanoscale science and technology	纳米科学[与]技术	奈米科技
nanosized crystals	纳米微晶	奈米晶體
nanostructure	纳米结构	奈米結構
nanotechnical	纳米技术[的]	奈米技術[的]
naphtha	石脑油	①石油腦②輕油
naphtha cracking	石脑油裂解	輕油裂解
naphtha pyrolysis	石脑油热解	輕油裂解
naphthene	环烷	環烷
naphthol dye	萘酚染料	萘酚染料
Nash pump	纳氏泵	納氏泵
natural aging	自然老化	自然老化
natural circulation evaporator	自然循环蒸发器	自然循環蒸發器
natural convection	自然对流	自然對流
natural cooling	自然冷却	自然冷卻
natural draft	自然通风	自然通風
natural draft cooling tower	自然通风冷却塔	自然通風冷卻塔
natural draught	自然通风	自然通風
natural evaporation	自然蒸发	自然蒸發
natural frequency	自然频率	自然頻率
natural gas	天然气	天然氣
natural gas hydrates	天然气水合物	天然氣水合物
natural gas liquefaction	天然气液化	天然氣液化
natural gasoline	天然汽油	天然汽油
natural gas reforming	天然气重整	天然氣重組
natural period	自然周期	自然週期
natural resources	自然资源	天然資源
natural rubber	天然橡胶	天然橡膠
natural soda	天然碱	天然鹼
natural varnish	天然清漆	天然清漆
needle number	针孔数	針孔數
needle valve	针阀	針形閥
negative adsorption	负吸附	負吸附
negative azeotrope	下降性共沸混合物	負共沸液
negative catalysis	负催化作用	負催化[作用]

英 文 名	祖国大陆名	台湾地区名
negative catalyst	负催化剂	負觸媒
negative chemical promoter	负化学促进剂	負化學促進劑
negative coupling	负偶合	負偶合
negative deviation	负偏差	負偏離
negative feedback	负反馈	負回饋
negative resistance	负电阻	負電阻
negative valued product	负值产物	負值產物
nematic liquid crystal	向列型液晶	向列型液晶
nematic state	向列态	向列液晶態
neomycin	新霉素	新黴素
neoprene	氯丁橡胶	新平[橡膠]
neoprene adhesive	氯丁橡胶黏合剂	新平黏合劑
neoprene rubber	氯丁橡胶	新平橡膠, 氯平橡膠
nephelite	霞石	霞石
nested control loop	多巢控制环路	巢式控制環路
net flux	净通量	淨通量
net heating value	净热值	淨熱值
net positive suction head (NPSH)	汽蚀余量, 净正吸压头	淨正吸高差
net present value	净现值	淨現值
net present worth	净现值	淨現值
net profit	净利润	淨利潤
net rate	净速率	淨速率
net weight	净重	淨重
net work	净功	淨功
network	网络	網路
neural network training	神经网络训练	神經網絡訓練
neuron	神经元	神經元
neutral fertilizer	中性肥料	中性肥料
neutral firebrick	中性耐火砖	中性耐火磚
neutral glass	中性玻璃	中性玻璃
neutralization	中和	中和
neutralization number	中和值	中和值
neutralization point	中和点	中和點
neutralization reaction	中和反应	中和反應
neutralization titration	中和滴定	中和滴定[法]
neutralization value	中和值	中和值
neutralizing agent	中和剂	中和劑
neutral refractory	中和耐火材料	中性耐火物
neutral size	中性施胶	中性膠料

英　文　名	祖国大陆名	台湾地区名
neutral soap	中性肥皂	中性肥皂
neutral sulfite process	中性亚硫酸盐法	中性亞硫酸[鹽]法
neutral sulfite semichemical process	中性亚硫酸盐半化学法	中性亞硫酸[鹽]半化學法
neutral zone	中和区	中和區
news-printing paper	新闻纸	新聞紙
Newtonian fluid	牛顿流体	牛頓流體
Newton method for convergence	牛顿收敛法	牛頓收斂法
Newton Raphson method	牛顿-拉弗森法	牛頓-拉弗森法
nichrome	镍铬合金	鎳鉻合金
nickel alumina catalyst	镍铝氧催化剂	鎳鋁氧觸媒
nickel carbonyl catalyst	羰基镍催化剂	鎳羰觸媒
nickel crucible	镍坩埚	鎳坩堝
nickel plating	镀镍	鍍鎳
nickel resistance bulb	镍电阻球	鎳電阻球
nickel steel	镍钢	鎳鋼
nickel storage battery	镍蓄电池	鎳蓄電池
niger	皂脚	皂腳
niger oil	皂脚油	皂腳油
nigrosine	苯胺黑	苯胺黑
niter	硝石	硝石
niter cake	硝饼	硝餅
niter cake furnace	硝饼炉	硝餅爐
niter oven	硝炉	硝爐
niter pot	硝锅	硝鍋
nitration	硝化作用	硝化
nitration process	硝化法	硝化程序, 硝化法
nitrator	硝化器	硝化器
nitrification	硝化作用	氮硝化
nitrifying bacteria	硝化菌	氮硝化菌
nitrile rubber	丁腈橡胶	腈橡膠
nitrocellulose	硝酸纤维素	硝化纖維素, 硝酸纖維素
nitrocellulose powder	硝酸纤维素火药	硝化纖維素無煙火藥
nitro dye	硝基染料	硝基染料
nitroexplosive	硝基火药	硝基火藥
nitrogen cycle	氮循环	氮循環
nitrogen fixation	固氮[作用]	氮固定
nitrogen guano	富氮海鸟粪	富氮海鳥糞

英 文 名	祖国大陆名	台湾地区名
nitrogen mustard gas	氮芥子气	氮芥子氣
nitrogenous fertilizer	氮肥	氮肥
nitrogen oxygen demand	氮氧需要量	氮氧需要量
nitrogen purge	氮气清除	氮氣沖洗
nitroglycerine explosive	硝化甘油炸药	硝化甘油炸藥
nitrosamine red	亚硝胺红	亞硝胺紅
nitroso dye	亚硝基染料	亞硝基染料
node	节点	節點
nodulizer	造粒机	製粒機
nodulizing kiln	造粒转窑	粒化轉窯
noise	噪声	①雜訊, 噪音②干擾
noise level	噪声水平	噪音水平
noise pollution	噪声污染	噪音污染
noise thermometer	噪声温度计	雜訊溫度計
nomenclature	命名法	命名[法]
nominal	公称	標稱
nominal pipe diameter	公称管径	標稱管徑
nominal size	公称尺寸	標稱尺寸
nominal stress	名义应力	標稱應力
nominal value	标称值	標稱值
nonactivated chemisorption	未活化学吸附	未活化學吸附
nonadiabatic reaction	非绝热反应	非絕熱反應
nonadiabatic reactor	非绝热反应器	非絕熱反應器
non-affinity adsorption	非亲和吸附	非親和吸附
noncatalytic reaction	非催化反应	非催化反應
nonchain reaction	非连锁反应	非連鎖反應
noncoking coal	非炼焦煤	非煉焦煤
noncompetitive inhibition	非竞争性抑制	非競爭性抑制[作用]
noncondensable gas	不凝气体	不冷凝氣體
noncrystalline	非晶性的	非晶
nondestructive test	非破坏性试验	非破壞性試驗
nondrying oil	非干性油	不乾性油
nonelementary reaction	非基本反应	非基本反應
non-equilibrium stage model	非平衡级模型	非平衡級模型
non-equilibrium system	非平衡系统	非平衡系統
non-equilibrium thermodynamics	非平衡热力学, 不可逆 过程热力学	不平衡熱力學
nonfilamentous bacteria	非丝状细菌	非絲狀細菌
nonhomogeneity	非均匀性	不均匀性

英　文　名	祖国大陆名	台湾地区名
nonideal flow	非理想流动	非理想流動
nonideal gas	非理想气体	非理想氣體
nonideal reactor	非理想反应器	非理想反應器
nonideal solution	非理想溶液	非理想溶液
nonideal surface	非理想表面	非理想表面
noninertial coordinate system	非惯性坐标系统	非慣性坐標系統
noninteracting system	无相互作用系统	無相互作用系統
noninvasive instrument	非侵入性仪器	非侵入性儀器
noninvasive sensor	非侵入性遥感器	非侵入性感測器
nonionic detergent	非离子洗涤剂	非離子清潔劑
nonionic surface active agent	非离子型表面活性剂	非離子界面活性劑
nonionic surfactant	非离子型表面活性剂	非離子界面活性劑
non-isothermal absorption	非等温吸收	非等溫吸收
nonisothermality	非等温性	非等溫性
nonisothermal reactor	非等温反应器	非等溫反應器
nonisotropic turbulence	各向异性湍流	非各向同性紊流
nonlinear controller	非线性控制器	非線性控制器
nonlinear control system	非线性控制系统	非線性控制系統
nonlinear control valve	非线性控制阀	非線性控制閥
nonlinearity	非线性	非線性
nonlinear kinetics	非线性动力学	非線性動力學
nonlinear programming	非线性规划	非線性規劃
nonlinear regression analysis	非线性回归分析	非線性回歸分析
nonlinear stability	非线性稳定性	非線性穩定性
nonlinear system	非线性系统	非線性系統
nonlinear viscoelasticity	非线性黏弹性	非線性黏彈性
nonmetal	非金属	非金屬
nonminimum phase lag	非最小相位落后	非最小相位落後
non-Newtonian fluid	非牛顿流体	非牛頓流體
nonoverlapping operation	非重叠操作	非重疊操作
nonporous membrane	非多孔膜	非多孔膜
nonporous pellet absorber	非多孔颗粒吸收器	非孔性粒吸收器
nonproductive period	辅助周期	不生產期
non-random two liquid equation	NRTL 方程, 非随机两液方程	NRTL 方程, 非随機兩液方程
nonreactive component	无反应性组分	無反應性成分
nonreactivity	无反应性	無反應性
nonregular solution	非正规溶液	非規則溶液
nonrenewable energy source	非再生能源	非再生性能源

英　文　名	祖国大陆名	台湾地区名
nonrepetitive polymer	非重复聚合物	非重複性聚合物
nonsegregated backmixing	非分离返混	非隔離逆混
nonsegregated mixing	非分离混合	非隔離混合
nonsegregated reactor	非分离反应器	非隔離反應器
non-spontaneous process	非自发过程	非自發過程
nonspontaneous reaction	非自发反应	非自發反應
nonsteady state process	非稳态过程	非穩態程序
nontronite	绿脱石	矽鐵石
nonuniform flow	非均匀流	非均匀流動
nonvolatility	不挥发性	不揮發性
nonwashing plate	非洗式板框	非洗式板框
nonwetting surface	非润湿表面	非潤溼表面
nonwoven fabric	无纺布	不織布
normal channel depth	正常渠深	正常渠深
normal concentration	规定浓度	當量濃度
normal condition	正常状态	常態
normal coordinates	法坐标	正規座標
normal distribution	正态分布	常態分布
normal distribution function	正态分布函数	常態分布函數
normalization	归一化	正規化
normal linearization	正常线性化	常態線性化
normal operation	正常操作	正常操作
normal reaction kinetics	正常反应动力学	正常反應動力學
normal shock wave	正激波	正震波
normal solution	正常溶液	當量溶液
normal stress	法向应力	正應力,法向應力
normal stress difference	法向应力差	正應力差
no slip condition	无滑动条件	無滑動條件
notation	记号	①記號②記法
notch	凹槽	凹槽,Ｖ形槽
notched weir	切口堰	Ｖ形堰
novobiocin	新生霉素	新生黴素
noxious gas	有害气体	有害氣體
nozzle	喷嘴	噴嘴
nozzle pressure	喷嘴压力	噴嘴壓力
NPSH(＝net positive suction head)	汽蚀余量,净正吸压头	淨正吸高差
NRTL equation	NRTL 方程,非随机两液体方程	NRTL 方程
NTU(＝number of transfer unit)	传递单元数	質傳單元數

英　文　名	祖国大陆名	台湾地区名
nuclear bomb	核弹	核彈
nuclear chemistry	核化学	核化學
nuclear energy	核能	核能
nuclear fuel	核燃料	核燃料
nuclear fusion	核聚变	核熔合,核聚变
nuclear physics	核物理学	核(子)物理學
nuclear radiation	核辐射	核輻射
nuclear reactor	核反应堆	核反應器
nucleate boiling	泡核沸腾	成核沸騰
nucleate crystallization	有核结晶	成核結晶
nucleation	成核,晶核生成	成核
nucleation crystallization	成核结晶	成核結晶
nucleic acid	核酸	核酸
nucleoside	核苷	核苷
nucleotide	核苷酸	核苷酸
nucleus	晶核	①核②晶核
nucleus formation	晶核形成	晶核形成
nuclide	核素	核種
null balance	零点平衡	零點均衡
null hypothesis	零假设	零假設
null potentiometer	零电位计	零點電位計
number analysis	数值分析	數值分析
number average molecular weight	数均分子量	數量平均分子量
number of heat transfer unit	传热单元数	熱傳單元數
number of mass transfer unit	传质单元数	質傳單元數
number of overall transfer units	总传质单元数	總質傳單元數
number of transfer unit (NTU)	传质单元数	傳送單元數
numerical analysis	数值分析	數值分析
numerical control	数字控制	數值控制
numerical method	数值方法	數值方法
numerical solution	数值解	數值解
Nusselt number	努塞特数	努塞特數
nylon	尼龙	耐綸,耐隆,尼龍
nystatin	制霉菌素	奈黴素,制黴菌素

O

英 文 名	祖国大陆名	台湾地区名
objective control	目标控制	目標控制
objective function	目标函数	目標函數
objective management	目标管理	目標管理
observability	可观测性	可觀測性
observed order	观测次序	測得階
obsolescence	退化	退化
occluded resin	吸着树脂	包藏樹脂
occlusion	包藏	①吸著②包藏
occurrence matrix	事件矩阵	事件矩陣
ocean pollution	海洋污染	海洋污染
ocher	赭石	赭石
octane enhancer	辛烷值增进剂	辛烷值增進劑
octane number	辛烷值	辛烷值
octane number test	辛烷值测试	辛烷值測試
odor	气味	氣味
odor pollutant	恶臭污染物	氣味污染物
off-line	离线	離線
offset	偏离	偏差
offset paper	胶版印刷纸	平板印紙
off-spec product	不合格产品	不合規格產品
ohmmeter	欧姆计	歐姆計
oil	油	油
oil atomizer	油喷雾器	噴油器
oil bath	油浴	油浴
oil cake	油饼	油餅
oiled paper	油纸	油紙
oil field	油田	油田
oil flotation	油浮选	油浮選
oil foot	油脚	油腳, 油渣
oil gas	油气	油氣
oil in water emulsion	水包油乳状液	水中油乳液
oil mordant	油媒染剂	油媒染劑
oil of vitriol	浓硫酸	礬油, 硫酸
oil paint	油脂涂料	油性漆
oil recovery	石油回收	石油回收

英　文　名	祖国大陆名	台湾地区名
oil sand	油砂	油砂
oil shale	油页岩	油頁岩
oil tank	储油罐	油槽
oil tanker	油船	油輪
oil-water separator	油水分离器	油水分離器
oil winterizing	油冬化	①油冬化②油低溫脱脂
ointment base	软膏基质	軟膏基
oleandomycin	竹桃霉素	安徽素, 奧連徽素
oleo-resin	油性树脂	油性樹脂
oligoclase	奥长石	鈉鈣長石
oligomer	低聚物	低聚合物
oligomerization	低聚反应	低[元]聚合[反應]
oligosaccharide	寡糖, 低聚糖	寡糖
olive-kernel oil	橄榄仁油	橄欖仁油
olive oil	橄榄油	橄欖油
olivine	橄榄石	橄欖石
once-through operation	单程操作	單程操作
once-through process	单程过程	單程程序
one-component system	单组分系统	單成分系統
one-dimensional model	一维模型	一維模型
one-way decoupling	单向解偶	單向解偶
on-line	在线	線上
on-line adaptation	在线适应	線上調適
on-line computer control	线上计算机控制	線上計算機控制
on-line tuning	线上调谐	線上調諧
on-off action	开关作用	開關作用
on-off control	通断控制	開關控制
Onsager reciprocal relation	昂萨格倒易关系	昂薩格倒易關係
opacifier	乳浊剂	乳白劑
opacifying agent	乳浊剂	失透劑
opacity	不透明度	乳白度
opalescent glass	乳白玻璃	乳光玻璃
opaque glass	乳浊玻璃	不透明玻璃
opaque glaze	不透明釉	不透明釉
open boundary	开式边界	開式邊界
open channel	明槽	明渠
open channel flow	明槽流	明渠流
open circuit	开路	開路
open circuit grinding	开路研磨	開路研磨

英　文　名	祖国大陆名	台湾地区名
open filter	敞式沙滤器	開敞過濾器
open impeller	开敞叶轮	敞動葉輪
open loop	开环	開環
open-loop control	开环控制	開環控制
open-loop pole	开环极点	開環極點
open-loop system	开环系统	開環系統
open-loop test	开环测试	開環測試
open-loop transfer function	开环转移函数	開環轉移函數
open-loop zero	开环零点	開環零點
open nozzle	开敞喷嘴	開敞噴嘴
open reactor	开式反应器	開敞反應器
open sand filter	开式过滤器	開敞砂濾器
open sequence reaction	开式序列反应	非連鎖反應
open system	敞开系统	開放系統
open vessel	开式容器	開放容器
operability	可操作性	可操作性
operating condition	操作条件	操作條件
operating curve	操作曲线	操作曲線
operating data	操作数据	操作數據
operating diagram	操作图	操作圖
operating flexibility	操作弹性	操作彈性
operating frequency	操作频率	操作頻率
operating limit	操作界限	操作界限
operating line	操作线	操作線
operating manual	操作手册	操作手冊
operating parameter	操作参量	操作參數
operating point	操作点	操作點
operating pressure	操作压力	操作壓力
operating principle	操作原理	操作原理
operating stability	操作稳定性	操作穩定性
operating temperature	操作温度	操作溫度
operating time	操作时间	操作時間
operating window	操作窗	操作窗
operational variable	操作变量	操作變量
operation amplifier	操作放大器	操作放大器
operation condition	操作条件	操作條件
operation cost	操作成本	操作成本
operation cycle	操作周期	操作周期
operation data	操作数据	操作數據

英　文　名	祖国大陆名	台湾地区名
operation diagram	操作图	操作圖
operation efficiency	操作效率	操作效率
operation flowsheet	操作流程图	操作流程圖
operation length	操作时间	操作期
operation line	操作线	操作線
operation mode	工况	操作方式
operation regeneration cycle	操作再生循环	操作再生循環
operation standard	操作标准	操作標準
opposing reactions	对峙反应	可逆反應
optical compensator	光学补偿器	光學補償器
optical density	光学密度	光學密度
optical disc	光碟	光碟
optical fiber	光纤	光纖(維)
optical glass	光学玻璃	光學玻璃
optical isomer	旋光异构体	光學異構物
optical pyrometer	光学高温计	光學高溫計
optical rotatory dispersion	旋光色散	旋光色散
optimal composition	最优组成	最適組成
optimal condition	最优工况	最適條件, 最適狀況
optimal control	最优控制	最適控制
optimal conversion	最优转化率	最適轉化率
optimal design	最优设计	最適設計
optimality	最优性	最適性
optimal operation condition	最优操作条件	最適操作條件, 最適操作狀況
optimal path	最优途径	最適途徑
optimal policy	最优策略	最適策略
optimal reflux	最优回流[量]	最適回流
optimal searching method	最佳搜寻法	最適搜尋法
optimal set point	最优设计点	最適設定點
optimal switching	最优开关	最適開關
optimal system	最优系统	最適系統
optimal temperature	最优温度	最適溫度
optimal tuning	最优调谐	最適調諧
optimal value	最优值	最適值
optimization	优化, 最优化	最適化
optimum control	最优控制	最適控制
optimum conversion	最优转化率	最適轉化率
optimum design	最优设计	最適設計

英　文　名	祖国大陆名	台湾地区名
optimum operation condition	最优操作条件	最適操作條件, 最適操作狀況
optimum path	最佳途径	最適途徑
optimum policy	最适策略	最適策略
optimum reflux	最佳回流	最適回流
optimum reflux ratio	最佳回流比	最適回流比
optimum searching method	最佳搜寻法	最適搜尋法
optimum set point	最优设计点	最適設定點
optimum switching	最优开关	最適切換
optimum system	最优系统	最適系統
optimum temperature	最适温度	最適溫度
optimum tuning	最佳调谐	最適調諧
optimum value	最优值	最適值
order of deactivation	去活化级	去活化階
order of magnitude	数量级	量階
order of reaction	反应级	反應階
ordinary differential equation	常微分方程	常微分方程式
ore dressing	选矿	選礦
ore flotation	矿浮选	礦浮選
organic acid	有机酸	有機酸
organic binder	有机黏合剂	有機黏合劑
organic builder	有机补助剂	有機補助劑
organic carbon	有机碳	有機碳
organic fertilizer	有机肥料	有機肥料
organic load	有机负荷量	有機物負載
organic overload	有机超负载	有機物超負載
organic sludge	有机污泥	有機污泥
organic solid	有机固体	有機固體
organisms	有机体	微生物
organization	组织	組織
organoclay	有机黏土	有機黏土
organo mercury	有机汞	有機汞
organo metallic compound	有机金属化合物	有機金屬化合物
organosol	有机增塑糊	有機溶膠
orientation	定向	定向, 定位, 方位
orientation distribution	定向分布	方向分布
orientation flat	定向平面	定向平面
oriented adsorption	定向吸附	位向吸附
orifice	孔口	孔口, 小孔

英　文　名	祖国大陆名	台湾地区名
orifice coefficient	孔口系数	孔口係數
orifice discharge	孔口排放	孔口排放
orifice flange	孔口法兰	孔口法蘭, 孔口凸緣
orifice flowmeter	孔板流量计	孔口流量計
orifice meter	孔板流量计	孔口[流量]計
orifice meter coefficient	孔板计系数	孔口計係數
orifice plate	孔板	孔口板
orifice ring	孔口环	孔口環
orifice tap	孔板计接头	孔口計接頭
orifice tube	锐孔管	孔口管
orlon	奥纶	奧綸
orthoflow process	正交流过程	正交流程序
orthogonal collocation	正交配置	正交配置
orthogonality	正交性	正交性
oscillating reaction	振荡反应	振盪反應
oscillating screen	振动筛	擺動篩
oscillation	振荡	振盪, 振動, 擺動
oscillation frequency	振荡频率	振盪頻率
oscillatory element	振荡元件	振盪元件
oscillograph	示波器	示波器
oscilloscope	示波器	示波器
Oslo evaporative crystallizer	奥斯陆蒸发结晶器	奧斯陸蒸發結晶器
osmometer	渗透压计	滲[透]壓[力]計
osmoscope	渗透试验器	滲透試驗器
osmosis	渗透[作用]	滲透
osmotic coefficient	渗透系数	滲透係數
osmotic diffusion	渗透扩散	滲透擴散
osmotic pressure	渗透压	滲透壓
outgassing	出气	除氣
outlet	出口	出口
outlet temperature	出口温度	出口溫度
outlet valve	出口阀	出口閥
output function	输出函数	輸出函數
output layer	输出层	輸出層
output pressure	输出压力	輸出壓力
output set	输出集	輸出集
output signal	输出信号	輸出信號
output variable	输出[变]量	輸出變數
outside diameter	外径	外徑

英　文　名	祖国大陆名	台湾地区名
outside reflux	外回流	[塔]外回流
outside reflux ratio	外回流比	[塔]外回流比
oven	烘箱	烘箱, 爐
oven gas	炉气	焦爐氣
overall absorption coefficient	总吸收系数	總吸收係數
overall coefficient	总系数	總係數
overall composition	总组成	總組成
overall efficiency	总效率	總效率
overall energy balance	总能量平衡	總能量均衡
overall entropy balance	总熵平衡	總熵均衡
overall fractional yield	总产量分率	總產量分率
overall heat balance	总热平衡	總熱量均衡
overall heat transfer coefficient	总传热系数	總熱傳係數
overall mass balance	总质量平衡	總質量均衡
overall mass transfer coefficient	总传质系数	總質傳係數
overall process	总过程	總程序
overall rate	总速率	總速率
overall recovery	总回收	總回收[率]
overall transfer function	总传递函数	總轉移函數
overall transfer unit	总传递单元	總傳送單位
overcapacity	生产能力过剩	超負荷
overdamped response	过阻尼应答	過阻尼應答
overdamped system	过阻尼系统	過阻尼系統
overdamping	过阻尼	過阻尼
over-driven buhrstone mill	过载磨石机	上動石磨[機]
overexpansion phenomenon	过膨胀现象	過膨脹現象
overflow	溢流, 上溢	溢流
overflow alarm	溢流警报	溢流警報
overflow pipe	溢流管	溢流管
overflow rate	溢流速率	溢流速率
overflow velocity	溢流速度	溢流速度
overflow weir	溢流堰	溢流堰
over gain	过增益	過增益
overhead condenser	塔顶冷凝器	頂部冷凝器
overhead cost	管理费用	管理費
overhead distillate	头馏分	頂部餾出物
overhead vapor	塔顶蒸气	頂部蒸氣
overheat	过热	過熱
overlapping operation	重叠操作	重疊操作

英　文　名	祖国大陆名	台湾地区名
overload	过载	超載, 超負荷
overpotential	超电势	過電位
overproduction	生产过剩	過度生產
over range protection	超限保护	超限保護
overrelaxation method	过松弛法	過鬆弛法
override control	超越控制	超越控制
oversaturation	过饱和	過飽和
overshoot	超调	超越量
overshooting	过调节	超越
oversize distribution curve	筛上物分布曲线	篩上物分布曲線
oxidant	氧化剂	氧化劑
oxidase	氧化酶	氧化酶
oxidation	氧化	氧化[反應]
oxidation catalyst	氧化催化剂	氧化觸媒
oxidation ditch	氧化渠	氧化渠
oxidation inhibitor	氧化抑制剂	氧化抑制劑
oxidation period	氧化期	氧化期
oxidation pond	氧化塘	氧化池
oxidation potential	氧化电位	氧化電位
oxidation process	氧化过程	氧化程序
oxidation reaction	氧化反应	氧化反應
oxidation reduction cell	氧化还原电池	氧化還原電池
oxidation reduction potential	氧化还原电位	氧化還原電位
oxidation reduction reaction	氧化还原反应	氧化還原反應
oxidation reduction titration	氧化还原滴定	氧化還原滴定[法]
oxidative degradation	氧化降解	氧化降解[反應]
oxidative pyrolysis	氧化热裂化	氧化熱解[反應]
oxidized oil	氧化油	氧化油
oxidized starch	氧化淀粉	氧化澱粉
oxo process	羰基合成	羰氫化法
oxydase	氧化酶	氧化酶
oxydo-reductase	氧化还原酶	氧化還原酶
oxygen analyzer	氧分析仪	氧分析儀
oxygenation	氧合作用	加氧[反應], 充氧
oxygen consumption rate	耗氧速率	耗氧速率
oxygen content	含氧量	含氧量
oxygen demand	需氧量	需氧量
oxygen depletion	氧量耗尽	氧之耗盡
oxygenizer	充氧器	充氧器

英　文　名	祖国大陆名	台湾地区名
oxygen supply	供氧	供氧
oxygen transfer	氧传递	氧傳遞
oxygen transfer coefficient	传氧系数	傳氧係數
oxygen transfer rate	传氧速率	傳氧速率
oxygen uptake	摄氧	攝氧[量]
oxygen uptake rate	摄氧速率	攝氧速率
oxygen yield coefficient	氧收率系数	氧收率係數
oxyluminescence	氧化发光	氧發光
ozone	臭氧	臭氧
ozone cracking	臭氧龟裂	臭氧紋裂
ozonization	臭氧化	臭氧化[反應]
ozonometer	臭氧计	臭氧計

P

英　文　名	祖国大陆名	台湾地区名
packaged equipment	成套设备	套裝裝置
package software	软件包	套裝軟體
packaging	包装	包裝
packed bed	填充床	填充床
packed bed reactor	填充床反应器	填充床反應器
packed column	①填料塔②填充柱	①填料塔②填充柱
packed column absorber	填料吸收器	填充塔吸收器
packed density	填充密度	填充密度
packed distillation column	填充蒸馏塔	填充蒸餾塔
packed extractor	填充式萃取塔	填充式萃取塔,填充式萃取器
packed height	填充高度	填充高度
packed scrubber	填充式洗涤器	填充洗氣器
packed spray tower	填充式喷淋塔	填充噴霧塔
packed tower	填料塔	填充塔
packing density	填充密度	填充密度
packing effect	敛集效应	緊束效應,質量效應
packing factor	填料因子	填充因數
packing fraction	充填率	填充分率
packing leather	密封革	墊皮
packing material	填充物,填料材料	填充物
packing paper	包装纸	包裝紙

英　文　名	祖国大陆名	台湾地区名
packing ring	①密封圈②填料环	①墊圈②填充環
packing support	填料支承板	填充物支架
padding	①垫料②浸染	①填料②浸染
padding liquor	浸染液	浸染液
padding machine	浸染机	浸染機
paddle	平桨	槳
paddle agitator	桨式搅拌器	槳式攪拌器
paddle flocculator	桨式絮凝器	槳式絮凝器
paddle impeller	桨式叶轮	槳式葉輪
paddle mixer	桨式混合器	槳式混合器
paddle stirrer	桨式搅拌器	槳式攪拌器
paddle wheel	叶轮	槳輪
paint	①油漆②涂料	①油漆②塗料
paint remover	脱漆剂	去漆劑
pair annihilation	[正负电子]对湮没	對偶互毀
pair creation	电子偶产生	對偶生成
pair glass	双层中空玻璃	雙層玻璃
pairing	对偶,成对	對偶,成對
pairing variables	对偶变量	對偶變數
pair production	电子偶生成	对偶生成
palladium catalyst	钯催化剂	鈀觸媒
pallet	平板架	托板
palm kernel oil	棕榈仁油	棕櫚仁油
palm nut oil	棕榈仁油	棕櫚實油
palm oil	棕榈油	棕櫚油
palm oil grease	油润滑脂	棕櫚油[潤]滑脂
palm wax	棕榈蜡	棕櫚蠟
panel board	①仪表板②配电盘	儀表板
panel spalling test	[耐火砖]格子散裂试验	屏列剝落試驗
pan feeder	加料盘	盤飼器
pan salt	锅盐	鍋鹽
pan seed	缸内晶种	罐內晶種
pan seeding	缸内晶种	罐內晶種
papain	木瓜蛋白酶	木瓜酶
paper chromatography	纸色谱法	紙層析[法]
paper electrophoresis	纸电泳	紙電泳法
paper partition chromatography	纸分配色谱	紙分配層析[法]
parabolic flume	抛物线溜槽	抛物線槽
parabolic valve	抛物线阀	抛物線閥

英　文　名	祖国大陆名	台湾地区名
paraboloid condenser	抛物线体聚光器	抛物線體聚光器
paraffin	石蜡	石蠟
paraffin conversion	石蜡转化	石蠟轉化
paraffin distillate	石蜡馏分	石蠟餾出物
paraffin oil	石蜡油	石蠟油
paraffin paper	石蜡纸	石蠟紙
paraffin slop	含蜡废油	含蠟污油
paraffin soap	石蜡皂	石蠟皂
paraffin wax	石蜡	石蠟
paraflow	并流,同向流	並流
paraflow heat exchange	并流热交换	並流熱交換
parahelium	仲氦	仲氦
parahydrogen	仲氢	仲氫
parallel circuit	并联电路	並聯電路
parallel compensation	并联补偿	並聯補償
parallel compound turbine	并列复式涡轮机	並列複式渦輪機
parallel deactivation	平行失活	平行失活
parallel feed	平行进料	平行進料
parallel operation	并行操作	並聯操作
parallel processes	并行过程	並聯程序
parallel processing	平行处理	平行處理
parallel reaction	平行反应	平行反應
parallel running	并联运行	並聯運轉
paramagnetic body	顺磁体	順磁體
paramagnetic effect	顺磁效应	順磁效應
paramagnetic oxygen analyzer	顺磁氧分析仪	順磁氧分析儀
paramagnetism	顺磁性	順磁性
parameter	参数	參數
parameter selection	参数选择	參數選擇
parameter sensitivity	参数灵敏度	參數靈敏度
parametric pump	参数泵	參數泵
parametric pumping	参数法分离	參數法分離
parasite	寄生物	寄生蟲,寄生物
parathion	对硫磷	巴拉松
parchment paper	羊皮纸	羊皮紙
parchmyn paper	仿羊皮纸	玻璃紙
parent cell	母细胞	母細胞
parent element	母体元素	母元素
parent nuclide	母体核素	母核種

英　文　名	祖国大陆名	台湾地区名
parison	型坯	型坯
parity	奇偶性	奇偶性
paromomycin	巴龙霉素	巴龍黴素
partial combustion	部分燃烧	部分燃燒
partial condensation	部分冷凝	部分冷凝
partial condenser	分凝器	部分冷凝器
partial conversion	部分转化	部分轉化
partial correlation	偏相关	部分相關
partial cracking	部分裂化	部分裂解
partial decomposition	部分分解	部分分解
partial decoupling	部分解耦	部分解偶
partial differential equation	偏微分方程	偏微分方程式
partial fraction	部分分式	部分分式
partial ionization	部分游离	部分游離
partially miscible system	部分互溶系统	部分互溶系統
partially segregated flow	部分分隔流动	部分隔離流動
partial methylation	部分甲基化[反应]	部分甲基化[反應]
partial miscibility	部分互溶	部分互溶
partial molar enthalpy	偏摩尔焓	偏莫耳焓
partial molar excess property	偏摩尔超额物性函数	部分莫耳過量性質
partial molar Gibbs free energy	偏摩尔吉布斯自由能	偏莫耳吉布斯自由能
partial molar internal energy	偏摩尔内能	部分莫耳内能
partial molar property	偏摩尔物性函数	部分莫耳性質
partial molar quantity	偏摩尔量	偏莫耳量
partial molar volume	偏摩尔体积	部分莫耳體積
partial oxidation	部分氧化	部分氧化
partial oxidation reaction	部分氧化反应	部分氧化反應
partial pressure	分压力	分壓
partial reaction	部分反应	部分反應
partial recycle process	部分再循环过程	部分循環程序
partial reflux	部分回流	部分回流
partial separation	部分分离	部分分離
partial vacuum	部分真空	部分真空
particle	粒子	粒子
particle board	碎料板	塑合板,碎屑膠合板
particle characteristics	颗粒特性	粒子特性
particle conversion	粒子转化	粒子轉化
particle cyclone separator	粒子旋风分离器	粒子旋風分離器
particle density	颗粒密度	粒子密度

英　文　名	祖国大陆名	台湾地区名
particle diameter	粒径	粒徑
particle kinetics	粒子动力学	粒子動力學
particle shape	颗粒形状	粒形
particle shape factor	颗粒形状因子	粒形因數
particle size	粒度	粒度,粒子大小
particle size analysis	粒度分析	粒度分析
particle size analyzer	粒度分析仪	粒度分析儀
particle size distribution	粒度分布	粒子大小分布
particle size separator	粒度分离器	粒度分離器
particle swarm	颗粒群	粒子群
particulate	微粒	顆粒,粒狀物
particulate bed	颗粒床	顆粒床
particulate fluidization	散式流态化	顆粒式流體化,散式流體化
particulate process	颗粒过程	顆粒程序
particuology	颗粒学	顆粒學
parting agent	脱模剂	脱模劑
parting line	分界线	分模線
partition chromatography	分配色谱法	分配層析[法]
partition coefficient	分配系数	分配係數
partition constant	分配常数	分配常數
partitioning	分隔	分隔
pascal	帕(压力单位)	帕斯卡(壓力單位)
passivation	钝化	鈍化
passive iron	钝化铁	不活性鐵,鈍鐵
passive state	钝态	鈍態
passivity	钝态	不活性,鈍性
paste	糊剂	糊
paste paint	厚漆	厚漆
paste resin	糊状树脂	糊狀樹脂
pasteurization	巴氏消毒法	低溫殺菌法,巴氏殺菌法
pasteurizer	巴氏灭菌器	低溫殺菌器
pastille	锭剂	研棒
pat	试饼	試餅
patch test	斑贴试验	布片試驗
patent	专利	專利
patent leather	漆革	黑漆皮,漆革
path function	路径函数	路徑函數

英　文　名	祖国大陆名	台湾地区名
path length	程长	路徑長度
path line	迹线	徑線
pathogen	病原体	病原體
path tracing	路径追踪	路徑追蹤
pattern	模式	型式, 方式
pattern recognition	模式识别	模式識別
pattern search	模式搜索	模式搜索
payback period	投资回收期	投資回收期
payout	支付	支付
payroll	工资单	薪資單
peach bloom glaze	桃红釉	桃紅釉, 桃花浪釉
peach-kernel oil	桃仁油	桃仁油
peak	峰值	尖峰
peak flow	峰流量	尖峰流量
peak flow rate	最高流速	尖峰流量
peak gain	峰增益	尖峰增益
peak gain ratio	峰增益比	尖峰增益比
peak load	峰负荷	尖峰負載
peak resonance	峰共振	尖峰共振
peak time	峰时间	尖峰時間
peanut oil	花生油	花生油
peat	泥炭	泥煤
peat bog	泥炭沼地	泥煤田
pebble	卵石	卵石
pebble bed	卵石层	卵石床
pebble heater	卵石加热器	卵石加熱器
pebble mill	砾磨机	卵石磨
pecan oil	胡桃油	胡桃油
Peclet number	佩克莱数	佩克萊數
pectin	果胶	果膠
pectinase	果胶酶	果膠酶
pecto-cellulose	果胶纤维素	果膠纖維素
pedestal	底座	底座, 拖架
peeling	剥离	剝離, 去皮
peep hole	观察孔	視孔
pegmatite	伟晶岩	偉晶花崗石
pelleted catalyst	粒状催化剂	成粒觸媒
pelletizer	造粒机	製粒機, 製丸機
pelletizing	造粒	造粒

英　文　名	祖国大陆名	台湾地区名
penalty function	罚函数	處罰函數
pendant drop test	悬滴试验	懸滴試驗
pendulum-type tension testing machine	摆式张力试验机	擺式張力試驗機
pendulum viscometer	摆式黏度计	擺式黏度計
penetrability	穿透性	穿透性
penetrant	渗透剂	浸透劑
penetrating agent	渗透剂	浸透劑
penetration index	针入度指数	穿透指數
penetration potential	渗透电位	穿透電位
penetration probability	穿透概率	穿透機率
penetration test	渗入测试	浸透試驗, 穿透試驗
penetration theory	穿透理论	穿透理論
penetrometer	针入度测定计	穿透計
Peng-Robinson equation	PR[状态]方程	PR 方程
penicillin	青霉素	青黴素, 盤尼西林
penicillinase	青霉素酶	青黴素酶
peppermint oil	薄荷油	薄荷油
peptidase	肽酶	肽酶
peptide	肽	肽
peptization	胶溶	解膠[作用]
peptizing agent	①胶溶剂②塑解剂	①助消化劑②解膠劑
peptone	胨	腖
peptonization	胨化	腖化[作用]
peracetic acid	过乙酸	過醋酸
percentage absolute humidity	百分绝对湿度	百分絕對濕度
percentage humidity	百分湿度	百分濕度
percolate	渗滤液	滲濾液
percolation	渗滤	滲濾
percolation extractor	渗滤器	滲濾器
perdistillation	透析蒸馏	透析蒸餾
perfect control	完全控制	完美控制
perfect decoupling	完全解偶	完全解偶
perfect fluid	理想流体	理想流體
perfect gas	理想气体	理想氣體
perfectly mixed reactor	完全混合反应器	完全混合反應器
perfect mixing	全混	完全混合
perfect solution	理想溶液	理想溶液
perforated brick	多孔砖	多孔磚
perforated false bottom	多孔假底	多孔假底

英　文　名	祖国大陆名	台湾地区名
perforated iron cathode	多孔铁阴极	多孔鐵陰極
perforated pipe	多孔管	多孔管
perforated plate	多孔板	多孔板
perforated plate column	孔板塔	多孔層板塔
perforated plate distillation column	孔板蒸馏塔	多孔板蒸餾塔
perforated plate tower	孔板塔	多孔層板塔
perforated screen	多孔板筛	多孔[板]篩
perforated tray	多孔塔板	多孔板
perforation	穿孔	①打洞②孔,洞
performance	性能	性能
performance characteristics	工作特性	性能特性
performance chart	操作性能图	性能圖
performance control	性能控制	性能控制
performance criterion	特性判据	性能準則
performance curve	性能曲线	性能曲線
performance equation	性能方程	性能方程式
performance index	性能指标	性能指數
performance number	特性数	性能值
performance test	性能试验	性能試驗
performance test report	性能试验报告	性能試驗報告
perform tray	网孔塔板	網孔塔板
perfume	香料	香料
perfume fixative	香料固定剂	香料固定劑
perfume fixing agent	香料固定剂	香料固定劑
perfume oil	芳香油	香料油
perfusion culture	灌注培养	灌注培養
periclase	方镁石	方鎂石
perilla oil	紫苏子油	蘇子油
period	周期	週期,期間
periodic coagulation	周期凝聚	週期凝聚
periodic flow	周期流	週期流動
periodic flow reactor	周期流反应器	週期流動反應器
periodicity	周期性	週期性
periodic kiln	间歇窑	週期窯,間歇窯
periodic load	周期负荷	週期負載
periodic operation	周期操作	週期操作
periodic oscillation	周期振荡	週期振盪
periodic phenomenon	周期现象	週期現象
periodic solution	周期解	週期解

英 文 名	祖国大陆名	台湾地区名
periodic table	元素周期表	週期表
peripheral speed	圆周速度	圓周速度
peripheral velocity	圆周速度	周邊速度
periscope	潜望镜	潛望鏡
perlite	珍珠岩	珠岩
permanent deformation	永久变形	永久變形
permanent disposal	永久清除	永久處理
permanent disturbance	永久扰动	永久擾動
permanent hardness	永久硬度	永久硬度
permanent loss	永久损失	永久損失
permanent press	永久定形	永久定形
permanent set	永久变形	永久定形
permanganate number	高锰酸盐值	過錳酸鹽數
permanganate value	高锰酸盐[滴定]值	過錳酸鹽值
permeability	①渗透性②磁导率	①透過性②磁導率
permeability apparatus	渗透仪	透過性測定儀
permeability test	渗透试验	透過性試驗
permeate	渗透物	滲透物
permeation flux	渗透通量	滲透通量
permutit	人造沸石	人造沸石
peroxide	过氧化物	過氧化物
peroxide decomposer	过氧化物分解剂	過氧化物分解劑
persistent accelerator	长效促进剂	長效催速劑
persistent agent	长效剂	長效劑
personal computer	个人计算机	個人計算機, 個人電腦
pertraction	渗透萃取	透[過]萃[取]法
perturbation	微扰	微擾
perturbation method	微扰法	微擾法
perturbation theory	微扰理论, 摄动理论	微擾理論
perturbation variable	微扰变量	微擾變數
perturbed hard chain theory	微扰硬链理论	微擾硬鏈理論
pervaporation	渗透蒸发	滲透蒸發
pesticide	杀虫剂	殺蟲劑
petrochemical	石油化学品	石油化學品, 石化品
petrochemical complex	石油化工厂	石[油]化[學]工業區
petrochemical industry	石油化学工业	石油化學工業, 石化工業
petrochemistry	石油化学	石油化學
petrol	车用汽油	石油

英　文　名	祖国大陆名	台湾地区名
petrolatum	矿脂	石蠟脂
petroleum	石油	石油
petroleum coke	石油焦	石油焦
petroleum engineering	石油工程	石油工程
petroleum ether	石油醚	石油醚
petroleum gas	石油气	石油氣
petroleum naphtha	石脑油	石油腦
petroleum processing engineering	石油加工工程	石油煉製工程
petroleum processing technology	石油加工技术	石油煉製技術
petroleum refining	石油炼制	石油煉製
phage	噬菌体	噬菌體
pharmaceutical chemistry	药物化学	藥物化學
pharmaceutical industry	制药工业	製藥工業
pharmaceutical preparation	药物制剂	藥劑
pharmaceutics	药剂学	藥劑學
pharmacokinetics	药动学	藥物動力學
pharmacology	药理学	藥物學, 藥理學
pharmacopoeia	药典	藥典
phase	①相②相位	①相②相位
phase advance	相位超前	相位超前, 相位領先
phase angle	相角	相位, 相角
phase angle locus	相角轨迹	相角軌跡
phase behavior	相特性	相行為
phase change	相变	相變化
phase crossover	相位交叉	相位交越
phase crossover frequency	相位交叉频率	相位交越頻率
phase crossover point	相位交叉点	相位交越點
phase diagram	相图	相圖
phase difference	相[位]差	相位差
phase equilibrium	相平衡	相平衡
phase lag	相位滞后	相位落後
phase lead	相位超前	相位領先, 相位超前
phase margin	相位裕量	相位邊限
phase plane	相平面	相平面
phase plane analysis	相平面分析	相平面分析
phase plane plot	相平面标绘	相平面圖
phase plane portrait	相平面图	相平面圖
phase relation	位相关系	相關係
phase rule	相律	相律

英　文　名	祖国大陆名	台湾地区名
phase shift	相移	相位位移
phase space	相空间	相空間
phase transfer	相转移	相轉移
phase transfer catalysis	相转移催化	相[間]轉移催化[作用]
phase transfer catalyst	相转移催化剂	相[間]轉移觸媒
phase transfer catalyzed reaction	相转移催化反应	相[間]轉移催化反應
phase transformation	相变	相變換
phase velocity	相速度	相速度
pH control	pH 控制	pH 值控制
PHC theory	微扰硬链理论	微擾硬鏈理論
pH electrode	pH 电极	pH 電極
phenol formaldehyde resin	酚醛树脂	酚甲醛樹脂
phenolic resin	酚醛树脂	酚數脂
phenolics	酚醛塑料	酚醛樹脂
phenolphthalein	酚酞	酚酞
phenol red	酚红	酚紅
phenol resin adhesive	酚树脂黏合剂	酚樹脂黏合劑
phenomenological coefficient	唯象系数	現象係數
phenomenological model	唯象模型	現象模式
pH meter	pH 计	pH 計, 酸度計
phosgenation process	光气化法	光氣法
phosgene	光气	光氣
phosgenite	角铅矿	角鉛礦
phosphate buffer	磷酸盐缓冲剂	磷酸鹽緩衝劑
phosphate builder	磷酸盐增效助剂	磷酸鹽補助劑
phosphate fertilizer	磷肥	磷肥
phosphate glass	磷酸盐玻璃	磷酸鹽玻璃
phosphate rock	磷酸盐岩	磷岩
phospholipid	磷脂	磷脂
phosphorescence	磷光	磷光
phosphorylation	磷酸化作用	磷酸化[反應]
phosphosilicate glass	磷硅酸盐玻璃	磷矽酸鹽玻璃
photoactivation	光活化	光活化[作用]
photocatalysis	光催化	光催化[作用]
photocatalyst	光催化剂	光觸媒
photocell	光电池	光電管
photochemical absorption law	光化学吸收律	光化學吸收律
photochemical activity	光化学活性	光化學活性
photochemical chlorination	光化学氯化	光化學氯化[反應]

英 文 名	祖国大陆名	台湾地区名
photochemical decomposition	光化学分解	光化學分解[反應]
photochemical degradation	光化学降解	光化學降解[反應]
photochemical efficiency	光化学效率	光化學效率
photochemical equivalent	光化学当量	光化學當量
photochemical excitation	光化学激化	光化學激化
photochemical induction	光化学诱导	光化學誘導
photochemical kinetics	光化学动力学	光化學動力學
photochemical ozonization	光化学臭氧化	光化學臭氧化[反應]
photochemical process	光化学过程	光化學程序
photochemical reaction	光化学反应	光化學反應
photochemical reactivity	光化学反应性	光化學反應性
photochemical reactor	光化学反应器	光化學反應器
photochemical threshold	光化学低限	光化學低限
photochemical yield	光化学产量	光化學產率
photochemistry	光化学	光化學
photochlorination	光氯化	光氯化[反應]
photocolorimetry	光电比色计	光比色法
photoconductive material	光电导材料	光導電材料
photoconductivity	光电导性	光電導性
photodecomposition	光[分]解[反应]	光分解[反應]
photodegradation	光降解	光降解[反應]
photodensitometer	光密度计	光密度計
photodetector	光检测器	測光器
photodissociation	光致离解	光解離
photoelasticity	光弹性	光彈性
photoelectric cell	光电池[管]	光電管
photoelectric colorimeter	光电比色计	光電比色計
photoelectric effect	光电效应	光電效應
photoelectricity	光电, 光电学	光電學
photoelectric polarimeter	光电旋光计	光電旋光計
photoelectric spectrophotometer	光电分光光度计	光電光譜儀
photoelectrometer	光电计	光電計
photoetching process	光刻蚀过程	照相蝕刻法
photoform glass	感光玻璃	感光玻璃
photogenic bacteria	发光菌	發光菌
photographic desensitizer	感光减敏剂	感光減敏劑
photographic emulsion	感光乳剂	感光乳液
photographic paper	照相纸	感光紙
photographic sensitizer	照相增感剂	感光敏化劑

英　文　名	祖国大陆名	台湾地区名
photolithography	光刻	照相石印法
photoluminescence	光致发光	光發光
photolysis	光解	光解[反應]
photometer	光度计	光度計
photometric analysis	光度分析	光測定分析[法]
photometric method	光度测定法	光測定法
photometry	光度学	光測定法
photon	光子	光子
photoneutron	光中子	光中子
photonic material	光子材料	光子材料
photonuclear reaction	光[致]核反应	光核子反應
photooxidation	光氧化	光氧化[反應]
photooxidative aging	光氧化老化	光氧化老化
photopolymer	光[致]聚合物	光聚合物
photopolymerization	光致聚合	光聚合[反應]
photoreaction	光反应	光反應
photoreceptor	光感受体	感光器
photoresist	光致抗蚀剂	光阻劑
photorespiration	光呼吸	光呼吸
photosensitivity	光敏性	感光性
photosensitized oxidation	光敏氧化	感光氧化[反應]
photosensitizer	光敏剂	光敏劑
photostabilization	耐光作用	光穩定化
photosynthesis	光合作用	光合作用
photothermal analysis	光热分析	光熱分析[法]
phototube	光电管	光電管
photoyellowing	光黄化	光黄化
pH-sensitive electrode	pH 敏感电极	pH 敏感電極
phthalocyanine blue	酞菁蓝	酞花青藍
phthalocyanine dye	酞菁染料	酞花青染料
phthalocyanine green	酞青绿	酞花青綠
phthalocyanine pigment	酞菁颜料	酞花青顏料
pH value	pH 值	pH 值, 酸度值
physical absorbent	物理吸收剂	物理吸收劑
physical absorption	物理吸收	物理吸收
physical adsorption	物理吸附	物理吸附
physical chemistry	物理化学	物理化學
physical constant	物理常数	物理常數
physical constraint	物理约束	物理限制

英　文　名	祖国大陆名	台湾地区名
physical law	物理定律	物理定律
physical model	物理模型	物理模型
physical parameter	物理参数	物理參數
physical process	物理过程	物理程序
physical promoter	物理促进剂	物理促進劑
physical property	物理性质	物理性質, 物性
physical realizability	实际可实现性	實際可真實化性
physical separation	物理分离	物理分離
physical system	物理系统	物理系統
physical test	物理试验	物理試驗
physical tracer	物理示踪剂	物理示蹤劑
physical transport step	物理传递步骤	物理輸送步驟
physical treatment	物理处理	物理處理
physical treatment process	物理处理过程	物理處理法
physical vapor deposition	物理气相沉积	物理蒸氣沈積法
physicochemical absorption	物理化学吸收	物理化學吸收
physisorption	物理吸附	物理吸附
phytosterol	植物甾醇	植固醇
pickle bath	酸洗槽	浸酸浴
pickle liquor	酸洗液	浸酸液
pickling	酸洗	浸酸, 酸洗
pickling bath	酸洗槽	浸酸浴
pickling process	酸洗法	浸酸法
pictorial flowsheet	图形流程图	圖形流程圖
PID (＝piping and instrument diagram)	管路-仪表流程图	管線及儀器圖
piezoelectric crystal	压电晶体	壓電晶體
piezometer ring	均压环	測壓環
pig iron	生铁	生鐵
pigment	颜料	顏料
pilot plant	中间试验装置	實驗工廠
pilot valve	导向阀	導引閥
pinch	箍	捏縮
pinch effect	箍缩效应	捏縮效應
pinch point	夹点	夾點
pinch technology	夹点技术	夾點技術
pine oil	松油	松香油
pine seed oil	松子油	松子油
pine tree oil	松油	松樹油
pin-point gate	针点浇口	針點澆口

英 文 名	祖国大陆名	台湾地区名
pipe	管	管
pipe bend	弯管	彎管
pipe bending machine	弯管机	彎管機
pipe diameter	管径	管徑
pipe filter	管过滤器	管過濾器
pipe fitting	管件	管配件
pipe flow	管流	管流
pipe friction	管路摩擦	管路摩擦
pipe joint	管接头	管接頭
pipeline	管线	管線
pipe line list	管线表	管線表
pipeline network	管路网络	管路網絡
pipe reducer	异径管	漸縮管
pipe relative roughness	管子相对粗糙度	管子相對粗糙度
pipe roughness	管子粗糙度	管粗糙度
pipe schedule number	管壁厚系列号	管分類號
pipe sealing	管接头密封	管密封
pipe size	管子尺寸	管子尺寸
pipe still	管式蒸馏釜	管餾器
pipe tap	锥管螺纹	管接頭
piping	配管,管路	配管,管路
piping and instrument diagram(PID)	管路-仪表流程图	管線及儀器圖
piping flowsheet	管道流程图	配管流程圖
piping standard	配管标准	配管標準
piping stress	管线应力	管線應力
piping system	管路系统	管路系統,配管系統
piston	活塞	活塞
piston actuator	活塞式执行机构	活塞引動器
piston flow reactor	活塞流反应器	塞流反應器
piston gage	活塞式压力计	活塞壓力計
piston meter	活塞流量计	活塞流量計
piston pump	活塞泵	活塞泵
piston valve	活塞阀	活塞閥
pit	凹坑	坑,槽,池
pitch	①管中心距②沥青	①節距②瀝青
pitching	添加酵母	添加酵母,投種
pitching tank	加酵母槽	種母槽
Pitot tube	皮托管	皮托管
pitting	麻点	麻點,凹痕

英　文　名	祖国大陆名	台湾地区名
pitting corrosion	点蚀	麻點腐蝕
pivot	枢轴	樞, 樞軸
plait	褶	褶, 疊
plait point	共溶点, 褶点	褶點
plane stress	平面应力	平面應力
planet agitator	行星式搅拌器	迴繞式攪拌器
planimeter	求积仪	測面計
plant asset	工厂资产	工廠資產
plant cost estimate	工厂成本估算	工廠成本估計
plant growth hormone	植物生长激素	植物生長激素
plant growth regulator	植物生长调节剂	植物生長調節劑
plant hormone	植物激素	植物激素
plant layout	工厂布置	工廠布置
plant location	厂址	廠址
plant site selection	厂址选择	廠址選擇
plasma	等离子体	電漿
plasma-enhanced etching	等离子增强蚀刻	電漿加強蝕刻[法]
plasma etching	等离子蚀刻	電漿蝕刻[法]
plaster	硬石膏	石膏, 硬膏
plaster mold	石膏模	石膏模
plaster of Paris	熟石膏	熟石膏
plastic	塑料	塑膠
plastic binder	塑料黏合剂	塑膠黏結劑
plastic cement	塑料黏结剂	塑膠膠合劑
plastic engineering	塑料工程	塑膠工程
plastic fluid	塑性流体	塑性流體
plastic foam	塑料泡沫	塑膠泡體, 塑膠泡沫
plasticimeter	塑料计	塑性試驗計
plastic index	塑性指数	塑性指數
plasticity	可塑性	塑性
plasticization	增塑作用	塑化
plasticizer	增塑剂	增塑劑, 塑化劑, 可塑劑
plastic material	塑性材料	塑性材料, 塑膠材料
plastic packing	塑料填料	塑膠填充物
plastic processing	塑料加工	塑膠加工
plastic state	黏流态	可塑狀態
plastic sulfur	弹性硫	塑性硫
plastic yield	塑性屈伏	塑性降伏
plastic yield point	塑性屈伏点	塑性降伏點

英　文　名	祖国大陆名	台湾地区名
plastigel	塑性凝胶	塑性凝膠
plastimeter	塑度计	塑性計
plastisol	增塑溶胶	塑料溶膠
plastograph	塑性计	塑性測定器
plastometer	塑度计	塑性計
plate	塔板	板
plate-and-frame filter press	板框式压滤机	板框壓濾機
plate-and-frame module	板框组件	板框組件
plateau	曲线平直部分	平線區
plateau effect	平坦效应	平線區效應
plate column	板式塔	層板塔
plate culture	平皿培养	平碟培養[菌]
plate efficiency	[塔]板效率	板效率
plate-fin heat exchanger	板翅换热器	板翅換熱器
plate heat exchanger	板式换热器	板式換熱器
plate number	板数	板數
plate tower	板式塔	層板塔
plate-type evaporator	板式蒸发器	板式蒸發器
plate type heat exchanger	平板式换热器	板式熱交換器
platform	平台	平臺
platformate	铂重整产品	鉑媒重組油
platform balance	台秤	檯秤
platforming	铂重整	鉑媒重組
platforming process	铂重整过程	鉑媒重組程序
plating	喷镀	①電鍍②壓紋
plating industry	电镀工业	電鍍工業
platinized asbestos	附铂石棉	附鉑石綿
platinized asbestos catalyst	附铂石棉催化剂	附鉑石綿觸媒
platinized carbon electrode	镀铂碳电极	鍍鉑碳[電]極
platinized pumice	附铂浮石	附鉑浮石
platinized silica gel catalyst	附铂硅胶催化剂	附鉑矽膠觸媒
platinum-alumina catalyst	铂氧化铝催化剂	附鋁氧觸媒
platinum black catalyst	铂黑催化剂	鉑黑觸媒
platinum-chromia catalyst	铂氧化铬催化剂	鉑鉻氧觸媒
platinum contact process	铂接触过程	鉑接觸法
platinum crucible	铂坩埚	鉑坩堝, 白金坩堝
platinum gauze	铂网	鉑網
platinum nickel catalyst	铂-镍催化剂	鉑鎳觸媒
platinum oxide catalyst	氧化铂催化剂	氧化鉑觸媒

英　文　名	祖国大陆名	台湾地区名
platinum resistance bulb	铂电阻球	鉑電阻球
platinum resistance thermometer	铂电阻温度计	鉑電阻溫度計
platinum sponge	铂海绵	鉑海綿
plenum chamber	充气室	充氣室
plot	图	圖
plug flow	平推流, 活塞流	塞流, 栓流
plug flow backmixing	平推流返混	塞流逆混
plug flow reactor	活塞流反应器	塞流反應器
plugging	堵塞	阻塞
plug valve	旋塞阀	塞閥
plumb bob float	铅锤式浮标	錘式浮標
plum kernel oil	梅仁油	梅仁油
plunger	柱塞	柱塞
plunger pump	柱塞泵	柱塞泵
plutonium reactor	钚反应器	鈽反應器
plywood	胶合板	合板
pneumatic actuator	气动执行装置	氣動引動器
pneumatic analog	气动模拟	氣動類比
pneumatic classifier	气力分级器	氣動類析器
pneumatic control	气动控制	氣動控制
pneumatic controller	气动调节器	氣動控制器
pneumatic control valve	气动调节阀	氣動控制閥
pneumatic conveying	气动输送	氣動輸送
pneumatic conveyor	气流输送器	氣動運送機, 氣運機
pneumatic device	气动装置	氣動裝置
pneumatic dryer	气流干燥器	氣流乾燥器
pneumatic electric relay	气动电流继电器	氣動電流替續器
pneumatic flow transmitter	气动式流动传送器	氣動式流動傳送器
pneumatic force meter	气动测力仪	氣動測力計
pneumatic liquid density gage	气动式液体密度计	氣動液體密度計
pneumatic liquid level gage	气动式液面计	氣動液面計
pneumatic mechanism	气动机理	氣動機構
pneumatic sensor	气动传感器	氣動感測器
pneumatic set controller	气动设定控制器	氣動設定控制器
pneumatic signal	气动信号	氣動信號
pneumatic stirring	气动搅拌	氣動攪拌
pneumatic system	气动系统	氣動系統
pneumatic transmission	气动传动	氣動傳送
pneumatic transmission line	气动传动线	氣動傳送線路

英　文　名	祖国大陆名	台湾地区名
pneumatic transmission receiver	气动传动接受器	氣動傳送接受器
pneumatic transmitter	气动传动器	氣動傳送器
pneumatic transport	气动输运	氣動輸送
pneumatic valve	气动阀	氣動閥
pocket grinder	袋式研磨机	袋式研磨機
point efficiency	点效率	點效率
point of contact	接触点	接觸點
point of steepest ascent	最陡上坡点	最陡上坡點
point of steepest descent	最陡下坡点	最陡下坡點
point selectivity	点选择性	點選擇性
point sink	点汇座	點匯座
point source	点源	點源
poise	泊	泊
poison	毒物	毒物
poisoning	中毒	中毒
polar coordinates	极坐标	極坐標
polarimeter	旋光计	旋光計
polariscope	偏光镜	旋光計
polarity	极性	極性
polarizability	极化率	極化率
polarization	①极化②偏振	①極化②偏光
polarization factor	极化因子	極化因子
polarograph	极谱仪	極譜儀
polarographic method	极谱法	極譜法
polarography	极谱法	極譜法
polaroid glass	偏振片	偏光玻璃
polar plot	极坐标图	極坐標圖
polar solvent	极性溶剂	極性溶劑
polishing	磨光	磨光
polishing pan	上光盘	上光機
pollutant	污染物	污染物
pollutant standards index	污染物标准指数	污染物標準指數
polluted water	污染水	受污染水
pollution	污染	污染
pollution abatement	污染减少	污染減量
pollution control	污染控制	污染控制,污染防治
pollution engineering	污染工程	污染工程
pollution source	污染源	污染源
polyacrylate	聚丙烯酸酯	聚丙烯酸酯

英　文　名	祖国大陆名	台湾地区名
polyacrylate resin	聚丙烯酸酯树脂	聚丙烯酸酯樹脂
polyacrylonitrile	聚丙烯腈	聚丙烯腈
polyalcohol	多元醇	多元醇
polyalkylation	多烷基化作用	多烷化[反應]
polyamide	聚酰胺	聚醯胺
polyamide fiber	聚酰胺纤维	聚醯胺纖維
polyamide resin	聚酰胺树脂	聚醯胺樹脂
polyblend	高分子共混物	聚摻合物
polycaprolactam	聚己内酰胺	聚己内醯胺
polycarbonate	聚碳酸酯	聚碳酸酯
polycarbonate resin	聚碳酸酯树脂	聚碳酸酯樹脂
polycondensation	缩聚反应	聚縮合[反應]
polyester	聚酯	聚酯
polyester fiber	聚酯类纤维	聚酯纖維
polyester resin	聚酯类树脂	聚酯樹脂
polyester rubber	聚酯类橡胶	聚酯橡膠
polyether	聚醚	聚醚
polyethylene	聚乙烯	聚乙烯
poly-fluid theory	多流体理论	多流體理論
polyformaldehyde	聚甲醛	聚甲醛
polyisobutylene	聚异丁烯	聚異丁烯
polyisoprene rubber	聚异戊二烯橡胶	聚異戊二烯橡膠, 聚異平橡膠
polymer	高分子, 大分子	聚合物, 高分子, 聚合體
polymer blend	共混聚合物	高分子摻合物, 聚合體摻合物
polymer degradation	聚合物降解	聚合物降解[反應]
polymer flooding	聚合物泛滥	聚合物氾流[法]
polymeric material	高分子材料	聚合材料, 高分子材料
polymerization	聚合[反应]	聚合[反應]
polymerization kinetics	聚合动力学	聚合[反應]動力學
polymerization mechanism	聚合机理	聚合[反應]機構
polymerization reaction	聚合反应	聚合反應
polymerization reaction engineering	聚合反应工程	聚合反應工程
polymerization reactor	聚合反应器	聚合反應器
polymerized oil	厚油	聚合油
polymer processing	聚合物加工	聚合物加工, 高分子加工
polymer processing engineering	聚合物加工工程	聚合物加工工程

英　文　名	祖国大陆名	台湾地区名
polymer processing technology	聚合物加工技术	聚合物加工技術
polymer stabilization	聚合物稳定剂	聚合物穩定化
polymetallic catalyst	多金属催化剂	多金屬觸媒
polymorphism	多态[现象]	多形性
polymyxin	多黏菌素	多黏菌素
polyolefin resin	聚烯烃树脂	聚烯烴樹脂
polypropylene	聚丙烯	聚丙烯
polypropylene fiber	聚丙烯纤维	聚丙烯纖維
polysaccharidase	多糖酶	多醣分解酶
polysaccharide	多糖	多醣
polystyrene	聚苯乙烯	聚苯乙烯
polysulfide polymer	多硫聚合物	多硫聚合物
polysulfide rubber	聚硫橡胶	多硫橡膠
polysulfone	聚砜	聚碸
polytetrafluoroethylene	聚四氟乙烯	聚四氟乙烯,鐵氟龍
polytrifluorochloroethylene	聚三氟氯乙烯	聚三氟氯乙烯
polytropic constant	多变常数	多變常數
polytropic expansion	多变膨胀	多變膨脹
polytropic index	多变指数	多變指數
polytropic process	多变过程,多方过程	多變程序
polytropic specific heat	多变比热	多變比熱
polytropism	多变性	多變性
polyurethane	聚氨基甲酸酯	聚胺甲酸酯
polyurethane rubber	聚氨酯橡胶	聚胺甲酸酯橡膠,PU 橡膠
polyvinyl acetate	聚乙酸乙烯酯	聚乙烯乙酯
polyvinyl alcohol	聚乙烯醇	聚乙烯醇
polyvinyl butyral resin	聚乙烯酯缩丁醛树脂	聚乙烯丁醛樹脂
polyvinyl chloride	聚氯乙烯	聚氯乙烯
polyvinyl chloride resin	聚氯乙烯树脂	聚氯乙烯樹脂
polyvinyl ether	聚乙烯基醚	聚乙烯醚
polyvinyl ether adhesive	聚乙烯醚黏合剂	聚乙烯醚黏合劑
polyvinylidene chloride	聚偏氯乙烯	聚偏二氯乙烯
polyvinylidene fluoride	聚偏氟乙烯	聚氟乙烯
polyvinyl pyrrolidone	聚乙烯吡咯烷酮	聚乙烯四氫吡咯酮
Ponchon-Savarit diagram	焓-浓图,P-S 图	P-S 圖
pool boiling	池沸腾	池沸騰
pool boiling curve	池沸腾曲线	池沸騰曲線
population balance	种群平衡	族群均衡

英　文　名	祖国大陆名	台湾地区名
population density function	种群密度函数	族群密度函數
population dynamics	种群动态	族群動態
population parameter	种群参数	雜群參數
pop valve	突开式安全阀	急洩閥
porcelain	瓷	瓷
porcelain crucible	瓷坩埚	瓷坩堝
porcelain enamel	搪瓷	搪瓷
porcelain enameled steel	搪瓷钢	搪瓷鋼
porcelain glaze	瓷釉	瓷釉
porcelain insulator	瓷绝缘子	瓷絕緣體
pore	孔隙, 细孔	孔隙, 細孔
pore diffusion	微孔扩散	細孔擴散
pore diffusion resistance	微孔扩散阻力	細孔擴散阻力
pore entrance	细孔入口	細孔入口
pore geometry	细孔形状	細孔[幾何]形狀
pore-mouth poisoning	孔口中毒	孔口中毒
pore orientation distribution	细孔方向分布	細孔方向分布
pore radius	孔半径	細孔半徑
pore shape factor	细孔形状因子	細孔形狀因素
pore size	孔径	孔徑, 細孔大小
pore size distribution	孔径分布	孔徑分布
pore structure	孔隙结构	細孔結構
pore volume	孔体积, 孔容	細孔體積, 孔隙體積
porosimeter	孔隙计	孔隙計
porosity	孔隙率	孔隙度
porosity apparatus	孔隙仪	孔隙儀
porosity effect	孔效应	孔隙度效應
porous barrier	孔膜隔板	多孔[性]障壁
porous catalyst	多孔性催化剂	多孔性觸媒
porous material	多孔性材料	多孔[性]材料
porous medium	多孔介质	多孔介質
porous membrane	多孔膜	多孔薄膜
porous packing	多孔填料	多孔填充物
porous pellet	多孔粒	多孔粒
porous pellet pulse reactor	多孔粒脉冲反应器	多孔粒脈衝反應器
porous plate	密孔板	多孔板
porous solid	多孔固体	多孔[性]固體
porous-wall tube	多孔壁管	多孔壁管
positive azeotrope	正共沸混合物	正共沸液

英　文　名	祖国大陆名	台湾地区名
positive deviation	正偏差	正偏離
positive displacement	正位移	正位移
positive displacement flowmeter	容积式流量计	正位移流量計
positive displacement meter	正位移计	正壓位移計
positive displacement pump	容积式泵, 排代泵	正位移泵
positive feedback	正反馈	正回饋
positive-negative azeotrope	正负共沸物	正負共沸液
postbaking	后焙, 后烘	後焙
post vulcanization	后硫化	後硫化
potable water	饮用水	飲用水
potash feldspar	钾长石	鉀長石
potash fertilizer	钾肥	鉀肥
potash glass	钾玻璃	鉀玻璃
potash-lead glass	钾铅玻璃	鉀鉛玻璃
potash-lime glass	钾钙玻璃	鉀鈣玻璃
potash soap	钾软皂	鉀皂
potash superphosphate	含钾过磷酸钙	含鉀過磷酸鈣
potency of accelerator	加速器效能	催速劑效能
potential	①势②电势③位势	①勢②電位③位能
potential barrier	势垒	勢障
potential correction	势修正	勢修正
potential deviation	势偏差	勢偏差
potential difference	势差	勢差
potential drop	电势落	勢降
potential energy	势能	位能
potential energy surface	势能面	位能面
potential flow	势流	勢流
potential gradient	电势梯度	勢梯度
potential head	位头	勢差, 位能差
potential line	势线	勢線
potentiometer	电位计	電位計, 電勢計
potentiometric analysis	电位分析[法]	電位分析
potentiometric method	电位法	電位法
potentiometric titration	电位滴定[法]	電位滴定[法]
poundal	磅达(英制力的单位)	磅達
pouring glazing	浇注釉	澆釉
powder	粉[体]	①粉②火藥
powder density	粉体密度	粉體密度
powder metallurgy	粉末冶金	粉末冶金[學]

英　文　名	祖国大陆名	台湾地区名
powder technology	粉体技术, 粉体工程	粉體技術
power	功率	功率
power consumption	功力消耗	功率消耗量
power efficiency	功率效率	功率效率
power factor	功率因数	功率因數
power function type equation	幂函数型方程	冪函數型方程
power generator	动力发动机	能量產生器, 發電機
power-law fluid	幂律流体	冪次律流體
power loss	功率损失	功率損失
power number	功率数	功率數
power plant	发电厂	①動力廠②發電廠
power plant cycle	发电厂循环	動力廠循環
power requirement	功率需要量	功率需要量
Poynting correction	坡印亭校正	波印亭校正
Poynting factor	坡印亭因子	波印亭因子
pozzolana	火山灰	火山灰
practical stability	实际稳定性	實用穩定性
Prandtl number	普朗特数	普朗特數
prebaking	预焙	預焙
precedence ordering	排序	排序
precipitate	沉淀物	沈澱物
precipitation	沉淀	沈澱
precipitation polymerization	沉淀聚合	沈澱聚合[反應]
precipitation reaction	沉淀反应	沈澱反應
precipitation titration	沉淀滴定[法]	沈澱滴定[法]
precipitator	①沉淀器②除尘器	①沈澱器②集塵器
precise fractionation	精密分馏	精密分餾
precision	精[密]度	精確度, 精密度
precision fractional distillation	精密蒸馏	精密分餾
precocity of accelerator	促进剂的早熟性	催速劑早熟性
precooler	预冷器	前冷卻器
precursor	前体	前體
predesign cost estimate	预设计成本估算	設計前成本估計
prediction	预测	預測
predictive control	预测控制	預測控制
predistribution	预分布	預分布
preevaporator	预蒸发器	預蒸發器
preexponential factor	指数前因子	指數前因數, 頻率因數
preferential adsorption	选择吸附	優先吸附

英 文 名	祖国大陆名	台湾地区名
preferential solubility	有择溶解度	優先溶解度
prefermentation period	预发酵期	預發酵期
prefiltration	预滤	預濾
preflash tower	预闪蒸塔	初餾塔
prefractionation	预分馏	初分餾
preheater	预热器	預熱器
preheat furnace	预热炉	預熱爐
preheating	预热	預熱
preheating evaporator	预热蒸发器	預熱蒸發器
preheating period	预热期	預熱期
preheating zone	预热区	預熱區
preliminary design	初步设计	初步設計
preliminary filter	初滤机	初濾機
preliminary sizing	粗筛选	初級篩選
preliminary treatment	预处理	初步處理
preload	预载荷	預負載
premix molding	预混模制物	預混模製[物]
present value	现值	現值
present worth	现值	現值
present worth factor	现值因子	現值因數
present worth method	现值法	現值法
preservation	防腐	防腐
preservative	防腐剂	防腐劑
press cake	压滤饼	壓濾餅
press mold	压模	壓模
pressure balanced valve	压力平衡阀	壓力均衡閥
pressure calibration	压力校准	壓力校正
pressure coefficient	压力系数	壓力係數
pressure compensation	压力补偿	壓力補償
pressure-composition diagram	压力-组成图	壓力-組成圖
pressure control	压力控制	壓力控制
pressure controller	压力控制器	壓力控制器
pressure cooker	加压蒸煮器	加壓蒸煮器
pressure correction	压力校正	壓力修正
pressure difference	压[力]差	壓力差
pressure diffusion	压力扩散	壓力擴散
pressure distillate	加压馏出物	加壓餾出物
pressure distillation	加压蒸馏	加壓蒸餾
pressure distribution	压力分布	壓力分布

英　文　名	祖国大陆名	台湾地区名
pressure drop	压降	壓[力]降
pressure effect	压力效应	壓力效應
pressure efficiency	压力效率	壓力效率
pressure energy	压力能	壓[力]能
pressure-enthalpy diagram	压焓图	壓力-焓圖
pressure equalizer	压力平衡器	均壓器
pressure filter	压滤机	壓濾機
pressure filtration	加压过滤	加壓過濾
pressure flotation	加压浮选	加壓浮選
pressure fractionation	加压分馏	加壓分餾
pressure gage	压力计, 压力表	壓力計
pressure gradient	压力梯度	壓力梯度
pressure head	压头	壓力高差
pressure indicator	压力指示计	壓力指示計
pressure leaf filter	加压叶滤机	加壓葉濾機
pressure loss	压力损失	壓力損失
pressure measurement	压力测量	壓力量測
pressure meter	压力计, 压力表	壓力計
pressure oxidation	加压氧化	加壓壓力
pressure range	压力范围	壓力範圍
pressure recorder	压力纪录器	壓力記錄器
pressure reducing valve	减压阀	減壓閥
pressure regulating valve	调压阀	壓力調節閥
pressure regulator	调压器	壓力調節器
pressure sand filter	加压砂滤器	加壓砂濾器
pressure screen	加压筛	壓力篩
pressure sensitive adhesive	压敏胶	壓感黏合劑
pressure signal	压力信号	壓力信號
pressure swing adsorption(PSA)	变压吸附	變壓吸附
pressure system	压力系统	壓力系統
pressure tap	测压孔	壓力接頭
pressure-temperature diagram	压力-温度图	壓力-溫度圖
pressure transducer	压力传感器	壓力轉換器
pressure transmitter	压力变送器	壓力傳送器
pressure type thermometer	压力式温度计	壓力式溫度計
pressure-vacuum valve	压力-真空阀	壓力真空閥
pressure vessel	压力容器	壓力容器, 壓力槽
pressure-volume diagram	压容图	壓力-體積圖
pressure-volume work	压力-容积功	壓力-體積功

英　文　名	祖国大陆名	台湾地区名
pretreatment	预处理	預處理
preventative treatment	预防处理	預防處理
prevention	预防	預防處理
prilling	成球	製粒
prilling tower	造粒塔	製粒塔
primary clarifier	初级澄清器	初級澄清器, 初級沈降槽
primary design variable	主要设计变量	主要設計變數
primary dimension	主要尺寸	主因次
primary distillation	初级蒸馏	初級蒸餾
primary feeder	主加料机	主飼機, 初級取食者
primary impact	主要冲击	主要衝擊
primary liquid	原液	原液
primary loop	主环路	主環路
primary measurement	一次测量	主量測[值]
primary normal stress coefficient	一级正应力系数	主正應力係數
primary nucleation	原核成核	初級成核
primary plant	原装置	初級工廠
primary plasticizer	主增塑剂	主塑化劑
primary product	主产品	主產物
primary purification	初级净化	初級純化
primary reaction	主反应	主反應
primary settler	初级沉降槽	初級沈降槽, 初級澄清器
primary standard	一级标准	一級標準
primary tower	初馏塔	初餾塔
primary treatment	一次处理	初級處理
primary variable	主变量	主要變數
primer	底漆	底漆
priming	起动	引動, 發火
principal axis	主轴	主軸
principal fermentation	主发酵	主發酵
principal stress	主应力	主應力
principle	原理	原理
principle of conservation of energy	能量守恒原理	能量守恆原理
principle of conservation of mass	质量守恒原理	質量守恆原理
principle of conservation of moment of momentum	动量矩守恒原理	動量力矩守恆原理
principle of conservation of momentum	动量守恒原理	動量守恆原理

英 文 名	祖国大陆名	台湾地区名
principle of corresponding state	对应态原理, 对比态原理	對應狀態原理
principle of detailed balance	精细平衡原理	明細均衡原理
principle of dimensional homogeneity	因次均一性原理	因次均勻原理
principle of entropy increase	熵增原理	熵增原理
principle of material objectivity	物质客观性原理	物質客觀性原理
principle of microscopic reversibility	微观可逆性原理	微觀可逆性原理
prior estimate	预估值	預估值
probabilistic process	随机过程	隨機程序
probability	概率	機率, 或然率
probability curve	概率曲线	機率曲線
probability density	概率密度	機率密度
probability density function	概率密度函数	機率密度函數
probability distribution	概率分布	機率分布
probability distribution function	概率分布函数	機率分布函數
probability of acceptance	认可概率	驗收機率
probability plot	概率标绘图	機率圖
probe	探头	探針, 探桿
process analysis	过程分析	程序分析
process application	过程应用	程序應用
process automation	过程自动化	程序自動化
process availability	过程有效能	程序可用性
process behavior	过程行为特性	程序行為
process calculation	工艺计算	程序計算
process capacity	过程能力	程序容量
process characteristics	过程特性	程序特性
process chart	工艺流程图	程序圖
process condition	工艺条件	程序條件
process control	过程控制	程序控制
process control pattern	过程控制模式	程序控制方式
process decomposition	过程分解	程序分解
process design	工艺过程设计	程序設計
process design engineer	工艺设计工程师	程序設計工程師
process designer	工艺设计师	程序設計師
process development	过程开发	程序開發
process dynamics	过程动态学	程序動態[學]
process element	过程组成元件	程序元件
process engineer	工艺工程师	程序工程師, 方法工程師

英文名	祖国大陆名	台湾地区名
process engineering	过程工程	程序工程,方法工程
process engineering check list	过程工程核查表	程序工程檢查報表,方 法工程檢查報表
process equipment	工艺设备	程序装置
process evaluation	过程评价	程序評估,程序評價
process feed	过程进料	程序進料
process flow diagram	工艺流程图	程序流程圖
process flowsheet	工艺流程图	程序流程圖
process gain	过程增益	程序增益
process identification	过程辨识	程序辨識
process identification	过程辨识	程序鑑別
process industry	过程工业	程序工業
processing	加工	處理,加工
processing engineering	加工工程	加工工程
processing industry	加工工业	加工工業
processing technology	加工技术	加工技術
processing variable	加工变量	加工變數
process instability	过程不稳定性	程序不穩定性
process integration	过程集成	整合程序
process integration	过程集成	程序整合
process layout	工艺布置图	程序布置
process load	工艺负荷	程序負載
process load change	工艺负荷变化	程序負載改變
process modeling	过程模型化	程序模式化
process optimization	过程优化	程序最適化
process overload	过程超载	程序超載
process parameter	工艺参数	程序參數
process performance	工艺性能	程序性能
process planning	工艺规划	程序規劃
process reaction curve	过程反应曲线	程序反應曲線
process reliability	过程可靠性	程序可靠性
process resilience	过程反弹	程序反彈
process response	过程响应	程序應答
process response curve	过程响应曲线	程序應答曲線
process safety	工艺安全	程序安全性
process-scale chromatography	工业色谱	工業色譜
process sensitivity	过程灵敏度	程序靈敏度
process simulation	过程模拟	程序模擬
process stability	过程稳定性	程序穩定性

英　文　名	祖国大陆名	台湾地区名
process steam	工艺蒸汽	程序蒸汽
process stream	工艺物流	程序流
process synthesis	过程综合, 过程合成	程序合成
process system	工艺系统	程序系统
process system analysis	过程系统分析	程序系统分析
process system engineering	过程系统工程, 化工系统工程	程序系统工程
process technology	工艺技术	程序技術
process time lag	过程时间滞后	程序時間落後
process unit	工艺单元	程序單元
process upset	过程失稳	程序失穩
process vacuum	工艺用真空[系统]	真空程序
process variability	过程变异性	程序變異性
process variable	过程变量	程序變數
process water	工艺用水	程序用水
producer	煤气发生炉	發生爐
producer gas	发生炉煤气	發生爐氣
product cost	产品成本	產品成本
product design	产品设计	產品設計
product development	产品开发	產品開發
product distribution	产物分布	產物分布
product distribution distortion	产品分布变异	產物分布畸變
product engineering	产品工程	產品工程
product inhibition	产物抑制	產物抑制[作用]
production capacity	生产能力	生產能力
production control	生产控制	生產管制
production cost	生产成本	生產成本
production data	生产数据	生產數據
production line	生产线	生產線
production rate	生产率	生產速率
production schedule	生产进度表	生產時程表
productivity	生产力	生產力
product planning	产品规划	產品規劃
product quality	产品质量	產品品質
product recovery	产品回收	產物回收
product separation	产品分离	產物分離
product separator	产品分离器	產物分離器
product shipment check list	产品运送检查[清]单	產品運送檢查報表
product specification	产品规格	產品規格

英 文 名	祖国大陆名	台湾地区名
product storage	产品储存	產品儲存
product stream	产品物流	產物流
profile	剖面[图]	剖面[圖]
profit	利润	利潤
profitability	可获利程度	獲利能力
profitability factor	利润因子	利潤因數
profitability index	利润指数	利潤指數
profit function	利润函数	利潤函數
profit margin	利润限度	利潤邊際
program evaluation and review technique	计划评审法	項目評審技術
programmable calculator	可编程计算器	可程式計算器
programmable controller	可编程调制器	可程式控制器
programmable logic control	可编程逻辑控制	可程式邏輯控制
programmed control	程序控制	程式化控制
programmed controller	程序控制器	程式化控制器
programming	规划	規劃
progressive conversion model	渐进转化模型	漸進轉化模型
progressive freezing	逐步冷冻法	漸進冷凍
project	项目	專案, 計劃
project engineer	设计主管工程师	專案工程師
project engineering	项目工程	專案工程
promoted catalyst	促进化催化剂	促進化觸媒
promoter	助催化剂	促進劑, 促進體
propagation	传播	①傳播②增長
propagation of chain reaction	链增长反应	連鎖反應[的]傳播
propagation of error	误差传递	誤差傳遞
propagation reaction	链增长反应	增長反應
propagator	繁殖槽	繁殖槽
propellant	推进剂	推進劑
propeller	螺旋桨	螺旋槳
propeller agitator	螺旋桨式搅拌器	螺槳攪拌器
propeller capacity	螺旋桨容量	推進器容量
propeller impeller	螺旋桨式叶轮	螺旋式葉輪
propeller mixer	螺旋桨式搅拌器	螺旋槳混合器
propeller type fan	螺旋桨风扇	螺旋槳風扇
property change	性质变化	性質變化
property relationship	性质关联式	性質關係[式]
property tax	财产税	財產稅
propionate cellulose	丙酸纤维素	丙酸纖維素

英　文　名	祖国大陆名	台湾地区名
proportional action	比例作用	比例作用
proportional band	比例带	比例帶
proportional band adjustment	比例带调整	比例帶調整
proportional control	比例控制	比例產物
proportional controller	比例控制器	比例控制器
proportional counter	正比计数器	比例計數器
proportional-derivative action	比例-微分作用	比例-微分作用
proportional-derivative control	比例-微分调节	比例-微分控制
proportional element	比例元件	比例元件
proportional gain	比例增益	比例增益
proportional-integral action	比例-积分控制	比例-積分作用
proportional-integral control	比例-积分控制	比例-積分控制
proportional-integral-derivative action	比例-积分-微分作用	比例-積分-微分作用
proportional plus reset action	比例-重调作用	比例-重整作用
proportional plus reset plus rate action	比例-重调-速率作用	比例-重整-速率作用
proportional sensitivity	比例灵敏度	比例靈敏度
propylene oligomer	丙烯低聚物	低聚丙烯
propylene tetramer	四聚丙烯	四聚丙烯
propylene trimer	三聚丙烯	三聚丙烯
protease	蛋白酶	蛋白酶
protein	蛋白质	蛋白質
proteinase	蛋白酶	蛋白酶
protein engineering	蛋白质工程	蛋白質工程
protein fractionation	蛋白质分级	蛋白質分級
proteolysis	蛋白水解	蛋白質水解
proteolytic enzyme	蛋白[水解]酶	蛋白解酶
protic solvent	质子溶剂	質子溶劑
protoplasm	原生质	原生質
protoplast	原生质体	原生體
prototype	原型	雛型
prototype experiment	原型试验	雛型實驗
protozoa	原生动物	原生動物
protruded corrugated sheet packing	压延孔板波纹填料	壓延孔板波紋填料
Prussian blue	普鲁士蓝	普魯士藍
PSA (＝pressure swing adsorption)	变压吸附	變壓吸附
pseudocritical constant	假临界常数	假臨界常數
pseudocritical method	假临界点法	假臨界點法
pseudocritical pressure	假临界压力	假臨界壓力
pseudocritical temperature	假临界温度	假臨界溫度

英　文　名	祖国大陆名	台湾地区名
pseudohomogeneous model	拟均相模型	假均相模型
pseudokinetics	假动力学	假動力學
pseudomechanism	虚拟机理	假機構
pseudoparameter	虚拟系数	虚擬係數
pseudoplastic fluid	假塑性流体	假塑性流體
pseudoplasticity	假塑性	假塑性
pseudoreaction	虚拟反应	假反應
pseudo-reduced constant	假对比常数	假對比常數
pseudosalt	假盐	假鹽
pseudo steady state	假稳态	假穩態
pseudovariable	虚拟变量	虚擬變量
p-styrene sulfonic acid	对苯乙烯磺酸	對苯乙烯磺酸
psychrometer	干湿球湿度计	溼度計
psychrometric chart	湿度图	濕度圖
psychrometric difference	湿度差	溫度計量差
psychrometric line	湿度线	溼度線
psychrometric ratio	干湿比	溼度計量比
psychrometry	湿度测定法	溼度測定法
psychrophile	嗜冷菌	嗜冷菌
psychrophilic bacteria	嗜冷菌	嗜冷菌
puddle	搅炼	攪煉, 煉鐵
pug mill	揉捏机	捏泥機
pulp	浆料, 淤浆	紙漿
pulp beater	打浆机	打漿機
pulping	制浆	製漿
pulping process	制浆法	製漿法
pulsating fluidized bed	脉动流化床	脈動流化床
pulsation	脉动	脈動
pulsation damper	脉动阻尼器	脈動阻尼器
pulsation velocity	脉动速度	脈動速度
pulse	脉冲	脈動
pulse disturbance	脉冲扰动	脈動擾動
pulsed sieve plate column	脉冲筛板塔	脈沖篩板塔
pulse flow technique	脉冲流技术	脈動流動法
pulse function	脉冲函数	脈動函數
pulse input	脉动输入	脈動輸入
pulse method	脉冲法	脈動法
pulse reactor	脉冲反应器	脈動反應器
pulse response	脉冲响应	脈沖響應

英　文　名	祖国大陆名	台湾地区名
pulsing flow	脉冲流	脈動流動法
pulverization	粉碎	粉碎
pulverized coal	粉煤	粉煤
pulverized fuel	粉状燃料	粉狀燃料
pulverizer	粉磨机	粉碎機
pulverizing mill	粉磨机	粉碎機
pumice	浮石	浮石
pump	泵	泵
pump characteristics	泵特性	泵特性
pump efficiency	泵效率	泵效率
pump horsepower	泵功率	泵功率
pumping loss	泵送损耗	泵抽損失
pumping work	泵送功	泵功
pump pressure drop	泵压降	泵壓[力]降
pump rotor	泵转子	泵轉子
pump series	连串泵组	串聯泵組
pump shaft	泵轴	泵軸
pure component	纯组分	純成份
purge gas	吹扫气	沖洗氣
purification	提纯	純化, 淨化
purification process	净化过程	純化程序
purified water	净化水	淨化水, 純化水
purity	纯度	純度
puzzolana cement	火山灰水泥	火山灰水泥
pycnometer	比重瓶	比重瓶
pyrazolone dye	吡唑啉酮染料	吡唑啡染料
pyrethrin	除虫菊酯	除蟲菊精
pyrex glass	派热克司硬质玻璃	派熱司玻璃
pyrite	黄铁矿	黃鐵礦
pyrogenic decomposition	高温分解	熱解, 高溫分解
pyrolusite	软锰矿	軟錳礦
pyrolysis	热解	熱解
pyrolysis furnace	热解炉	裂解爐, 熱解爐
pyrolysis oil	热解油	熱解油
pyrolysis process	热解过程	熱解程序, 裂解程序
pyrometallurgical process	火法冶金过程	高溫冶金法
pyrometer	高温计	高溫計
pyrophyllite	叶蜡石	葉蠟石
pyrrhotine	磁黄铁矿	磁黃鐵礦

Q

英　文　名	祖国大陆名	台湾地区名
q-line	进料状态线	q 線, 進料狀態線
qualitative analysis	定性分析	定性分析
quality control	质量控制	品質管制
quality specification	质量规格	品質規格
quantimet	图象分析仪	圖像分析儀
quantitative analysis	定量分析	定量分析
quantitative flow diagram	物量流程图	計量流程圖
quantity meter	[累计]总量表	總量表
quantum effect	量子效应	量子效應
quartz	石英	石英
quartz glass	石英玻璃	石英玻璃
quartzite	石英岩	石英石
quartz tube	石英管	石英管
quasi-chemical approximation	准化学近似, 类化学近似	似化學近似
quasi-chemical solution model	准化学溶液模型, 类化学溶液模型	似化學溶液模型
quasi-linearization	拟线性化	似線性化
quasi-static process	准静态过程	似靜態過程
quasistatic thermograrimetry	准静态热重量分析法	似靜態熱重量法
quasi-stationary state	准静态	似定態
quasi steady state	准定常态	似穩態
quenched system	急冷系统	抑止系統
quenching	骤冷	驟冷, 淬火
quenching oil	骤冷油	驟冷油
quenching water	骤冷水	驟冷水
quick lime	生石灰	生石灰
quick-opening valve	速启阀	速啟閥
quiescent bed	静止床	靜止床

R

英　文　名	祖国大陆名	台湾地区名
racemic mixture	外消旋混合物	消旋混合物

英 文 名	祖国大陆名	台湾地区名
racemization	外消旋化	消旋反應
radial bearing	径向轴承	徑向軸承
radial dispersion	径向分散	徑向分散
radial dispersion coefficient	径向分散系数	徑向分散係數
radial distribution function	径向分布函数	徑向分布函數
radial flow reactor	径向反应器	徑向反應器
radial flow turbine	径向流涡轮机	沿徑流渦輪(機)
radial grooved filter plate	辐射沟槽滤板	輻射溝狀濾板, 輻溝濾板
radial mixing	径向混合	徑向混合
radial stress	径向应力	徑向應力
radial velocity	径向速度	徑向速度
radiant beam	辐射束	輻射束
radiant emissivity	辐射系数	輻射發射係數
radiant energy	辐射能	輻射能
radiant energy emission	辐射能发射	輻射能發射
radiation	辐射	輻射
radiation bonnet	辐射帽	輻射帽
radiation chemistry	辐射化学	輻射化學
radiation constant	辐射常数	輻射常數
radiation damage	辐射损伤	輻射損害
radiation density	辐射密度	輻射密度
radiation dose	辐射剂量	輻射劑量
radiation dosimeter	辐射剂量计	輻射劑量計
radiation energy	辐射能	輻射能
radiation hazard	辐射危害	輻射危害
radiation heat	辐射热	輻射熱
radiation heat transfer	辐射传热	輻射熱傳
radiation intensity	辐射强度	輻射強度
radiation law	辐射定律	輻射定律
radiation level	辐射能级	輻射能階
radiation loss	辐射损耗	輻射損失
radiation meter	辐射测量仪	輻射計
radiation of heat	热辐射	熱輻射
radiation penetration	辐射穿透	輻射穿透
radiation pressure	辐射压[强]	輻射壓[力]
radiation protection	辐射防护	輻射防護
radiation pyrometer	辐射高温计	輻射高溫計
radiation safety	辐射安全	輻射安全性

英 文 名	祖国大陆名	台湾地区名
radiation screen	辐射隔屏	辐射屏
radiation section	辐射段	辐射段
radiation shield	辐射屏蔽	辐射遮蔽
radiation sterilization	辐射消毒	辐射滅菌
radiator	①辐射器②散热器	①辐射器②散熱器
radical scavenger	自由基捕获剂	自由基捕捉劑
radioactive contamination	放射性污染	放射性污染
radioactive decay	放射性衰变	放射性衰變
radioactive disintegration	放射性蜕变	放射性蛻變
radioactive dust	放射性尘	放射性塵
radioactive element	放射性元素	放射性元素
radioactive fallout	放射性落尘	放射性落塵
radioactive fission	放射性裂变	放射性分裂
radioactive indicator	放射性指示剂	放射性示蹤劑
radioactive isotope	放射性同位素	放射性同位素
radioactive nucleus	放射性核	放射性核
radioactive nuclide	放射性核素	放射性核種
radioactive poison	放射性毒物	放射性毒
radioactive pollutant	放射性污染物	放射性污染物
radioactive source	放射源	放射[性]源
radioactive tracer	放射性示踪物	放射性示蹤劑
radioactive waste	放射性废物	放射性廢棄物
radioactivity	放射性	放射性
radiochemistry	放射化学	放射化學
radiotracer	放射性示踪剂	放射性示蹤劑
raffinate	萃余物, 抽余液	萃餘物
raining solid reactor	淋粒反应器	淋粒反應器
rammer process	夯锤法	鎚結法
ramp disturbance	斜坡扰动	斜坡擾動
ramp function	斜坡函数	斜坡函數
ramp response	斜坡响应	斜坡應答
random copolymer	无规共聚物	隨機共聚物
random degradation	无规降解	無規降解[反應]
random diffusion	无规扩散	隨機擴散
random distribution	无规分布	隨機分布
random disturbance	随机扰动	隨機擾動
random error	随机误差	隨機誤差
random fluctuation	随机波动	隨機波動
random load change	随机负荷变化	隨機負載變化

英　文　名	祖国大陆名	台湾地区名
randomness	随机性	隨機性
random number	随机数	隨機數
random packing	散堆填料	隨機填充
random process	随机过程	隨機程序
random sampling	随机抽样	隨機取樣
random search	随机搜索	隨機搜索
random signal	随机信号	隨機信號
random surface renewal	随机表面更新	隨機表面更新
random variable	随机变量	隨機變數
rangeability	量程可调范围	範圍性
rangeability of control valve	控制阀范围	控制閥範圍性
Rankine cycle	兰金循环	蘭金循環
Raoult's law	拉乌尔定律	拉午耳定律
rape oil	菜子油	菜子油, 菜油
rape seed oil	菜子油	菜子油
rapid filter	快速过滤器	快速濾器
rapid sand filtration rate	快速砂滤速率	快速砂濾[法]
Rasching ring	拉西环	拉西環
rate of evaporation	蒸发率	蒸發速率
rate of return on investment	投资收益率	投資收益率
ratio-flow control	流量比例调节	比值流量控制
rational function	有理函数	有理函數
rational number	有理数	有理數
raw gas	原料气	原料氣體
raw material	原料	原料
raw material check list	原料检查报表	原料檢查報表
raw material cost	原料成本	原料成本
raw material storage	原料储罐	原料儲存
raw sewage	未处理污水	原污水
raw sludge	未处理淤泥	原污泥
raw wastewater	未处理废水	原廢水
raw water	未经净化水	原水
Rayleigh number	瑞利数	瑞利數
Raymond mill	雷蒙磨	雷蒙磨
rayon	人造丝	嫘縈
rayon pulp	人造丝浆	嫘縈紙漿
rayon staple	人造丝短纤维	嫘縈棉
rayon tow	人造丝束	嫘縈束
RDC (＝rotating disc contactor)	转盘接触器	轉盤塔

英　文　名	祖国大陆名	台湾地区名
reachability matrix	可及矩阵	可及矩陣
reactant	反应物	反應物
reactant complex	反应络合物	反應物錯合物
reactant ratio	反应物比例	反應物比例
reacting phase	反应相	反應相
reaction	反应	反應
reaction affinity	反应亲和力	反應親和力
reaction chamber	反应室	反應室
reaction coordinates	反应坐标	反應坐標
reaction curve	反应曲线	反應曲線
reaction curve method	反应曲线法	反應曲線法
reaction cycle	反应循环	反應循環
reaction cycle time	反应周期	反應周期
reaction diagram	反应图	反應圖
reaction diffusion network	反应扩散网络	反應擴散網路
reaction engineering	反应工程	反應工程
reaction kettle	反应釜	反應釜
reaction kinetics	反应动力学	反應動力學
reaction mechanism	反应机理	反應機制
reaction network	反应网络	反應網絡
reaction network analysis	反应网络分析	反應網路分析
reaction order	反应级数	反應級數
reaction pair	反应对	反應對
reaction path	反应途径	反應途徑
reaction probability	反应概率	反應機率
reaction process	反应过程	反應程序
reaction progress variable	反应进行变量	反應進行變數
reaction rate	反应速率	反應速率
reaction rate constant	反应速率常数	反應速率常數
reaction rate theory	反应速率理论	反應速率理論
reaction sensitivity	反应灵敏度	反應靈敏度
reaction specificity	反应专一性	反應特異性
reaction step	反应步骤	反應步驟
reaction system	反应系统	反應系統
reaction temperature	反应温度	反應溫度
reaction time	反应时间	反應時間
reaction tower	反应塔	反應塔
reaction-type distributor	反应式分配器	作用式分配器
reaction velocity	反应速度	反應速度

英　文　名	祖国大陆名	台湾地区名
reaction zone	反应区	反應區
reactivation	再活化	再活化
reactive center	反应中心	反應中心
reactive distillation	反应蒸馏	反應蒸餾
reactive extraction	反应萃取	反應萃取
reactive ion etching	反应离子蚀刻	反應性離子蝕刻[法]
reactive process	反应过程	反應性程序
reactive sputtering etching	反应溅射蚀刻法	反應性噴濺蝕刻[法]
reactivity	反应性	①反應性②反應度
reactor	反应器	反應器
reactor analysis	反应器分析	反應器分析
reactor capacity	反应器规模	反應器容量
reactor configuration	反应器构型	反應器組態
reactor control	反应器控制	反應器控制
reactor core	堆芯活性区	反應器核心, 反應器爐心
reactor design	反应器设计	反應器設計
reactor dynamics	反应器动态学	反應器動態[學]
reactor engineering	反应器工程	反應器工程
reactor kinetics	反应器动力学	反應器動力學
reactor model	反应器模型	反應器模式
reactor network	反应器网络	反應器網絡
reactor operation	反应器操作	反應器操作
reactor optimization	反应器优化	反應器最適化
reactor parameter	反应器参数	反應器參數
reactor performance	反应器操作特性	反應器性能
reactor poisoning	反应器中毒	反應器中毒
reactor safety	反应器安全	反應器安全性
reactor setup	反应器装配	反應器配置
reactor stability	反应器稳定性	反應器穩定性
reactor system	反应器系统	反應器系統
reactor theory	反应器理论	反應器理論
reactor train	反应器组	反應器組
ready-mixed paint	调合漆	調和漆
reaeration	再曝气	再曝氣
reaggregation	再聚集	再聚集[作用]
real composition	真实组成	真實組成
real gas	真实气体	真實氣體
realizability	可实现性	可真實化性

英　文　名	祖国大陆名	台湾地区名
real stage	实际阶段	實際階
rear bearing	后轴承	後軸承
reboiler	再沸器, 重沸器	再沸器
reboiler condenser	再沸冷凝器	再沸冷凝器
reboiler dynamics	再沸器动力学	再沸器動能
recalcination	再锻烧	再燃燒[反應]
recalcining process	再锻烧过程	再燃燒程序
receiver	受槽	接受器
recessed plate	凹槽式滤板	帶框濾板
reciprocal plot	倒数标绘	倒數圖
reciprocal relation	倒数关系	倒數關係
reciprocal rule	倒数法则	倒數法則
reciprocating	往复式	往復運動
reciprocating compressor	往复式压缩机	往復[式]壓縮機
reciprocating feeder	往复加料器	往復飼機
reciprocating law	倒易律	互反律
reciprocating piston compressor	往复式活塞压缩机	往復式活塞壓縮機
reciprocating plate column	往复板[式]塔	往復板萃取塔
reciprocating plate extraction column	往复板[式]萃取塔	往復板萃取塔
reciprocating rake	往复耙	往復耙
reciprocating reactor	往复循环反应器	循環反應器
reciprocating screen	往复筛	往復篩
reciprocating vacuum pump	往复式真空泵	往復真空泵
recirculating ratio	循环比	循環比
recirculation	再循环	循環
recirculation factor	循环因子	循環因數
recirculation flow	循环流动	循環流動
recirculation pattern	循环模式	循環型式
recirculation ratio	循环比	循環比
recirculation reactor	循环反应器	循環反應器
reclaimed gypsum	回收石膏	再生石膏
reclaimed material	再生材料	再生材料
reclaimed rubber	再生胶	再生橡膠
reclaiming tank	回收槽	回收槽
reclaiming unit	回收装置	回收裝置
reclamation	回收	回收
recombinant DNA	重组 DNA	重组 DNA
recombination	重组	重組
recorder	记录器	記錄器, 錄音機

英　文　名	祖国大陆名	台湾地区名
recorder-controller	记录控制器	記錄控制器
recording barometer	记录式气压计	記錄氣壓計
recording chart	记录图表	記錄紙
recording equipment	记录设备	記錄裝置
recording material	记录材料	記錄材料, 錄製材料
recoverable stress	可恢复应力	可復應力
recovery	回收[率]	回收
recovery curve	回收率曲线	回收曲線
recovery factor	回收系数	回收因數
recovery system	回收系统	回收系統
recovery time	恢复时间	回收時間
recovery tower	回收塔	回收塔
recrystallization	再结晶	再結晶
rectangular coordinates	直角坐标	直角坐標
rectangular pulse	矩形脉冲	矩形脈動
rectangular weir	矩形堰	矩形堰
rectification	精馏	精餾
rectification section	精馏段	精餾段
rectifier	①精馏器②整流器	①精餾器②整流器
rectifier column	精馏塔	精餾塔
rectifying column	精馏塔	精餾塔
rectifying plate	精馏塔板	精餾板
rectifying section	精馏段	精餾段, 增濃段
rectifying still	精馏釜	精餾器, 整流器
rectifying tower	精馏塔	精餾塔
rectifying tray	精馏板	精餾板
recursion	递归	遞回
recursion formula	递归公式	遞回公式
recursive function	递归函数	遞回函數
recursive relation	递归关系式	遞回關係[式]
recycle	循环	循環
recycle feedstock	循环加料	循環原料
recycle oil	循环油	循環油
recycle process	循环过程	循環程序
recycle ratio	循环比	循環比
recycle stream	循环物流	循環流動
recycle synthesis gas	循环合成气	循環合成氣
recycle system	循环系统	循環系統
recycling	循环	循環

英　文　名	祖国大陆名	台湾地区名
redeposition	再沉积	再沉積
red lead	红铅	紅鉛, 紅丹
Redlich-Kwong equation(RK equation)	RK 方程	RK 方程
redox reaction	氧化还原反应	氧化還原反應
red phosphorus	赤磷	赤磷
reduced coordinates	对比坐标	對比坐標
reduced density	对比密度	對比密度
reduced equation of state	对比状态方程	對比狀態方程式
reduced mass	约化质量	折合質量
reduced model	降阶模型	縮減模式
reduced molecular weight	对比分子量	折合分子量
reduced pressure	对比压力	對比壓力
reduced property	对比性质	對比性質, 折合性質
reduced saturated vapor pressure	对比饱和蒸汽压	對比飽和蒸汽壓
reduced temperature	对比温度	對比溫度
reduced turbidity	对比浊度	對比濁度
reduced variable	对比变量	對比變數
reduced viscosity	对比黏度	折合對比黏度
reduced volume	对比体积	對比體積
reducer	异径管	漸縮管
reducer resistance	异径管阻力	漸縮管阻力
reducing agent	还原剂	還原劑
reducing coupling	缩径管接头	漸縮接頭
reducing elbow	异径弯头	漸縮彎頭
reducing fitting	异径管件	漸縮管件
reducing gear	减速齿轮	減速齒輪
reducing tee	异径三通管	漸縮 T 型管, 漸縮三通管
reducing valve	减压阀	減壓閥
reductant	还原剂	還原劑
reduction	还原	①還原[反應]②降低
reduction potential	还原电位	還原電位
reduction ratio	破碎比	磨碎比
reduction reaction	还原反应	還原反應
reduction value	还原值	還原值
reduction zone	还原区	還原帶
reductive bleaching	还原性漂白	還原漂白
reductive dehalogenation	还原性脱卤	還原性脱鹵[反應]
redundancy	冗余[度]	冗餘
redundant equation	冗余方程	冗餘方程

英　文　名	祖国大陆名	台湾地区名
re-entrainment	二次夹带	二次夾帶
reference condition	参比条件	參比條件
reference electrode	参比电极	參考電池
reference input	参比输入	參考輸入
reference junction	参考接点	參考接點
reference standard	参比标准	參考標準
reference state	参比态, 参考态	參考狀態
reference temperature	参考温度	參考溫度
reference value	参考值	參考值
refill	再装满	再填充物
refinement	精炼	精煉
refiner	精制机	精製機
refinery	炼油厂	煉油廠, 精煉廠
refinery gas	炼厂气	煉油氣
refining	精炼	精煉
reflectance spectroscopy	反射光谱法	①反射光譜學②反應光譜法
reflecting condenser	反射聚光器	反射聚光器
reflection spectrophotometer	反射分光光度计	反射分光光度計
reflectivity	反射率	反射係數
reflectometer	反射计	反光計
reflux	回流	回流
reflux accumulator	回流罐	回流累積量
reflux column	回流塔	回流塔
reflux condenser	回流冷凝器	回流冷凝器
reflux distillation	回流蒸馏	回流蒸餾
reflux drum	回流槽	回流槽
reflux operation	回流操作	回流操作
reflux ratio	回流比	回流比
reflux tube	回流管	回流管
reformed gasoline	重整汽油	重組汽油
reformer	重整器	重組器
reforming	重整	重組[反應]
reforming catalyst	重整催化剂	重組觸媒
reforming reaction	重整反应	重組反應
refraction temperature coefficient	折射温度系数	折射溫度係數
refractive index	折射率	折射率
refractometer	折射计	折射計
refractory	耐火材料	耐火物

英　文　名	祖国大陆名	台湾地区名
refractory brick	耐火砖	耐火磚
refractory porcelain	耐火瓷	耐火瓷
refractory shale	耐火页岩	耐火頁岩
refractory surface	耐火表面	耐火表面
refrigerant	冷冻剂	冷媒, 冷凍劑
refrigerating coil	冷冻盘管	冷凍旋管
refrigerating fluid	冷冻液	冷凍流體
refrigerating ton	冷冻吨	冷凍噸
refrigeration	冷冻	冷凍
refrigeration capacity	冷冻能力	冷凍能力
refrigeration cycle	制冷循环	冷凍週期
refuse	废物	廢棄物
refuse pit	废物坑	廢棄物坑
regain	回收	回收
regenerated cellulose	再生纤维素	再生纖維素
regenerated fiber	再生纤维	再生纖維
regenerated rubber	再生橡胶	再生橡膠
regeneration	再生	再生
regeneration furnace	再生炉	再生爐
regenerative cycle	再生循环[周期]	再生循環
regenerative heat	再生热	再生熱
regenerative heat exchanger	蓄热式换热器	再生熱交換器
regenerative oscillation	再生周期运动	再生振盪
regenerative system	再生系统	再生系統
regenerator	再生器	再生器
region	区[域]	區[域]
region of asymptotic stability	渐近稳定区	漸近穩定[域]
region of attraction	吸引力区[域]	吸引區[域], 漸近穩定區[域]
region of practical stability	实际稳定区	實用穩定區[域]
regression	回归	回歸
regression analysis	回归分析	回歸分析
regression coefficient	回归系数	回歸係數
regression equation	回归方程	回歸方程式
regression line	回归线	回歸線
regular solution	正规溶液	正規溶液
regular solution theory	正规溶液理论	規則溶液理論
regulating box	调节箱	調節箱
regulator	①调节器②调节剂	①調節器②調節劑

英　文　名	祖国大陆名	台湾地区名
regulatory control	调节控制	調節控制
reheater	再热器	再熱器
reheating	再加热	再熱
reinforced plastics	增强塑料	強化塑膠
reinforcing filler	增强填充料	增強填料
reinforcing material	增强材料	增強材料
relative absorptivity	相对吸收率	相對吸收率
relative abundance	相对丰度	相對含量, 相對豐度
relative adsorptivity	相对吸附能力	相對吸附度
relative blackness	相对黑度	相對黑度
relative deviation	相对偏差	相對偏差
relative error	相对误差	相對誤差
relative gain	相对增益	相對增益
relative humidity	相对湿度	相對濕度
relative rate constant	相对速率常数	相對速率常數
relative roughness	相对粗糙度	相對粗糙度
relative saturation	相对饱和度	相對飽和
relative stability	相对稳定性	相對穩定性
relative supersaturation	相对过饱和度	相對過飽和
relative vapour pressure	相对蒸气压	相對蒸氣壓
relative velocity	相对速度	相對速度
relative velocity factor	相对速度因子	相對速度因數
relative viscosity	相对黏度	相對黏度
relative volatility	相对挥发度	相對揮發度
relaxation method	松弛法, 弛豫法	鬆弛法
relaxation modulus	弛豫模量	鬆弛模數
relaxation phenomenon	松弛现象	鬆弛現象
relaxation technique	松弛法	鬆弛法
relaxation test	松弛试验	鬆弛試驗
relaxation time	弛豫时间	鬆弛時間
relaxed state	松弛状态	鬆弛狀態
relay	继电器	替續器, 繼電器
relay valve	继动阀	替續閥
release agent	脱模剂	脱模劑
release valve	放气阀	釋放閥
reliability	可靠性	可靠性
relief device	泄放装置	釋放裝置
relief paint	浮雕漆	浮彫漆
relief valve	安全减压阀	釋放閥

英 文 名	祖国大陆名	台湾地区名
remote control	遥控	遙控
renewable energy source	可再生能源	再生能源
renewable resources	可再生资源	可再生資源
renormalization	再归一化	再正規化
rent	租金	租金
reoxygenation	再充氧作用	再充氧
repair	修理	修理
repellent	拒虫剂	驅蟲劑, 排斥劑
repetitive operation	重复操作	重覆操作
repetitive polymer	重复性聚合物	重覆性聚合物
replacement	置换	置換, 替代
replacement cost	更换费用	置換成本
replacement evaluation	更换评估	汰換評估
replacement value	更换价值	汰換價值
repolymerization	再聚合	再聚合[反應]
report	报告	報告
repress	再压	再壓
reprocessing	再处理	再處理
reprocessing plant	[核燃料]后处理工厂	再處理廠
reproducibility	再现性	重複性, 再現性
reproduction	复制[品]	繁殖
reproduction factor	再现因素	繁殖因數
reproduction rate	再生速率	繁殖速率
reproductive cycle	生殖周期	繁殖周期
repulsive potential	推斥势	推斥勢
rerun column	再蒸馏塔	重餾塔
rerunning tower	再蒸馏塔	重餾塔
rerun yield	再蒸馏收率	重餾產率
research	研究	研究
reservoir	贮槽	貯槽, 蓄水池
reset	复位	重整
reset action	复位作用	重整作用
reset corner	复位角	重整角
reset rate	复位速率	重整速率
reset response	复位响应	重整應答
reset time	返位时间	重整時間
reset windup	复位终结	重整繞緊
residence time	停留时间	滯留時間

英　文　名	祖国大陆名	台湾地区名
residence time distribution density function	停留时间分布密度函数	滯留時間分布密度函數
residence time distribution(RTD)	停留时间分布	滯留時間分布
residual analysis	残差分析	殘留物分析
residual contribution	残余贡献, 剩余贡献	剩餘貢獻
residual enthalpy	残余焓, 剩余焓	殘留焓
residual entropy	残余熵, 剩余熵	殘留熵
residual error	残差	殘留誤差
residual fraction	残余馏分	殘留餾分
residual gas	残余气体	殘留氣體
residual product	残油	殘留產物
residual property	残余性质, 剩余性质	殘留性質
residual strain	残余应变	殘留應變
residual term	残余项, 剩余项	殘留項
residual volume	残余体积, 剩余体积	殘留體積
residue	残液, 釜液	殘渣, 殘留物
residuum	渣油	蒸餘物, 殘渣油
resilience	弹性	反彈
resin	树脂	樹脂
resin soap	树脂皂	樹脂皂
resin tannage	树脂鞣法	樹脂鞣製
resistance	①阻力②电阻	①阻力②電阻
resistance coefficient	阻力系数	阻力係數
resistance furnace	电阻炉	電阻爐
resistance thermometer	电阻温度计	電阻溫度計
resolution	分离度	解析度
resolving power	分辨能力	解析能力
resonance	共振	共振
resonance frequency	共振频率	共振頻率
resonance integral	共振积分	共振積分式
resonance peak	共振峰	共振尖峰
resonant frequency	共振频率	共振頻率
resonant peak	共振峰	共振尖峰
resources cost	资源成本	資源成本
respiration	呼吸	呼吸
respiratory chain	呼吸链	呼吸鏈
response curve	响应曲线	應答曲線
response speed	响应速度	應答速率
response surface	响应面	應答面

英　文　名	祖国大陆名	台湾地区名
response time	响应时间	應答時間
restrainer	抑制剂	制止器, 壓制劑
resuspension	再悬浮	再懸浮
retardation factor	阻滞因子	阻滞因數
retardation time	推迟时间	減速時間, 遲延時間
retarded motion	阻迟运动	阻滞運動
retarder	阻滞剂	阻滞劑
retarding force	减速力	減速力
retarding torque	减速扭矩	減速扭矩
retentate	渗余物	滲餘物
retention	截留, 保留	停留
retention period	保留期间	停留期間
retention time	保留时间	停留時間
retention volume	保留体积	停留體積
retort	甑, 干馏釜	甑
retort process	甑馏法	甑[蒸]餾法
retrogradation	逆反作用	退減[作用]
retrograde condensation	逆反冷凝	降壓冷凝
return action	反回作用	反作用
return bend	回弯管	回彎管
return on investment	投资收益率	投资收益率
return rate	返回率	報酬率
return trap	回流阱	回流阱
reverse current	反向电流	逆流
reverse diffusion	反扩散	逆擴散
reversed micelle	可逆胶囊	反微胞
reverse extraction	反萃取	逆萃取
reverse flow	反流	逆向流動
reverse fractionation	逆分馏法	逆分餾[法]
reverse micelle extraction	反胶团萃取	逆膠團萃取
reverse osmosis	反渗透	逆滲透
reverse reaction	逆反应	逆反應
reverse-running pump	反向泵	反向泵
reversibility	①可逆度②可逆性	①可逆度②可逆性
reversible adiabatic flow	可逆绝热流动	可逆絕熱流動
reversible cell	可逆电池	可逆電池
reversible change	可逆变化	可逆變化
reversible chemical reaction	可逆化学反应	可逆化學反應
reversible decomposition potential	可逆分解电位	可逆分解電位

英　文　名	祖国大陆名	台湾地区名
reversible diffusion	可逆扩散	可逆擴散
reversible mechanism	可逆机理	可逆機構
reversible precipitation	可逆沉淀	可逆沈澱
reversible process	可逆过程	可逆過程
reversible reaction	可逆反应	可逆反應
reversible work	可逆功	可逆功
revolution per minute (rpm)	每分钟转数	每分轉數
revolving screen	回转筛	迴轉篩
Reynolds number	雷诺数	雷諾數
rheodestruction	流变破坏	流變破壞
rheological property	流变性质	流變性質
rheology	流变学	流變學
rheometer	流变仪	流變儀
rheometry	流变测定法	流變測定法
rheopectic fluid	震凝性流体	搖變增黏流體
rheopexy	震凝性	觸變性
ribbon mixer	螺带混合机	螺帶混合機
ribonucleic acid(RNA)	核糖核酸	核糖核酸
ribose	核糖	核糖
rice oil	米糠油	米糠油
rich gas	富气	富氣
rich phase	富相	富相
rich solution	浓溶液	濃溶液
Rideal mechanism	里迪尔机理	里迪爾機構
rigidity	刚度	剛性
rigidity modulus	刚性模数	剛性模數
rigorous method	严格法	嚴格法
ringing pole	振铃极点	振鈴極點
ring joint	环形接头	環線接頭
ring piezometer	环形压力计	環形測壓計
ring roll mill	环滚磨机	環輥磨機
ring roll pulverizer	环滚粉碎机	環輥粉碎機
ring sprayer	环形喷雾器	環狀噴霧器
ring type grinder	环形研磨机	環形研磨機
ring type manometer	环形压力计	環形壓力計
ripple tray	波纹塔板	波動穿流板
rippling	浪动的	波動
riser	提升管	①上升管②竪板
riser reactor	提升管反应器	上升管反應器

英　文　名	祖国大陆名	台湾地区名
risk	风险	風險
risk analysis	风险分析	風險分析
risk earning rate	风险收益率	風險收益率
risk factor	风险因子	風險因素
risk function	风险函数	風險函數
ristocetin	瑞斯托菌素	利徽素
RK equation (= Redlich-Kwong equation)	RK 方程	RK 方程式
RNA (= ribonucleic acid)	核糖核酸	核糖核酸
roaster	焙烧炉	①焙燒爐②風箱
robot	机器人	機器人
robust process control	鲁棒过程控制	韌性程序控制
robust stabilizability	鲁棒稳定化度	韌性穩定化度
robust stabilization	鲁棒稳定化	韌性穩定化
rocket	火箭	火箭
rocket engine	火箭引擎	火箭引擎
rocket fuel	火箭燃料	火箭燃料
rock sugar	冰糖	冰糖
rodenticide	杀鼠剂	殺鼠劑
rod mill	棒磨机	棒磨機
roll crusher	辊式破碎机	輥式破碎機
roller	滚柱	輥, 滾筒
roller bearing	滚子轴承	滾筒軸承, 輥軸承
roller mill	轧制机	輥磨機
roller mill press	辊压机	輥壓機
roller printing machine	滚筒印花机	滾筒印花機
root locus	根轨迹	根軌跡
root locus angle	根轨迹角	根軌跡角
root locus method	根轨迹法	根軌跡法
root locus plot	根轨迹图	根軌跡圖
root locus technique	根轨迹方法	根軌跡法
root-mean-square deviation	均方根偏差	均方根偏差
root-mean-square error	均方根误差	均方根誤差
root-mean-square fluctuating velocity	均方根振荡速度	均方根速度
Roots blower	罗茨鼓风机	羅次鼓風機
rosin	松香	松香, 松脂
rosined soap	松香皂	松香皂
rosin oil	松香油	松香油
rosin size	松香胶	松香膠料

英　文　名	祖国大陆名	台湾地区名
rosin sizing	松香施胶	松香上膠
rotameter	转子流量计	浮子流量計
rotary blower	回转鼓风机	旋轉鼓風機
rotary compressor	回转压缩机	旋轉壓縮機
rotary continuous filter	旋转连续过滤机	旋轉續濾器
rotary cooler	旋转冷却器	旋轉冷卻器
rotary crusher	旋转压碎机	旋轉壓碎機
rotary crystallizer	旋转结晶器	旋轉結晶器
rotary disc filter	转盘过滤机	轉盤濾機
rotary disc meter	转盘流量计	轉盤流量計
rotary displacement pump	旋转排代泵	旋轉排量泵
rotary distributor	旋转分配器	旋轉分配器
rotary drum filter	转鼓过滤机	轉桶濾機
rotary dryer	旋转干燥器	旋轉乾燥器
rotary efficiency	旋转效率	旋轉效率
rotary feeder	旋转进料器	旋轉飼機
rotary filter	旋转过滤机	旋濾機
rotary filter press	旋转压滤机	旋轉壓濾機
rotary kiln	回转窑	旋轉窯
rotary pump	回转泵	旋轉泵
rotary roaster	回转焙烧炉	旋轉焙燒爐
rotary screen	旋转筛	旋轉篩
rotary shaker	旋转摇床	旋轉搖動器
rotary sieve	旋转筛	旋轉篩
rotary sieve shaker	旋转筛摇床	旋轉搖篩器
rotary vacuum drum filter	转筒真空过滤机	轉桶真空過濾機
rotary vacuum pump	旋转真空泵	旋轉真空泵
rotary valve	旋转阀	旋轉閥
rotary viscometer	旋转黏度计	旋轉黏度計
rotary washer	旋转洗涤机	旋轉洗滌機
rotating-basket reactor	旋筐反应器	旋籃反應器
rotating disc contactor(RDC)	转盘塔	轉盤塔
rotating-disc process	生物转盘法	[旋轉]生物盤法
rotating-drum absorber	转筒吸收器	轉桶吸收器
rotating drum dryer	滚筒干燥器	轉桶乾燥器
rotating extractor	旋转萃取器	旋轉萃取器
rotating sprinkler	旋转式喷头	旋轉噴灑器
rotational energy	转动能	旋轉能
rotational energy level	旋转能级	旋轉能階

英　文　名	祖国大陆名	台湾地区名
rotational flow	有旋流	旋轉流動
rotationality	旋转度	旋轉度
rotational partition function	转动配分函数	轉動分配函數
rotation viscometer	旋转黏度计	旋轉黏度計
rotatory power	旋光能力	旋光能力
rotenone	鱼藤酮	魚藤酮
rotifer	轮虫	輪虫
rotor	转子	轉子
rotor blade	转子叶片	轉子葉片
roughing bed	粗滤床	粗濾床
roughing filter	粗滤床	粗濾床
roughness	粗糙度	粗糙度
roughness factor	粗糙因子	粗糙因數
roughness scale	粗糙标度	粗糙標度
rounded orifice	圆孔口	圆孔口
route	途径	途徑
royalty	专利权使用费	權利金
rpm(=revolution per minute)	每分钟转数	每分轉數
RTD (=residence time distribution)	停留时间分布	滯留時間分布
rubber	橡胶	橡膠
rubber hose	胶管	橡皮軟管
rubber latex	橡胶胶乳	橡膠乳膠
rubber-like liquid	橡胶态液体	彈性液體
rubber-lined pipe	橡胶衬里管	橡膠襯管
rubber mill	橡胶磨	橡膠磨碾機
rubber tube	橡皮管	橡皮管
rubber tubing	橡皮管	橡皮管
rubbery state	橡胶状态	橡膠狀態
run	运转	運轉
runaway	失控	失控
run-down tank	溢流槽	半成品槽
runner	浇道	澆道
rupture stress	断裂应力	斷裂應力
rust	锈	銹
rust remover	除锈剂	除銹劑
rutile	金红石	金红石

S

英　文　名	祖国大陆名	台湾地区名
saccharide	糖类	糖类
saccharification	糖化[反应]	糖化
saccharifying agent	糖化剂	糖化劑
saccharimeter	糖量计	糖量計
saccharin	糖精	糖精
saccharomyces	酵母菌属	酵母菌
saccharose	蔗糖	蔗糖
saddle azeotrope	鞍点共沸物	鞍式共沸液
safety check list	安全检查清单	安全檢查報表
safety device	安全装置	安全裝置
safety door	安全门	安全門
safety explosive	安全炸药	安全炸藥
safety factor	安全系数	安全因數
safety glass	安全玻璃	安全玻璃
safety margin	安全裕度	安全邊界
safety relief area	安全泄压面积	安全釋放面積
safety-relief valve	安全泄压阀	安全釋放閥
safety testing	安全性试验	安全性試驗
safety valve	安全阀	安全閥
safety vent	安全放空	安全通氣孔
safe working pressure	安全操作压力	安全操作壓力
safe working stress	安全工作应力	安全工作應力
sag flow	淌流	淌流
sag flow board	淌流板	淌流板
salicylic acid	水杨酸	柳酸, 水楊酸
salmon oil	鲑鱼油	鲑油
SALS (=small angle light scattering)	小角光散射	小角度散射
salt	盐	鹽
salt catcher	盐捕集器	受鹽器
salt cellulose	盐纤维素	鹽纖維素
salted hide	盐皮	鹽皮, 腌皮
salt effect	盐效应	鹽效應
salt-fog resistance	防盐雾性	防鹽霧性
salt-free polyelectrolyte solution	无盐聚电解质溶液	無鹽聚電解質溶液
salt glaze	盐釉	鹽釉

英 文 名	祖国大陆名	台湾地区名
salting	腌制	醃製
salting in	盐溶	鹽溶
salting out	盐析	鹽析
salting out agent	盐析剂	鹽析劑
saltpeter	硝石	硝石
salvage value	残[余价]值	殘餘價值
sample	样品	樣品
sampled data	采样数据	取樣數據
sampled data control	采样控制	取樣數據控制, 抽樣數據控制
sampled-data control system	采样控制系统	取樣數據控制系統, 抽樣數據控制系統
sampled-data signal	采样数据信号	取樣數據信號, 抽樣數據信號
sampled-data system	采样数据系统	取樣數據系統, 抽樣數據系統
sample system	样品系统	樣品系統
sampling	取样	取樣, 抽樣
sampling delay	取样滞后	取樣延遲
sampling error	取样误差	取樣誤差
sampling system	取样系统	取樣系統, 抽樣系統
sampling tube	取样管	取樣管
sandal oil	檀香油	檀香油
sandarac	桧树胶	檜樹膠
sand-bed filter	砂滤器	砂濾器
sand-blast	喷砂	噴砂
sand filtration	砂滤法	砂濾法
sand mill	砂磨	砂磨
sand paper	砂纸	砂紙
sandwich panel	夹心板	三夾板
sandwich structure	夹层结构	夾層結構
sanitary wastewater	生活污水	衛生污水, 家庭污水
saponification	皂化	皂化反應
saponification number	皂化值	皂化值
saponified acetate rayon	皂化乙酸人造丝	皂化乙酸人造絲
saponified cellulose acetate	皂化纤维素乙酸酯	皂化纖維素乙酸酯
saponified polymer	皂化聚合物	皂化聚合物
saponifying agent	皂化剂	皂化劑
sapphire	蓝宝石	藍寶石

英　文　名	祖国大陆名	台湾地区名
saran	聚偏氯乙烯纤维	纱隆, 聚偏氯乙烯樹脂
saran pipe	萨纶管	紗隆管
sardine oil	沙丁鱼油	沙丁魚油
SAS (＝small angle scattering)	小角散射	小角散射
satellite plant	卫星工厂	衛星工廠
satin weave	纤维编织	纖維編織
saturated air	饱和空气	飽和空氣
saturated color	饱和色	飽和色
saturated humidity	饱和湿度	飽和溼度
saturated hydrocarbon polymer	饱和烃聚合物	飽和碳氫聚合物
saturated liquid	饱和液	飽和液
saturated polyester	饱和聚酯	飽和聚酯
saturated solution	饱和溶液	飽和溶液
saturated state	饱和态	飽和態
saturated steam	饱和蒸汽	飽和蒸汽
saturated vapor	饱和蒸气	飽和蒸氣
saturated vapor pressure	饱和蒸气压	飽和蒸氣壓
saturating paper	油纸	油紙
saturation	饱和	飽和, 飽和度
saturation curve	饱和曲线	飽和曲線
saturation limit	饱和极限	飽和極限
saturation pressure	饱和压力	飽和壓力
saturation temperature	饱和温度	飽和溫度
saturator	饱和器	飽和器
SAXD (＝small angle X-ray diffraction)	小角 X 射线衍射	小角 X 射線繞射
SAXS (＝small angle X-ray scattering)	小角 X 射线散射	小角 X 射線散射
scalar	标量	純量
scale	污垢, 结垢	積垢
scale coefficient	结垢系数	結垢係數
scale down	缩小	縮小
scale factor	标度因子	結垢係數
scale fiber	有鳞[片]纤维[黏胶]	有鱗[片]纖維[黏膠]
scale formation	结垢形成	積垢形成
scale indicator	标度指示器	標度指示器
scale up	放大	按比例放大
scale-up problem	放大问题	放大問題
scale-up rule	放大法则	放大法則
scaling	定标, 比例换算	定標, 比例化, 標度化
scaling factor for heat transfer	传热污垢因子	熱傳積垢因素

英　文　名	祖国大陆名	台湾地区名
scaling law	标度律	①比例定律②放大定律
scanning calorimetry	扫描量热法	掃描量熱法
scanning electron microscope (SEM)	扫描电子显微镜	掃描電子顯微鏡
scanning electron microscopy	扫描电子显微镜法	掃描電子顯微鏡法
scanning rate	扫描速度	掃描速度
scarf joint	嵌接	嵌接接頭
scattering	散射	散射
scattering angle	散射角	散射角
scattering cell	散射池	散射池
scattering coefficient	散射系数	散射係數
scattering of light	光散射	光散射
scattering pattern	散射形式	散射形式
scattering volume	散射体积	散射體積
scavenger	①净化剂②捕获剂	①淨化劑,清除劑②捕獲劑
schedule number	管壁厚系列号	分類號
scheduling computer control	排程计算机控制	排程計算機控制
schematic diagram	示意图	示意圖
schematics	示意图	示意圖
Schiff base	席夫碱	希夫鹼
Schiff base polymer	席夫碱聚合物	希夫鹼聚合物
Schmidt number	施密特数	西門諾夫數
Schob elastometer	肖伯弹力试验机	肖伯彈力試驗機
Schob resilience	肖伯回弹性	肖伯回彈性
schonite	软钾镁矾	硫酸鉀鎂礦
Schopper tensile tester	朔佩尔张力试验机	肖伯張力試驗機
Schulz-Zimm distribution	舒尔茨-齐姆分布	舒爾茨-齊姆分布
scission	断键	斷鍵
SCLC (＝space charge limited current)	空间电荷限制电流	空間電荷限制電流
sclerometer	硬度计	硬度計,邵氏硬度計
scope	范围	範圍
scorching	①焦烧②过早硫化	①焦燒②過早硫化
scorching test	焦烧试验	焦燒試驗
scorch-resisting treatment	抗焦处理	防焦處理
scorch retarder	防焦剂	防焦劑
Scott flexometer	斯科特挠度计	斯科特曲度計
scouring	洗毛	洗毛
scraper	刮板	刮刀
scraper chiller	刮刀式冷却器	刮刀式冷凍器

英　文　名	祖国大陆名	台湾地区名
scraper conveyor	刮板输送机	刮運機
scraper heat exchanger	刮板式换热器	刮刀式熱交換器
scrap rubber	废橡胶, 杂胶	廢橡膠, 雜膠
scratch	刮痕	刮痕
scratch hardness	刮痕硬度	刮痕硬度
scratch resistance	抗刮性	抗刮性
screen	①筛网②屏蔽	①篩子②屏蔽
screen analysis	筛析	篩析
screen aperture	筛孔	篩孔
screen bank	筛组	篩組
screen changer	过滤网调换装置	過濾網調換裝置
screen conveyor	筛网输送机	篩式運送機, 篩運機
screen effectiveness	屏蔽有效度	篩有效度
screen filter	筛滤器	篩濾機
screening	筛选	篩選
screening device	筛选装置	篩選裝置
screening efficiency	筛选效率	篩選效率
screening test	筛选测验	篩轉試驗
screening trommel	转筒筛	轉筒篩
screen mill	筛磨	篩曆機
screen opening	筛孔	篩孔
screen pack	过滤网组[合]	過濾網組[合]
screen plate	筛板	篩板
screw	螺杆	螺桿
screw core pin	螺杆芯孔销	螺桿蕊栓
screw dislocation	螺型位错	螺旋型差排
screw efficiency	螺杆效率	螺桿效率
screw extractor	螺杆出料机	螺桿出料機
screw extruder	螺杆压出机	螺桿擠出機
screw extruder reactor	螺杆挤出反应器	螺桿擠出反應器
screw extrusion	螺旋挤出	螺旋擠出
screw preplasticating type injection molding machine	螺杆式预塑化注压成型机	螺桿式預塑化成型機
screw pump	螺杆泵	螺旋泵
screw structure	螺旋结构	螺旋結構, 螺桿結構
screw type injection molding machine	螺杆式注塑成型机	螺桿式注塑成型機
scroop	丝鸣[绸料摩擦声]	絲鳴
scrubber	洗涤器	洗氣器, 洗滌器
scrubbing tower	洗涤塔	洗氣塔

英　文　名	祖国大陆名	台湾地区名
scum	浮渣	浮渣
scum-collecting device	浮渣收集装置	浮渣收集設備
sea island cotton	海岛型棉	海島棉
seal	密封	密封
sealant	密封胶	密封劑
seal cement	密封胶合剂	密封膠合劑
sealed tube	密封管	密閉管
sealer	热合机	熱合機, 封焊機
seal gum	密封胶	密封膠
sealing	①热合②封口	①熱合②封合
sealing compound	密封胶	密封劑
sealing joint	热合接头	熱合接頭
sealing liquid	密封液	密封液
sealing machine	封口机	封口機
sealing oil	密封油	密封油
sealing property	密封性能	密閉性
sealing wax	火漆	封蠟, 火漆
seal leg	密封腿[料腿]	封柱
seal oil	海豹油	海豹油
seal pot	密封罐	密封罐
seamless pipe	无缝管	無縫管
seam welder	缝焊机	縫焊機
seam welding	缝焊	縫焊
seaweed fiber	海藻纤维	海藻纖維
sebacic acid	癸二酸	癸二酸
secant modulus	正割模量	正割係數
secondary accelerator	助促进剂	助加速劑
secondary amine	二级胺	二級胺
secondary antioxidant	助抗氧剂	助抗氧劑
secondary cell wall	次生胞壁	次生胞壁
secondary clarifier	二级澄清器	二級澄清器, 二級沉降槽
secondary creep	次级蠕变	次級蠕變
secondary crystallization	后期结晶	後期結晶
secondary dispersion	二次色散	二次色散
secondary feeder	二级加料器	二級進料器
secondary gluing	二次胶合	二次膠合
secondary loop	副环路	副環路
secondary measurement	次级测量值	次量测值
secondary normal stress coefficient	次正应力系数	次正應力係數

英 文 名	祖国大陆名	台湾地区名
secondary nucleation	二次成核作用	二次成核作用
secondary nucleation	二次成核	第二次成核
secondary oil recovery	二次石油回收	二次石油回收
secondary plasticizer	辅助增塑剂	助增塑剂
secondary polymerization reaction	二级聚合反应	二級聚合反應
secondary product	次级产品	次產物
secondary reaction	二次反应	次要反應, 後續反應
secondary relaxation	次级松弛	次級鬆弛
secondary relaxation temperature	次级弛豫温度	次級鬆弛溫度
secondary settler	二级沉降槽	二級沉降槽
secondary standard	二级标准	二級標準
secondary transition	次级转变	次級轉變
secondary treatment	二级处理	二級處理
secondary trickling filter	二级滴滤池	二級滴濾池
second evaporator	二次蒸发器	二次蒸發器
second law of thermody namics	热力学第二定律	熱力學第二定律
second order crystallization	二级结晶	二級結晶
second-order fluid	二阶流体	二階流體
second order ionization potential	二级电离电位	二級電離
second order nucleation	二级成核作用	二級成核作用
second-order response	二阶响应	二階應答
second-order system	二阶系统	二階系統
second order transition	二级转变	二級轉變
second order transition temperature	二级转变温度	二級轉變溫度
second virial coefficient	第二位力系数, 第二维里系数	第二位力係數
sectioning	切片	切片
sediment	沉积物	沉積物
sedimentary clay	沉积黏土	沉積黏土
sedimentary kaolin	沉积高岭土	沉積高嶺土
sedimentary rock	沉积岩	沉積岩
sedimentation	沉降	沉積
sedimentation average molecular weight	沉降平均分子量	沉積平均分子量
sedimentation basin	沉淀池	沉積池
sedimentation bottle	沉淀瓶	沉積瓶
sedimentation chamber	沉淀槽	沉積槽
sedimentation coefficient	沉降系数	沉積係數
sedimentation constant	沉降常数	沉積常數
sedimentation equilibrium	沉降平衡	沉積平衡

英 文 名	祖国大陆名	台湾地区名
sedimentation equilibrium method	沉降平衡法	沉積平衡法
sedimentation method	沉降法	沉積法
sedimentation pan	沉降盘	沉積盤
sedimentation potential	沉积电位	沉降電位
sedimentation rate	沉积速率	沉積速率
sedimentation tank	沉积槽	沉積槽
sedimentation velocity	沉降速度	沉積速度
sedimentation velocity method	沉降速度法	沉積速度法
seed cotton	籽棉	種棉
seed crystal	晶种	種晶
seeded crystallization	晶种结晶	晶種結晶
seed grain	晶种粒	結晶種粒
seed hair	晶种纤维	晶種纖維
seeding	加晶种	①加種晶②加種菌
seeding polymerization	种子聚合	晶種聚合
seed mixer	晶种混合槽	晶種混合槽
segment	链节,线段	鏈段
segmental-bed reactor	分段床反应器	分段床反應器
segmental Brownian motion	链段布朗运动	鏈段布朗運動
segmental copolymer	多嵌段共聚物	多段共聚物
segmental downtake calandria	弓形降液管	弓形下導排管
segmental friction factor	链段摩擦因子	鏈段摩擦因子
segmental jump frequency	链段跃迁频率	鏈段躍遷頻率
segmental motion	链段运动	鏈段運動
segmental orifice	弓形孔口	弓形孔口
segmental orifice plate	弓形孔口板	弓形孔口板
segment anisotropy	链段各向异性	鏈段各向異性
segment copolymer	多嵌段共聚物	多段共聚物
segment copolymerization	嵌段共聚合	多段共聚合
segment-density distribution	链段密度分布	鏈段密度分布
segmented copolymer	多嵌段共聚物	多段共聚物
segment-interaction parameter	链段相互作用参数	鏈段交互作用參數
segment rotation	链段旋转	鏈段旋轉
segregated flow	隔离流动	隔離流動
segregated reactor	隔离流反应器	隔離流反應器
segregation	离析	離析
selected diffraction	选区衍射	選擇繞射
selection rule	选择定则	選擇法則
selective absorbent	选择吸收剂	選擇性吸收劑

英 文 名	祖国大陆名	台湾地区名
selective absorption	选择吸收	選擇性吸收
selective azeotropic distillation	选择性共沸蒸馏	選擇性共沸蒸餾
selective control	选择性控制	選擇性控制
selective controller	选择性控制器	選擇性控制器
selective cracking	选择裂化	選擇性裂解
selective evaporation	分馏, 蒸馏	選擇性蒸發
selective extraction	选择性萃取	選擇性萃取
selective fermentation	选择性发酵	選擇性發酵
selective flotation	选择性浮选	選擇性浮選
selective hydrogenation	选择加氢	選擇性氫化
selective leaching	选择性浸取	選擇性瀝取
selective poisoning	选择性中毒	選擇[性]中毒
selective polymerization	选择聚合	選擇聚合
selective rectification	选择精馏	選擇[性]精餾
selective solvent	选择性溶剂	選擇性溶劑
selectivity	选择性	①選擇性②選擇度
selectivity coefficient	选择性系数	選擇性係數
selectivity dispersion curve	选择性分散曲线	選擇性分散曲線
selectivity ratio	选择性比	選擇性比
selector	选择器	選擇器
selenium polymer	硒聚合物	硒聚合物
self-adaptive control	自适应控制	自適應控制
self-adhesive	自黏着剂	自黏著劑
self-adhesive tape	自黏胶带	自黏膠帶
self ageing	自老化	自老化
self-balancing bridge	自平衡电桥	自均衡電橋
self-balancing potentiometer	自平衡电位计	自均衡電位計
self-bonded fiber	自黏合纤维	自黏合纖維
self-crosslinking acrylate rubber	自交联丙烯酸酯橡胶	自交聯丙烯酸酯橡膠
self-crosslinking acrylic resin	自交联丙烯酸树脂	自交聯丙烯酸樹脂
self-cure	自固化	自塑
self-cure epoxy resin	自固化环氧树脂	自塑化環氧樹脂
self-curing	自动硫化	自動硫化
self-curing adhesive	自固化黏合剂	自塑化黏合劑
self-diffusion	自扩散	自擴散
self extinguish ability	自熄性	自熄性
self-extinguishing	自动灭火	自熄
self-heating property	自动加热性	自熱性
self-ignition	自燃	自燃

英 文 名	祖国大陆名	台湾地区名
self-ignition temperature	自燃温度	自燃溫度
self-initiation	自引发	自發
self-oligomerization	自身低聚化	自身低聚化
self-polymerization	自聚合	自聚合
self-priming pump	自起动泵	自引泵
self-propagation	自增长	自增長
self purification	自净化	自淨化[作用]
self regulation	自调整	自調節
self-reinforcing polymer	自增强聚合物	自增強聚合物
self-sustained reaction	自持续反应	自續反應
self sustaining	自持续	自續
self-termination	自终止	自終止
self-tuning controller	自校正控制器	自調諧控制器
self-tuning regulator	自校正调节器	自調諧調節器
self-vulcanizing	自动硫化	自動硫化
semi-anthracite	半无烟煤	半無煙煤
semi-automatic press	半自动压机	半自動壓縮機
semi-automatic system	半自动系统	半自動系統
semi-batch operation	半分批式操作	半批式操作
semi-batch process	半分批法	半批次法
semi-batch reactor	半分批反应器	半分批反應器
semi-batch selectivity	半分批选择性	半批式選擇性
semi-bituminous coal	半烟煤	半煙煤
semi-boiled soap	含水皂	含水皂
semi-boiling process	半煮法	半煮法
semi-chemical pulp	半化学纸浆	半化學紙漿
semi-chemical pulping process	半化学制浆法	半化學製漿法
semiconducting polymer	高分子半导体	高分子半導體
semiconductive coating	半导体涂料	半導體塗料
semiconductive polymer	半导体聚合物	半導體聚合物
semiconductor	半导体	半導體
semiconductor catalyst	半导体催化剂	半導體觸媒
semi-continuous kiln	半连续窑	半連續窯
semi-continuous polymerization	半连续聚合	半連續聚合
semicontinuous process	半连续过程	半連續程序
semi-crystal	半水晶	半晶體
semi-crystalline polymer	半结晶聚合物	半結晶聚合物
semi-cure	半硫化	半硫化
semi-curing	半硫化	半硫化

英　文　名	祖国大陆名	台湾地区名
semi drying oil	半干性油	半乾性油
semi-dull	半无光	半無光
semi-durable adhesive	半耐久性黏合剂	半耐久性黏合劑
semi-enclosed impeller	半封闭叶轮	半封閉葉輪
semi-flexible chain	半挠性链	半撓性鏈
semiflow reactor	半流动式反应器	半流動式反應器
semifractionating	半分馏	半分餾
semigloss coating	半光涂料	半光澤塗料
semi-hard board	半硬质纤维板	半硬質纖維板
semi-interpenetrating polymer network	半互穿透聚合物网络	半互穿聚合物網絡
semi-isolated fluidized catalyst bed	半封闭流化态催化剂床	半封閉流化態觸媒床
semi-logarithmic paper	半对数坐标纸	半對數坐標紙
semi-microanalysis	半微量分析	半微量分析
semi-octagonal cell	半八角形池	半八角形池
semi-opaque	半透明的	半透明的
semi-paste paint	半糊状漆	半糊狀漆
semi-pearl polymerization	半悬浮聚合	半懸浮聚合
semipermeable membrane	半透膜	半透膜
semipermeable molecule	半透性分子	半透性分子
semipolar bond	半极性键	半極性鍵
semireinforcing furnace black	半补强炉黑	半補強爐黑
semi-rigid foam	半硬泡沫	半硬性泡沫
semi-strong accelerator	半强性促进剂	半強促進劑
semisynthetic	半合成物	半合成物
semi-synthetic fiber	半合成纤维	半合成纖維
SEM (＝scanning electron microscope)	扫描电子显微镜	掃描式電子顯微鏡
sensible heat	显热	顯熱
sensing element	敏感元件	感測元件
sensitivity	灵敏度	靈敏度
sensitivity analysis	灵敏度分析	靈敏度分析
sensitivity estimate	灵敏度估计	靈敏度估計
sensitized paper	感光纸	感光紙
sensitizer	敏化剂	敏化劑
sensitizing dye	增感染料	感光染料
sensor	传感器	感測器
separated application adhesive	组分分涂黏合剂	分施膠黏劑
separated flow	分离流	分離流
separation	分离	分離
separation cell	分离池	分離池

英　文　名	祖国大陆名	台湾地区名
separation cyclone	旋风分离器	分離旋風器
separation factor	分离因子	分離因數
separation point	分离点	分離點
separation process	分离过程	分離程序
separation specification table	分离技术规格表	分離規格表
separation technology	分离技术	分離技術
separative membrane	分离膜	分離膜
separator	分离器	分離器
sephadex gel	交联葡聚糖[凝胶]	交聯葡凝膠, 交聯葡凝糖
septum	隔膜	隔膜, 隔板
sequence	序列	序列
sequence control	顺序控制	順序控制
sequence length	序列长度	序列長度
sequence-length distribution	序列长度分布	序列長度分布
sequencing on-off operation	逐次开闭操作	逐次開閉操作
sequencing valve	顺序阀	順序閥
sequential analysis	序贯分析	順序分析
sequential copolymer	序列共聚物	序列共聚物
sequential operation	顺序操作	順序操作
sequential polymerization	序列聚合	序列聚合
sequential sampling	序贯抽样	順序取樣
sequestering agent	螯合剂	螯合劑, 鉗合劑
sequestration	多价螯合作用	螯合作用, 鉗合作用
serial processes	重串过程	串聯程序
sericin	丝胶蛋白	絲膠
series compensation	重串补偿	串聯補償
series flow	重串流动	串流
series reactions	重串反应	串行反應
series reactors	重串反应器	串聯反應器
serimeter	①验丝计②生丝强伸力 　试验计	①驗絲計②強力延伸試 　驗器
serration	锯齿形	鋸齒狀缺口, 鱗片
serum albumin adhesive	血清白蛋白黏合剂	血清蛋白黏合劑
serum globulin	血清球蛋白	血清球蛋白質
serviceable life	使用期	適用期, 耐用期
servo control	伺服控制	伺服控制
servo mechanism	伺服机构	伺服機構
servomechanism compensation	伺服机构补偿	伺服機構補償
servomechanism-type problem	伺服问题	伺服問題

英　文　名	祖国大陆名	台湾地区名
servo problem	伺服问题	伺服問題
sesame oil	芝麻油	芝麻油
set point	设定值	設定點
set point control	设定值控制	設定點控制
set point disturbance	设定值扰动	設定點擾動
set point response	设定值响应	設定點應答
set point tracking	设定值跟踪	設定點追蹤
set point variation	设定值变动	設定點變動
set time	凝固时间	①凝固時間②固化時間③變定時間
setting temperature	凝固温度	凝固溫度
settler	沉降器	沈降器, 沈降槽
settling	沉降	沈降
settling basin	沉降池	沈降池
settling chamber	沉降室	沈降室
settling column	沉降柱	沈降柱
settling column analysis	沉降柱分析	沈降柱分析
settling filtration	沉降过滤	沈降過濾
settling tank	沉降槽	沈降槽
settling time	沉降时间	沈降時間, 安定時間
settling velocity	沉降速度	沈降速度
settling zone	沉降区	沈降區
setup cost	装置建立成本	配置成本
set-up effect	开始[硫化]效应	開始[硫化]效應
sewage disposal	污水处理	污水處理
sewage raw	原污水	原污水
sewer pipe	污水管	污水管
sexamer	六聚物	六聚物
shadowing	屏蔽	屏蔽
shaft	轴	軸
shaft brake horsepower	轴制动马力	軸制動馬力
shaft work	轴功	轉軸功
shaker	震动器	搖動器
shaking conveyor	振动输送机	搖運機
shaking screen	摇动筛	搖動篩
shale	页岩	頁岩
shale oil	页岩油	頁岩油
shallow-bed reactor	浅床反应器	淺床反應器
shape birefringence	形状双折射	形狀雙折射

英　文　名	祖国大陆名	台湾地区名
shape factor	形状系数	形種因數
shape-selective catalyst	择形催化剂	形狀選擇觸媒
shape stability	形状稳定性	形狀穩定性
shared electron pair	共价电子对	共用電子對
sharkskin	鲨鱼皮	鯊魚皮
sharp edged orifice	锐缘孔口	銳緣孔口
sharp separation	清晰分离	清晰分離
shaving	削匀	修裹
shear	剪切	剪力
shear compliance	剪切柔量	剪切柔量
shear creep	剪切蠕变	剪切蠕變
shear deformation	剪切变形	剪切形變
shear elasticity	剪切模量	切變模量
shear flexure	剪切挠曲	剪切撓曲
shear force	剪切力	剪切力
shearing	剪切	剪切
shearing disc viscometer	剪切圆盘式黏度计	剪切圓盤式黏度計
shearing stiffness	剪切刚度	剪切剛度
shearing strength	剪切强度	抗切強度, 抗剪強度
shear modulus	剪切模量	剪力係數
shear rate	剪切速率	切變速率
shear-rate thinning	切速稀化	切速稀化
shear relaxation	剪切松弛	剪切弛豫, 剪切鬆弛
shear storage modulus	剪切储能模量	剪切儲能模量
shear strain	剪切应变	剪切應變
shear strength	剪切强度	剪斷強度
shear stress	剪应力	剪應力
shear test	剪切试验	剪切試驗
shear thickening	剪切增稠	剪力增長
shear-thickening fluid	剪切增稠流体	剪切增稠流體
shear-thinning fluid	剪切稀化流体	剪切淺稠流體
shear velocity	剪切速度	剪速度
shear viscosity	剪切黏度	切變黏度, 剪切黏性
shear viscosity coefficient	剪切黏滞系数	切變黏切係數
sheathing paper	绝热[沥青]纸	絕熱紙
sheet blowing method	薄膜吹制法	薄膜吹製法
sheeter line	刨纹	切削條痕
sheeting	压片	薄層
sheet molding	片状成型	模壓成型

英　文　名	祖国大陆名	台湾地区名
sheet molding compound	片状成型料	模壓成型板材
sheet polymer	片型聚合物	片狀聚合物
shelf drier	柜式干燥机	箱形乾燥器
shelf dryer	柜式干燥机	箱形乾燥器
shellac	紫胶	蟲膠
shellac varnish	紫胶清漆	蟲膠清漆
shellac wax	紫胶蜡	蟲膠蠟
shell-and-tube heat exchanger	列管换热器	殼管熱交換器
shell-and-tube reactor	列管式反应器	殼管反應器
shell balance	壳平衡	殼均衡
shell molding	壳模铸造	殼形鑄造, 殼膜法
shell molding resin	壳模铸造[用]树脂	殼形鑄造樹脂
shell-side heat transfer coefficient	壳程传热系数	殼側熱傳係數
shell still	简单[壳式]蒸馏釜	鍋餾器
shielding	屏蔽	屏障
shielding length	屏蔽长度	屏蔽長度
shift conversion	[水煤气]变换反应	[水煤氣]轉化[反應]
shift converter	变换炉	[水煤氣]轉化器
shift factor	平移因子	平移因子, 移位因子
shifting-order reaction	变阶反应	變階反應
shift reaction	变换反应	[水煤氣]轉化反應
shift reactor	[水煤气]变换反应器	[水煤氣]轉化反應器
shikimic acid	莽草酸	莽草酸
ship bottom paint	船底漆	船底塗料
shock coagulation	冲击凝固	沖擊凝固
shock elasticity	冲击弹性	沖擊彈性
shock load	冲击载荷	震動負載
shock proof lacquer	防震涂料	防震塗料
shock wave	冲击波	震波
shock wave polymerization	冲击波聚合	沖擊波聚合
shock wave reactor	冲击波反应器	沖擊波反應器
Shore durometer	肖氏硬度计	邵氏硬度計
Shore elastometer	肖氏弹性计	邵氏彈性計
Shore hardness	肖氏硬度	邵氏硬度
short-chain branch	短支链	短支鏈
short-chain branching	短链支化	短鏈支化
shortcut design method	简捷设计法	簡捷設計法
shortcut method	简捷法	簡捷法
shortening oil	起酥油	酥脆油

英　文　名	祖国大陆名	台湾地区名
short fiber	短纤维	短纖維
short fiber composite	短纤维复合材料	短纖維複合材料
short fiber reinforced plastic	短纤维增强塑料	短纖維增強塑料
short fiber reinforcement	短纤维增强	短纖維增強
short nipple	双头螺纹短接头	短絲接管
short range	近程	近距離
short-range interaction	近程相互作用	近程
short-range intrachain crankshaft movement	近程链内曲柄运动	内近程鏈間曲軸運動
short-range intramolecular interaction	近程分子内相互作用	分子内近程相互運動
short-range order	近程有序	[鏈分子排列的]近程有序
short-range structure	近程结构	近程結構
short ripening	短期熟成	短期成熟
shortstopped polymerization	速止聚合	速止聚合
shortstopper	速止剂	速止劑
shortstopping agent	速止剂	速止劑
shot capacity	注射量	射出容量
shot cycle	注射周期	射出週期
shot rate	注射速率	射出速率
shot size	注射量	注射量
shot volume	注射体积	注射體積
shot volume controller	注射体积控制器	注射體積控制器
shot weight	注射量	注射量
shredder	撕碎机	撕碎機
shredding	水淬	水淬
shredding device	撕碎装置	撕碎裝置
shrinkage block jigs	防缩楦模	縮裂
shrinkage crack	收缩裂纹	收縮開裂
shrinkage curve	收缩曲线	收縮曲線
shrinkage jigs	防缩模	防縮模
shrinkage rate	收缩率	收縮率
shrinkage temperature	防缩温度	防縮溫度
shrinkage tension tester	收缩张力试验机	收縮張力試驗機
shrink back	回缩	退縮
shrink fixture	防缩器	防縮器
shrinking agent	收缩剂	收縮劑
shrinking machine	[热]收缩机	[熱]收縮機
shrinking power	收缩能力	收縮能力
shrinking stress	收缩应力	收縮應力

英　文　名	祖国大陆名	台湾地区名
shrink mark	缩痕	收縮標誌, 收縮皺紋
shrink resistant finish	防缩整理	防縮整理
shrunk glass	高硅氧玻璃	耐熱玻璃
shutdown control	停车控制	停工控制
shutdown inspection	停工检查	停工檢查
shutdown period	停工期	停工期
shutdown schedule	停工日程表	停工時程
shutdown time	停工时间	停工時間
shut-off head	关闭高差	關閉高差
side branch	侧[链]支	側支
side capacitance	旁容量	旁容量
side chain motion	侧链运动	側鏈運動
side chain radical	侧链基	側鏈基
side cooler	侧冷却器	側冷卻器
sidecut	侧馏分	側流餾出物
sidecut distillate	侧馏分	側流餾出物
side effect	副效应	副效應
side group	侧基	側基
side reaction	副反应	副反應
side reflux	侧回流	側回流
siderite	菱铁矿	菱鐵礦
side stream	侧流	側流, 支流
sidestream column	侧线塔	側流塔
side stream stripper	侧线[馏分]汽提塔	側流汽提塔
side stripper	侧线汽提塔	側流汽提塔
side tube	支管	支管
Sieber number	齐白值(100 份纸浆消程的有效氯)	西伯值
sieve	筛	篩
sieve analysis	筛析	篩析
sieve plate	筛板	篩板
sieve plate column	筛板塔	篩板塔
sieve plate tower	筛板塔	篩板塔
sieve shaker	振筛器	搖篩器
sieve test	筛分试验	篩分試驗
sieve tray	筛板	篩板, 篩盤
sieving	筛分	篩[分]
sifter	筛	篩
sifting machine	筛选机	篩粉機

英　文　名	祖国大陆名	台湾地区名
sight box	窥箱	窥箱
sight glass	视镜	窥鏡
sight hole	视孔	視孔
signal flow diagram	信号流程图	信號流程圖
signal flow graph	信号流程图	信號流程圖
signal smoke	信号烟	信號煙
signal transfer lag	信号传递滞后	信號傳送落後
sign convention	符号规定	符號規定
silane	硅烷	矽甲烷
silane adhesion promoter	硅烷类增黏剂	矽烷類增黏劑
silane coupling agent	硅烷偶联剂	矽烷類偶聯劑
silanol	甲硅烷醇	矽烷醇
silazane polymer	硅氮烷聚合物	矽氯烷聚合物
silica	二氧化硅	矽石, 二氧化矽
silica aerogel	硅补强剂	矽土氣凝膠
silica brick	硅砖	矽磚
silica flour	石英细粉	石英粉
silica gel	硅胶	矽膠凝體
silica glass	石英玻璃	矽石玻璃
silica refractory	硅质耐火材料	矽石耐火物
silica rock	硅岩	矽岩
silicate adhesive	硅酸盐类黏合剂	矽酸鹽類黏合劑
siliceous clay	硅质黏土	矽質黏土
siliceous earth	硅藻土	矽藻土
siliceous fireclay	硅质耐火黏土	矽質耐火黏土
siliceous fireclay brick	硅质耐火黏土砖	矽質耐火黏土磚
siliceous limestone	硅石灰石	矽質石灰石
silicon carbide	碳化硅	碳化矽
silicon carbide fiber	碳化硅纤维	碳化矽纖維
silicone adhesive	有机硅黏合剂	矽氧樹脂黏合劑
silicone elastomer	有机硅弹性体	矽酮彈性體
silicone enamel	有机硅瓷漆	聚矽氧瓷漆
silicone grease	硅[润滑]脂	矽酮脂, 聚矽氧脂
silicone nitrile rubber	氰硅橡胶	矽腈橡膠
silicone oil	硅油	矽油
silicone plastic	有机硅塑料	矽[酮]塑膠
silicone polymer	有机硅聚合物	矽氧聚合物
silicone release	有机硅脱模剂	矽[酮]脱膜劑
silicone resin	有机硅树脂	矽氧樹脂

英　文　名	祖国大陆名	台湾地区名
silicone rubber	硅橡胶	矽橡膠
silicone rubber adhesive	硅橡胶黏合剂	矽膠黏著劑
silicone rubber compound	硅橡胶混炼胶	矽氧橡膠聚合物
silicone rubber foam	泡沫硅橡胶	矽氧橡膠泡沫
silicon nitride	氮化硅	氮化矽
silicon-nitrogen polymer	硅氮聚合物	矽氮聚合物
silicon rectifier	硅整流器	矽整流器
silicon rubber	硅橡胶	矽橡膠
silicon wafer	硅片	矽晶片
silk	蚕丝	絲
silk degumming	生丝精练	生絲精練
silk fiber	丝纤维	絲纖維
silk rubber	绢丝橡胶	絹絲橡膠
silk scouring	生丝精练	絲綢精練
silk spinning	丝纺	絲紡
sillimanite	硅线石	矽線石
siloxane	硅氧烷	矽氧烷
silver necking	银颈	銀頸
silver number	银值	銀值
silvichemical	林产化学品	林材化學品
similarity	相似性	相似性, 模擬
similarity solution	相似性解	相似解
similarity theory	相似理论	相似性理論
similarity transformation	相似变换	相似變換
similar law	相似定律	相似定律
simile paper	模造纸	模造紙
simple cubic lattice	简单立方晶格	簡單立方晶格
simple elongation	简单伸长	簡單伸長, 單向伸長
simple fluid	简单流体	簡單流體
simple interest	单利	單利
simple pole	单极点	單極點
simple reaction	简单反应	簡單反應
simple shear	简单剪切	單純剪力
simple shear deformation	简单剪切形变	簡單剪切形變
simple shear flow	简单剪切流动	簡單剪切流動
simplex algorithm	单纯形算法	簡式運算[法], 簡式演算[法]
simple xanthating machine	简式黄原酸化机	簡式黃化機
simplex method	单纯形法	簡式法

英　文　名	祖国大陆名	台湾地区名
simplexometer	简易胶乳比重计	簡易膠乳比重計
simplex reciprocating pump	单缸往复泵	單缸往復泵
simple zero	单零点	單零點
simulated meat	人造肉	人造肉
simulation	模拟, 仿真	模擬, 模型化
simulation model	模拟模型	模擬模式, 模擬模型
simulator	模拟器	模擬器
simultaneous absorption	同时吸收	同時吸收
simultaneous interpenetrating network	同步互穿网络	同步互穿網絡
simultaneous operation	同时操作	同時操作
simultaneous polymerization	同时聚合	同時聚合
simultaneous reactions	同时反应	併發反應
sine response	正弦响应	正弦應答
sine wave	正弦波	正弦波
sine wave generator	正弦波发生器	正弦波產生器
single acting pump	单动泵	單動泵
single-bridged polymer	单桥聚合物	單橋聚合物
single cavity mold	单腔模具	單穴模
single cell protein	单细胞蛋白	單細胞蛋白
single channel analyzer	单道分析仪	單波道分析儀
single component	单组分	單成分
single crystal	单晶	單晶
single-crystal fiber	单晶纤维	單晶纖維
single distributed component	单分布组分	單分布成分
single effect evaporator	单效蒸发器	單效蒸發器
single flighted screw	单螺线螺杆	單絲螺桿
single impression mold	单型腔模	單模
single-input-single-output system	单输入单输出系统	單輸入單輸出系統
single liquid approximation	单一液体近似法	單一液體近似法
single-loop control	单环路控制	單環路控制
single pass conversion	单程转化率	單程轉化率
single-pass heater	单程加热器	單程加熱器
single-phase reactor	单相反应器	單相反應器
single-phase system	单相系统	單相系統
single-point determination	单点测定法	單點測定法
single reaction	单反应	單一反應
single roll crusher	单辊压碎机	單輥壓碎機
single-screw extruder	单螺杆挤出机	單軸壓出機
single-screw mixer	单螺杆混合机	單軸混合器

英　文　名	祖国大陆名	台湾地区名
single-site mechanism	单部位机理	單部位機構
single split point	单分隔点	單分隔點
single spread	单面涂布	單面塗佈
single stage centrifugal pump	单级离心泵	單級離心泵
single stage centrifuge	单级离心机	單級離心機
single stage distillation	单级蒸馏	單階蒸餾
single stage extraction	单级萃取	單階萃取
single stage operation	单级操作	單階操作
single stage pilot valve	单级导向阀	單級導引閥
single-stage process	单级过程	單階程序
single-stage resin	一步法[酚醛]树脂	一步法[酚醛]樹脂
single-strand polymer	单股聚合物	單股聚合物
single-stroke preforming press	单冲程预塑机	單衝程預塑機
single substrate reaction	单底物反应	單受質反應
single thread tester	单纱强力试验机	單紗強度試驗儀
single twist	单丝加捻	單捻線, 單紗捻度
single unit depreciation	单元折旧率	單一貶值
single yarn	单纱, 单丝	單捻紗
singular point	奇点	奇點
sinking funding	债务基金	償債基金, 債務基金
sink mark	凹痕	凹痕
sinter	烧结	燒結
sintered glass	烧结玻璃	燒結玻璃
sintered glass filter	烧结玻璃过滤器	燒結玻璃過濾器
sintered material	烧结料	燒結物料
sinter forming	烧结成型	燒結成型
sintering	烧结	燒結法
sintering zone	烧结带	燒結區
sinter membrane	烧结膜	燒結膜
sinusoidal analysis	正弦分析	正弦分析
sinusoidal deformation	正弦式变形	正弦式變形
sinusoidal disturbance	正弦扰动	正弦擾動
sinusoidal response	正弦响应	正弦應答
sinusoidal signal	正弦信号	正弦信號
sinusoidal wave	正弦波	正弦波
siphon barometer	虹吸气压计	虹吸氣壓計
siphon gauge	虹吸压力计	虹吸壓力計
siphon oiler	虹吸加油器	虹吸加油器
siphon pipe	虹吸管	虹吸管

英　文　名	祖国大陆名	台湾地区名
site-model theory	位置模型理论	位置模型理論
size analysis	粒度分析	粒度分析
size distribution	粒度分布	粒度分布
size enlargement	粒度增大	粒子增大
size factor	尺寸因子	大小因數
size ratio	尺寸比	尺寸比
size reduction	粉碎	碎解, 磨碎, 粉碎
size reduction ratio	粉碎比	磨碎程度
size separation	粒度分离	粒度分離
sizing agent	上胶剂	上膠劑
sizing machine	上浆机	上漿機
skein	绞丝	絲球
skeining	绕成绞	成絞
skeletal vibration	骨架振动	骨架振動
skewness	扭曲度	偏斜度
skim	撇去浮渣	浮渣
skim milk	脱脂奶	脱脂牛乳
skimming machine	乳油分离机	乳油分離機
skimming process	原油拔顶(蒸出轻馏分) 　　过程	粗餾程序
skim rubber	胶清橡胶	膠清橡膠
skin	皮	表皮
skin and core effect	皮心效应	皮蕊效應
skin-core structure	皮心结构	皮蕊結構
skin effect	表皮效应	表面效應
skin friction	表面摩擦	表面摩擦
skinning	结皮[现象]	結皮現象
skin temperature	表面温度	表面溫度
skiving machine	削皮机	削皮機
slabber	切块机	切塊機
slab glass	光学玻璃板	光學玻璃板
slab grating	木浆除滓机	木漿除滓機
slack coal	煤屑	碎煤
slacking test	风蚀试验	風蝕試驗
slack mercerization	松弛丝光	鬆弛絲光
slack variable	松弛变量	假擬變數
slack wax	含残油软石蜡	粗蠟
slag	炉渣	熔渣
slag action	熔蚀作用	熔蝕作用

英　文　名	祖国大陆名	台湾地区名
slag cement	炉渣水泥	熔渣水泥
slag wool	渣棉	礦渣, 溶渣
slaking	湿化	水化
slat conveyor	板式输送机	板式運送機, 板運機
slat packed tower	百叶板填充塔	填板塔
sleeve coupling	套筒联轴节	套筒連結器
sleeve joint	套管接头	套筒接頭
slenderness	细长度	細長, 微小
slenderness ratio	高径比	細長度, 長径比
slice	薄片	①薄片②平切
sliced film	平切薄膜	切片薄膜
sliced sheet	平切片材	切片
slicer	切片机	切片機
slicing	切片	切片
slide core	滑动型芯	滑動心
slide valve	滑阀	滑閥
slide vane pump	滑板式泵	滑葉泵
slidewire	滑线	滑線
slime	①黏泥②阳极泥	①黏土②陽極泥
slime control	腐浆防治	腐漿控制
slimy fermentation	黏滞发酵	黏液發酵
sling psychrometer	手摇干湿度计	搖轉濕度計
slip	滑动	①滑動②泥漿
slip band	滑移带	滑動區
slip casting process	注浆成型过程	注漿法
slip factor	滑移系数	滑動因數
slip friction	滑动摩擦	滑動摩擦
slip glaze	泥釉	泥釉
slippage factor	滑动因子	滑動因子
slip velocity	滑移速度	滑動速度
slit	割缝	狹縫
slit fiber	切膜纤维	切膜纖維
slitter	切条机	切條機
slit yarn	切膜扁丝	切膜絲
slop	①废液②污油	①廢液②污油
slot liquid seal	沟缝液体密封	溝縫液體密封
slot velocity	沟缝速度	溝縫速度
slow reaction	缓慢反应	緩慢反應
slow release	缓释	緩釋

英　文　名	祖国大陆名	台湾地区名
slow release capsule	缓释胶囊	緩釋膠囊
slow release control	缓释控制	緩釋控制
slow release device	缓释器件	緩釋裝置
slow release drug	缓释药物	緩釋藥物
slub yarn	竹节丝	竹節絲
sludge blanket	污泥层	污泥層
sludge bulking	污泥蓬松现象	污泥蓬鬆[現象]
sludge density index	污泥密度指数	活泥密度指數
sludge digestion	污泥消化	污泥消化
sludge disposal	污泥处理	污泥處理
sludge disposal process	污泥处理过程	污泥處理法
sludge filtration	污泥过滤	污泥過濾
sludge furnace	污泥焚烧炉	污泥焚燒爐
sludge incinerator	污泥焚烧炉	污泥焚燒爐
sludge process	淤泥法	污泥法
sludge promoter	淤渣生成促进剂	污泥促進劑
sludge recycle	淤渣再循环	污泥循環
sludge thickener	污泥浓缩器	污泥增稠器
sludge thickening	污泥浓缩	污泥增稠
sludge volume index	污泥体积指数	污泥體積指數
slug flow	节涌流, 弹状流, 团状流	塊狀流動, 汽胞流動, 塞流
slug flow reactor	节涌流反应器	塞流反應器
slugging tablet making	压片	乾壓製錠
slurry	浆料, 淤浆	漿體, 泥漿
slurry coating	水浆涂料	泥漿塗料
slurry piping system	浆料配管系统	漿體配管系統
slurry polymerization	淤浆聚合	漿狀聚合
slurry process	淤浆法	淤漿法
slurry reaction	浆料反应	漿體反應
slurry reaction kinetics	浆料反应动力学	漿體反應動力學
slurry reactor	浆料反应器	漿體反應器
slush molding	①中空铸型法②搪塑	①中空模型②冷凝模塑
slush pulp	纸浆[粕]液	漿[粕]液
small angle light scattering (SALS)	小角光散射	小角光散射
small angle scattering (SAS)	小角散射	小角散射
small angle X-ray diffraction (SAXD)	小角 X 射线衍射	小角 X 射线繞射
small angle X-ray scattering (SAXS)	小角 X 射线散射	小角 X 射线散射
smectic phase	晶态	層列型液晶相

英　文　名	祖国大陆名	台湾地区名
smectic state	近晶相	層列型液晶態
smectic structure	近晶型结构	層列型液晶結構
smelter	冶炼厂	①熔煉爐②冶煉廠
smelter gas	冶炼炉气	熔煉爐氣
smelter hearth	冶炼炉	熔煉爐
smeltery	冶炼厂	熔煉廠
smelting	熔炼	熔煉
smelting furnace	熔炼炉	熔煉爐
smelting pot	熔炼坩埚	熔煉坩堝
smelting process	冶炼过程	熔煉法
smoke agent	烟雾剂	發煙劑
smoke bomb	烟幕弹	煙幕彈
smoke candle	烟雾烛[缸]	發煙罐
smoked sheet	烟片[橡胶]	煙片[橡膠]
smoke generator	烟雾发生器	發煙器
smoke grenade	烟雾手榴弹	煙幕手榴彈
smokeless fuel	无烟燃料	無煙燃料
smokeless powder	无烟火药	無煙火藥, 無煙葯
smokeless propellant	无烟推进剂	無煙推進劑
smoke point	烟点	發煙點
smoke pot	发烟罐	發煙罐
smoke shell	发烟弹	煙幕彈
smoke tanning	烟鞣制	煙鞣製
smoking	熏制	燻燒
smoking test	发烟试验	發煙試驗
smoothed signal	修匀信号	修匀信號
smoothing	平滑	平滑化, 修匀
smoothing constant	修匀常数	修匀常數
smooth surface	平滑表面	平滑表面
snail	蜗结	渦輪, 渦形板
snake-cage amphoteric ion exchange resin	蛇笼型两性离子交换树脂	蛇籠型兩性離子交換樹脂
snake-cage polyelectrolyte	蛇笼型聚电解质	蛇籠型聚電解質
snapback fiber	弹性纤维	彈性纖維, 鬆緊纖維
snarl	卷缩	卷縮, 扭結
snubber roll	缓冲辊	緩衝輪
soakage	浸湿法	浸濕法
soaking factor	均热因子	均熱因數
soak test	浸泡试验	浸泡試驗

英　文　名	祖国大陆名	台湾地区名
soap content	含皂量	皂含量
soap extraction	皂液萃取	皂液萃取
soap fastness	耐皂洗[色]牢度	耐臭性
soap flake	皂片	皂片
soap micelle	皂胶束	皂膠粒
soap paste	皂糊	皂糊
soap resistance	耐皂性	耐皂性
social cost	社会成本	社會成本
soda ash	苏打灰	鹼灰, 純鹼, 碳酸鈉
soda cellulose	碱纤维素	鹼纖維素
soda feldspar	钠长石	鈉長石
soda glass	钠玻璃	鈉玻璃
soda lime	碱石灰	鹼石灰
soda-lime feldspar	钠钙长石	鈉鈣長石
soda-lime glass	钠钙玻璃	鈉鈣玻璃
soda process	烧碱法(制浆)	鹼法
soda pulp	烧碱法浆	苛性鈉紙漿
soda soap	钠[硬]皂	鈉皂
sodium alginate	藻酸钠	薄酸鈉
sodium-butadiene rubber	丁钠橡胶	丁鈉橡膠
sodium cellulose xanthate	纤维素黄酸钠	黃酸纖維素鈉
sodium ethylene sulfonate polymer	乙二磺酸钠聚合物	乙烯基磺酸鈉聚合物
sodium iron tartrate	酒石酸铁钠	酒石酸鐵鈉
sodium lamp	钠灯	鈉燈
sodium polyacrylate	聚丙烯酸钠	聚丙烯酸鈉
sodium polybutadiene rubber	丁钠橡胶	鈉聚丁二烯橡膠
sodium polymerization	钠[引发]聚合作用	鈉[引發]聚合作用
sodium rubber	丁钠橡胶	鈉橡膠
soft acid	软酸	軟酸
soft agent	软化剂	軟化劑
soft coal	烟煤	煙煤
soft detergent	软性洗涤剂	軟性清潔劑
softened water	软化水	軟化水
softener	软化剂	軟化劑
softening	软化	軟化
softening agent	软化剂	軟化劑
softening point	软化点	軟化點
softening temperature	软化温度	軟化溫度
soft fiber	软纤维	軟纖維

英　文　名	祖国大陆名	台湾地区名
soft glass	软玻璃	軟玻璃
soft grease	软润滑脂	軟潤滑脂
soft-mud process	软泥法	軟泥法
soft paraffin	软石蜡	軟石蠟
soft polymer	软质聚合物	軟質聚合物
soft rubber	软质橡胶	軟橡膠
soft segment	软段	軟段
soft segment	软[链]段	柔性鏈段
soft solid materials	半固体物料	半固體物料, 塑性物質
software	软件	軟體
soft water	软水	軟水
soft wax	软质蜡	軟蠟
softwood	软木	軟木
softwood flour	软木粉	軟木粉
soft X-ray	软性 X 射线	弱 X 射線
soggy	欠硫	欠硫, 硫化不足
Sohio process	索亥俄法	索亥俄法
soil	土壤	土壤
soil acidity	土壤酸度	土壤酸度
soil bearing pressure	土壤承受压力	泥土承受壓力
soil cement	土水泥	土壤水泥
soil conditioner	土壤改良剂	土壤改良劑
soil contamination	土壤污染	土壤受污
soil corrosion	土壤腐蚀	土壤腐蝕
soil fertility	土壤肥力	土壤肥度
soil improver of polymer	聚合物土壤改良剂	聚合物土壤改良劑
soil mechanics	土壤力学	土壤力學
soil organism	土壤微生物	土壤微生物
soil pollution	土壤污染	土壤污染
soil sol	土壤溶胶	土壤溶膠
soil stabilizer	土壤稳定剂	土壤穩定劑
soil sterilant	土壤消毒剂	土壤消毒劑
soil treatment	土壤处理	土壤處理
soil yeast	土壤酵母	土壤酵母
sol	溶胶	膠溶體
solar aquafarming	太阳能水耕	太陽能水耕
solar cell	太阳能电池	太陽電池
solar collector	太阳能集热器	太陽能收集器
solar constant	太阳常数	太陽常數

英　文　名	祖国大陆名	台湾地区名
solar distillation	太阳能暴晒蒸馏	太陽能蒸餾
solar energy	太阳能	太陽能
solar engine	太阳能发动机	太陽引擎
solar evaporation	暴晒蒸发	太陽能蒸發
solar oil	太阳油	太陽油
solar pond	太阳能贮池	太陽能貯池
solar radiation	太阳辐射	太陽輻射
solar spectrum	太阳光谱	太陽光譜
solation	溶胶化	溶膠
solder	[低温]焊料	軟焊料, 焊錫
soldering	钎焊	軟焊
soldering paste	钎焊膏	焊膏
solenoid valve	电磁阀	電磁閥
sol fraction	溶胶部分	溶膠部份
sol-gel transformation	溶胶-凝胶转换	溶膠-凝膠轉化
solid	固体	固體
solid-catalyzed reaction	固体催化反应	固體催化反應
solid concentration	固体浓度	固體濃度
solid content	固相含量	固體含量
solid filled plastic	固体填充塑料	固體填充塑料
solid flux	固体通量	固體通量
solid flux theory	固体通量理论	固體通量理論
solid fuel	固体燃料	固體燃料
solidification	凝固	凝固
solidification heat	固化热	固化熱
solidification point	凝固点	固化點
solidification process	凝固过程	固化程序
solidification rate	固化速率	固化速率
solidification rate parameter	固化速率参数	固化速率參數
solidified alcohol	固化酒精	固化酒精
solidified gasoline	固化汽油	固化汽油
solidifying point	凝固点	固化點
solid lattice	固体点阵	晶體晶格
solid-liquid separator	固液分离器	固液分離器
solid lubricant	固体润滑剂	固體潤滑劑
solid mechanics	固体力学	固體力學
solid overload	固体超载	固體超載
solid particle	固体粒子	固體粒子
solid phase	固相	固相

英　文　名	祖国大陆名	台湾地区名
solid phase polycondensation	固相缩聚	固相縮聚
solid phase polymerization	固相聚合	固相聚合
solid phase reaction	固相反应	固相反應
solid phase synthesis	固相合成	固相合成
solid polymerization	固体聚合法	固相聚合
solid propellant	固体推进剂	固體推進劑
solid-solid reaction	固固相反应	固固反應
solid-solid transition	固-固相转变	固-固相轉變
solid solution	固溶体	固溶體
solid solution alloy	固溶体合金	固溶體合金
solid state	固态	固態
solid state chemistry	固态化学	固態化學
solid-state diffusion	固态扩散	固態擴散
solid state physics	固态物理	固態物理
solid-state polyelectrolyte	固态高分子电解质	固態高分子電解質
solid-state polymerization	固相聚合	固態聚合
solid-state radiation-initiated polymerization	辐射引发的固态聚合	固態輻射引發聚合
solid suspension	固体悬浮体	固體懸浮物
solid transport	固体输送	固體輸送
solidus	固相线	固相線
solid waste	固体废物	固體廢棄物
solid waste disposal	固体废弃物处理	固體廢棄物處理
solid waste landfill	固体废弃物掩埋场	固體廢棄物掩埋場
sol rubber	溶胶橡胶	溶橡膠
solubility	①溶解度②可溶性	①溶解度②可溶性
solubility analysis	溶度分析	溶度分析
solubility barrier	溶解度隔板	溶解度障壁
solubility curve	溶线	溶度曲線
solubility fractionation	溶度分级	溶度分級
solubility limit	溶度极限	溶度極限
solubility parameter	溶度参数	溶度參數
solubility product	溶度积	溶度積
solubility product constant	溶[解]度积常数	溶[解]度積常數
solubility-temperature curve	溶解度-温度曲线	溶解度-溫度曲線
solubility test	溶解度试验	溶度試驗
solubilization	增溶作用	增溶作用
solubilization chromatography	增溶色谱法	增溶色譜法
solubilizer	增溶剂	增溶劑

英　文　名	祖国大陆名	台湾地区名
solubilizing agent	增溶剂	助溶剂
solubilizing reaction	增溶反应	增溶反應
soluble gum	可溶性胶	溶性膠
soluble polymer	可溶性聚合物	可溶性聚合物
soluble resin	可溶性树脂	可溶樹脂
soluble RNA (=sRNA)	可溶性核糖核酸	可溶性核糖核酸, 可溶性 RNA
soluble starch	可溶性淀粉	可溶澱粉
soluble starch	可溶性淀粉	可溶澱粉
solute	溶质	溶質
solution	溶液	溶液
solution birefringence	溶液双重折射	溶液雙折射
solution casting	溶液浇铸	溶液澆鑄
solution chemistry	溶液化学	溶液化學
solution polymerization	溶液聚合	溶液聚合
solution pressure	溶解压	溶解壓力
solution spinning	溶液纺丝	溶液紡絲
solution styrene-butadiene rubber	溶液丁苯橡胶	溶液丁苯橡膠
solution tension	溶解张力	溶解張力
solvability	可解性	可溶解性, 溶劑合性
solvate	溶剂合物	溶劑合物
solvating effect	溶剂化效应	溶劑化效應
solvating plasticizer	溶剂化增塑剂	溶劑化塑化劑
solvating power	溶剂化能力	溶劑化能力
solvation	溶剂化	溶合作用
solvation effect	溶剂化效应	溶劑合效應
solvation energy	溶剂化能	溶劑合能
solvent	溶剂	溶劑
solvent assisted dyeing	溶剂助染	溶劑助染劑
solvent based adhesive	溶剂基黏合剂	溶劑基黏合劑
solvent based coating	溶剂基涂料	溶劑基涂料
solvent cast process	溶剂流铸法	溶劑流鑄法
solvent cement	溶剂胶浆	液狀凝固劑
solvent cracking	溶剂裂化	溶劑裂解
solvent crazing	溶剂裂纹法	溶劑銀紋
solvent deashing	溶剂去灰分法	溶劑去灰分[法]
solvent deashing process	溶剂去灰分过程	溶劑去灰分程序
solvent dyeing	溶剂染色	溶劑染色
solvent effect	溶剂效应	溶劑效應

英　文　名	祖国大陆名	台湾地区名
solvent etching	溶剂蚀刻	溶劑腐蝕
solvent extraction	溶剂萃取	溶劑萃取
solvent-free basis	无溶剂基准	無溶劑基準
solvent front	溶剂锋面	溶劑鋒面
solvent-gradient elution	溶剂梯度洗脱	溶劑梯度洗脱
solvent hold-up	溶剂滞留量	溶劑滯留
solvent impregnated resin	浸渍树脂	浸漬樹脂
solvent interaction parameter	溶剂相互作用参数	溶劑相互作用參數
solvent lamination	溶剂复合	溶劑複合
solventless adhesive	无溶剂黏合剂	無溶劑黏合劑
solventless coating	无溶剂涂料	無溶劑塗料
solventless paint	无溶剂漆	無溶劑漆
solvent naphtha	溶剂滤油	溶劑油
solvent process	溶剂法	溶劑法
solvent-refined coal	溶剂精炼煤	溶劑精煉煤
solvent refining	溶剂精制	溶劑精煉[法]
solvent resistance	耐溶剂性	耐溶劑, 耐溶性
solvent-segment interaction	溶剂-链节相互作用	溶劑敏性黏合劑
solvent sensitive adhesive	溶剂感敏性黏合剂	易溶膠黏性
solvent separator	溶剂分离器	溶劑分離器
solvent spinning	溶剂纺丝	溶劑紡絲
solvent spun fiber	溶纺纤维	溶紡纖維
solvent strength	溶剂浓度	溶劑濃度
solvent test	溶剂试验	溶劑試驗
solvent tolerance	溶剂最大容限	溶劑容許溶解力
solvent type adhesive	溶剂型黏合剂	溶劑型黏合劑
solvent type plasticizer	溶剂型增塑剂	溶劑型增塑劑
solvent welding	溶剂黏接	溶劑黏接
solvolysis	溶剂分解作用	溶劑分解
solvolysis reaction	溶剂化反应	溶劑分解反應
solvus	固溶线	[固]溶線
sonic flow	声速流动	音速流動
sonic velocity	声速	音速
sonoluminescence	声致发光	聲發光
sonolysis	超声波分解	超音波分解
sophistication	复杂性	摻雜
sorbent	吸着剂	吸著劑
sorbitan monooleate	山梨糖醇酐单油酸酯	去水山梨糖醇三硬脂酸酯

英　文　名	祖国大陆名	台湾地区名
sorption	吸附	吸附, 吸著
sorption hysteresis	吸附滞后现象	吸附滯後現象
sorption isotherm	吸附等温线	吸附等溫線
sorption of water	吸湿	吸濕, 水吸著
sorption ratio	吸附比	吸附比
sorter	[纤维长度]分析器	[纖維長度]分析器
sorting	[纤维]分级	[纖維]分級
sound-insulating adhesive	隔声黏合剂	隔音黏合劑
sound-insulating material	隔声材料	隔音材料
sound-proof coating	防声涂料	防聲塗料
source	源	源
sour fermentation	酸败发酵	酸敗發酵
sour gas	酸气	酸氣
souring	酸败	酸腐
sour water	酸水	酸水
sour-water stripping	酸水汽提	酸水汽提
soybean fiber	大豆纤维	大豆蛋白纖維
soybean oil	大豆油	黄豆油
soybean protein	大豆蛋白质	大豆蛋白質
space charge	空间电荷	空間電荷
space charge limited current (SCLC)	空间电荷限制电流	空間電荷限制電流
space conditioning	空间空调	空間空調
space cooling	空间冷却	空間冷卻
space factor	空间系数	空間因數
space group	空间群	空間群
space heating	环流空间供暖	空間加熱
space lattice	空间点阵	空間晶格
space network polymer	立体网状聚合物	立體網狀聚合物
space polymer	立体聚合物	立體聚合物
space time	时空	空間時間
space velocity	空间速率	空間速度
spacing	间隔	間隔, 距離
spall	散裂	散裂
spallation	散裂	散裂
spalling	散裂	散裂
spalling resistance	抗散裂强度	抗散裂強度
spalling test	散裂试验	散裂試驗
spalling test panel	散裂试验板台	散裂試驗板台
span	跨度	寬度, 跨距

英　文　名	祖国大陆名	台湾地区名
spandex	弹力纤维	彈性纖維
spandex fiber	弹力纤维	彈性纖維, 鬆緊纖維
spar	晶石	晶石
sparged gas	鼓泡气	噴佈氣
sparged reactor	鼓泡反应器	氣泡反應器
sparger	鼓泡器	噴佈器
spark discharge	火花放电	火花放電
spark discharge detector	火花放电检测器	火花放電檢測器
spark erosion	火花电蚀	火花電蝕, 電火花腐蝕
sparse matrix	稀疏矩阵	零散矩陣
spar varnish	桅杆清漆	晶石清漆
spatial configuration	立体排列	立體排列
spatial oscillation	空间振荡	空間振盪
spatial structure	立体结构	立體結構
spatial velocity	空间速度	空間速度
spatter	飞溅	濺鍍
special adhesive	特种黏合剂	特殊膠黏劑
special coating	特种涂料	特殊塗佈
speciality chemical	专用化学品	特用化學品
speciality polymer	特殊性能高分子	特殊性能高分子
speciation	物种形成	物種形成, 物種演變
species	物种	物種
specific adhesion	特性黏合	特性黏合, 比黏合
specification	规格	規格
specification of equipment	设备规格表	設備規格
specific cake resistance	比滤饼阻力	比濾餅阻力
specific damping capacity	比阻尼容量	比衰減容量
specific designation	特定牌号	特定名稱
specific enthalpy	比焓	比焓
specific entropy	比熵	比熵
specific extinction coefficient	比消光系数	比消光係數, 吸光係數
specific gravity	比重	比重
specific gravity bottle	比重瓶	比重瓶
specific growth rate	比生长速率	比生長速率
specific heat	比热	比熱
specific humidity	比湿度	比濕度
specific impulse	比冲量	比衝量, 比推力
specific inductive capacity	介电常数	比介電常數
specific insulation resistance	比绝缘电阻	比絕緣電阻

英　文　名	祖国大陆名	台湾地区名
specificity	特异性	特定性, 特異性
specific polarization	比极化度	比極化度, 比偏光度
specific rate	比速率	比速率
specific rate coefficient	比速率系数	比速率係數
specific reaction rate	比反应速率	比反應速率
specific refractivity	折射系数	比折射率
specific resistance	比电阻	比阻力
specific rigidity	比刚性	比剛性
specific rotation	旋光率	比旋光度
specific rotatory power	比旋光度	光轉偏極係數
specific speed	比转速	特定速率
specific strength	比强度	比強度, 強度係數
specific surface	比表面	比面, 表面係數
specific surface area	比表面积	比表面積
specific tenacity	比强度	比強度, 比抗張力
specific thrust	比推力	比推力
specific viscosity	比黏度	比黏度
specific volume	比容[积]	比容
specific weight	比重	比重
speck	污点	污點
speck dyeing	斑染	斑染
spectral analysis	光谱分析	光譜化學分析
spectral transmittance	透光率	分光透射率
spectrochemistry	光谱化学	光譜化學分析
spectrogram	光谱图	光譜圖, 譜圖
spectrograph	摄谱仪	攝譜儀
spectrometer	分光计	分光計
spectrometry	光谱测定法	光譜法
spectrophotometer	分光光度计	分光光度計, 光譜儀
spectroscope	分光镜	分光鏡
spectroscopic analysis	光谱分析	光譜分析
spectroscopy	光谱学	①光譜學②光譜法
spectroturbidimetric titration	分光浊度滴定	分光濁度滴定
spectrum	谱	光譜, 譜
speed	速率	速率
Spencer method	斯潘塞[分级]法	斯彭瑟[分級]法
spent caustic	废碱	廢鹼
spent liquor	废液	廢液
spent lye	废碱液	廢鹼液

英　文　名	祖国大陆名	台湾地区名
spermaceti oil	鲸蜡油	鲸油
spermaceti wax	鲸蜡	鲸蠟
sperm oil	鲸蜡油	抹香鲸油, 鲸蜡油
spew	①溢料缝②毛刺	①溢料②毛邊
spew area	溢料面	溢料面
spew groove	溢料槽	溢料槽
spew line	溢料线	溢料線
sphere	球体	球體
spherical clusters	球状团簇	球狀團簇
spherical coordinates	球形坐标	球形坐標
spherical polar coordinates	球形极坐标	球極坐標
spherical tank	球形油罐	球形槽
sphericity	球形度	球形度
spherulite	球晶	球晶
spherulite structure	球晶结构	球晶結構
spider	纤维纲	支架
spin	自旋	自轉, 自述
spin coating	旋涂	旋轉塗佈
spin coupling	自旋偶合	自旋偶合
spin coupling constant	自旋偶合常数	自旋偶合常數
spin decoupling	自旋去偶	自旋去偶
spindle oil	锭子油	錠子油
spin-drawing	旋转延伸	旋轉延伸
spin echo	自旋回波	自轉回聲, 自轉回波
spin echo method	自旋回波法	自旋回波法, 自轉回聲法
spin-lattice relaxation	自旋晶格弛豫	自旋晶格弛豫, 自旋晶格鬆弛
spin-lattice relaxation time	自旋晶格弛豫时间	自旋-格子鬆弛时间
spinnability	可纺性	可紡性
spinneret	喷丝嘴	噴絲嘴
spinning	①纺丝②抽丝	①紡絲②抽絲
spinning acid	纺丝酸	紡絲酸
spinning bath	纺丝浴	紡絲浴
spinning bath stretch	纺丝浴拉伸	紡絲浴拉伸
spinning cabinet	纺丝信道	紡絲通道
spinning cake	[纺丝]丝饼	[紡絲]絲餅
spinning can	纺丝罐	紡絲罐
spinning cell	纺丝仓	紡絲倉, 紡絲[機]部位
spinning channel	纺丝甬道	紡絲甬道

英　文　名	祖国大陆名	台湾地区名
spinning draft	纺丝头拉伸	紡絲頭拉伸
spinning head	喷丝头	噴絲頭
spinning machine	纺丝机	紡絲機
spinning nozzle	纺丝头	紡絲帽, 噴絲帽
spinning oil	纺丝油	紡絲油
spinning pot	纺丝罐	紡絲罐
spinning solution	纺丝溶液	紡絲溶液
spinodal	亚稳单相极限线	亞穩均相極限線
spin-spin coupling	自旋-自旋偶合	自旋耦合
spin-spin relaxation	自旋-自旋弛豫	旋-旋傳遞
spin-spin relaxation time	自旋-自旋弛豫时间	旋-旋傳遞時間
spin welding	旋转焊接	旋轉焊接
spiral	螺旋	螺旋
spiral coil heater	螺旋管加热器	旋管加熱器
spiral condenser	螺旋板式冷凝器	螺旋冷凝器
spiral conveyor	螺旋运输机	螺旋運送機
spiral flow	螺旋流	螺旋流動
spiral gear pump	螺旋齿轮泵	螺旋齒輪泵
spiral plate heat exchanger	螺旋板换热器	螺旋板熱交換器
spiral polymer	螺旋形高分子	螺旋形高分子
spiral pressure spring	螺旋压力弹簧	螺旋壓力彈簧
spiral ring	螺旋环	螺旋環
spiral separator	螺旋形分离器	螺旋分離器
spiral structure	螺旋形结构	螺旋形結構
spirit	酒精	酒精
spirit varnish	醇质清漆	酒精清漆
spiropolymer	螺旋聚合物	螺旋聚合物
spitzkasten	角锥[沉淀]池	錐形選粒器
splash disc	防溅盘	防濺盤
splash plate	防溅挡板	防濺板
split feed adsorption	分馈吸附	分饋吸附
split fibre	膜裂纤维	裂散纖維
split flow	分流	分叉流動
split mold	分瓣模	組合模, 瓣合塑模
split-range control	分程调节	分開範圍控制
splitter	分流器	分裂機, 分離機
splitting	分裂	裂解
splitting yarn	裂膜纱	裂膜紗
sponge	海绵	海綿

英 文 名	祖国大陆名	台湾地区名
sponge gum	海绵胶	海綿膠
sponge iron	海绵铁	海綿鐵
sponge nickel	海绵镍	海綿鎳
sponge plastic	多孔塑料	海綿塑膠
sponge platinum	海绵铂	海綿鉑
sponge rubber	泡沫橡胶	海綿狀橡膠
spontaneous coagulation	自然凝结	自然凝固
spontaneous combustion	自燃	自燃
spontaneous crystallization	自发结晶	自發結晶
spontaneous emulsification	自发乳化	自發乳化
spontaneous evaporation	自然蒸发	自然蒸發
spontaneous nucleation	自发形核	自發成核
spontaneous polarization	自发极化	自然極化
spontaneous process	自发过程	自發程序
spontaneous reaction	自发反应	自發反應
sporadic nucleation	自发成核[作用]	自發成核[作用]
spot	斑点	斑漬, 斑點
spot analysis	斑点分析	斑點分析
spot gluing	点胶合	點膠合, 局部膠合
spot reaction	斑点反应	斑點反應
spot survey	定点观察	定點察視
spot test	斑点试验	斑點試驗
spot welder	点焊机	點焊機
spot welding	点焊	點焊
spouted bed	喷动床	噴流床
spouting velocity	喷流速度	噴流速度
spray	喷雾	①噴霧②噴霧劑
spray atomizer	喷雾雾化器	噴霧[霧化]器
spray chamber	喷雾室	噴霧室
spray chamber contactor	喷雾室接触器	噴室接觸器
spray coating process	喷涂法	噴塗法
spray column	喷雾塔	噴霧塔
spray condenser	喷雾冷凝器	噴霧冷凝器
spray cooled crystallizer	喷雾冷却结晶器	噴霧冷卻結晶器
spray cooling	喷雾冷却	噴霧冷卻結晶器
spray drier	喷雾干燥器	噴霧乾燥器
spray drying	喷雾干燥	噴霧乾燥[法]
spray drying process	喷雾干燥法	噴霧乾燥法
sprayed rubber	喷雾[制]橡胶	噴霧[法]橡膠

英　文　名	祖国大陆名	台湾地区名
sprayer	喷雾器	①噴霧器②噴漆器
spray gun	雾化器	噴槍
spraying	喷涂	噴霧
spraying drier	喷雾干燥器	噴霧乾燥器
spraying glazing	喷釉	噴釉
spray nozzle	雾化喷嘴	噴嘴
spray paint	喷漆	噴漆
spray pond	喷水池	噴水池
spray separator	喷雾分离器	噴霧分離器
spray tower	喷粉塔	噴霧塔
spread	①涂布②涂布量	①塗布②塗布量
spreadable life	可涂期	使用期
spread coater	刮涂机	刮塗機
spreading calender	涂胶压延机	塗膠壓延機, 擦膠壓延機
spring actuator	弹簧执行机构	彈簧引動器
spring balance	弹簧秤	彈簧秤
spring balanced type bell gauge	弹簧平衡型钟式压力计	彈簧均衡鐘型壓力計
spring constant	弹簧常量	彈簧常數
spring force	弹簧力	彈簧力
spring hanger	弹簧吊架	彈簧吊架
springless actuator	无弹簧执行机构	無彈簧引動器
spring loaded area meter	弹簧载荷面积计	彈簧負載面積計
spring loaded pressure regulator	弹簧载荷调压器	彈簧負載調壓器
spring loaded reducing valve	弹簧载荷减压阀	彈簧負載減壓閥
spring-mass-damper system	弹簧-质量-阻尼系统	彈簧-質量-阻尼系統
springness	弹力性	彈簧性
spring safety valve	弹簧式安全阀	彈簧安全閥
spring scale	弹簧秤	彈簧秤
spring type hardness tester	弹簧式硬度试验仪	彈簧硬度試驗儀
spring wood	春材	春材, 早材[木材]
sprinkler	洒水器	灑水器
sprinkler head	水喷头	灑水頭
sprue ejector	注残料顶杆	注殘料頂桿
spue	压铸硫化	壓鑄硫化
spun-colored	纺前染色	紡前染色
spun-dyed yarn	纺前染色丝	紡紗
spun yarn	短纤纱	短纖紗, 細紗, 精紡紗
spur gear pump	正齿轮泵	正齒輪泵
sputtering	溅射	噴濺

英　文　名	祖国大陆名	台湾地区名
sputtering deposition	溅射镀	噴濺沈積[法]
sputtering etching	溅射蚀刻	噴濺蝕刻[法]
square edge orifice	直边孔口	方緣孔口, 直邊孔口
square root extractor	开方器	開方器
square wave	方波	正方形波
squeeze pump	挤压泵	擠壓泵
squeezer	压榨机	擠壓機
sRNA (＝soluble RNA)	可溶性核糖核酸	可溶性 RNA, 可溶性核糖核酸
stabiliser	稳定剂	穩定劑
stability	稳定性	穩定性
stability analysis	稳定性分析	穩定性分析
stability condition	稳定条件	穩定條件
stability criterion	稳定性判据	穩定準則
stability limit	稳定极限	穩定極限
stability of steady state	稳态稳定性	穩態穩定性
stability poisoning	稳定性受损	持久中毒
stability region	稳定区域	穩定區[域]
stability test	稳定性试验	穩定性試驗
stabilizability	可稳性	①穩定性②穩定化度
stabilization	稳定[作用]	穩定[作用]
stabilization basin	稳定化池	穩定化槽, 穩定化池
stabilization pond	稳定化池	穩定化池
stabilization tank	稳定化槽	穩定化槽
stabilizator	①稳定器②稳定剂	①穩定器②穩定劑
stabilized gasoline	稳定汽油	穩定汽油
stabilized reformate	稳定重整油	穩定重組油
stabilizing agent	稳定剂	穩定劑
stabilizing ingredient	稳定成分	穩定成份
stable emulsion	稳定乳液	穩定乳液, 穩定乳態
stable isotope	稳定同位素	穩定同位素
stable operating condition	稳定操作条件	穩定操作條件
stable region	稳定区域	穩定區[域]
stable response	稳定响应	穩定應答
stable solution	稳定溶液	①穩定解②安定溶液
stable state	稳态	穩定狀態
stable steady state	稳定稳态	穩定穩態
stack gas	烟道气	煙道氣
stacking	堆垛	堆垛

英 文 名	祖国大陆名	台湾地区名
stacking effect	堆垛效应	堆垛效應
stack-up reactor	迭式反应器	叠式反應器
stage calculation	逐级计算	分階計算
stage contacting	逐级接触	分階接觸
stage contactor	逐级接触器	分階接觸器
stage efficiency	级效率	階效率
stage ejector	多级喷射器	多級射出器
stage model	分级模型	分階模式
stage operation	分段操作	分階操作
stage process	多级过程	分階程序
stage stripping	多级汽提	分階汽提
stage system	多级系统	分階系統
stagewise model	多级模型	分階模式
staggered air heater	拐折空气加热器	交錯空氣加熱器
staggered form	交叉式构型	對位交叉型
staggered tubes	交错管排	交錯管
stagnant boundary layer	静止边界层	静止邊界層
stagnant film	滞止膜	[停]滯膜
stagnant phase	滞止相	[停]滯相
stagnation point	驻点	停滯點
stagnation pressure	驻点压力	停滯壓力
stagnation property	驻点性质	停滯性質
stagnation temperature	驻点温度	停滯溫度
stain	①着色②生锈	①著色②生鏽
staining	着色	著色
staining method	着色法	著色法
staining test	着色试验	著色試驗
stainless steel	不锈钢	不鏽鋼
stainless-steel fiber	不锈钢纤维	不鏽鋼纖維
staircase reaction	逐级反应	逐級反應
stalactite	钟乳石	鐘乳石
stamp battery	捣矿杵组	捣礦杵組
stamp mill	捣磨机	捣碎機
stamp pad ink	打印墨	打印墨
standard	标准	標準
standard atmospheric pressure	标准大气压	標準大氣壓[力]
standard cell	标准电池	標準電池
standard cellulose	标准纤维素	標準纖維素
standard compound	标准化合物	標準化合物

英　文　名	祖国大陆名	台湾地区名
standard conditions	标准状态	標準狀態
standard consistency	标准稠度	標準稠度
standard deviation	标准差	標準偏差
standard enthalpy change of formation	标准生成焓变化	標準生成焓改變量
standard enthalpy change of reaction	标准反应焓变化	標準反應焓改變量
standard error	标准误差	標準誤差
standard gage	标准量规	標準規
standard gate	标准浇口	標準澆口
standard heat of combustion	标准燃烧热	標準燃燒熱
standard heat of formation	标准生成热	標準生成熱
standard heat of reaction	标准反应热	標準反應熱
standardization	标准化	標準化
standardization flow-sheet	标准化流程图	標準化流程圖
standard linear solid	标准线性固体	標準線性固體
standard nomenclature	标准命名法	標準命名法
standard nozzle	标准喷嘴	標準注嘴
standard opal glass	标准乳浊玻璃	標準乳濁玻璃
standard operation	标准操作	標準操作
standard operation procedure	标准操作步骤	標準操作步驟
standard orifice	标准孔口	標準孔口
standard reaction condition	标准反应条件	標準反應條件
standard screen	标准筛	標準篩
standard screwed elbow	标准螺纹弯头	標準螺旋彎頭
standard sieve	标准筛	標準篩
standard solution	标准溶液	標準溶液
standard state	标准态	標準狀態
standard state fugacity	标准态逸度	標準狀態逸壓
standard substance	基准物	標準物, 基準物
standard symbol	标准符号	標準符號
standard temperature	标准温度	標準溫度
standard temperature and pressure	标准温度和压力	標準溫度和壓力
standard test	标准试验	標準試驗
standard testing sieve	标准试验筛	標準試驗篩
standard thermometer	标准温度计	標準溫度計
stand oil	聚合油	熟油, 亞麻仁油, 聚合油
staple	人造短纤维	①人造短纖維②纖維長度
staple diagram	[纤维]长度分布图	[纖維]長度分布圖
staple fiber	切段[定长]纤维	定長短纖維

英 文 名	祖国大陆名	台湾地区名
staple fiber	短纤维	短纖維, 定長纖維, 人造棉
staple glass fiber	常产玻璃纤维	玻璃短纖維
staple length	纤维长度	纖維長度
staple rayon	人造棉	嫘縈棉
staple yarn	棉状纱	棉狀紗
starch	淀粉	澱粉, 漿糊
starch adhesive	淀粉黏合剂	澱粉黏合劑
starch based polymer	淀粉基聚合物	澱粉基聚合物
starch iodide paper	淀粉碘化物试纸	澱粉碘化物試紙
starch iodine paper	淀粉碘试纸	澱粉碘試紙
starch paper	淀粉试纸	澱粉試紙
starch paste	浆糊	漿糊
star polymer	星形聚合物	星形聚合物
starter	起动器	起動機, 起動器
starved joint	缺胶接头	缺膠接頭
starved line	缺胶层	缺膠層
state	状态	狀態
state function	状态函数	狀態函數
state of cure	硫化状态	熟化狀態
state property	物态参量	狀態性質
state variable	状态变量	狀態變數
static compliance	静态柔量	靜態柔量
static dielectric constant	静电介电常数	靜電介電常數
static electrification	带静电[作用]	帶靜電[作用]
static eliminator	静电消除器	靜電消除器
static error	静态误差	靜態誤差
static error coefficient	静态误差系数	靜態誤差係數
static fatigue	静态疲劳	靜態疲勞
static friction	静摩擦	靜摩擦
static head	静压头	靜力高差
static mixer	静态混合器	靜態混合器
static modulus	静态模量	靜態模量
static optimization	静态优化	靜態最適化
static pressure	静压	靜壓[力]
statics	静力学	靜力學
static state	静态	靜態
static test	静态试验	靜態試驗
static thermogravimetry	静态热重分析法	靜態熱重量法

英　文　名	祖国大陆名	台湾地区名
stationary growth phase	生长静止期	生長靜止期
stationary phase	静止期	靜止相
stationary platen	固定台	固定台
stationary point	驻点	穩定點
stationary state	定态	固定狀態
stationary state approximation	稳态近似	穩態近似法
statistical analysis	统计分析	統計分析
statistical chain	统计链	統計鏈
statistical coil	统计线团	統計線團
statistical copolymer	统计[结构]共聚物	統計[結構]共聚物
statistical design	统计设计	統計設計
statistical error	统计误差	統計誤差
statistical estimation	统计估计	統計估計
statistical homogeneity	统计均匀性	統計均勻性
statistical mechanics	统计力学	統計力學
statistical segment	统计链段	統計鏈段
statistical theory	统计理论	統計理論
statistical thermodynamics	统计热力学	統計熱力學
statistical unit	统计单元	統計單元
statistical weight matrix	统计权重矩阵	統計權重矩陣
statistics	统计学	統計學
statistics of rupture	断裂统计学	斷裂統計學
stator	定子	定子
Staudinger's viscosity law	斯托丁格黏度定律	斯托丁格黏度定律
steady flow	稳定流	穩流
steady motion	定常运动	恆穩運動, 穩定運動
steady periodic operation	稳态周期操作	平穩週期操作
steady-state	定态	穩態
steady state analysis	稳态分析	穩態分析
steady state approximation	稳定近似	穩態近似法
steady-state compliance	稳态柔量	穩態柔量
steady-state creep	稳态蠕变	穩態蠕變
steady-state diffusion rate	稳态扩散速率	穩態擴散速率
steady state error	稳态误差	穩態誤差
steady state flow	稳态流动	穩態流動
steady state gain	稳态增益	穩態增益
steady-state model	稳态模型	穩態模式
steady-state multiplicity	多重稳态性	多穩態性
steady state operation	稳态操作	穩態操作

英 文 名	祖国大陆名	台湾地区名
steady-state performance	稳态性能	穩態性能
steady state process	稳态过程	穩態程序
steady state response	稳态响应	穩態應答
steam	水蒸气	蒸汽, 水蒸汽
steam bath	蒸汽浴	蒸汽浴
steam boiler	蒸汽锅炉	蒸汽鍋爐
steam chamber	蒸汽室	蒸汽室
steam coil	蒸汽旋管	蒸汽旋管, 蒸汽盤管
steam condensate	蒸汽冷凝水	蒸汽冷凝液
steam condenser	蒸汽冷凝器	蒸汽冷凝器
steam cracking	蒸汽裂解	蒸汽裂解
steam cure	①蒸汽熟化②蒸汽硫化	①蒸汽固化②蒸汽硫化
steam distillation	水蒸气蒸馏	蒸汽蒸餾
steam-driven pump	蒸汽驱动泵	蒸汽驅動泵
steam drum	汽包	蒸汽鼓
steam economizer	蒸汽省热器	蒸汽省熱器
steam economy	蒸汽经济	蒸汽經濟
steam ejector	蒸汽喷射器	蒸汽射出器, 蒸汽抽氣機
steam engine	蒸汽机	蒸汽機
steam hammer	汽锤	蒸汽鎚
steam header	蒸汽总管	蒸汽集管[箱]
steaming	汽蒸	汽蒸, 蒸煮
steam jacket	蒸汽夹套	蒸汽套鍋
steam jet ejector	蒸汽喷射泵	噴汽抽氣機
steam jet pump	蒸汽喷射泵	蒸汽抽氣泵
steam power plant	蒸汽动力发电厂	蒸汽發電廠
steam purge	蒸汽吹洗	蒸汽清除
steam quality	蒸汽品质	蒸汽乾度, 蒸汽品質
steam reforming	蒸汽转化	蒸汽重組
steam regeneration	蒸汽再生法	蒸汽再生[法]
steam relief valve	蒸汽释放阀	蒸汽釋放閥
steam separator	蒸汽水分离器	蒸汽水分離器
steam sterilization	蒸汽灭菌	蒸汽滅菌
steam sterilizer	蒸汽消毒器	蒸汽滅菌器
steam stripping	汽提	蒸汽汽提
steam table	水蒸气图表	蒸汽表
steam trap	疏水器	蒸汽阱, 袪水器
steam tube	蒸汽管	蒸汽管
steam turbine	蒸汽涡轮	蒸汽渦輪

英　文　名	祖国大陆名	台湾地区名
steam valve	蒸汽阀	蒸汽閥
steam valve capacity	水蒸气阀容量	水蒸汽閥容量
steel	钢	鋼
steel fiber	钢纤维	鋼纖維
steel glass	强力玻璃	強力玻璃
steel pipe	钢管	鋼管
steepest ascent method	最速上升法	最陡上升法
steepest descent method	最速下降法	最陡下降法
steeping	①浸渍②浸染	①浸漬②浸染
steeping press	浸渍压榨机	浸漬壓榨機
stem	主干	主桿
stem fiber	韧皮纤维	韌皮纖維
stenter	展幅机	展幅機, 拉幅機
step addition polymer	逐步加成聚合物	逐步加成聚合物
step analysis	阶梯分析	階梯分析
step copolymerization	逐步共聚合	逐步共聚合
step disturbance	阶梯干扰	階梯擾動
step function	阶跃函数	階梯函數
step function response	阶跃函数响应	階梯函數應答
step input	阶跃输入	階梯輸入
step ladder polymer	分段梯形聚合物	分段梯形聚合物
step motor	步进电机	步進馬達
step reaction polymerization	逐步聚合	逐步聚合
step response	阶跃响应	階梯應答
stepwise addition polymerization	逐步加成聚合	逐步加成聚合[反應]
stepwise polymerization	逐步聚合	逐步聚合
stepwise reaction	逐步反应	逐步反應
stereoblock copolymer	立构嵌段共聚物	立構嵌段共聚物
stereocenter	立构中心	立構中心
stereochemistry	立体化学	立體化學
stereohybridization	立构杂化作用	立構雜化作用
stereoisomer	立体异构体	立體異構體
stereomicrography	立构显微照相法	立構顯微照相法
stereorandom copolymer	立构无规共聚物	無規立構共聚物
stereoregularity	立构有规性	立構規整性
stereoregular polymer	有规立构聚合物	有規立構聚合物
stereoregular polymerization	立构规整聚合	定向聚合, 立構規整聚合
stereorepeating unit	立构重复单元	立構重復單元
stereo rubber	立构规整橡胶	立構規整橡膠

英 文 名	祖国大陆名	台湾地区名
stereoselective polymerization	立构有择聚合	立構選擇聚合
stereoselectivity	立体选择性	立體選擇性
stereosequence	立体序列	立體序列
stereosequence distribution	立体序列分布	立體序列分布, 立構序列分布
stereospecific catalyst	立体有择催化剂	有規立構催化劑
stereospecific configuration	立体有规构型	有規立構組態
stereospecific copolymerization	有规立构共聚	有規立構共聚
stereospecificity	立体专一性	立構規整性
stereospecific polymer	立构规整聚合物	立構規整聚合物
stereospecific polymerization	立构规整聚合	立構規整聚合
stereospecific reaction	立体定向反应	立體特定反應
stereospecific rubber	有规立构橡胶	有規立構橡膠
stereosymmetrical homopolymer	立体对称均聚物	立體對稱均聚物
stereotacticity	立构规整度	立構規整度
stereotactic polymer	立构规整聚合物	立構規整聚合物
stereotactic polymerization	立构规整聚合	立構規整聚合
steric effect	空间效应	位阻效應, 空間效應
steric factor	空间因子	位阻因素
steric hindrance	位阻	位阻[現象], 空間障礙
steric regularity	立构规整性	立構規整性
steric restriction	空间障碍	空間障礙
steric strain	空间张力	立構張力, 立構應變
sterilization	灭菌	滅菌
sterilizer	灭菌器	滅菌器
sterol	甾醇, 固醇	固醇
stibinosiloxane polymer	锑基硅氧烷聚合物	銻硅氧烷聚合物
stickiness	黏性	黏性
sticky stage	黏态	黏態, 黏性階段
stiff chain	硬性链	硬性鏈, 剛性鏈
stiffener	硬化剂	硬化劑, 硬挺劑
stiffening agent	硬化剂	硬化劑, 硬挺[整理]劑
stiff equation	刚性方程	剛性方程式
stiff flow	难流动[性]	難流動[性]
stiff-mud process	硬泥法	硬泥法
stiffness modulus	刚性模量	剛性模量
stiffness test	刚度试验	勁度試驗
stiffness-weight ratio	比刚度	比剛度, 剛性-重量比
stilbene	芪	二苯乙烯

英　文　名	祖国大陆名	台湾地区名
stilbene dye	芪染料	二苯乙烯染料
still	蒸馏釜	蒸餾器
stimulus-response technique	受激响应技术	刺激應答法
stirred flow reactor	搅拌流反应器	攪拌流動反應器
stirred fluidized bed	搅拌流化床	攪拌流體化床
stirred reactor	搅拌反应器	攪拌反應器
stirred tank	搅拌罐	罐式攪拌
stirred-tank contactor	搅拌槽接触器	攪拌槽接觸器
stirred tank reactor	搅拌罐式反应器	罐式攪拌反應器
stirrer	搅拌器	①攪拌器②攪拌棒
stirring	搅拌	攪拌
stitch bonded fabric	缝编织物	縫編織物
stochastic analysis	随机分析	隨機分析
stochastic feature	随机特性	隨機特質
stochastic process	随机过程	概率過程
stochastic stability	随机稳定性	隨機穩定性
stock	原料	原料
stock dyeing	散纤维染色	散纖維染色
stock pile	储料堆	儲料
stoichiometric calculation	化学计量计算	化學計量計算
stoichiometric coefficient	化学计量系数	化學計量係數
stoichiometric compatibility	化学计量兼容性	化學計量相容性
stoichiometric mixture	化学计量混合物	化學計量混合物
stoichiometric number	化学计量数	化學計量數
stoichiometry	化学计量学	①化學計量學②化學計量法
stokes	斯托克斯(动力黏度单位)	斯托克斯(動力黏度單位)
Stokes approximation	斯托克斯近似法	斯托克斯近似法
Stokes line	斯托克斯线	斯托克斯線
stoneware	炻器	缸器, 陶石器
stopped flow method	停止流动法	止流法
stopping agent	阻化剂	終止劑
storage	储存器	儲存
storage battery	蓄电池	蓄電池
storage capacity	储存量	儲存容量
storage cell	蓄电池	蓄電池
storage equipment	储存设备	儲存裝置
storage equipment flowsheet	储存装置流程图	儲存裝置流程圖

英 文 名	祖国大陆名	台湾地区名
storage facility	储存设施	儲存設施
storage loss	储存损失	儲存損失
storage stability	贮存稳定性	貯存穩定性
storage tank	储槽	儲存槽
storage temperature	储存温度	儲存溫度
storage vessel	储槽	儲存容器
stored energy function	储能函数	儲能函數
Stormer's viscometer	斯氏黏度计	斯氏黏度計
stoving	烘干	烘乾
straight-chain polymer	直链聚合物	直鏈聚合物
straight dipping process	单纯浸渍法	單純浸漬法
straight-line depreciation	直线折旧	直線折舊[法]
straight run	直馏油	直餾油
straight-run distillation	直馏法	直餾
strain adhesive	应变黏合剂	應變黏合劑
strain birefringence	应变双折射	應變雙折射
strain ellipsoid	应变椭球	應變橢圓
strain energy	应变能	應變能
strain energy function	应变能函数	應變能函數
strain gauge	应变仪	應變儀
strain hardening	应变硬化	應變硬化
straining	粗滤	粗濾
straining process	粗滤过程	粗濾程序
strain optical coefficient	应变光学系数	應變光學係數
strain softening	应变软化	應變軟化
strain tensor	应变张量	應變張量
stratification	分层	成層[作用]
stratified flow	分层流	分層流動
stratified plastic	层压塑料	層壓塑料
stratum reticular	网状层	網狀層, 網狀組織
straw pulp	草纸浆	草紙漿
stray light	杂散光	雜散光
streak line	条纹线	煙線
stream function	流函数	流線函數
streaming birefringence	流动双折射	流動雙折射
streaming potential	泳动电势	流動電位
streamline	流线	流線
streamline coordinates	流线坐标	流線坐標
streamlined body	流线型物体	流線形物體

英　文　名	祖国大陆名	台湾地区名
streamlined filter	流线式过滤器	流線式濾器
streamlined flow	层流, 滞流	層流, 直線流動
streamlined valve	流线型阀	流線形閥
streamline flow	层流	流線流動, 層流
streamline motion	流线运动	流線運動
streamlining	流线型化	流線化
strength	强度	強度
strengthening agent	补强剂	補強劑
strength-to-weight ratio	比强度	比強度, 強度-重量比
streptomycin	链霉素	鏈黴素
stress	应力	應力
stress analysis	应力分析	應力分析
stress birefringence	应力双折射	應力雙折射
stress concentration	应力集中	應力集中
stress concentration factor	应力集中系数	應力集中因子
stress corrosion	应力腐蚀	應力腐蝕
stress cracking	应力开裂	應力開裂
stress crazing	应力银纹	應力銀紋
stress crystallinity	应力结晶性	應力結晶性
stress-deformation curve	应力-形变曲线	應力-形變曲線
stress distribution	应力分布	應力分布
stressed shell	预应力外壳	預應力外殼
stress ellipsoid	应力椭球体	應力橢球
stress graphitization	应力石墨化	應力石墨化
stress history	应力史	應力史
stress-induced crystallization	应力诱导结晶	應力誘導結晶
stress-induced growth	应力诱导生长	應力誘導生長
stress-induced orientation	应力诱导取向	應力誘導取向
stress-induced polarization	应力诱导极化	應力誘導極化
stress intensity factor	应力强度因子	應力強度因子
stress optical coefficient (SOC)	应力光学系数	應力光學係數
stress power	应力功率	應力功率
stress relaxation	应力弛豫	應力弛豫, 應力松弛
stress relaxation curve	应力弛豫曲线	應力弛豫曲線, 應力鬆弛曲線
stress relaxation modulus	应力弛豫模量	應力弛豫模量, 應力鬆弛模量
stress relaxometer	应力弛豫试验机	應力弛豫試驗機, 應力鬆弛試驗機

英　文　名	祖国大陆名	台湾地区名
stress relief test	应力消除试验	應力消除試驗
stress softening	应力软化	應力軟化
stress-strain behavior	应力-应变行为	應力-應變行為
stress-strain curve	应力-应变曲线	應力-應變曲線
stress-strain curve relation	应力-应变曲线关系式	應力-應變曲線關係式
stress-strain loop	应力-应变滞后圈	應力-應變滯后圈
stress-strain response	应力-应变响应	應力-應變響應
stress tensor	应力张量	應力張量
stress-transfer mechanism	应力传递机理	應力傳遞機構
stress whitening	应力致白	應力致白
stress wrinkless	无应力皱纹	無應力皺紋
stretchability	拉伸性	拉伸性
stretched membrane	拉伸膜	拉伸膜
stretch forming	拉伸成型	拉伸成型
stretching	拉伸	拉伸
stretch orientation	拉伸取向	拉伸取向
stretch rate	拉伸速率	伸展速率
stretch ratio	拉伸比	拉伸比
stretch spinning	拉伸纺丝	拉伸紡絲
stretch yarn	弹力丝	彈力絲
stringiness	起黏丝性	黏絲性,黏稠性
strip chart	长条记录纸	長條記錄紙
stripe	条纹	①條紋②加條紋
strippable coating	可剥性涂料	可剝塗料
stripper	①汽提塔②解吸塔③脱模机	汽提塔
stripper plate	脱模板	脱模板
stripping	提馏	汽提,脱除
stripping agent	①汽提剂②解吸剂③褪色剂④脱模剂	①汽提劑②解吸劑③褪色劑
stripping column	汽提塔	汽提塔
stripping factor	解吸因子	汽提因數
stripping section	提馏段	汽提段
stripping steam	汽提用蒸汽	汽提蒸汽
stripping still	汽提塔	汽提器
stripping tower	汽提塔	汽提塔,脱除塔
stroboscope	频闪仪	頻閃轉速計
stroke	冲程	衝程
strong viscose rayon	强力黏胶纤维	强力黏膠人造絲

英　文　名	祖国大陆名	台湾地区名
structural adhesive	结构黏合剂	結構黏合劑
structural birefringence	结构双折射	結構雙折射
structural composite	结构复合材料	結構複合材料
structural defect	结构缺陷	結構缺陷
structural disorder	结构无序	結構無序
structural fatigue	结构疲劳	結構疲勞
structural foam plastic	结构泡沫塑料	結構泡沫塑料
structural stability	结构稳定性	結構穩定性
structural unit	结构单元	結構單元
structural viscosity	结构黏度	結構黏度
structure analysis	结构分析	結構分析
structured fluid	有规结构流体	結構化流體
structured model	结构化模型	結構化模式
structured packing	整装填料	結構填充物
stuffing	①填料②填充剂③上脂	①填料②填充劑③上脂
stuffing box	填料箱	填料箱
stuffing box seal	填料函式密封	填料箱密封
styrenated oil	苯乙烯基化油	苯乙烯基化油
styrene	苯乙烯	苯乙烯
styrene acrylate copolymer coating	苯乙烯-丙烯酸酯共聚涂料	苯乙烯-丙烯酸酯塗料
styrene acrylonitrile copolymer	苯乙烯-丙烯腈共聚物	苯乙烯-丙烯腈共聚物
styrene butadiene random copolymer	丁苯无规共聚物	丁苯無規共聚物
styrene divinylbenzene copolymer	苯乙烯-二乙烯基苯共聚物	苯乙烯-二乙烯基苯共聚物
styrene maleic anhydride copolymer	苯乙烯-顺丁烯二酸酐共聚物	苯乙烯-順丁烯二酸酐共聚物
styrene methyl methacrylate resin	苯乙烯-甲基丙烯酸甲酯树脂	苯乙烯-甲基丙烯酸甲酯樹脂
styrene oxide polymer	氧化苯乙烯聚合物	氧化苯乙烯聚合物
styrene resin	苯乙烯树脂	苯乙烯樹脂
styrene rubber	苯乙烯橡胶	苯乙烯橡膠
subatmospheric pressure	负压	真空壓力
subbituminous coal	次烟煤	次煙煤
subcell	亚晶胞	亞晶胞
subchain motion	链段运动	鏈段運動
subcooled boiling	过冷沸腾	過冷沸騰
subcooled liquid	过冷液体	過冷液體
subcooled state	过冷状态	過冷狀態

英　文　名	祖国大陆名	台湾地区名
subcooling	过冷	過冷
suberaldehyde polymer	辛二醛聚合物	辛二醛聚合物
sublayer	次层	次層
sublimate	升华物	昇華物
sublimate drying	升华干燥法	昇華乾燥[法]
sublimation	升华	昇華
sublimation apparatus	升华装置	昇華裝置
sublimer	升华器	①昇華器②昇華材料
submerged combustion	浸没燃烧	水中燃燒
submerged condenser	浸没式冷凝器	沉式冷凝器
submerged pump	液下泵	沉式泵
submicrocrack	亚微裂纹	亞微裂紋
submicrofracture	亚微观断裂	亞微觀斷裂
submicroscopic	亚微观的	亞微觀的
submicroscopic micelle	亚微观胶束	亞微觀膠束
submicroscopic structure	亚微观结构	亞微觀結構
submolecule	亚分子	亞分子
suboptimal control	次优控制	次最適控制
suboptimal controller	次优控制器	次最適控制器
suboptimality	次优性	次最適性
subprogram	子程序	次程式
subroutine	子程序	次程式
subsonic nozzle	亚声速喷嘴	次音速噴嘴
substituent	①取代基②替代物	①取代基②替代物
substituent constant	取代基常数	取代基常數
substituent uniformity	取代基均匀度	取代基均勻度
substitute natural gas	替代天然气	替代天然氣
substitution reaction	取代反应	取代反應
substitution rule	取代法则	代換法則
substrate	底物,基质	底物,基質,基材
substrate inhibition	底物抑制	受質抑制[作用]
substrate material	基质材料	基質材料
substrate specificity	基质专一性	基質專一性
subunit	亚单元	亞單元
subunit structure	亚单元结构	亞單元結構
successive polymerization	逐次聚合	逐次聚合
successive reaction	逐次反应	逐次反應
successive tear strength	逐次撕裂强度	連續撕裂強度
sucrose	蔗糖	蔗糖

英　文　名	祖国大陆名	台湾地区名
suction filter	吸滤器	吸濾器
suction head	吸取高差	吸取高差
suction lift	吸引升液器	吸引上升
suction line	吸引管线	吸入管線
suction potential	吸引势	吸入勢
suction pressure	吸入压力	吸入壓[力]
suction pump	空吸泵	吸取泵
suction velocity	吸入速度	吸入速度
sudden expansion	突然膨胀	突然膨脹
sugar	糖	糖
sugarcane	甘蔗	甘蔗
sugarcane wax	甘蔗蜡	蔗蠟
suitability	适合性	適合性
sulfated oil	硫酸化油	硫酸化油
sulfate process	硫酸盐制纸浆法	硫酸鹽法, 牛皮紙漿法
sulfate pulp	硫酸盐纸浆	硫酸鹽紙漿, 牛皮紙漿
sulfate resistant cement	耐硫酸水泥	抗硫酸鹽水泥
sulfating	硫酸化	硫酸化
sulfation	硫酸盐化	硫酸化[反應]
sulfidation	硫化过程	加硫反應
sulfidity	硫化度	硫化度
sulfite process	亚硫酸盐法	亞硫酸鹽法
sulfite pulp	亚硫酸盐浆	亞硫酸鹽紙漿
sulfonate	磺酸盐	磺酸鹽
sulfonated oil	磺化油	磺酸化油
sulfonating agent	磺化剂	磺酸化劑
sulfonation	磺化	磺酸化[反應]
sulfonator	磺化器	磺酸化器
sulfur	硫	硫
sulfur bleach	硫漂白	硫[黃]漂白
sulfur chloride vulcanization	氯化硫溶液硫化	氯化硫溶液硫化
sulfur-containing polymer	含硫聚合物	含硫聚合物
sulfur crosslinking	硫交联	硫交聯
sulfur donor	给硫体	給硫體
sulfur donor agent	给硫剂	給硫劑
sulfur dye	硫化染料	硫化染料
sulfur elimination	脱硫	脫硫
sulfuric acid ester	硫酸酯	硫酸酯
sulfuric acid process	硫酸法	硫酸法

英　文　名	祖国大陆名	台湾地区名
sulfurization	硫化	硫化[反應]
sulfurless cure	无硫硫化	無硫硫化
sulfur vulcanization	硫[黄]硫化	硫[黄]硫化
sulphidizing	黄化	黄化, 硫化
sulphonic acid ionomer	磺酸型离子交联聚合物	磺酸離聚體, 磺酸型離子交聯聚合物
summer	加法器	相加器
summer oil	夏季油	夏油
summer wood	晚材	夏材
summing point	加和点	和點
sum over state	状态和	狀態和
sun-checking agent	抗日光龟裂剂	抗日光龜裂劑, 抗晒劑
suncrack	晒裂	晒裂
sun-cracking	晒裂	晒裂
sun-crazing	晒裂	晒裂
sun-discoloration	日晒变色	日晒變色
sunflower oil	向日葵油	向日葵油
supercalender	多辊压延机	強度軋光機
supercentrifuge	高速离心机	超速離心機
supercomputer	超级计算机	超級計算機
superconducting alloy	超导合金	超導合金
superconducting carbonitride	超导碳氮化物	超導碳氮化物
superconducting characteristic	超导特性	超導特性
superconducting generator	超导发电器	超導發生器
superconducting state	超导态	超導態
superconducting thin film	超导薄膜	超導薄膜
superconducting transition temperature	超导转变温度	超導轉變溫度
superconductive polymer	超导聚合物	超導聚合物
superconductivity	超导性	超導性
superconductor	超导体	超導體
supercooled liquid	过冷液	過冷液
supercooled vapor	过冷蒸气	過冷蒸氣
supercooling	过冷	過冷
supercritical extraction	超临界萃取	超臨界萃取
supercritical flow	超临界流	超臨界流動
supercritical gas	超临界气体	超臨界氣體
super drawing	超拉伸	超拉伸
super duty refractory	特级耐火材料	超強耐火物, 特級耐火物
superelasticity	超弹性	超彈性

英　文　名	祖国大陆名	台湾地区名
superficial velocity	表观速度	表觀速度
superfluid	超流体	超流體
superheat	过热[量]	過熱量
superheated state	过热状态	過熱狀態
superheated water	过热水	過熱水
superheater	过热器	過熱器
superheating	过热	過熱
super ion-conductive polymer	超离子导电聚合物	超離子導電聚合物
superlattice	超晶格	超晶格
superminicomputer	超小型计算机	超級小型計算機
supermolecular order	超分子有序	超分子有序
supermolecular structure	超分子结构	超分子結構
supermolecular texture	超分子织态结构	超分子織[態結]構
supermolecular transition	超分子转变	超分子轉變
supermolecule	超分子	超分子
supernatant	上清液	上澄液
superoxide	超氧化物	超氧化物
superphosphate	过磷酸钙	重過磷酸鈣
superpolymer	超高聚物	超聚合物
superposition	叠加	疊加
superposition principle	叠加原理	疊加原理
supersaturated solution	过饱和溶液	過飽和溶液
supersonic nozzle	超声速喷嘴	超音速噴嘴
supervision	监督	監督
supervisory control	监督控制	監督控制
supply	供应	供應
supply pressure	供给压力	供給壓[力]
supported bimetallic catalyst	有载体双金属催化剂	受載雙金屬觸媒
supported catalyst	有载体催化剂	受載觸媒
supported metal catalyst	有载体金属催化剂	受載金屬觸媒
supporting film	支持膜	支持膜
support plate	支承板	支撐板
support plate	托板	托板
supramolecular structure	超分子结构	超分子結構
surface abrasion	表面磨蚀	表面磨蝕, 表面磨耗
surface absorber	表面吸收器	表面吸收器
surface acid site	表面酸性部位	表面酸性部位
surface-active agent	表面活性剂	表面活性劑
surface activity	表面活性	表面活性, 界面活性

英　文　名	祖国大陆名	台湾地区名
surface adsorption	表面吸附	表面吸附
surface area	表面积	表面積
surface bonding	表面粘接	表面黏接
surface catalysis	表面催化	表面催化[作用]
surface charge	表面电荷	表面電荷
surface chemistry	表面化学	界面化學, 表面化學
surface coating	表面涂层	表面塗層
surface complex	表面络合物	表面錯合物
surface concentration	表面浓度	表面濃度
surface condenser	表面冷凝器	表面冷凝器
surface condition	表面状况	表面狀況
surface conductance	表面电导	表面傳導
surface coverage	表面覆盖度	表面覆蓋率
surface crack	表面裂纹	表面裂紋
surface cultivation	表面培养	表面培養
surface density	表面密度	表面密度
surface diffusion	表面扩散	表面擴散
surface dislocation	表面位错	表面移位
surface drag	表面曳引	表面阻力
surface emissivity	表面发射率	表面發射係數
surface energy	表面能	表面能
surface energy of fracture	断裂表面能	斷裂表面能
surface engineering	表面工程	表面工程
surface fermentation	表面发酵	表面發酵
surface flow	表面流动	表面流動
surface force	表面力	表面力
surface free energy	表面自由能	表面自由能
surface grafting	表面接枝	表面接枝
surface hardening	表面硬化	表面硬化
surface imperfection	表面缺陷	表面缺陷
surface layer	表面层	表面層
surface modification	表面改性	表面改性
surface modified fiber	表面改性纤维	表面改性纖維
surface modifier	表面改性剂	表面改性劑
surface moisture	表面水分	表面水分
surface morphology	表面形态	表面形態
surface phenomenon	表面现象	表面現象
surface polymerization	界面聚合	界面聚合
surface porosity	表面孔隙度	表面孔隙度

英 文 名	祖国大陆名	台湾地区名
surface potential	表面电位	表面電位, 表面勢
surface preparation	表面制备	表面製備
surface pressure	表面压力	表面壓[力]
surface profile	表面轮廓	表面輪廓
surfacer	二道底漆	二道底漆
surface rate	表面速率	表面速率
surface reaction resistance	表面反应阻力	表面反應阻力
surface renewal factor	表面更新因数	表面更新因數
surface renewal theory	表面更新理论	表面更新理論
surface resistance	表面电阻	表面電阻
surface resistivity	表面电阻率	表面電阻率
surface rheology	表面流变学	表面流變學
surface roughening treatment	表面粗化处理	表面粗化處理
surface roughness	表面粗糙度	表面糙度
surface science	表面科学	表面科學
surface sizing	表面施胶	表面上膠
surface structure	表面结构	表面結構
surface temperature	表面温度	表面溫度
surface tensiometer	表面张力计	表面張力計
surface tension	表面张力	表面張力
surface transcrystallinity	表面横向结晶	表面橫列結晶, 表面跨晶結晶
surface treatment	表面处理	表面處理
surface viscometer	表面黏度计	表面黏度計
surface viscosity	表面黏度	表面黏度
surface winder	表面卷取机	表面捲取機
surfactant	表面活性剂	表面活性劑
surfactant flooding	表面活性剂液泛	界面活性劑泛流[法]
surge tank	缓冲罐	緩衝槽
surge wave	涌波	湧波
surroundings	[热力学]环境	外圍
surrounding temperature	环境温度	外圍溫度
survival rate	生存率	生存率
survival theory	生存理论	生存理論
suspended catalyst	悬浮催化剂	懸浮觸媒
suspended growth	悬浮生长	懸浮生長
suspended-growth system	悬浮生长系统	懸浮生長系統
suspending agent	助悬剂	懸浮劑
suspension	①悬浮②悬浮液	懸浮

英　文　名	祖国大陆名	台湾地区名
suspension adhesive	悬浮黏合剂	懸浮黏合劑
suspension colloid	悬浮胶体	懸膠體
suspension load	悬浮物负载	懸浮物負載
suspension polymerization	悬浮聚合	懸浮聚合
suspension stabilizer	悬浮稳定剂	懸浮液安定劑
suspension system	悬浮系统	懸浮系統
suspension wire	悬浮线	懸線
suspensoid	悬浮体	懸溶膠體
sustained oscillation	自持振荡	持續振盪
sweating	发汗(增塑剂渗出)	發汗(增塑劑滲出)
sweating process	发汗工艺	發汗法
sweat out	发汗(增塑剂渗出)	發汗(增塑劑滲出)
sweep diffusion	吹扫扩散	掃掠擴散
sweetener	甜味剂	低硫天然甜味料
sweetening process	①加糖增浓法②脱臭过程③脱硫过程	脱臭程序, 脫硫程序
sweet gas	低硫天然气	脱臭氣
swell	溶胀	膨脹, 潤脹
swelling agent	溶胀剂	溶脹劑
swelling capacity	溶胀量	溶脹量
swelling power	溶胀能力	溶脹能力
swelling value	溶胀值	溶脹值
switch	开关	開關
switching line	切换线	切換線
symbiosis	①类聚效应②共生	共生, 共棲
symbol	符号	符號
symmetric membrane	对称膜	對稱膜
symmetry	对称性	對稱性
symmetry center	对称中心	對稱中心
synchro	同步机	同步儀
synchro-control transformer	同步控制变压器	同步控制變壓器
synchro-transmitter	同步传送器	同步傳送器
synchrotron	同步加速器	同步加速器
syncrude	合成原油	合成原油
syndiotactic addition	间同立构加成	間規立構加成
syndiotacticity	间同立构规整度	間規異構性
syndiotactic placement	间同立构键接	間規立構鍵接
syndiotactic polymer	间同立构聚合物	間規立構聚合物
syndiotactic polymerization	间同立构聚合	間規聚合

英　文　名	祖国大陆名	台湾地区名
syndiotactic polypropylene	间规聚丙烯	間規聚丙烯
syndiotactic sequence	间同立构序列	間規序列
syndiotactic triad	间同立构三[单]元组	間規立構三[單]元組
syndiotactic unit	间同立构单元	間規立構單元
syneresis	脱水收缩	脱水收縮
synergism	增效作用	增效, 綜效
synergist	增效剂	增效劑
synergistic	增效性的	增效性的
synergistic additive	增效添加剂	增效填加劑
synergistic effect	协同效应	增效效應
synergistic flame retardant	增效性阻燃剂	增效燃劑
synergistic mechanism	增效机理	增效機構
synergistic mixture	增效混合剂	增效混合劑
synergistic stabilizer	增效性稳定剂	增效穩定劑
synfuel	合成燃料	合成燃料
syngas	合成气	合成氣
syntan	合成鞣剂	合成鞣劑
synthesis	合成	合成
synthesis gas	合成气	合成[煤]氣
synthesis process	合成过程	合成程序
synthesis reaction	合成反应	合成反應
synthesis reactor	合成反应器	合成反應器
synthesizer	合成器	合成器
synthetic ammonia	合成氨	合成氨
synthetic boundary cell	合成界面池	合成界面池
synthetic chemistry	合成化学	合成化學
synthetic coating material	合成涂布材料	合成塗布材料
synthetic coloring agent	合成色素	合成色素
synthetic crude	合成原油	合成原油
synthetic crude oil	合成原油	合成原油
synthetic detergent	合成洗涤剂	合成洗滌劑
synthetic drying oil	合成干性油	合成乾性油
synthetic elastomer	合成弹性体	合成彈性體
synthetic fat	合成脂肪	合成脂肪
synthetic fatty acid	合成脂肪酸	合成脂肪酸
synthetic fiber	合成纤维	合成纖維
synthetic fiber paper	合成纤维纸	合成纖維紙
synthetic foam	合成泡沫塑料	合成泡沫塑料
synthetic fuel	合成燃料	合成燃料

英　文　名	祖国大陆名	台湾地区名
synthetic gas	合成气	合成[煤]氣
synthetic gasoline	合成汽油	合成汽油
synthetic gem	合成宝石	合成寶石
synthetic glycerol	合成甘油	合成甘油
synthetic gypsum	合成石膏	合成石膏
synthetic high polymer	合成高分子	合成高分子
synthetic indigo	合成靛	合成靛
synthetic latex	合成胶乳	合成膠乳
synthetic leather	合成革	合成革
synthetic macromolecule	合成高分子	合成高分子
synthetic natural gas	合成天然气	合成天然氣
synthetic natural rubber	合成天然橡胶	合成天然橡膠
synthetic nitrogenous fertilizer	合成氮肥	合成氮肥
synthetic oil	合成油	合成油
synthetic paper	合成纸	合成紙
synthetic plastic	合成塑料	合成塑膠
synthetic plasticizer	合成增塑剂	合成增塑劑
synthetic polymer	合成聚合物	合成聚合物
synthetic polypeptide	合成多肽	合成多肽
synthetic process	合成过程	合成程序
synthetic reaction	合成反应	合成反應
synthetic resin	合成树脂	合成樹脂
synthetic resin adhesive	合成树脂黏合剂	合成樹脂黏合劑
synthetic resin cement	合成树脂胶泥	合成樹脂膠泥
synthetic resin varnish	合成树脂涂料	合成樹脂塗料
synthetic rubber	合成橡胶	合成橡膠
synthetic rubber adhesive	合成橡胶黏合剂	合成橡膠黏合劑
synthetic rubber latex	合成橡胶胶乳	合成橡膠膠乳
synthetic tannin	合成单宁	合成單寧
synthetic tanning material	合成鞣料	合成鞣料
synthetic wood	合成木材	合成木材
synthetic wool	合成羊毛	合成羊毛
synthon	合成子	合成纖維
syphon seal	虹吸式密封	虹吸式密封
syringe	注射器	注射器
syrup	糖浆剂	①糖漿②漿
system	系统, 体系	系統
system analysis	系统分析	系統分析
system boundary	系统边界	系統邊界

英　文　名	祖国大陆名	台湾地区名
system dynamics	系统动力学	系統動態
system engineering	系统工程	系統工程
system function	①系统函数②系统功能	系統函數
system optimization	系统优化	系統最適化
system parameter	系统参数	系統參數
system performance	系统性能	系統性能
system stability	系统稳定性	系統穩定性

T

英　文　名	祖国大陆名	台湾地区名
table salt	食盐	食鹽
tablet	片剂	錠, 片劑
tabletting	压片	壓片
tachometer	转速计	轉速計, 轉數計
tachometer constant	转速计常数	轉速計常數
tail gas	尾气	尾氣
tailing	拖尾	尾料, 尾渣
tail liquid	尾部液体	稀溶液
tail pipe	尾管	尾管
take off point	接取点	脫離點
talc	滑石	滑石
talcum powder	爽身粉	滑石粉
tall oil	妥尔油	松油
tall-oil rosin	妥尔油松香	松香
tall-oil soap	妥尔油皂	松香油皂
tallow	牛脂	牛脂
tandem compound turbine	串联复式涡轮机	串列複式渦輪機
tangential force	切向力	切線力
tangential stress	切向应力	切線應力
tank crystallizer	槽式结晶器	槽式結晶器
tanker	油轮	油輪
tank furnace	槽炉	槽爐
tank reactor	槽式反应器	槽式反應器
tanks-in-series model	多釜串联模式	串聯槽模式
tannage	鞣制	鞣製
tannery	鞣革厂	製革廠
tannic acid	单宁酸	單寧酸, 鞣酸

英 文 名	祖国大陆名	台湾地区名
tannin	单宁	單寧, 鞣酸
tanning	制革	鞣製
tap density	振实密度, 夯实密度	振實密度
tapered aeration	渐减曝气	漸減通氣, 漸減曝氣
tapered pipe	锥形管	錐形管
tap water	自来水	自來水
tar	焦油	溚
tar acid	焦油酸	溚酸
tar distillate	焦油馏出液	溚溜出物
target efficiency	靶效率	靶效率
target tube	靶管	靶管
tariff	关税	關稅
tar oil	焦油	焦油, 溚油
tar sand	沥青砂	溚砂
tartar	酒石	酒石
tartar emetic	吐酒石	吐酒石
tartaric acid	酒石酸	酒石酸
tautomer	互变异构体	互變異構物
tautomerism	互变异构	①互變異構現象②互變異構性
tawing paste	白鞣酸	白鞣酸, 明礬糊
tax	税	稅
tax rate	税率	稅率
T-die	T 字模	T 字模
tear gas	催泪毒气	催淚毒氣
tearing	断开	撕裂
tearing strength	抗撕强度	撕裂強度
tearing stress	撕裂应力	撕裂應力
technical evaluation	技术评估	技術評估
technical service	技术服务	技術服務
technology	技术	技術
technology transfer	技术转移	技術轉移
tee	三通	T 型管
tee equivalent pipe length	T 型管件当量长度	T 型管件相關管長
teflon	聚四氟乙烯	鐵氟龍, 聚四氟乙烯
telemetering	遥测	遙測
temperature	温度	溫度
temperature bulb	温泡	溫度球莖
temperature calibration	温度校正	溫度校正

英　文　名	祖国大陆名	台湾地区名
temperature-composition diagram	温度组成图	溫度-組成[關係]圖
temperature control	温度控制	溫度控制
temperature controller	温度控制器	溫度控制器
temperature correction	温度校正	溫度修正
temperature difference	温差	溫差
temperature difference driving force	温差推动力	溫差驅動力
temperature distribution	温度分布	溫度分布
temperature-entropy diagram	温熵图	溫度-熵[關係]圖
temperature gage	温度计	溫度計
temperature gradient	温度梯度	溫度梯度
temperature-humidity chart	温湿图, T-H 图	溫濕圖
temperature indicating controller	温度指示控制器	溫度指示控制器
temperature indicator	温度指示计	溫度指示計
temperature level	温度水平	溫階
temperature measuring element	测温元件	測溫元件
temperature profile	温度[分布]剖面[图]	溫度剖面圖
temperature programmer	程序升温图	溫度規劃圖
temperature programming	程序升温	溫度規劃
temperature progression	温度进展	溫度進展, 溫度歷程
temperature range	温度范围	溫度範圍
temperature recorder	温度记录器	溫度記錄器
temperature regulator	温度调节器	溫度調節器, 調溫器
temperature runaway	飞温	溫度失控
temperature scale	温标	溫標
temperature schedule	温度程序	溫度排程
temperature swing adsorption	变温吸附	變溫吸附
temperature transmitter	温度变送器	溫度傳送器
tempered glass	钢化玻璃	回火玻璃, 強化玻璃
tempering	回火	回火
tempering air	回火空气	回火空氣
tempering coil	调温旋管	調溫旋管
tenacity	韧度	韌度
tenorite	黑铜矿	黑銅礦
tensile strength	抗拉强度	抗拉強度
tensiometer	张力计	張力針
tension	张力	張力
tension testing machine	张力试验机	張力試驗機
tensor	张量	張量
terminal	终端	①終端機②[末]端

英 文 名	祖国大陆名	台湾地区名
terminal condition	界限条件	最終條件
terminal falling velocity	终端下降速度	最終下降速度
terminal velocity	终端速度	終端速度
termination reaction	终止反应	終止反應
termolecular reaction	三分子反应	三分子反應
ternary azeotrope	三元共沸物	三元共沸液
ternary system	三元系[统], 三组分系统	三成分系統, 三元系
terpene	萜	萜
terpene oil	萜油	萜油
terramycin	土霉素	土黴素, 地靈黴素
tertiary chemical clarifier	三级化学澄清器	三級化學澄清器
tertiary oil recovery	三次采油	三次石油回收
tertiary treatment	三级处理	三級處理
tester	测试仪	試驗器
testing	试验	試驗法
test point	测试点	測試點
test procedure	测试步骤	測試步驟
test tube	试管	試管
tetracycline	四环素	四環素, 四環黴素
tetraethyl lead	四乙铅	四乙基鉛
tetramer	四聚物	四聚[合]物
textile	纺织品	①紡織品②纖維
textile fiber	纺织纤维	紡織纖維
textile finishing	织物整理	纖物整理
thawing	融化	融化
theorem	定理	定理
theoretical air	理论空气量	理論空氣[需要]量
theoretical model	理论模型	理論模式
theoretical plate	理论[塔]板	理論板
theoretical plate number	理论板数	理論板數
theoretical stage	理论级	理論階
theoretical tray number	理论板数	理論板數
theory	理论	理論
thermal alkylation	热烷化	熱烷化[反應]
thermal analysis	热分析	熱分析[法]
thermal autoxidation	热自氧化	熱自氧化
thermal balance	热[量]平衡	熱量均衡
thermal behavior	热行为	熱行為
thermal black	热炭黑	熱碳黑

英　文　名	祖国大陆名	台湾地区名
thermal boundary layer	热边界层	熱邊界層
thermal capacity	热容量	熱容量
thermal characterization	热特性	熱示性[法]
thermal conductivity	导热系数, 导热率	熱傳導係數
thermal conductivity coefficient	导热系数	熱導係數
thermal conductivity detector	热导检测器	熱傳導係數偵檢器
thermal constitutive equation	热本构方程	熱本質方程式
thermal contact resistance	接触热阻	熱接觸阻力
thermal cracking	热裂化	熱裂解
thermal cycle	热循环	熱循環
thermal deactivation	热失活	熱衰化
thermal decomposition	热分解	熱分解
thermal degradation	热降解	熱降解
thermal depolymerization	热解聚	熱解聚合
thermal devolatilization	热脱挥发分	熱去揮發物作用
thermal diffusion	热扩散	熱擴散
thermal diffusion coefficient	热扩散系数	熱擴散係數
thermal diffusivity	热扩散系数, 导温系数	熱擴散係數
thermal dissipation	热耗散	熱散逸
thermal dissociation	热离解	熱解離
thermal efficiency	热效率	熱效率
thermal endurance	耐热性	耐熱性
thermal energy	热能	熱能
thermal energy equation	热能方程式	熱能方程式
thermal equilibrium	热平衡	熱平衡
thermal expansion	热膨胀	熱膨脹
thermal expansion coefficient	热膨胀系数	熱脹係數
thermal fission	热裂变	熱裂变
thermal insulation	隔热	隔熱
thermal method	热方法	熱分析法
thermal pollution	热污染	熱污染
thermal polymerization	热聚合	熱聚合[反應]
thermal precipitation	热沉降	熱沉降
thermal radiation	热辐射	熱輻射
thermal reforming	热重整	熱重組
thermal resistance	热阻	熱阻力
thermal shock	热震	熱震
thermal shock test	热冲击试验	熱震實驗
thermal source	热源	熱源

英　文　名	祖国大陆名	台湾地区名
thermal stability	热稳定性	熱安定性
thermal strain	热应变	熱應變
thermal stress	热应力	熱應力
thermal synthesis	热合成	熱合成
thermal technique	热方法	熱分析法
thermal unit	热单位	熱單位
thermal value	热值	熱值
thermal value test	热值试验	熱值試驗
thermal wave	热波	熱波
thermal well	热套管	熱套管
thermistor	热敏电阻器	熱阻體, 熱阻器, 電熱調節器
thermite	铝热剂	鋁熱劑
thermoacoustimetry	热声强测量法	熱測音法
thermoammeter	热安培计	熱偶安培計
thermoanalysis	热分析	熱分析[法]
thermoanalytical method	热分析法	熱分析法
thermoanalytical microscopy	热分析显微镜	熱分析顯微鏡
thermoanalyzer	热分析器	熱分析儀
thermoatomic process	热原子法	熱原子法
thermobalance	热天平	熱天平
thermobattery	热电池	熱電池
thermochemistry	热化学	熱化學
thermo-conductivity cell	热传导率测定槽	熱傳導係數測定槽
thermocouple	热电偶	熱電偶
thermocouple pyrometer	热电偶高温计	熱電偶高溫計
thermocouple reference junction	热电偶参考接点	熱電偶參考接點
thermocouple type anemometer	热电偶式流速计	熱電偶式流速計
thermocurrent	热电流	熱電流
thermodiffusion	热扩散	熱擴散
thermodynamic analysis	热力学分析	熱力分析
thermodynamic analysis of process	过程热力学分析	程序熱力分析
thermodynamic characteristic function	热力学特性函数	熱力學特性函數
thermodynamic consistency	热力学一致性	熱力一致性
thermodynamic consistency test	热力学一致性检验	熱力一致性試驗[法]
thermodynamic constitutive equation	热力学本构方程	熱力本質方程式
thermodynamic diagram	热力学图	熱力圖
thermodynamic efficiency	热力学效率	熱力效率
thermodynamic equilibrium	热力学平衡	熱力學平衡

英　文　名	祖国大陆名	台湾地区名
thermodynamic flux	热力学通量	熱力學通量
thermodynamic force	热力学力	熱力學力
thermodynamic function	热力学函数	熱力學函數
thermodynamic probability	热力学概率	熱力擴散
thermodynamic property	热力学性质	熱力學性質
thermodynamic property table	热力学性质表	熱力性質表
thermodynamic relation	热力学关系	熱力關係
thermodynamics	热力学	熱力學
thermodynamic system	热力学系统	熱力學系統
thermodynamic temperature	热力学温度	熱力學溫度
thermodynamic temperature scale	热力学温标	熱力溫標
thermo-economics	热经济学	熱經濟學
thermoelectric effect	温差电效应	熱電效應
thermoelectricity	热电	熱電
thermoelectric junction	热电偶接点	熱電[偶]接點
thermoelectric potentiometer	热电式电位计	熱電[式]電位計
thermoelectric pyrometer	热电高温计	熱電高溫計
thermoelectric thermometer	热电温度计	熱電溫度計
thermoelectrometer	热电计	熱電計
thermoelectrometry	热电法	熱電法
thermofixation	热固定	熱固定
thermoforming	热成型	熱重組, 熱壓成型
thermogalvanometer	热电流计	熱電流計
thermogelling	热胶凝	熱膠凝
thermogravimetric analysis	热重量分析	熱重量分析[法]
thermogravimetric analyzer	热重量分析器	熱重量分析儀
thermogravimetry	热重法	熱重量分析法
thermokinematics	热运动学	熱運動學
thermoluminescence	热释发光	熱發光
thermomagnetic analysis	热磁分析	熱磁分析[法]
thermomagnetism	热磁分析	熱磁分析[法]
thermomechanical analysis	热机械分析	熱機械法
thermomechanometry	热机械法	熱機械法
thermometer	温度计	溫度計
thermometric titration	热滴定[法]	溫度滴定[法]
thermometric titrimetry	温度滴定法	溫度滴定[分析]法
thermometry	计温学	測溫法
thermonuclear reaction	热核反应	熱核反應
thermooptometry	热光学法	熱光學法

英 文 名	祖国大陆名	台湾地区名
thermoosmosis	热渗透	熱滲透[作用]
thermooxidative aging	热氧化老化	熱氧化老化
thermooxidative degradation	热氧化降解	熱氧化降解
thermopane	双层隔热玻璃板	雙層隔熱玻璃板
thermoparticulate analysis	热颗粒分析	熱顆粒分析[法]
thermophilic bacteria	嗜热细菌	嗜熱菌
thermophoresis	热电泳	熱泳法
thermophotometry	热光分析法	熱光[分析]法
thermophysical property	热物理性质	熱物理性質
thermopile	热电堆	熱電堆
thermoplastic	热塑性塑料	熱塑性塑膠
thermoplasticity	热塑性	熱塑性
thermoplastic polymer	热塑性聚合物	熱塑性聚合物
thermoplastic resin	热塑性树脂	熱塑性樹脂
thermopolymerization	热聚合	熱聚合[反應]
thermoreflectometry	热反射法	熱反應[分析]法
thermo regulating valve	热调节阀	熱調閥
thermoregulator	温度调节器	調溫器
thermoset	热固性塑料	熱固物
thermosetting	热固化	熱固性
thermosetting resin	热固性树脂	熱固性樹脂
thermo siphon	热虹吸管	熱虹吸
thermo siphon reboiler	热虹吸式再沸器	熱虹吸再沸器
thermosonimetry	热声法	熱聲[分析]法
thermostability	热稳性	熱安定性
thermostat	恒温器	恆溫器
thermostatics	热静力学	靜熱力學
thermostatic trap	恒温阱	恆溫阱
thermotropic liquid crystal	热致液晶	熱致性液晶
thermotropism	向热性	向熱性
thermowell	温度计套管	熱套管
thickener	增稠器,浓密机	增稠器,稠厚器,增稠器
thickening	增稠	增稠
thickening agent	增稠剂	增稠劑
thickening capacity	增稠能力	增稠能力
thickening process	增稠过程	增稠程序
thickening sludge	污泥增稠	污泥增稠
thick slurry process	浓浆法	濃漿法
Thiele modulus	蒂勒模数	蒂勒模組

英　文　名	祖国大陆名	台湾地区名
thin-boiling starch	稀糊化淀粉	稀糊化澱粉
thin-film evaporator	薄膜蒸发器	薄膜蒸發器
thin layer chromatography	薄层色谱法	薄層分析[法]
thinner	稀释剂	減黏劑
thiohydrogenolysis	硫氢解反应	硫氫解[反應]
thiohydrolysis	硫代水解	硫化氫解[反應]
thioindigo	硫靛蓝	硫靛藍
thiokol	聚硫橡胶	多硫橡膠
thiokol polymer	聚硫聚合物	多硫聚合物
thiol	硫醇	硫醇
thiolmodified rubber	硫醇改性橡胶	硫醇改質橡膠
thiostrepton	硫链丝菌肽	硫鏈絲菌
thiourea resin	硫脲树脂	硫脲樹脂
third law of thermodynamics	热力学第三定律	熱力學第三定律
third virial coefficient	第三位力系数, 第三维里系数	第三維里係數
thixotrope	触变胶	①觸變膠②流動減黏膠
thixotropic fluid	触变性流体	觸變流體
thixotropic property	触变性质	觸變性質
thixotropy	触变性	觸變性
thorite	硅酸钍矿	矽酸釷礦
three-mode control	三式控制	三式控制
three-mode controller	三式控制器	三式控制器
three-phase fluidization	三相流态化	三相流態化
three-phase fluidized bed	三相流化床	三相流化床
three-phase reactor	三相反应器	三相反應器
threshold	阈[值]	低限
threshold concentration	浓度极限	低限濃度
threshold energy	阈能	低限能量
threshold frequency	阈频[率]	低限頻率
threshold limit value	阈限值	低限值
throat	喉管	喉道
throat velocity	喉道流速	喉道流速
throttle valve	节流阀	節流閥
throttling	节流	節流
throttling band	节流带	節流帶
throttling calorimeter	节流量热计	節流卡計
throttling device	节流装置	節流裝置
throttling process	节流过程	節流程序

英 文 名	祖国大陆名	台湾地区名
throttling valve	节流阀	節流閥
throughput	通过量, 产量	產量
throwing	拉坯	拉坯
throwing power	均镀能力	電鍍能力
tie line	结线, 系线	連結線
time and motion study	工时学	工時學
time constant	时间常量	時間常數
time control	时间控制	時間控制
time delay	时延	時間遲延, 時間延遲
time dependence	时间关连	時間關連[性]
time derivative	时间导数	時間導數
time domain	时域	時間領域
time invariant system	非时变系统	非時變系統
time invarying system	不变时间系统	不變時間系統
time lag	时滞	時間落後
time line	时间线	時間線
time minimum control	时间最短控制	時間最短控制
timer	计时器	計時器
time ratio method	时间比率法	時間比率法
time scale change factor	对标换算因数	時間換算因數
time scaling	时间标度	時間標度
time schedule	时程表	時程表
time series	时间序列	時間序列
time series model	时间序列模型	時間序列模式
time sharing	分时	分時
time switch	定时开关	定時開關
time value of money	资金的时间价值	貨幣的時間價值
time variant system	时变系统	時變系統
time varying system	时变系统	時變系統
tinfoil	锡箔	錫箔
tin mordant	锡媒染剂	錫媒染劑
tinning	镀锡	鍍錫
tin plate	镀锡板	馬口鐵
tin plating	电锡镀	鍍錫
tin soap	锡皂	錫皂
tip speed	浆尖速度	尖端速度
tire	轮胎	輪胎
tire cord	轮胎帘线	輪胎簾布
tire yarn	轮胎砂	輪胎砂

英　文　名	祖国大陆名	台湾地区名
tissue	组织	組織
titania	二氧化钛	氧化鈦
titania porcelain	钛瓷	鈦瓷[器]
titania whiteware	钛白陶	鈦白陶[器]
titanium alloy	钛合金	鈦合金
titanium enamel	钛搪瓷	鈦搪瓷
titanium sponge	海绵钛	鈦海綿
titanium white	钛白	鈦白
titration	滴定	滴定[法]
titrimeter	滴定计	微滴定
titrimetric method	滴定[分析]法	滴定[分析]法
titrimetry	滴定[分析]法	滴定[分析]法
toilet soap	香皂	香皂
tolerance	容差	公差, 容許度
tolerance interval	容许间隔	容許間隔
top phase	顶相, 上相	頂相
topping	拔顶蒸馏	直餾
top plate	顶板	頂板
torque	转矩	扭矩
torque rheometer	转矩流变仪	扭矩流變儀
torque tube	转矩管	扭犛管
torr	托(压力单位)	托(壓力單位)
torsion	扭转	扭轉, 扭力
torsional braid analysis	扭辫分析	扭編分析
torsional creep	扭转蠕变	扭轉潛變
torsional flow	扭转流动	扭轉流動
torsional stress	扭转应力	扭轉應力
torsion balance	扭力天平	扭力天平
torsion viscometer	扭力黏度计	扭力黏度計
tortuosity	曲折因子	扭曲度
tortuosity factor	弯曲因子	扭曲[度]因數
total capital investment	总资本投资额	總資本投資額
total derivative	全导数	全導數
total emissivity	总发射系数	總發射係數
total energy	总能	總能
total hardness	总硬度	總硬度
total head	总压头	總高差
total organic carbon	总有机碳	總有機碳
total pressure	总压力	總壓[力]

英 文 名	祖国大陆名	台湾地区名
total pressure method	总压法	總壓法
total recycle process	全部循环过程	全部循環程序
total reflux	全回流	全回流
total solid	总固体	總固體[量]
toughened glass	钢化玻璃	韌化玻璃
toughened polystyrene	韧性聚苯乙烯	韌化聚苯乙烯
toughness	韧度	韌度
tower	塔器	塔
tower hold-up	塔贮留量	塔貯留量
tower packing	塔填充物	塔填充物
toxic chemical	毒性化学品	毒性化學品
toxic gas	毒气	毒氣
toxicity	毒性	毒性
toxic material	有毒材料	有毒材料, 毒物
toxic smoke	毒烟	毒煙
toxin	毒素	毒素
T-piece	三通	三通
trace element	痕量元素	微量元素
tracer	示踪剂	示蹤劑, 示蹤器
tracer chemistry	示踪化学	示蹤[劑]化學
tracer curve	示踪曲线	示蹤劑應答曲線
tracer information	示踪信息	示蹤劑資訊
tracer isotope	示踪同位素	示蹤同位素
tracer response curve	示踪响应曲线	示蹤劑應答曲線
tracer technique	示踪技术	示蹤劑法
tracing contour	追踪等值线	追蹤等能線
tracing paper	描图纸	描圖紙
tracing steam	随管加热蒸汽	隨管加熱蒸汽
tragacanth gum	黄芪胶	紫雲英樹膠
trailing vortex	拖尾旋涡	拖尾旋渦
transaddition	反式加成	反式加成[反應]
transalkylation	烷基转移	轉烷化[反應]
transcrystallization	横向结晶	橫向結晶
transesterification	酯交换	轉酯化[反應]
transfer	传递	傳送, 轉移
transfer coefficient	传递系数	傳送係數
transfer function	传递函数	轉移函數, 轉換函數
transfer lag	传递滞后	傳送落後
transfer mold	传递成型	轉移模製

英　文　名	祖国大陆名	台湾地区名
transfer unit	传递单元	傳送單元
transform	变换	變換式
trans-form	反式[异构体]	反式
transformation	转化	變換, 轉換
transformer	变压器	變壓器
transient behavior	瞬态行为	暫態行為
transient diffusion	瞬态扩散	暫態擴散
transient phenomenon	瞬态现象	暫態現象
transient response	瞬态响应	暫態應答
transient response analysis	瞬态响应分析	暫態應答分析
transient state	暂态, 瞬态	暫態
transistor	晶体管	電晶體
transition element	过渡元素	過渡元素
transition flow	过渡流	過渡流動
transition length	过渡长度	過渡長度
transition metal	过渡金属	過渡金屬
transition point	转变点	轉變點
transition range	转变范围	轉變範圍
transition region	过渡区	過渡區[域], 轉變區[域]
transition state	过渡态	過渡狀態
transition-state theory	过渡态理论	過渡狀態理論
transition temperature	转变温度	轉移溫度
translational energy	平移位能	平移能
translational partition function	平动配分函数	移動配分函數
translation of function	函数平移	函數平移
translucence	半透明度	①半透明度②半透明性
transmethylation	甲基转移作用	轉甲基[反應]
transmission lag	传动滞后	傳送落後
transmission line	传输线路	傳送線路
transmissivity	透射率	透射係數
transmittance	透光率	①透光度②透射係數
transmitter	变送器	傳送器, [信號]發設機
transmitter gain	变送器增益	傳送氣增益
transparency	透明度	透明度
transparent glaze	透明釉	透明釉
transparent soap	透明皂	透明皂
transpiration	蒸腾	蒸散[作用], 發汗[現象]
transpiration cooling	发散冷却	蒸散冷卻
transportation	运输	運輸, 輸送

英 文 名	祖国大陆名	台湾地区名
transportation lag	运输滞后	輸送落後
transport diffusivity	传递扩散系数	輸送擴散係數
transport disengaging height	分离高度	分離高度
transport phenomenon	传递现象, 输运现象	輸送現象
transport property	传递性质	輸送性質
transverse fin	横向翅片	橫向鰭片
trap	疏水阱	①阱②閘③分離器
trapezoidal weir	梯形堰	梯形堰
traveling screen	移动筛	移動篩
tray	塔板	盤, 板
tray drier	箱式干燥器	盤乾燥器
tray efficiency	板效率	板效率
tray spacing	塔板间距	板間距
treatment device	处理设备	處理設備
treatment plant	处理工厂	處理工廠
treatment process	处理过程	①處理程序②處理法
trench	管沟	溝
triangular coordinates	三角坐标	三角座標
triangular diagram	三角图	三角圖
triangular notch	三角缺口	三角凹槽
triaxial stress	三轴应力	三軸應力
tricarboxylic acid cycle	三羧酸循环	三羧酸循環
trickle bed	滴流床, 涓流床	滴流床
trickle-bed reactor	滴流床反应器	滴流床反應器
trickling	滴流	滴流
trickling filter	滴滤池	滴濾池
trickling filtration	滴滤	滴濾[法]
tridiagonal matrix	三对角矩阵	三對角矩陣
trimer	三聚体	三聚物
trimolecular reaction	三分子反应	三分子反應
triple point	三相点	三相點
triple salt	三合盐	三合鹽
tripper	①倾卸装置②跳开装置	跳動裝置
trip wire	拉发线	引線
trommel	圆筒筛	礦石篩, 轉筒篩
trona	天然碱	碳酸鈉石
trough	料槽	槽
true boiling point	真沸点	真沸點
true boiling point curve	实沸点蒸馏曲线	真沸點曲線

英　文　名	祖国大陆名	台湾地区名
true density	真密度	真密度
true mean	实平均	實平均
true rate of return	实报酬率	實報酬率
true variance	真方差	實變異
tube[side]pass	管程	管程
tube bank	管排	管排
tube bundle	管束	管束
tube drier	管式干燥器	管式乾燥器
tube furnace	管式炉	管式爐
tube gage	管式气压计	管式氣壓計
tube mill	管磨机	管磨機
tube sheet	管板	管板
tube still	管式蒸馏釜	管餾器
tubing machine	制管机	製管機
tubular condenser	管式冷凝器	管式冷凝器
tubular evaporator	管式蒸发器	管式蒸發器
tubular heater	管式加热器	管式加熱器
tubular module	管式组件	管式組件
tubular reactor	管式反应器	管式反應器
tubular still	管式蒸馏釜	管餾器
tumbler mixer	转鼓混合机	轉鼓混合機
tumbling mill	滚磨机	滾磨機
tung oil	桐油	桐油
tuning	调谐	調諧
tuning parameter	调谐参数	調諧參數
tunnel drier	隧道干燥器	隧式乾燥器
tunnel kiln	隧道窑	隧式窯
turbidimeter	浊度计	濁度測定法
turbidity	浊度	濁度
turbine	涡轮	渦輪機
turbine agitator	涡轮搅拌器	渦輪攪拌器
turbine centrifugal pump	涡轮离心泵	渦輪離心泵
turbine flowmeter	涡轮流量计	渦輪[式]流量計
turbine impeller	涡轮叶轮	渦輪[式]葉輪
turbine pump	涡轮泵	渦輪泵
turbo-blower	涡轮鼓风机	渦輪鼓風機
turbo-compressor	涡轮压缩机	渦輪壓縮機
turbo-dryer	涡轮干燥机	渦輪乾燥機
turbo-grid tray	穿流栅板	渦輪格子板

英　文　名	祖国大陆名	台湾地区名
turbo-machine	涡轮机	渦輪機
turbo-pump	涡轮泵	渦輪泵
turbulence promoter	湍流促进器	紊流促進器
turbulence scale	湍流标度	紊流標度
turbulent boundary layer	湍流边界层	紊流邊界層
turbulent core	湍流核心	紊流核心
turbulent diffusion	湍流扩散	紊流擴散
turbulent diffusivity	湍流扩散系数	紊流擴散係數
turbulent-energy spectrum	湍流能谱	紊流能譜
turbulent flow	湍流, 紊流	紊流
turbulent flow reactor	湍流反应器	紊流反應器
turbulent fluidized bed	湍动流化床	湍動流化床
turbulent motion	湍流运动	紊流運動
turbulent relaxation phenomenon	湍流松弛现象	紊流鬆弛現象
turbulent shear stress	湍流剪应力	紊流剪應力
turbulent stress	湍流应力	紊流應力
turndown ratio	操作弹性	操作彈性
turnover number	转化数	觸媒變率, 酶變率
turnover ratio	周转率	週轉率
turpentine oil	松节油	松節油
tuyere distributor	风帽分布板	風帽分布板
twin screw extruder	双螺杆挤出机	雙螺桿擠出機
two-cell stabilization pond	双室稳定化池	雙室穩定化池
two-cycle engine	二行程引擎	二行程引擎
two-dimensional chromatography	二向色谱法	二維層析[法]
two-dimensional model	二维模型	二維模式
two-film theory	双膜理论	雙薄膜理論
two-fluid theory	两流体理论	兩流體理論
two-liquid theory	两液体理论	兩液體理論
two-mode control	双式控制	雙式控制
two-phase flow	两相流	兩相流動
two-phase flow pattern	两相流动型式	兩相流動型式
two phase heat transfer	二相传热	兩相熱傳
two-phase model	两相模型	兩相模式
two-phase region	两相区	兩相區[域]
two-phase system	两相系统	兩相系統
two-position actuator	双位驱动器	雙位引動器
two-position controller	双位控制器	雙位控制器
two-resistance theory	双阻力理论	雙阻力理論

英　文　名	祖国大陆名	台湾地区名
two-stage cascade system	两段串级系统	二段串級系統
two-stage distillation	两段蒸馏	二段蒸餾
two-stage pilot valve	两级导向阀	二级導引閥
two tier approach	双层法, 联立模块法	雙層法
Tyler standard sieve	泰勒标准筛	泰勒標準篩
tylosin	泰乐菌素	泰黴素
tyrocidin	短杆菌酪肽	短桿菌酪
tyrothricin	短杆菌素	短桿菌素, 混合短桿菌

U

英　文　名	祖国大陆名	台湾地区名
ultimate disposal	最终处理	最終處理
ultimate period	最终周期	最終週期
ultimate periodic response	最终周期应答	最終週期應答
ultimate periodic solution	最终周期解	最終週期解
ultimate sensitivity	极限灵敏度	極限靈敏度
ultimate stress	极限应力	極限應力
ultra accelerator	超加速剂	超催速劑
ultracentrifugation	超速离心	超速離心
ultracentrifuge	超[速]离心机	超離心機
ultrafilter	超滤机	超過濾器
ultrafiltration	超滤	超過濾
ultragrinder	超细研磨机	超細研磨機
ultramarine	群青	群青
ultramarine blue	群青蓝	群青
ultramarine brown	群青棕	群棕
ultramarine green	群青绿	群綠
ultramarine violet	群青紫	群紫
ultramarine yellow	群青黄	群黄
ultra microchemistry	超微量化学	超微量化學
ultra microscope	超显微镜	超顯微鏡
ultra microscopy	超显微法	超顯微法
ultrapurification	超净化	超淨化, 超純化
ultrasonic agglomeration	超声附聚	超曰波黏聚
ultrasonic cleaning	超声净化	超音波清潔
ultrasonic flow	超声波流动	超音波流動
ultrasonic meter	超声波计	超音波計

英　文　名	祖国大陆名	台湾地区名
ultrasonic precipitation	超声波沉淀	超音波沈澱
ultrasonics	超声波学	超音波學
ultrasonic vibrator	超声波振动器	超音波振動器
ultrasonic viscometer	超声波黏度计	超音波黏度計
ultrathin film	超薄膜	超薄膜
ultraviolet absorber	紫外线吸收剂	紫外線吸收劑
ultraviolet absorption	紫外线吸收	紫外線吸收
ultraviolet colorimeter	紫外线比色计	紫外線比色計
ultraviolet curing	紫外线熟化	紫外線熟化
ultraviolet fluorescence	紫外线荧光	紫外線螢光
ultraviolet generator	紫外线发生器	紫外線產生器
ultraviolet lamp	紫外灯	紫外線燈
ultraviolet light	紫外光	紫外光
ultraviolet microscope	紫外线显微镜	紫外線顯微鏡
ultraviolet microscopy	紫外线显微法	紫外線顯微法
ultraviolet ray	紫外线	紫外線
ultraviolet spectrophotometer	紫外[线]分光光度计	紫外線分光光度計, 紫外線光譜儀
umber	棕土	富錳棕土
unbound water	非结合水分	非結合水
uncompetitive inhibition	反竞争抑制	不競爭抑制[作用]
unconstrained optimization	无约束优化	無約束最佳化
uncoupling	解偶联	非偶合, 解偶
undamped natural frequency	无阻尼自然频率	無阻尼自然頻率
undamped response	无阻尼响应	無阻巴應答
undamped system	无阻尼系统	無阻尼系統
undercoat	底漆	底塗
under cure	欠硫	低硫化
under cured rubber	欠硫化橡胶	低硫化橡膠
underdamped response	欠阻尼响应	欠阻尼應答
underdamped system	欠阻尼系统	欠阻尼系統
underdamping	欠阻尼	欠阻尼
underdrainage system	排水系统	排水系統
under-driven buhrstone mill	下传动式石磨机	底動石磨[機]
underflow	底流	底流
underflow rate	底流速率	底流速率
underflow velocity	底流速度	底流速度
underloading	负荷不足	負荷不足
undershoot	低于额定值	①低越量②低越[現象]

英　文　名	祖国大陆名	台湾地区名
uniaxial stress	单轴应力	單軸向應力
unicellular growth	单细胞生长	單細胞生長
unidirectional reaction	单向反应	單向反應, 不可逆反應
uniflux tray	S 形塔板	長條泡罩板
uniform coking	均匀结焦	均勻結焦
uniform conversion	均匀转化	均勻轉化
uniform conversion model	均匀转化模型	均勻轉化模型
uniform crystal	均匀晶体	均勻晶體
uniform distribution	均匀分布	均勻分布
uniform flow	均匀流	均勻流動
uniformity	均匀性	均勻性
uniformity coefficient	均匀系数	均勻係數
uniform loading	均匀载荷	均勻負荷
uniform poisoning	均匀中毒	均勻中毒
uniform stream	均匀流动	均勻流
unilateral slit	单向狭缝	單向狹縫
unimolecular reaction	单分子反应	單分子反應
union	活[管]接头	管套節
union of asymptotic stability	渐近稳定性联合	漸近穩定聯合域
uniqueness of steady state	稳态唯一性	穩態唯一性
unique steady state	单一稳态	單一穩態
unit computation	单元计算	單元計算
unit cost	单位成本	單位成本
unit cost estimation	单位成本估计	單位成本估計
unit feedback	单位回馈	單位回饋
unit operation	单元操作	單元操作
unit process	单元过程	單元程序, 單元方法
universal condenser	通用冷凝器	通用冷凝器
universal gas constant	普适气体常量	通用氣體常數
universal indicator	通用指示剂	通用指示計
universal joint	万向接头	通用接頭
universal velocity distribution	泛速度分布	泛速度分布
unleaded gasoline	无铅汽油	無鉛汽油
unloading	卸载	卸載
unmixed flow reactor	无混合流动反应器	無混合流動反應器
unox process	纯氧活性污泥法	純氧活性污泥法
unpacked-tube reactor	空管反应器	空管反應器
unreacted core model	未反应核模型	未反應核模型
unsaponified matter	不皂化物	未皂化物

英　文　名	祖国大陆名	台湾地区名
unsaturated fatty acid	不饱和脂肪酸	不飽和脂肪酸
unsaturated hydrocarbon	不饱和烃	不飽和烴
unsaturated polyester	不饱和聚酯	不飽和聚酯
unsaturation	不饱和	不飽和
unstable operating point	不稳定操作点	不穩定操作點
unstable response	不稳定响应	不穩定應答
unstable state	非稳态	不穩定狀態
unstable system	不稳定系统	不穩定系統
unsteady flow	非定常流	非穩態流動
unsteady state	非定态	非恆穩狀態
unsteady state heat transfer	非定态传热	非穩態熱傳
unsteady state operation	非定态操作	非穩態操作
unsteady state process	非定态过程	非穩態程序
unstructured model	非结构模型	末結構化模式
unsupported catalyst	无载体催化剂	非受載觸媒
upflow column	上流塔	上流塔, 上流柱
upgrading	提升	[品質]提升, 升級
upper bound	上限	上限
upper control limit	上控制限	管制上限
upper limit	上极限	上限
upstream	上游	上游
upstream pressure	上游压力	上游壓力
uptake	摄取	攝取, 升道
uptake rate	摄取速率	攝取速率
uranite	云母铀矿	瀝青鈾礦
urea	尿素	尿素
urease	脲酶	尿素酶
urethane	尿烷	胺甲酸乙酯
utility	公用设施	公用設施
utility check list	公共设施检查报表	公用設施檢查報表
utility flowsheet	公共设施流程图	公用設施流程圖
utility service	公用服务事业	公用設施
U tube	U 型管	U 形管
U-tube heat exchanger	U 形管换热器	U 形管熱交換器
U type crystallizer	U 型结晶器	U 型結晶器
UV spectrophometer	紫外光谱仪	紫外線光譜儀

V

英　文　名	祖国大陆名	台湾地区名
vaccine	疫苗	疫苗
vacuometer	真空计	真空計
vacuum	真空	真空
vacuum column	真空蒸馏塔	真空蒸餾塔
vacuum concentration	真空浓缩	真空濃縮
vacuum crystallization	真空结晶	真空結晶
vacuum crystallizer	真空结晶器	真空結晶器
vacuum dehydrator	真空脱水器	真空脱水器
vacuum desiccator	真空干燥器	真空乾燥器
vacuum discharge	真空放电	真空放電
vacuum distillation	真空蒸馏, 减压蒸馏	真空蒸餾
vacuum drum filter	真空鼓式过滤器	真空濾桶
vacuum drying	真空干燥	真空乾燥
vacuum drying apparatus	真空干燥器	真空乾燥器
vacuum drying oven	真空干燥箱	真空乾燥箱
vacuum evaporation	真空蒸发	真空蒸發
vacuum evaporator	真空蒸发器	真空蒸發器
vacuum filter	真空过滤机	真空過濾機
vacuum flash vaporization	真空闪蒸	真空驟汽化
vacuum flotation	真空浮选	真空浮選
vacuum forming	真空成型	真空成形
vacuum fractionator	真空分馏器	真空分餾器, 真空分餾塔
vacuum gage	真空计	真空計
vacuum grease	真空润滑油	真空潤滑脂
vacuum head	真空高差	真空高差
vacuum manometer	真空压力计	真空壓力計
vacuum molding	真空成型	真空成型
vacuum oil	真空油	真空油
vacuum oven	真空烘箱	真空烘箱
vacuum pan	真空锅	真空罐
vacuum press	真空压制机	真空壓製機
vacuum pressing	真空压制	真空壓製
vacuum pressure	真空压	真空壓[力]
vacuum process	真空过程	真空程序

英　文　名	祖国大陆名	台湾地区名
vacuum pump	真空泵	真空泵
vacuum relief	真空解除	真空釋放
vacuum retort	真空甑	真空甑
vacuum seal	真空密封	真空密封
vacuum shelf drier	真空盘架干燥器	真空廂乾燥器
vacuum still	真空蒸馏釜	真空蒸餾器
vacuum tar	真空焦油	真空殘渣油, 真空浴
vacuum tray drier	真空盘架干燥器	真空盤乾燥器
vacuum tube	真空管	真空管
vacuum valve	真空阀	真空閥
valve actuator	阀驱动器	閥引動器
valve body	阀体	閥主體
valve capacity	阀容量	閥容量
valve characteristics	阀特性	閥特性
valve control	阀控制	閥控制
valve equivalent pipe length	阀当量管长	閥相當管長
valve gain	阀增益	閥增益
valve hysteresis	阀滞后	閥遲滯
valve identification code	阀鉴定编码	閥鑑定簡碼
valve loss	阀损失	閥損失
valve plug	阀塞	閥塞
valve positioner	阀门定位器	閥定位器
valve rangeability	阀范围性	閥範圍性
valve resistance	阀阻力	閥阻力
valve response	阀响应	閥應答
valve sequencing	阀序列	閥定序
valve stem	阀杆	閥桿
valve tray	浮阀塔板	閥板
vanadium contact process	钒接触法	釩接觸法
vancomycin	万古霉素	萬古黴素
van der Waals adsorption	范德瓦耳斯吸附	凡得瓦吸附
van der Waals constant	范德瓦耳斯常数	凡得瓦常數
van derWaals equation	范德瓦耳斯方程	凡得瓦方程式
van der Waals equation of state	范德瓦耳斯状态方程	凡得瓦狀態方程式
van der Waals force	范德瓦耳斯力	凡得瓦力
vane flow	叶片式流动	葉輪流動
vane type blower	叶片式鼓风机	葉輪式鼓風機
vanilla	香草	香草
vanillin	香草醛	香草精

英　文　名	祖国大陆名	台湾地区名
van Laar equation	范拉尔方程	凡得拉方程
vanner	淘矿机	播洗器
van't Hoff law	范托夫定律	凡特荷夫定律
vapor	蒸气	蒸氣
vapor bypass	蒸气旁路	蒸氣傍路
vapor compression	蒸气压缩	蒸氣壓縮
vapor deposition	气相沉积	蒸氣沉積
vapor diffusion pump	蒸气扩散泵	蒸氣擴散泵
vaporization	汽化	汽化
vaporization curve	汽化曲线	汽化曲線
vaporizer	汽化器	汽化器
vapor-liquid equilibrium	汽液平衡	汽液平衡
vapor-liquid equilibrium ratio	汽液平衡比	汽液平衡比
vapor-liquid separator	汽液分离器	汽液分離器
vapor phase association	汽相缔合	汽相結合
vapor pressure	蒸气压	蒸氣壓
variability	变异性	變異性
variable	变量	變數
variable controlled	受控变量	受控變數
variable cost	变动成本	變動成本
variable orifice meter	可变孔板流量计	可變孔口計
variable-speed pump	变速泵	變速泵
variable transformer	调压变压器	可變變壓器
variant	变体	變異體
variational calculus	变分学	變分學, 變分法
varnish	清漆	清漆
vaseline	凡士林	凡士林
vat dye	还原染料	染缸
vat leaching	桶式浸取	槽式浸取
vegetable oil	植物油	植物油
velocity boundary layer	速度边界层	速度邊界層
velocity boundary layer thickness	速度边界层厚度	速度邊界層厚度
velocity-distance lag	速度距离滞后	速度距離落後
velocity distribution	速度分布	速度分布
velocity gradient	速度梯度	速度梯度
velocity head	速度压头	速度高差
velocity meter	速度计	速度計
velocity potential	速度势	速度勢
velocity profile	速度[分布]剖面[图]	速度剖面圖

英　文　名	祖国大陆名	台湾地区名
vena contracta	流颈, 缩脉	收縮口
vena contracta tap	流颈接头	收縮口壓力接頭
vent	①排空②排气孔	通氣孔, 排氣孔
vent and drain seal	排气及排泄口密封	通氣及排洩口密封
ventilation	通风	通風
ventilation requirement	通风需求	通風需求
ventilator	通风机	通風器
vent line	排气管线	排收管線
vent pipe	排气管	通風管
vent ratio	通风比例	通風比率
venture analysis	风险分析	風險分析
venture profit	风险利润	風險利潤
Venturi caliper	游标卡尺	游標卡尺
Venturi flowmeter	文丘里流量计	文氏流量計, 細腰流量計
Venturi meter	文丘里流量计	文氏計, 細腰計
Venturi nozzle	文丘里喷嘴	文氏噴嘴, 細腰噴嘴
Venturi scale	游标尺	游標尺
Venturi scrubber	文丘里洗涤器	文式洗滌器
Venturi tube	文丘里管	文氏管, 細腰管
vertical digester	立式蒸煮器	直立式蒸煮器
vertical retort	竖甑	豎甑
vertical sieve tray	垂直筛板	垂直篩板
vertical tube evaporator	竖管式蒸发器	豎管蒸發管
very large scale integrated circuit	超大型集成电路	超大型積體電路
vibrated fluidized bed	振动流化床	振動流體化床
vibrating screen	振动筛	振動篩
vibrational energy	振动能	振動能
vibrational energy level	振动能级	振動能階
vibration partition function	振动配分函数	振動分配係數函數
vibrator	振动器	振動器
vibratory feeder	振动给料器	振動飼機
vibrometer	测振仪	振動計
view factor	视因数	視因數
vinyl chloride monomer	氯乙烯单体	氯乙烯單體
vinyl compound	乙烯基化合物	乙烯系化合物
vinylon	维纶	維尼綸
vinyl pipe	聚氯乙烯管	聚氯乙烯管
vinyl polymer	烯类聚合物	乙烯系聚合物

英　文　名	祖国大陆名	台湾地区名
vinyl resin	乙烯系树脂	乙烯系樹脂
viomycin	紫霉素	群紫
virial coefficient	位力系数	維里係數
virial equation	位力方程	維里方程
virus	病毒	病毒
visbreaking	减黏裂化	減黏
viscoelastic fluid	黏弹性流体	黏彈性流體
viscoelasticity	黏弹性	黏彈性
viscometer	黏度计	黏度計
viscometric flow	测黏流动	測黏[度]流動
viscometry	黏度测定法	黏度測定法
viscoplastic fluid	黏塑性流体	黏塑性流體
viscoplasticity	黏塑性	黏塑性
viscose rayon	黏胶人造丝	黏液嫘縈
viscose staple	黏胶人造短纤维	黏液嫘縈棉
viscosimeter	黏度计	黏度計
viscosimetry	黏度测定法	黏度測定法
viscosity	黏度	黏度
viscosity average molecular weight	黏度平均分子量	黏度平均分子量
viscosity blending chart	黏度掺和图	黏度摻配圖
viscosity coefficient	黏度系数	黏度係數
viscosity index	黏度指数	黏度指數
viscosity number	黏度数	黏度值
viscosity test	黏度试验	黏度試驗
viscous damper	黏性减震器	黏性制振器
viscous dissipation	黏性耗散	黏性散逸
viscous flow	黏性流动	黏性流動
viscous force	黏性力	黏性力
viscous motion	黏性运动	黏性運動
viscous resistance	黏性阻力	黏性阻力
viscous stress	黏性应力	黏性應力
viscous sublayer	黏性次层	黏性次層
vitamin	维生素	維生素, 維他命
vitreous china	玻化瓷器	玻化瓷器
vitreous phase	玻化相	玻化相
vitrification	玻璃固化	玻化
vitrification period	玻璃固化期	玻化期
vitrified brick	玻璃固化砖	玻化磚
V notch	V 型缺口	V 形凹槽

英　文　名	祖国大陆名	台湾地区名
void	空隙	空隙
voidage	空隙率	空隙度
void fraction	空隙分数	空隙分率
void volume	空隙体积	空隙體積
volatile suspended solid	挥发性悬浮固体	揮發性懸浮固體
volatility	挥发度	揮發度
volatilization	挥发	揮發[作用]
voltage	电压	電壓
voltameter	库仑计	電量計
voltmeter	伏特计	電壓計, 伏特計
volume	体积	體積, 容積
volume average velocity	体积平均速度	體積平均速度
volume expansivity	体膨胀率	體積膨脹係數
volume fraction	体积分率	體積分率
volume index	体积指数	體積指數
volume mean diameter	体积平均直径	體積平均直徑
volume modulus	体积模数	體積[彈性]模數
volume shrinkage	体积收缩	體積收縮
volume surface diameter	体积表面直径	體積表面[相當]直徑
volumetric average boiling point	体积平均沸点	體積平均沸點
volumetric efficiency	体积效率	體積效率
volumetric flow rate	体积流率, 体积流量, 体积流速	體積流量
volumetric hourly space velocity	[小时]体积空[间]速[度]	單位體積空間時速
volumetric oxygen transfer coefficient	容积传氧系数, 体积传氧系数	體積傳氧係數
volumetric titration	容量滴定法	容量滴定[法]
volumetry	容量分析法	容積分析法
volume work	体积功	體積功
volute	蜗壳	渦卷形
volute pump	涡壳泵	渦卷泵
von Karman analogy	冯卡门类比	馮卡門類比
von Karman boundary layer theory	冯卡门边界层理论	馮卡門邊界層理論
von Karman integral method	冯卡门积分法	馮卡門積分法
von Karman number	冯卡门数	馮卡門數
von Karman vortex street	冯卡门涡街	馮卡門渦列
von Weiman equation	冯韦曼方程式	馮韋曼方程式
vortex	涡旋	漩渦
vortex agitator	涡旋搅拌器	漩渦攪拌器

英 文 名	祖国大陆名	台湾地区名
vortex breaker	涡流消除器	碎漩涡器
vortex cavity	旋涡空穴	漩涡空洞
vortex line	涡线	漩涡線
vortex motion	涡旋运动	漩涡運動
vortex potential	涡旋势	漩涡勢
vortex shedding	涡旋脱落	漩涡分離
vortex sheet	涡片	漩涡面
vortex trail	涡旋尾迹	漩涡尾跡
vorticity	涡度	漩涡度
votator apparatus	套管冷却结晶器	套管冷卻結晶器
vulcanite	硬橡胶	硬橡膠
vulcanizate	硫化橡胶	硫化橡膠
vulcanization	硫化	橡膠硫化反應
vulcanization coefficient	硫化系数	硫化係數
vulcanized fiber paper	硫化纤维纸	硬化紙
vulcanized oil	硫化油	硫化油
vulcanizer	硫化器	硫化器
vulcanizing agent	硫化剂	硫化劑
vulcanizing chamber	硫化室	硫化室
vulcanizing pan	硫化锅	硫化鍋
vycor glass	高硅氧玻璃	耐熱玻璃

W

英 文 名	祖国大陆名	台湾地区名
wafer	晶片	晶片
wage	工资	工資
wake	尾流, 尾涡	尾流
wall effect	壁效应	壁效應
wall thickness	壁厚	壁厚
wall turbulence	壁湍流	壁紊流
warpage	翘曲	翹曲[變形]
warping	翘曲	①翹曲②[織物]整經
washable ointment base	可洗软膏基	可洗軟膏基
wash column	洗涤塔	洗滌塔
washing rate	洗涤速率	洗滌速率
washings	洗涤液	洗滌水
washing soda	洗涤碱	洗滌, 碳酸鈉

英　文　名	祖国大陆名	台湾地区名
washing tower	洗涤塔	洗滌塔
washout	冲洗	沖盡
waste	废弃物	廢棄物, 廢料
waste acid	废酸	廢酸
waste disposal	废物处理	廢棄物處理
waste effluent	废液	廢液
waste fuel	废燃料	廢燃料, 廢油
waste gas	废气	廢氣
waste heat	废热	廢熱
waste heat boiler	废热锅炉	廢熱鍋爐
waste heat drier	废热干燥器	廢熱乾燥器
waste heat recovery	废热回收	廢熱回收
waste management	废物管理	廢棄物管理
waste treatment	废物处理	廢棄物處理
waste water	废水	廢水, 污水
waste water treatment	废水处理	廢水處理
water absorbency	吸水率	吸水率
water analysis	水质分析	水質分析
water base paint	水性漆	水性漆
water bath	水浴	水浴
water conditioning	水调理	水調理
water content	含水量	含水量
water cooled crystallizer	水冷结晶器	水冷結晶器
water cooler	水冷却器	水冷卻器
water cooling	水冷却	水冷卻
water curtain	水帘	水簾
water gas	水煤气	水煤氣
water gas shift reaction	水煤气变换反应	水煤氣轉化反應
water glass	水玻璃	水玻璃
water hammer	水锤	水鎚
water hardness	水质硬度	水質硬度
water head	水位压头	水位高差
water heater	水加热器	水加熱器
water in oil emulsion	油包水乳状液	油中水乳液
water jacket	水套	水套
water of crystallization	结晶水	結晶水
water paint	水性漆	水性漆
water permeability	透水性	透水性
water polisher	水磨光机	水高純化器

英　文　名	祖国大陆名	台湾地区名
water pollutant	水污染物	水污染物
water pollution	水污染	水污染
water pollution abatement	水污染降低	水污染降低
water pollution control	水污染控制	水污染控制
water pollution engineering	水污染工程	水污染工程
waterproof	不透水的	防水布, 防水材料
water proof cement	防水水泥	防水水泥
water proof grease	防水润滑脂	防水潤滑脂
water proofing	防水处理	①防水[處理]②防水劑
water purification	水净化	水淨化, 水純化
water quality	水质	水質
water quality criterion	水质判据	水質準則
water repellence agent	抗水剂	拒水劑
water repellent	抗水[作用]	防水劑
water seal	水封	水密封
water separator	水分离器	水分離器
water softener	软水剂	①水軟化劑②水軟化器
water softening	水软化	水軟化
water-soluble ointment base	水溶性软膏基	水溶性軟膏基
water tolerance	耐水度	①耐水性②耐水度
water treatment	水处理	水處理
water valve	水阀	水閥
wavelength	波长	波長
wave mechanics	波动力学	波動力學
wave motion	波状运动	波形運動
wave number	波数	波數
wave propagation	波传播	波之傳播
wavy flow	波状流	波狀流
wax	蜡	蠟
wax distillate	蜡馏出物	蠟餾出物
wax fractionation	含蜡分馏	蠟分餾
wax oil	含蜡油	蠟油
wax paper	蜡纸	蠟紙
wax residuum	含蜡渣油	蒸餘蠟
waxy crude	含蜡原油	含蠟原油
waxy oil	含蜡油	含蠟油
weak acid	弱酸	弱酸
weak base	弱碱	弱鹼
weak electrolyte	弱电解质	弱電解質

英　文　名	祖国大陆名	台湾地区名
weak reversibility	弱可逆性	弱可逆性
wearing ring	耐磨圈	耐磨圈
weathering	风化	風化
weathering test	耐候性试验	風化試驗, 耐候性試驗
weaving machine	织机	織機
Weber	韦[伯](磁通量单位)	韋伯
Weber number	韦伯数	韋伯數
wedge	楔	楔形, 楔形體
weed killer	除草剂	除草劑
weeping	漏液	滴流
weeping hole	漏液孔	滴流孔
Wegstein method	韦格斯坦法	魏格斯坦法
weighing feeder	喂料装置	稱量飼機
weight average boiling point	重均沸点	重量平均沸點
weight average molecular weight	重均分子量	重量平均分子量
weight distribution	权重分布	重量分布
weighted mean	加权平均[值]	加權平均
weighted residual	加权残差	加權剩餘
weighted temperature difference	加权温差	加權溫差
weight factor	权因子	權因數
weight fraction	重量分率	重量分率
weight function	权函数	權函數, 權數
weighting factor	加权因子	加權因數
weighting function	加权函数	加權函數
weight loss	重量损失	重量損失
weight mean diameter	重量平均直径	重量平均直徑
weightometer	自动皮带秤	稱重計
weir	堰	堰
weir height	堰高	堰高
Weisz modulus	韦斯模数	魏斯模數
welding	焊接	熔接
well type nozzle	井式喷嘴	井型噴嘴
wet and dry bulb hygrometer	干湿球湿度	乾濕球濕度計
wet-bulb depression	湿球下降	濕球下降
wet bulb temperature	湿球温度	濕球溫度
wet-bulb thermometer	湿球温度计	濕球溫度計
wet combustion	湿式燃烧	濕式燃燒
wet gas	湿气	濕[天然]氣, 含油氣
wet gas meter	湿气计	濕氣計

英　文　名	祖国大陆名	台湾地区名
wet grinding	湿磨	濕磨
wet oxidation	湿式氧化	濕式氧化[法]
wet pan mill	湿辗机	濕輾機
wet process	湿法过程	濕製程
wet screening	湿筛选	濕篩選
wet seal holder	湿式密闭容器	濕式密閉容器
wet separation	湿法分离	濕法分離
wet spinning	湿纺	濕紡[絲]
wet steam	湿蒸汽	濕蒸汽
wet surface	湿表面	濕[表]面
wettable agent	可湿性助剂	助濕劑
wetted area	湿面积	潤濕面積
wetted perimeter	润湿周边	潤濕周邊
wetted surface area	润湿表面积	潤濕表面積
wetted wall column	湿壁塔	濕壁管
wetted-wall tower	湿壁塔	濕壁塔
wetting agent	润湿剂	潤濕劑
wetting effect	润湿效应	潤濕效應
wet well	湿井	濕井
whale oil	鲸油	鯨油
whale tallow	鲸脂	鯨脂
whipping	打泡	攪打
white carbon	白炭黑	白碳, 白煙
white cement	白水泥	白水泥
white gold	白金	白金[含鉛金]
white liquor	白液	白液, 燒鹼液
white metal	白合金	白合金
white phosphorus	白磷, 黄磷	黃磷, 白磷
white water	白水	白水
wild stream	未控流	未控流
wilkinite	膨润土	膨土, 漿土
willemite	硅锌矿	矽鋅礦
Wilson equation	威尔逊方程	威爾遜方程
Wilson method	威尔逊方法	威耳生方法
wind tunnel	风洞	風洞
windup	卷绕	繞緊
winterizing	冬化	低溫化, 冬化
wire glass	夹丝玻璃	夾網玻璃
W/O emulsion ointment base	水/油乳剂软膏基	水/油乳劑軟膏基

英　文　名	祖国大陆名	台湾地区名
Wohl expansion	沃尔展开式	沃爾展開式
wollastonite	硅灰石	矽灰石
wood alcohol	木精	木精
wood creosote	木杂酚油	木雜酚油
wood-derived chemical	木材化学品	木材化学品
wood distillation	木材蒸馏	木材蒸餾
wood extractive	木材提取物	木材萃取物
wood gas	木煤气	木煤氣
wood preservative	木材防腐剂	木材防腐劑
wood pyrolysis	木材热解	木材熱解
wood rosin	木松香	松香
wood saccharification	木材糖化	木材糖化
wood spirit	木精	木精, 甲醇
wood sugar	木糖	木糖
wood tar	木焦油	木溚
wool alcohol	羊毛脂醇	羊毛醇
wool fat	羊毛脂	羊毛脂
wool fiber	毛纤维	毛纖維
wool grease	羊毛脂	[粗]羊毛脂
wool oil	羊毛油	羊毛油
wool wax	羊毛蜡	羊毛蠟
work function	功函数	功函數
work index	功指数	功指數
working capital	流动资金	營運資金
working fluid	工作流体	工作流體
working stress	工作应力	工作應力
work requirement	功需要量	功需要量
worksheet	工作单	工作單
wort	麦芽汁	麥芽汁
wrought iron	熟铁	熟鐵
Wulff-Bock crystallizer	摆动连续结晶槽	擺動連續結晶槽

X

英　文　名	祖国大陆名	台湾地区名
xanthation	黄原酸化[反应]	黄酸化
xerogel	干凝胶	乾凝膠
X-ray diffraction	X 射线衍射	X 射線繞射

英 文 名	祖国大陆名	台湾地区名
X-ray fluorescence	X 射线荧光	X 射線螢光
X-ray spectrum	X 射线谱	X 射線譜
xylene	二甲苯	二甲苯
xylose	木糖	木糖
xylulose	木酮糖	木酮糖

Y

英 文 名	祖国大陆名	台湾地区名
yarn	纱	紗, 線
yarn dyeing machine	纱染机	紗染機
Yate's algorithm	耶特算法	耶特算法
yeast	酵母	酵母
yellow cake	黄饼	黃餅
yellow lead	黄丹	密陀僧
yield	收率	①產率, 產量②降伏
yield coefficient	收率系数	產率係數
yield limit	屈伏极限	降伏極限
yield per path	单程收率	單程產率
yield point	屈服点	降伏點
yield strength	屈服强度	降伏強度
yield stress	屈服应力	降伏應力

Z

英 文 名	祖国大陆名	台湾地区名
zeolite	沸石	沸石
zeolite catalyst	沸石催化剂	沸石催化劑
zero-flux surface	零通量表面	零流通量表面
zero frequency gain	零频率增益	零頻率增益
zero memory system	零记忆系统	零記憶系統
zero-order reaction	零级反应	零階反應
zero-pressure state	零压状态	零壓狀態
zeroth law of thermodynamics	热力学第零定律	熱力學第零定律
zinc alloy	锌合金	鋅合金
zinc blende	闪锌矿	閃鋅礦
zinc bronze	锌青铜	鋅青銅
zinc chrome	锌铬黄	鋅鉻黃

英　文　名	祖国大陆名	台湾地区名
zinc-crown glass	锌冕玻璃	鋅冕玻璃
zinc flower	锌华	鋅華
zinc green	锌绿	鋅綠
zincite	红锌矿	紅鋅礦
zinc plating	镀锌	鍍鋅
zinc spar	菱锌矿	菱鋅礦
zinc sponge	锌海绵	鋅綿
zinc storage battery	锌蓄电池	鋅蓄電池
zinc white	锌白	鋅白
zinc yellow	锌铬黄	鋅黃
zircon	锆石	鋯英石
zirconia	二氧化锆	氧化鋯
zirconia glass	氧化锆玻璃	鋯玻璃
zirconium oxide refractory	氧化锆耐火材料	氧化鋯耐火物
zircon porcelain	锆石瓷	鋯瓷
zircon refractory	锆英石耐火材料	鋯英石耐火物
zone electrophoresis	区带电泳	帶域電泳
zone freezing	区域冷冻	帶域冷凍
zone melting	区域熔炼	帶域熔化
zone melting method	区域熔炼法	帶域熔化法
zone refining method	区域精制	帶域純化法
zone settling	区域沉降	帶域沉降
Z transformation	Z 转换	Z 轉換
zymase	酿酶	解醣酶